The Birds of the West Midlands

The Birds of the West Midlands

Graham R. Harrison, *Editor*
Alan R. Dean
Alan J. Richards
David Smallshire

West Midland Bird Club, 1982

To all members of the West Midland Bird Club, both past and present, without whose records this book could not have been written.

First published in 1982 by the West Midland Bird Club,
Hon. Secretary: A. J. Richards,
PO Box 1, Studley, Warwickshire B80 7JG.
All rights reserved. No part of this book may be
reproduced, stored in a retrieval system, or
transmitted, in any form or by any means,
electronic, mechanical, photocopying,
recording or otherwise, without the prior
permission of the West Midland Bird Club.

Printed in Great Britain by
Ebenezer Baylis and Son Ltd
The Trinity Press, Worcester, and London

Contents

Drawings by R. A. Hume
Maps, histograms and graphs drawn by Janet Harrison
Ringing information compiled by P. L. Ireland

List of Photographs

Acknowledgements

During the compilation of this book a considerable amount of help has been forthcoming from a great many people, far too numerous to mention each by name. The authors, however, would like to express their gratitude to them all. We would particularly like to acknowledge the major contributions made by Michael Warren with his excellent jacket illustration, Rob Hume whose delightful drawings do so much to enliven the text, Philip Ireland who analysed and prepared notes on all the ringing recoveries, and Janet Harrison who meticulously prepared all the many maps, histograms and graphs.

For photographs we are indebted to Roy Blewitt, Charlie Brown, Arthur Cundall, Chris Fuller, Bill Jones, Tom Leach, Bill Smallshire, Peter Wakely, Geoff Ward, Mick Wilkes, the Nature Conservancy Council and the City of Birmingham Planning Department, who all made their high quality work freely available. They are individually acknowledged with each photograph, but we should like to thank them all and in particular Geoff Ward for his help with printing and Charlie Brown who gathered much of the material together.

To Tony Blake, Harry Green, Frank Gribble and Joe Hardman we owe an especial debt for their constructive comments on various sections of the manuscript. We are most grateful to Eunice Holder who typed so much of the manuscript, and to two of the authors' wives, Jenny Richards and Judy Smallshire, who also helped with the typing as well as giving assistance and encouragement in other ways.

We would like to thank the British Trust for Ornithology for kindly allowing us to quote Ringing, Common Bird Census and Sites Register data; the Wildfowl Trust for providing and permitting us to use monthly wildfowl counts and T. & A. D. Poyser for permission to publish data from the BTO Atlas. Maps are based on the Ordnance Survey Map with the permission of the Controller of Her Majesty's Stationery Office, Crown Copyright reserved. Dr. E. R. Austin, M. J. Austin, King Edward's School, Birmingham, and S. C. Nichols provided comprehensive data from their Common Bird Census plots, Barbara Jones gave valuable help on pest species, A. E. Coleman provided additional Mute Swan data, Andy Lowe loaned a card index of collated records from early WM *Bird Reports*, George and Maurice Arnold provided collated wildfowl counts and Clive Richards prepared the jacket typography. The Countryside Commission, Forestry Commission, Ministry of Agriculture, Fisheries and Food and the Nature Conservancy Council all supplied most useful information. We also received much help and understanding from the staff of Ebenezer Baylis and Son Ltd., the printers, for which we are extremely grateful.

Finally, we must acknowledge the vast contribution made by all those who have contributed records over the years and to thank previous editors of the WM *Bird Report*, whose past efforts greatly facilitated the assembly and analysis of all data. We are most grateful to the Committee of the West Midland Bird Club, under the chairmanship of Tony Blake, for its constant encouragement and support and lastly, but by no means least, we must thank the Royal Society, the Foyle Trust and all those members and friends of the West Midland Bird Club whose generosity through interest-free loans and donations enabled the publication of this book to be financed at a most difficult time.

Graham R. Harrison
Alan R. Dean
Alan J. Richards
David Smallshire

As Editor, I would also like to add my personal thanks to my co-authors, who, apart from their own very substantial contributions to the systematic list, proffered valued comments and opinions throughout. In addition, Dave Smallshire undertook a variety of tasks including the revision of many initial drafts and assistance with the gathering together of photographs, Alan Dean worked conscientiously through volumes of wildfowl data and Alan Richards ably handled fund raising, promotion and publicity almost singlehanded. Finally, I should like to thank my wife, Janet, for her considerable help, encouragement and tolerance.

Graham R. Harrison

Introduction

The publication of *The Birds of the West Midlands* by the West Midland Bird Club is an act of faith. To publish such a book commercially would have meant that the price would be too high to reach all the ornithologists who would like a copy. The initial impetus for this new work arose from the West Midland Bird Club's intention to commemorate its 50th Anniversary in 1979 in three ways – with a special celebration, the acquisition of a nature reserve and the publication of a book on the birds of the West Midland Region. We have now met all three objectives.

Unlike some areas, the Region's current county avifaunas are not too outdated. Smith's *The Birds of Staffordshire* is the oldest, having been published in 1938, but this was up-dated by Lord and Blake in their *Birds of Staffordshire* in 1962, whilst Harthan's *The Birds of Worcestershire* (1946) was updated by the author himself in 1961. However, the *Notes on the Birds of Warwickshire* written by our President, Tony Norris, in 1947 has never been revised. It contains few references to the Tame Valley, now producing such a wealth of records, and, of course, was published long before the creation of Draycote Water and Brandon Marsh. In addition the WMBC prepared its own *Atlas of Breeding Birds of the West Midlands* in 1970, a pioneering venture which at least brought up-to-date our knowledge of those species which nest in the Region. Nevertheless, the rate of change over the past twenty years and the growth in knowledge amply justified a complete reappraisal.

To a large extent this is a reflection of the success of the WMBC. Through an interesting and varied programme of indoor and field meetings, and the publication of regular *Bulletins* and an *Annual Report*, the Club has engendered interest in ornithology throughout the Region and this is reflected in its membership figures – a mere 30–40 in the 1930s, rising to 100 soon after the war, 500 in the 1950s, 1,000 in 1969 and 2,000 in the early 1970s.

Today the Club's activities are not just centred on its traditional Birmingham base, but have spread much further afield, with branches at Kidderminster and in Staffordshire and a local group at Solihull.

Nor have the scientific aspects of ornithology been neglected. Through its Research Committee the Club has undertaken various projects, both self-initiated and as part of national inquiries for organisations like the British Trust for Ornithology. Amongst these might be mentioned an early ringing project using a Heligoland trap, observation of visible migration during the late-1940s and early-1950s, and censusing by means of line transects, as well as the pioneer *Atlas* referred to above.

Over the years the Club has had a number of distinguished members. To mention names seems almost invidious, but those of Horace Alexander, Talbot Clay, Anthony Harthan, Lord Hurcomb, Cecil Lambourne, Tony Norris and Brunsden Yapp are perhaps best known.

The outcome of an expanding membership under such eminent guidance has been the assimilation of a wealth of data on the regional bird-life. This book brings all this data together and presents it in an informative and attractive manner. It is unusual for a single publication to deal with the avifauna of an entire region and the opportunity has been taken to thoroughly analyse the records of those species, such as the divers, which are too numerous for national analysis, yet too scarce to show significant trends at individual county level. Furthermore, in a Region that is visited by few ornithologists, it is possible to estimate the bias that results from increased observer activity. The number of people contributing records has remained fairly constant since 1968, so any changes in bird-life since then are likely to be real.

For these reasons this book should not only appeal to those who live, work, visit or watch birds in the West Midlands, but also to those in a much wider area who have an interest in ornithology generally. Inevitably some questions remain unanswered, but it is fairly safe to say that many of the answers will be forthcoming from members of the WMBC during the next fifty years.

A. R. M. Blake,
Chairman,
West Midland Bird Club,
October 1981.

1 A Profile of the West Midlands

The West Midlands is perhaps the most misunderstood and maligned region of Britain. To a northerner it is part of the lowland south, to a southerner part of the hilly north. Mention it and most people immediately think of Birmingham and motor cars. To those knowing no better, its unfortunate tag of "Black Country" conjures up a lurid picture of smoky factories and serried ranks of grimy terraced houses. The Region is, however, infinitely wider and more attractive than this narrow perspective implies and its unenviable reputation is by no means deserved. Indeed it is so wide and variable that it is arguable whether it can justifiably claim to be an entity at all. Its diverse nature is reflected in the many definitions of its boundaries even between Government departments, let alone between different organisations.

For the purpose of this book, however, the Region has been defined as the historical counties of Staffordshire, Warwickshire and Worcestershire. Together these form a roughly triangular area of 7,350 km², extending for 141 km from the southern tip of the Pennines in the north to the Cotswolds in the south and 102 km from the Welsh Massif in the west to the Northamptonshire Uplands in the east.

Today the bird-life of this entire area is recorded by members of the West Midland Bird Club (WMBC), though this has not always been so. Fifty years ago, when the Birmingham Bird Club, as it was known then, was formed, only Warwickshire and Worcestershire were covered. South Staffordshire, bounded by the Trent, Sow and the Stafford-Wellington railway, was added in 1935 and it was another ten years before the name was changed to the Birmingham and West Midland Bird Club. Finally, in 1948, the remainder of Staffordshire was annexed, though it was 1959 before Birmingham was eventually dropped from the Club's name.

During these fifty years the Region's landscapes and habitats have been constantly evolving in response to man's intervention. With their gift of flight, its birds have been swift to respond to changes in their environment, sometimes colonising new areas and sometimes deserting hitherto favoured haunts. To fully appreciate the avifaunal changes of the Region, therefore, it helps to look at the national context of the Region and how it has developed socially, physically and economically.

The West Midlands is the very heart of England. It is further from the sea than anywhere else in Britain, yet even at its centre it is no more than 140 km from the estuaries of the Mersey, Severn and Wash. To birds such distances are nothing and there is a surprising maritime flavour to the avifauna which serves as a salutary reminder of just how small an island Britain is. Common Scoter and Scaup, for example, are annual visitors and there are few years when an odd shearwater, petrel or Gannet is not recorded.

Although typically British in its variability, the Region's climate is nonetheless also influenced by its geographical position. Situated midway between the west and east coasts, it lies predominantly under the maritime influence of prevailing south-westerly or westerly airstreams and experiences the mild, damp weather associated with depressions and the North Atlantic Drift. However it escapes the worst of the heavy rainfall that affects more westerly districts. Rainfall is fairly evenly distributed throughout the year, with most in winter, but varies with altitude from 600 mm per annum in the Avon Valley to 1,300 mm on the moorlands of North Staffordshire. Except on these northern moors, there is a significant summer moisture deficiency everywhere and this affects both crops and the natural vegetation. The seasonal temperature range is more extreme than in coastal areas, with mean January temperatures between 0° and 2°C and July ones between 19° and 21°C in most areas, but falling as low as 15°C on the high moors.

Both the Severn and Avon valleys enjoy early springs with plants flowering a week or more earlier than in the Region as a whole. Fruit and vegetable growers in the Vale of Evesham have exploited this climatic advantage, which no doubt also contributes to the presence of several species of bird, like the Nightingale, which here are on the edge of their range. It is here too that Marsh Warblers thrive and Red-backed Shrike, Wryneck and Cirl Bunting lingered longest. From time-to-time, however, the Region comes under the influence of northerly or easterly airstreams, which respectively bring colder arctic or drier continental climates. Hard-weather movements of birds and the arrival of storm-driven vagrants often accompany such conditions. The inland situation leads to a high diurnal temperature range and a propensity for night frosts, but relative freedom from high winds. Combined with low rainfall, this means that a hard winter is more likely to manifest itself in extreme cold and prolonged frost than in deep snowdrifts, although snowfall on the northern moors can be considerable. On average frost occurs on 111 days each year on the moors compared to less than 100 days in the Avon valley, where snow cover is fairly exceptional. Such local variations in the winter climate affect both the availability of food and the survival of birds during severe weather.

The past fifty years have also brought long-term fluctuations and certain climatic extremes worthy of record. The summers of the 1930s and 1940s were generally drier and warmer than average, with a drought in 1933, though that of 1931 was notably wet. Of all years, however, 1947 made an indelible impression on the record books. The winter was especially severe, with prolonged cold persisting well into March and snow lying for a record fifty-three days. Bird mortality in such arctic conditions was very high and many species took several years to fully recover their numbers. When the thaw finally came it was so rapid that flooding was the worst for many years. By way of recompense, however, the summer of 1947 was exceptionally dry and hot, setting several new meteorological records. The 1950s and 1960s were characteristically variable. For example, spring was early in 1957, but late in 1958; 1958 also brought a wet summer and that of 1960 was amongst the wettest ever experienced, yet the intervening one was unusually dry. Arctic conditions returned again in 1963 and many birds perished during a protracted cold spell in the course of which temperatures plummeted into the negative range of the Fahrenheit scale. Since then, however, there has been a succession of mild winters, notably during the 1970s; combined with cold, late springs; variable summers and latterly an unusual prevalence for gales and even hurricanes in 1976 and 1977. Many early broods suffered in 1967, when May was the wettest for two-hundred years, and the summer of 1968 was the wettest for over thirty-years. Spring was very late in 1975, with snow falling on the moors in the first few days of June, but was then followed by a long, hot summer and the onset of the worst drought for two-hundred and fifty years, which was to last through the summer of 1976 – the driest and warmest of the century – only to end abruptly with the wettest September on record. Finally 1979 brought a third severe winter, all coincidentally on a sixteen year cycle.

Physically the Region resembles a lemon-squeezer, with the Birmingham Plateau at the centre surrounded by the three major river basins of the Severn, Avon and Trent and a perimeter of hills and uplands which rises through central Staffordshire to reach its peak on the northern moors and includes an outer rim embracing both the Malverns and the Cotswolds.

In subsequent chapters this structure is used as the basis for detailed descriptions of the Region and its habitats. There is a major watershed across the Region, with the Trent flowing north-eastwards into the Humber estuary and the Severn and Avon flowing south-westwards into the Bristol Channel. Small areas also drain north-westwards into the Mersey and south-eastwards into the Thames, but these are relatively insignificant. Centuries ago, when these rivers were virtually the only water in the Region, they figured prominently in historical records and must have been much more important to birds than they are today. Their subsequent pollution, along with the creation of man-made waters, may have diminished their specific importance, but their valleys are still broadly followed by many species of migrating birds, especially those crossing from the Wash to the Bristol Channel or those heading due north or south along the Severn or Tame. Significant steps have been taken in the past decade to reduce pollution and fish life is returning to many rivers like the Tame and with it birds too. Kingfishers appear now in some most unlikely places, whilst the clear, unpolluted streams of the extreme north and west still play host to Dippers and the occasional pair of Common Sandpipers.

Considering its distance from the sea, it is perhaps surprising that most of the Region lies below 200 m, with the lowest points of the Severn and Avon valleys no more than 10 m above sea-level. For the most part, the topography is one of wide river valleys and level or gently undulating, low plateaux. Steep hillsides and narrow valleys are characteristic of the extreme north and west, however, and there are some significant, if not extensive, uplands which reach 500 m in the north of Staffordshire. In general though it is the many subtle variations in scenery, soils and farming pattern stemming from the diversity of underlying rocks that give the Region its true character.

In the extreme north-east Carboniferous Limestone has been steadily eroded by streams into an area of exceptional scenic beauty, which stands in sharp contrast to the neighbouring bleak and windswept moorland of the Millstone Grit country, where the poor, acidic soils are suitable only as sheep or cattle pastures. Elsewhere in the Region, though, soils are generally productive. Small, isolated outcrops apart, the oldest rocks are found along the western flank, beyond the Severn, where soft red marls of the Old Red Sandstone yield rich, fertile soils in picturesque, undulating countryside dotted with hop yards and wooded hillsides. Here the landscape is dominated by the Malvern Hills – a bare, rugged outcrop formed by the remarkably sharp folding of pre-Cambrian rocks and rising to 425 m.

The newest rocks, excluding glacial deposits, occur along the southern flank, beyond the Avon, where the Lower Lias beds form a flat, featureless plain terminating abruptly to the south-east in the Jurassic scarp of the Cotswolds. Arable farming with long leys, often grazed by sheep, predominates on the heavy clay soils and intensive horticulture is now the trade mark of the Vale of Evesham, which was once epitomised by orchards. Gone too are the lofty hedgerow elms, which once were synonymous with this landscape. Extensive patches of boulder clay and glacial gravels have infiltrated the upper Avon valley, forming small hillocks of poor ground which are frequently wooded, especially on their steeper slopes.

Between these two age extremes Permo-triassic rocks are widespread, with Keuper beds underlying much of the Region's heartland north of the Avon and east of the Severn. The associated marls and sandstones have weathered into rich, red soils set in a gently undulating, but rather featureless landscape. What little relief there is can again often be traced to a thin, but widespread, covering of glacial drift. This is the true Midlands plain – pleasant countryside which conjures up an image of cattle pastures and fox coverts; of sturdy hedgerow oaks and Hereford cattle – though in truth arable fields and Friesian cattle are more in evidence today. Wherever Bunter beds occur, the ground is usually higher and the gravelly soils are thirsty and agriculturally poor. Heaths, usually relics of the once-widespread Royal hunting forests, are characteristic of such soils, with Cannock Chase and Sutton Park the best known examples. More recently though, some areas, notably Cannock Chase and Enville, have been afforested with conifers. Typically interesting bird communities are found in these places, including species like Nightjar, Woodcock, Grasshopper Warbler and Stonechat.

The Triassic deposits also embrace older rock masses, which in glacial times stood as islands in a lake basin. Although not vast in area, for centuries these rock masses have been exploited for their economic value. Indeed they were the very foundation of the Region's industrial growth. Although very small igneous intrusions occur at Nuneaton and Rowley Regis and there are outcrops of Silurian Limestone, Cambrian and pre-Cambrian rocks, it is the Carboni-ferous Coal Measures that occur in north Staffordshire, south Staffordshire and east Warwickshire that have undeniably been of greatest importance. In these three areas, which cradled the Region's great industrial centres of the Potteries, Birmingham and the Black Country, and Coventry respectively, the landscape has been drastically disfigured by urbanisation.

Today the West Midlands has a thriving industrial heart within an extensive rural hinterland, but this has not always been so. Until the late eighteenth century its principal economic centres were its market towns, like Worcester, Coventry and Lichfield, and only in the last two hundred years have values shifted to the extent that the hitherto uninviting Birmingham Plateau has become a hive of activity (Cherry 1975). Although initially based on the exploitation of coal and iron, this activity soon developed into a bewildering array of metal-based and engineering industries which have become justifiably renowned throughout the world. Indeed it is this strong economic base more than any other factor which provides the justi-fication for regarding the Region as a homogeneous unit.

Accompanying this industrial growth has been a rapid population expansion. In 1801 there were a mere 0·6 million people living in the Region. By 1901 this had risen to 2·6 millions; by 1921 to 3·1 millions and by 1971 to 4·6 millions – an increase of almost 50 per cent in the last fifty years alone. Inevitably such rapid growth led to poverty and over-crowding, but the squalid housing of the eighteenth and nineteenth centuries has been largely replaced by the new estates and suburbs of the twentieth century. To achieve this, thousands of hectares have been lost to building and over the past fifty years the percentage of urbanised land has risen from 10 per cent to 16 per cent.

Contrary to popular belief the process of urbanisation has not always been to the detriment or exclusion of birds. House Sparrows and feral pigeons have adapted so well as to become almost domesticated, yet in truth the tall buildings of a city centre

16

1. Even within sight of the centre of Birmingham, mature suburban gardens may have more trees and a richer bird-life than some country districts.

by courtesy of the City of Birmingham Planning Department

are not so very different from the pigeons' ancestral cliff homes; and man, by accident or design, provides sufficient food to sustain both species. With these species have come predators like Kestrel and Tawny Owl, whilst the winter warmth, which city and industrial buildings retain at night, attracts enormous roosts of Starlings and large numbers of Pied Wagtails. Even amidst the most desolate of asphalt and concrete jungles, the odd churchyard, square or park, with its grass, shrubs or trees, affords sanctuary for a few Blackbirds, whilst bombing during the last war and continuous redevelopment since have enabled weeds to colonise the inner city and finches to move in and feed on their seeds. More than anything, though, the Black Redstart has become the urban speciality of the Region over the past

decade or two, as indeed it has elsewhere in Britain.

Even richer is the bird-life of the suburbs. Here the abundant food supply is fully exploited from the ground right into the sky, with the tiny Wren searching every crevice of the rockery; Starlings systematically advancing across the lawn and, especially since the introduction of smoke control, increasing numbers of House Martins and Swifts hawking a myriad of flying insects. Indeed the density of birds in the suburbs often exceeds that of the open countryside and some species, like the Blackbird, are more successful here than in their more natural habitats. Within the suburbs bird-life varies with the age and density of the housing, being richest where density is lowest and the gardens most mature. The wealthier Victorian suburbs,

17

for example, with their spacious, secluded gardens, trees and dense shrubberies, harbour Blackbird, Song Thrush, Robin, Dunnock, Chaffinch and Greenfinch amongst others, whilst any conifers are always likely to attract Coal Tit, Collared Dove and Goldcrest. In the newer suburbs, wider, tree-lined roads and grass verges have consumed space at the expense of gardens, which have become smaller. Trees are less mature and the old-fashioned shrubbery has given way to lawns, herbaceous borders and vegetable plots, where less welcome visitors like Woodpigeons and Bullfinches move in to attack the cabbages or strip the fruit buds. Wherever mature trees have been kept in the interests of amenity, they are all too often within a sea of mown grass, which restricts their wildlife value except as perches for the ever-watchful opportunists like the Carrion Crow and Magpie.

Even in winter, the suburban garden is rich in invertebrates and berries, but once natural food becomes really scarce or unobtainable anything provided by man is quickly found. This auxilliary food supply, which is being provided in ever-increasing quantity, sustains not only the urban species like the House Sparrow and Starling, but proves irresistible also to Blue and Great Tits, which are less adapted to suburban life. Indeed, since the hard winter of 1963, it has even drawn new species like the Siskin into gardens and there is also a growing propensity for Blackcaps to over-winter in gardens. Some interesting adaptations have also emerged. Tits have discovered the rich prize of cream beneath the milk-bottle top and the Greenfinch has learnt to obtain food from suspended nut-bags. Growing interest in birds has also encouraged more people to provide nest-boxes or to put out water for drinking and bathing and overall the suburbs have become somewhat safer than the natural woodland-fringe habitat that so many species normally favour.

Open spaces in the suburbs vary considerably. Playing fields and recreation grounds attract flocks of Black-headed Gulls and Woodpigeons and a few Pied Wagtails, but little else. Parks can be richer, with the typical suburban species augmented wherever mature trees occur by the occasional Stock Dove, Jay, Nuthatch or woodpecker. If nature is allowed to override tidy-mindedness, even Blackcap and Willow Warbler may attempt to breed as the understorey develops. Most parks have a lake too, though all too frequently islands are lacking and the steep, artificial banks offer little prospect of wildfowl nesting, except perhaps for a few Mallard and Moorhen or the occasional Mute Swan or Canada Goose.

The demands of an urban society are not just confined to its towns and cities, however, and in the West Midlands man's activities beyond the urban fringe have most influenced the avifauna. In particular the canal system has been profoundly beneficial, adding not just a linear habitat for waterfowl, warblers, wagtails and Reed Buntings, but providing also several feeder reservoirs, which for many years have been a principal haunt in the Region for wildfowl and passage terns and waders. The canal system, too, has provided a corridor which enables birds to penetrate right into the heart of the urban area and Mallard, Tufted Duck and Coot have all been noted on a small basin close to the centre of Birmingham. Loss of traffic to the railways brought a steady decline in canal usage. Maintenance was neglected, lengths were abandoned and gradually less disturbed, more natural habitats emerged. Kingfishers and Grey Herons took to the quieter stretches as the rivers became more polluted; Mallard, Moorhen and even grebes and Reed Warblers moved in as emergent vegetation became established; Sedge and Willow Warblers, Whitethroats and finches colonised the scrub which developed along embankments and cuttings;

and, as bankside alders and willows have matured, wintering flocks of Siskin and Redpoll have arrived too. Since the last war interest in fishing has grown and more recently there has been an upsurge in pleasure cruising, and these have brought a new lease of life to the canals. Sadly, though, this has been at the expense of Kingfishers and Grey Herons, which are seen less often, whilst the clearance of scrub to keep banks and towpaths open has led to a reduction in Sedge Warblers and Lesser Whitethroats. Nonetheless, with so many hedges removed from farmland in recent years, the canals have extended their role as wildlife corridors right across the countryside and, with more and more land drainage, they have become increasingly important as feeding areas for hirundines and wagtails.

The quest for quicker transport has been one of the more interesting facets in the development of an industrial society. The canals were quickly ousted by the railways, which provided a similar linear habitat but without water. Since the passing of steam, however, embankments and cuttings have not been subjected to accidental, indiscriminate burning and scrub has steadily invaded the grassland to the benefit of species like the Linnet. It is not uncommon to see a Green Woodpecker attacking a railway fence or Goldfinches feeding on embankments or disused sidings, whilst both Stonechat and Black Redstart have been observed perching on gradient posts near the centre of Birmingham. Since the "Beeching axe" in 1963, many railways have been abandoned and their ensuing colonisation by scrub has made them safe sites for feeding, nesting and roosting. Some have even become rich enough to be designated as nature reserves.

The demise of the railways was a direct result of the greater flexibility and independence afforded by road transport. During the last twenty years or so, a national motorway network has been built

and again the West Midlands is its hub. Despite the many casualties sustained, roads are important for birds. Unsprayed verges are a rich feeding ground for finches and buntings and in certain areas the only remaining hedges and trees are those along the roadside. Even the tarmac is not unproductive, as it provides much needed grit for roughage. Most birds, particularly the small passerines, are found along country lanes, but main road verges are often frequented by Rooks or Carrion Crows, Chaffinches are commonly seen in roadside lay-bys and the Kestrel has become symbolic of the motorway. At night-time too, the headlights occasionally illuminate a Barn Owl perched on a lay-by litter bin ready to pounce on any unsuspecting, scavenging rodent.

Man's ultimate achievement in transport has been to emulate birds through flight. Surprisingly though, none of the Region's commercial airports has developed a reputation amongst bird-watchers as in other areas and it seems their bird-life is fairly mundane. Problems with bird-strikes have not been encountered, though the presence of Canada Geese on the new lake at the National Exhibition Centre is known to have caused concern to the authorities at Birmingham airport. Generally the plethora of abandoned military airfields – a legacy from the last war – is of greater interest. Most have now reverted to agriculture and are being increasingly ploughed, but where grasslands remain their short swards and lack of hedges make attractive feeding grounds for Lapwing, Golden Plover, Starling, Skylark and Meadow Pipit and rich hunting territories for owls, many of which roost or nest in the derelict buildings.

Transport apart, it is the services demanded by an urban society which have the most noticeable impact beyond the confines of the built-up area. Industrial and population growth, coupled with higher standards of hygiene, have produced a phenomenal increase in the demand for water and the past fifty years have witnessed the construction of several water-supply reservoirs of ever-increasing size. The most recent of these, Blithfield and Draycote, are on a scale hitherto unknown in the West Midlands and both have become important focal points for wintering wildfowl and gulls, and breeding and passage birds of all kinds. With depths suitable for divers, grebes and diving duck to feed; sufficient area for safe roosting; and fluctuating water-levels that expose muddy margins for waders, especially during autumn passage, the water-supply reservoirs have added a new dimension to the Region's avifauna and many species considered rare fifty-years ago are regular today.

The demand for water has been matched by that for power, and giant power stations have sprung up along the Severn and Trent valleys, where they can take advantage of an abundant supply of water for cooling. Abstracting water for cooling and then returning it to the rivers, artificially raises the temperature and this helps to keep the rivers open longer during freezing conditions. Wildfowl, waders and gulls all take advantage of this as well as exploiting the ash lagoons associated with many stations, whilst the surrounding scrub and rough pasture can hold species like Whinchat and Meadow Pipit. More recently Black Redstarts have colonised some of the buildings, whilst Kestrel and Rook have nested on switch-gear and both Cormorants and Shags have used cooling towers for roosting.

The increasing demand for clean water has also created the problem of its disposal after use. Fifty years or more ago this was done through old-fashioned sewage farms, with their settling lagoons, and in those days these were one of the Region's primary haunts for dabbling duck and migrating waders. Clinical modernisation has steadily diminished their importance, but the sheer size and seclusion afforded by some installations ensures their continuing attraction to birds, notably wagtails.

Since the last war, most of the Region's bird-watchers have diverted their attention from the sewage farms to the gravel pits. As the post-war building boom of the 1950s and 1960s gathered momentum, engineering and architecture entered their concrete eras, creating a virtually insatiable demand for sand and gravel. River valleys and the pebble beds of the higher plateau became scarred with a rash of quarries, where dredgers or draglines gouged great holes and conveyors clattered endlessly for miles before ascending some grotesque lattice steelwork to spew forth their contents onto great conical mounds. Incongruous as such surroundings were, they proved most attractive to birds and today gravel pits are amongst the richest habitats in the West Midlands. More often than not deposits were below the water-table and could only be worked by dredging or pumping. Dredging created instant lakes, which were quickly colonised by ducks, Mute Swans and Canada Geese, whilst pumping resulted in dry pits with conical spoil heaps and patches of bare gravel, mud or shallow water, which provided migrant waders with an ideal substitute for their fast disappearing sewage-sludge lagoons. Little Ringed Plover, in particular, became synonymous with gravel pits after it first nested here in 1952, but many other species found conditions to their liking too. Sand Martins soon established colonies in the quarry faces and those excavations that were back-filled with silt from the quarry washing-plants quickly developed marginal aquatic vegetation or willow scrub, where warblers breed and thousands of hirundines and wagtails roost each autumn. During the initial stages, the surrounding, disturbed land was colonised by an assortment of weeds, which became a paradise for feeding finches, pipits and larks. Subsequently the growth of scrub and its succession to trees

2. In their early stages, gravel pits often hold wildfowl and waders, particularly breeding Little Ringed Plover.

Bill Smallshire

3. As nature heals the scars of extraction, gravel pits often mature into a haven for passerines like Reed and Sedge Warblers.

J. V. & G. R. Harrison

has further increased diversity. Once working has ceased and pumping stops, there are usually two options for a gravel pit. Either the excavations can be left to fill naturally with ground water, in which case they are an obvious attraction to wildfowl, gulls and passage waders or terns, or they can be filled with one of industrial man's by-products such as household waste or power-station ash and returned to agriculture. Examples of both will be found throughout the West Midlands.

Reference to household waste highlights another of society's more pressing problems, namely disposing of the growing mountain of domestic rubbish. Dumping it into holes in the ground has been the standard practice for many years now and Starlings, corvids and gulls have become familiar scavengers at such tips. Gulls especially have exploited this source of food. Fifty years ago the large gulls were a rare sight so far inland, occurring only after storms, but today they are commonplace on most tips throughout the winter and cause some consternation by resorting in their thousands to roost on reservoirs at night. The Control of Pollution Act 1974 introduced new measures to regulate the operation of tips and, amongst other things, the code of practice seeks to reduce the number of scavenging gulls by restricting their food supply. Those studying gulls at rubbish tips report little evidence of a drop in numbers, however, though there has been a noticeable decline in the size of most roosts over the past few years. Perhaps it is simply that licensing has led to fewer tips, particularly since some local authorities initially favoured incineration, though this practice is now being abandoned because of its cost.

There are other influences stemming from an urban society, like mining and quarrying, which are locally important, but despite its industrial strength and overwhelming urban population, the West Midlands is still predominantly agricultural in terms of land use, with two-thirds of its area devoted to crops and grassland. Since 1929 some 70,000 ha of farmland have been sacrificed to urban development and this has undoubtedly caused a contraction in the range of many farmland birds. Of much greater significance, however, are the more subtle changes in rural bird communities that have paralleled changes in farm husbandry.

The annual agricultural statistics produced by the Ministry of Agriculture, Fisheries and Food give a useful insight into the broad changes in crop areas during the last fifty years, as Table 1 shows.

To place these changes into context,

Table 1: Percentage of Total Area under Crops and Grass of Selected Crops in the Region.

Crop	1929	1939	War	1949	1959	1973
Arable	26·2	23·1	54·2	52·3	49·3	53·4
Temporary grass*	5·8	4·9	10·9	15·0	18·8	15·3
Wheat	3·6	5·0	16·2	9·0	7·6	10·7
Barley	0·9	0·3	2·0	3·1	5·9	16·6
Oats	5·3	3·5	9·1	6·5	4·1	2·3
Permanent grass	73·8	77·0	45·8	47·7	50·7	46·6

*comprises clover and short-term leys, mainly the latter in recent decades

The Ministry of Agriculture, Fisheries and Food has kindly permitted this table to be published and reference to be made to its publication "Agriculture in the West Midlands" 1979.

though, it is necessary to delve a little deeper into history. The agricultural landscape as we know it today was largely fashioned during the reign of George III (1760–1820), when the Enclosure Acts were passed. Prior to then, the Midland scene had been one of open countryside and adjacent woodland "wastes", but, within the space of sixty years or so, it was divided by the Land Commissioners into two-to-six hectare fields with elm or thorn hedges. According to Hoskins (1955) this had a marked effect on bird life. Certainly, with woodland diminishing at the same time, these new hedgerows provided a ready replacement for scrub species displaced from the woodland fringe. Enclosure must have led to fewer Skylarks, partridge and Lapwing, but it has been suggested (Murton 1971) that overall the bird population increased three-fold as a result. Initially the main beneficiaries were Dunnock, Greenfinch, Whitethroat and Yellowhammer, but later, as the sapling ash, elm and oak rose from the hedgerows, other species like the Chaffinch profited too.

Widespread importation of cheap grain from the American prairies caused an agricultural depression in the late 1870s and 1880s which, apart from a brief respite resulting from the ploughing campaign of the First World War, continued until the outbreak of the Second World War. Capital was lacking for farm improvement and labour could not be sustained, so maintenance standards fell. As a result much marginal land was converted to permanent pasture or in extreme cases permitted to revert to scrub. Buildings became decrepit; hedges neglected and overgrown; and ditches blocked by rank vegetation. On some of the heavy lands south of the Avon, for example, there had been little change since the enclosures and even today remnants of the old ridge-and-furrow system are widespread in this area.

In short, much of the land was being inefficiently farmed and for many birds the countryside of fifty-years ago, when three-quarters of the land was under permanent grass devoted to dairying and stock rearing, had perhaps reached its zenith. The small fields of around four hectares were bordered by stock-proof hedges and much of the grassland was ancient and hosted a much richer flora and invertebrate fauna than the more productive and popular short-term leys of the post-war era. Furthermore, late haymaking was a much safer affair for ground-nesting birds than the more hectic, current fashion of cutting grass for silage two or three times between May and August.

The ploughing campaign of the Second World War began a revolution based on an ambitious mechanisation programme that was to bring fundamental and lasting benefits to an ailing agricultural industry. Until the outbreak of hostilities, the horse had remained the powerhouse of the farm, but with the war-time labour shortage it was quickly ousted by the tractor. This opened up completely new horizons by permitting cultivation of the heaviest land and allowing this to be sustained even when import restrictions were lifted after the war. Since then, agricultural technology has developed apace and today this is one of Britain's most efficient industries. The ploughing campaign, more than any other factor, was responsible for the decline of 45 per cent in permanent grasslands between 1929 and 1973, which is the last year for which data are available for the old county of Worcestershire. Since then the area of permanent grassland has declined still further. During the war some 170,000 ha of permanent pasture were ploughed and most of this was put down to short-term leys (55,000 ha), wheat (51,000 ha), oats (30,000 ha) and barley (15,000 ha). Import restrictions meant that home-grown lamb and mutton had to be eaten and this, combined with the loss of pasture, resulted in a drop in the Region's sheep population from 738,000 to less than 300,000 in 1944.

Since the war, however, flocks have usually totalled around 6–700,000. By contrast, cattle numbers rose steadily from 369,000 in 1929 to over 600,000 by 1973, with about half of these in Staffordshire.

Accompanying the post-war explosion in machine usage was the demise of the working horse, from 37,317 in 1929 to only 3,696 in 1958. This led to a considerable reduction in the area devoted to oats, from 49,000 ha in 1942 to only 12,000 ha in 1969. Even more dramatic, though, was the almost exponential post-war increase in barley growing, from a mere 1,850 ha in 1939 to over 90,000 ha by 1969. This has been concentrated on the lighter, sandier soils, especially those of south Staffordshire, north Worcestershire and mid-Warwickshire. By contrast, the winter wheat crop favours heavy land, such as the Lias Clays of the south of the Region. The area under wheat more than tripled in the early years of the war, but then declined with the easing of import restrictions and about 50,000 ha has been the post-war norm. The general increase in arable land after 1939 must have resulted in a great overall increase in the amount of invertebrates made available to birds by cultivation. To what extent this influenced bird population changes, for example the dramatic increase in wintering Black-headed Gulls, is largely unknown, but the effects must have been considerable. More efficient land usage, denser plant stands, greater use of pesticides and more efficient harvesting have led to a great reduction in the length of time cereal stubbles are left unploughed and the quantity of spilt grain, weed seeds and invertebrates to be found – all of which makes them poorer for birds.

Traditionally a fallow, or break crop, such as field beans was used to "rest" the land after cereals or root crops, but in recent years there has been a trend towards continuous cereals along with minimal cultivation on heavier land. However, one very lucrative break crop – oilseed rape – has exploded in popularity from 162 ha in 1973 to 2,300 ha in 1978. The brilliant yellow of a flowering rape crop in June has become a familiar sight in Warwickshire and very recently in other areas too. A tall, dense crop that is rarely traversed by machinery, rape is a good habitat for birds. Reed Buntings, now breeding widely in dry situations, find rape eminently suitable for nesting. Young rape plants are grazed by Woodpigeons in winter; the ripening seeds may be taken from their pods by flocks of Linnets or other finches; pigeons and doves take the spilt seed after harvest; and the crop holds many insects which provide food for warblers in summer. Rarely has such a beneficial situation been experienced by farmland birds. Indeed, most agricultural change has been to their detriment, though fortunately in the West Midlands the soils and climate generally favour mixed farming, so its countryside has escaped the worst deprivations of more arable-intensive areas. None the less, change there has been, of which removal of hedgerows, improvements to drainage and use of chemicals have most affected bird-life. Modern machinery operates most effectively in regularly-shaped fields of fifteen-to-twenty hectares rather than the two-to-six hectares prescribed under the Enclosure Acts, so during the 1950s and 1960s anything from a quarter to a half of the Region's hedgerows disappeared in the name of progress. Once many farms had a pond – often a marl pit excavated for fertiliser or to provide bricks for the farmstead – but these too were filled and extensive drainage improvements reduced the incidence of flooding and enabled traditionally damp, waterside meadows to be brought into cultivation. Ironically, this has resulted in the water table being lowered and the severe drought of 1976 forced some farmers to replace their vanished farm ponds with new storage reservoirs. Often these have been stocked with trout to create a fishery and many have become valuable habitats for birds.

The classic declines of Corncrake and Peregrine, attributed to mechanised hay-cutting and the use of persistent insecticides respectively, are well-known and are reflected in WMBC annual reports. A few Common Bird Census (CBC) plots in the Region have tried to monitor breeding densities since 1962, but farmland is generally much neglected by bird-watchers and there is little information on the population changes of many farmland birds, breeding or otherwise, or reasons for any status changes. Many questions are without answers. For example, what are the effects of hedgerow removal and flail-mower cutting on breeding and berry-eating birds? Has the increased passage of machinery over the ground reduced the nesting success of ground-nesting birds? What caused the dramatic decline in Rook numbers in the last two decades? Many species are dependent on earthworms as a major food item, but how have their populations been altered by changes in husbandry? What are the effects, direct or insidious, of the vast quantities of pesticides which have been liberally scattered over crops? Alas, husbandry changes so rapidly nowadays that it may never be possible to answer such questions fully, even after concerted research effort.

Our knowledge of where farmland species occur is considerably greater than that of why they are there. As arable crops now account for about half of agricultural land, let us deal with the birds associated with these crops first. Lapwing and Skylark are the typical breeding birds of large cereal fields, with occasionally Grey Partridge, Curlew, Yellow Wagtail, rarely Quail and, in increasing numbers, Corn Bunting in barley and Reed Bunting in wheat. Yellow Wagtails breed in potatoes and sugar beet on the lighter soils and in field beans and oilseed rape on the heavier lands. Most crops are kept clean with insecticides, so birds feeding on insects are confined to the diminishing hedgerows, but in young cereals, stubble and cultivated land, the ground and soil invertebrates and seeds attract feeding partridges, Lapwing, Golden Plover, Carrion Crow, Rook, Jackdaw, Black-headed Gull, Little Owl, pigeons, Pied Wagtail, Meadow Pipit, Starling, Wheatear, finches and thrushes and the increase in winter wheat has certainly helped to sustain large numbers of Lapwing, Fieldfare and Redwing in particular.

Grassland bird populations are dependent primarily on the number of livestock and the species diversity of the sward. Generally speaking, low stocking rates and agriculturally poorer swards are favoured by breeding birds such as partridges, Lapwing, Redshank, Curlew, Snipe, Skylark, Yellow Wagtail and Reed Bunting – the actual species present depending on the wetness of the field in question. Most have declined as more and more land has been drained, but Moorhen, Sedge Warbler and Reed Bunting have all shown themselves able to exist well away from water. Many of the foregoing, along with Golden Plover, Black-headed Gull, corvids, thrushes, Starling and Meadow Pipit, prefer short, well-grazed grass for feeding. Rough grassland may be hunted over by owls, and grass left to be cut for hay or silage may on rare occasions attract Quail or Corncrake. After mowing the wealth of exposed invertebrates attracts flocks of Rooks and Starlings.

Hedgerows still remain a characteristic part of the farmland scene over virtually all of the Region. As breeding haunts, they hold many of the common woodland species in addition to partridges, Whitethroat, Dunnock, Yellowhammer, Linnet, Reed Bunting and, when at their thickest, Turtle Dove, Magpie, Lesser Whitethroat and Bullfinch. When hedges are left uncut and allowed to bear fruit, then hips, haws, elderberries and the like provide food for

4. Much of the Region is still agricultural and its fields, hedges and copses harbour a good variety of birds.   *J. V. & G. R. Harrison*

winter thrushes, Greenfinch, Bullfinch and Blackcap. Where standard trees have been left in the older hedgerows, there may be breeding Kestrel, Little Owl, Barn Owl, Rook, Carrion Crow, Jackdaw, Tree Sparrow, Chaffinch and, rarely, Hobby. Mechanical cutting to a proper shape keeps a hedge neat, tidy and functionally stock-proof, but mitigates against species like Turtle Dove and Lesser Whitethroat that prefer tall or thick growth. It also precludes the emergence of saplings which are so important as song posts for species like the Chaffinch. Couple to this the practices of spraying into hedge bottoms, which eliminates so much of the bottom growth in which many species nest and feed, and straw or stubble burning, which each year accidentally destroys many hedgerows, and the picture is indeed a

gloomy one. As a final, devastating blow, a particularly virulent attack of Dutch elm disease in the late 1970s has killed virtually every elm in the Region. In parts of Warwickshire and Worcestershire over 80 per cent of all trees were elm and have had to be felled. Rooks in particular, which were low in numbers before the outbreak of disease, must have suffered from it, since, in 1973, half of them were nesting in elms. Hedgerows are a sub-optimal habitat for woodland birds and Murton (1974) has suggested that because of this they have become a red herring in the real conservation issues. In the West Midlands, where the stock of woodland is itself below the national average, it is hard to subscribe entirely to this view.

Where woodland remnants survive on farms, or copses have been planted, many typical woodland species occur and even quite small plantations may add to the farmland avifauna birds like Redpoll, Goldcrest and Sparrowhawk – the latter making a welcome recovery now that the dangers of certain "persistent" insecticides have been recognised and the offending products restricted. Most woodland on the farm, of course, is kept as cover for Pheasants and these are jealously guarded. Allied to the copses and small woodlands are orchards, which were once widespread throughout the Avon and Severn valleys. Throughout the Region the area of orchards has more than halved in recent decades, smaller trees are used to facilitate harvesting and there is more human activity, so orchards have generally become a poorer place for birds, with fewer natural holes for owls and less suitable trees for woodpeckers. In the Vale of Evesham especially, the loss of orchards has been enormous and the once famous plum has almost entirely vanished in the transition from a fruit growing district into an intensive horticultural unit, where glasshouses and polythene cloches are as common as orchards used to be.

The pond is another traditional farmland habitat that has disappeared on a scale similar to the hedgerow. Many ponds hold Moorhens, breeding Mallard and even Sedge Warblers in more overgrown situations, and all are important to a whole variety of birds for drinking and bathing. Last, but not least there are a number of species which can be regarded as typical of farm buildings themselves. Classically the Swallow is *the* bird of farmsteads, although the Barn Owl of dark old barns is an equally welcome visitor. Pied Wagtails and House Martins often nest around the farmstead too, and Collared Doves and House Sparrows make full use of an easy supply of grain around the out-buildings. Here again, though, there have been changes. For example, intensive rearing and the indoor housing of stock have led to cleaner farmyards and consequently fewer insects for the low-flying Swallow to hawk. Many old buildings have been demolished to make way for modern structures that offer fewer nest and roost sites for owls and, with the virtual disappearance of the old rick-yard, many grain-eating species have been forced to look elsewhere for winter food.

Whereas most species have a highly evolved ecological separation in their natural woodland habitat, in a sub-optimal farmland habitat they are often competing for the same food. Inevitably some have proved more successful than others and unfortunately a number of the most successful species have come to be regarded as pests. The Woodpigeon is partial to clover and brassicas and countless thousands have been shot, whilst the Collared Dove's habit of congregating round grain stores has earned it not a little disrepute. Bullfinches are a locally severe problem, taking the buds of plums, apples, pears, gooseberries, damsons and blackcurrants in roughly descending order of preference. Skylarks may damage sugar beet and brassicas, but corvids can be a worse problem. Rooks, especially, dig up recently sown grain and, along with Mallard, House Sparrows and sometimes Jackdaws, may take cereal grain prior to harvest. Coots, Moorhens, Mute Swans and, in particular, Canada Geese graze grass and cereals and the latter species suffers control measures because of this. In the Packington area feral Grey-lag Geese are also a nuisance.

The other major habitat of the West Midlands is woodland. Vast forests, such as those of Cannock Chase, Malvern Chase and Wyre, once covered most of the Region, but these have been steadily eroded by man and today, for example, only the wealth of half-timbered buildings in Worcestershire and Warwickshire bear testimony to the extent of forests like those of Feckenham and Arden – though in truth the latter was never more than a well-wooded district. Man's onslaught on the forests was begun initially to clear ground for agriculture and to provide wood for kindling, building and fencing, but later more and more areas were felled by charcoal burners for the furnaces of the Black Country and the Potteries – a process which continued well into the eighteenth century until charcoal was finally ousted by coke. All too frequently animals were introduced under the ancient rights of grazing and pannage and these prevented natural regeneration, so steadily the landscape changed from extensive woods and forest, with just a few clearings, into its position fifty years ago of largely open countryside with small, fragmented woodlands. During the past fifty years the Region's stock of woodland has increased slightly, but there have been some important changes in its management, notably a drastic decline in coppicing and a spread in conifer plantations, and these have changed its structure profoundly.

Today very little primary woodland, that is land that has been wooded for at least two thousand years, remains, but what there is shows well the influence of climate and soils. In the West Midlands the climate suits broad-leaved rather than coniferous trees, with the dominant species determined by the underlying rocks and the diversity of species by the management regime. For example, the ashwoods that clothe the limestone dales of north-east Staffordshire contrast sharply with the sessile oakwoods and birch scrub of the neighbouring moors. The ash, normally sub-dominant to either oak or beech, is here the dominant species because the soils are too shallow to support oak. Its lighter foliage creates a more open canopy through which sunlight can penetrate, so the shrub layer is well developed and the field layer is also rich. Willow Warbler and Chaffinch are the

commonest species in these ashwoods, but Blackcap find the mixture of mature trees and a good shrub layer much to their liking too.

Sessile oak is dominant on the steeper slopes of the Old Red Sandstone west of the Severn as well as on the moors. In both places the combination of a higher rainfall and well-drained soil is much to its liking. It forms a close canopy which excludes light, so the shrub layer is more sparse and the field layer less varied, being frequently dominated in the lower valleys by grass- and in the uplands by heather-communities. Chaffinch and Willow Warbler are again commonly found, together with tits, but the characteristic birds of sessile oakwoods are Pied Flycatcher, Wood Warbler, Redstart, Tree Pipit and, to a lesser extent, Woodcock. In general the Willow Warbler prefers the birch scrub that is frequently found around the higher perimeter, where it may be extremely abundant. In winter the more exposed sessile oakwoods, notably those on the moors, are partially deserted by species such as the Chaffinch and Robin and only a few wandering parties of tits and roosting thrushes remain.

Beechwoods are not common in the West Midlands, though stands of beech are sometimes encountered in mixed woods. Those on the Jurassic outcrops of the Cotswolds have been largely cleared, though here and there a graceful hanger still clothes the steepest scarp slopes. Sadly, though, these appear to be too fragmented to support the characteristic avifauna found in the beechwoods of southern England and to find this one has to look to some of the smaller woods south-west of Birmingham or around the Potteries. The interlocking mosaic of beech leaves obliterates sunlight almost completely, so that apart from an odd evergreen holly or yew, the shrub and field layers are often non-existent and the wood is carpeted with leaf litter. In such situations Chaffinch, Blackbird and Great Tit are usually the commonest breeding species, but conditions can suit Nuthatch and Wood Warbler too. Beechwoods come more into their own in winter, however, when beech-mast figures prominently in the diet of many birds. The mast crop varies from year-to-year, but overall appears to follow a biennial cycle and in good years foraging flocks of tits and finches, especially Brambling and Chaffinch, gather to feed in considerable numbers and small parties of the elusive Hawfinch may also be encountered.

If there is one tree that exemplifies the West Midlands it must be the pedunculate oak. This is the dominant woodland tree throughout most of the Region, occurring right across the heavy clays of the drift-covered Keuper marl and on the Coal measures and Permian beds. In north-east Warwickshire, though, where the drift is composed of a chalky boulder clay, the pedunculate oak is supplanted by the ash as the dominant species. In the main, pedunculate oakwoods are small and fragmented, having so often been cleared to make way for agriculture or urban development. That they contain a much wider diversity of species than the sessile oakwoods is largely because many of them were managed until recently as coppice-with-standards, some since medieval times. This has produced short trees with spreading crowns at well-spaced intervals, between which light can penetrate freely. Furthermore oak comes into leaf late, so conditions are perfect for secondary growth and both shrub and field layers are rich. Rotational cropping, particularly of hazel, on a five-to-fifteen year cycle also ensured a varied age structure within the understorey and this added to diversity, with flowers in the field layer flourishing immediately after coppicing, but gradually being shaded out or prevented from flowering as the stools regrew. Indeed, it is woodlands like these, with their carpets of bluebells in spring, glades of dappled shade in summer and abundance of haws and holly berries each autumn, that conjure up the popular, poetic image of woodland. It is in such woods, too, that Wren, Robin, Blackbird, Dunnock, Blue Tit, Song Thrush and Great Tit join the Willow Warbler and Chaffinch amongst the commoner breeding species and where large numbers and a great variety of birds come to feed and roost in winter. Oaks are especially important to Jays and Wood-pigeons, which feed on acorns, but it is the diversity that draws such a wide range of finches, many of which specialise on particular food plants. Management is important too. Rotational cropping of a coppice used to benefit the Garden Warbler, which was once as common as the Blackcap, and was most important for the Nightingale, which here is on the edge of its range and consequently exhibits fairly fastidious habitat preferences. Coppiced woods, of course, are managed habitats and Simms (1971) has suggested that perhaps they are more typical of scrub than true woodland. Nevertheless, they added much to the Region's avifauna and their continued loss through clearance or reversion to scrub or high forest as a result of lack of management is to be regretted.

Many of the smaller woods referred to so far are in the form of farmland spinneys, copses or coverts. For centuries they provided a ready supply of timber for the repair of fences or buildings, or as a source of fuel. With new building materials, more efficient fuels and less fencing because of larger fields, however, there is little demand for timber on the farm today. Coppicing, furthermore, is labour-intensive and therefore expensive to a mechanised industry. In short, the need for small woodlands on the farm has all but disappeared. Most are now neglected and at best maintained only as game coverts, but the spiralling cost of land is forcing more and more farmers to look again at their small woods to see whether they can be brought into productive farming. With modern machinery the answer is increasingly yes, and there is a real danger that these woods will pass into history within the foreseeable future. In the Severn and Avon valleys, particularly, the situation has been exacerbated through so many spinneys being devastated by Dutch elm disease.

More woodland has survived on the large estates, often because of the strong sporting interests, and the more enlightened attitude of keepers these days has enabled species like Sparrowhawk and Buzzard to stage a come-back. Even on the estates, though, the position of woodlands is precarious, since capital transfer tax is punitive and encourages owners either to abandon their woods or else to fell them and replace their hardwoods with the fastest growing softwoods, so that these in turn can be cropped on the tightest possible rotation. More and more woodland is passing to commercial forestry enterprises, which share the same basic objectives as the Forestry Commission, but without the same obligation to have regard for amenity and wildlife.

Today in the West Midlands, woodland is extensive only on the sandstones and Bunter pebble beds, especially at Wyre Forest and Cannock Chase. Here the natural vegetation comprises oak-birch woodland and dry heath, but because the thirsty, acidic soils have so little agricultural value, large areas have been afforested with conifers. Birch grows both in wet ground, where its rotting stumps will attract Willow Tit and Great Spotted Woodpecker, and in dry situations with a field layer of bracken or heather, where it is usually an intermediate in the natural succession to oak woodland. Scattered birches in a heathland setting are a favourite haunt of Redpoll, Willow Warbler, Tree Pipit and sometimes Nightjar

too, whilst in winter their seed draws flocks of tits, Redpoll, Siskin and Bullfinch, which have to indulge in acrobatics in order to feed from the slender twigs. Perhaps it is as well that birch thrives on such impoverished soils and regenerates so freely, since it has little or no commercial value and must otherwise have surely declined in a Region where land is always at a premium. Sadly the oaks in these situations are less healthy. The oldest are stag-headed and full of natural cavities that house owls, Jackdaws, Redstarts, Tree Sparrows and the like, but, despite willing assistance from a thriving population of Jays, they are frequently slow to regenerate. Often the oaks show characteristics intermediate between sessile and pedunculate and the bird-life in these woods is equally mixed.

In wetter situations, such as meres, mosses, disused gravel pits and the old marl pits of the Severn and Avon valleys, willow and alder woods occur and these too are often rich in bird-life, with breeding warblers, Lesser Spotted Woodpecker and Willow Tit; and flocks of Redpoll and Siskin in winter.

Excluding primary woodland, many of today's woods were planted when the large estates were laid out in the eighteenth century. A further surge of planting occurred between 1820 and 1880, to replace timber felled during the Napoleonic wars, and more recently another since 1950 to replenish the depleted stock of woodland after two world wars. Statistics provided by the Forestry Commission and reproduced in Table 2 give some insight into the changes in woodland over the past fifty years and acknowledgements are due to the Commission for permission to publish these.

Direct comparisons between the three surveys are complicated by subtle differences in sample size and structure and by small changes in the classification of woodland, but broad trends can be clearly

Table 2: Extent and Structure of Woodland in the Region.

	1924	1947	1965
Woodland as a percentage of total land area	4·3	4·4	4·6
Structure of woodland as a percentage	100·0	100·0	100·0
High forest	50·2	52·9	65·7
Coniferous	11·4	13·6	32·5
Hardwood	23·6	34·9	33·2
Mixed*	15·2	4·4	
Coppice and coppice-with-standards	26·9	9·7	0·2
Scrub	5·6	17·5	32·2
Felled, devastated or uneconomic	17·3	19·9	1·9

* The 1965 survey classified high forest as mainly coniferous or mainly hardwood and the mixed category was dropped.

discerned. The actual area of woodland increased from 4·3 per cent of the land area in 1924 to 4·6 per cent in 1965. This was a fairly small increase however and the proportion of wooded land in the Region is still below the national average, especially in Warwickshire where the cover is less than 3 per cent. Changes in woodland structure are much more evident, especially the decline of coppice and the increase in the area of coniferous forest since 1947. For example, in 1924 a quarter of all woodland was classed as coppice or coppice-with-standards, but by 1947 only 10 per cent fell into this category and by 1965 it had all but disappeared. With the cessation of coppicing, these types of woodland, depending on age and species, tended either to degenerate to scrub or to develop towards high forest. This is reflected in the 1947 figures, where there has been some addition to both the area of broad-leaved high forest and of scrub. A substantial part of the increase in the area of the latter category also arose from fellings between the wars and during the last war. Between 1947 and 1965 coppiced woodland declined by a further 10 per cent and scrub increased by 15 per cent. The size of this latter figure is also indicative of a general decline in the management of

many small, privately-owned woodlands. The reasons for the decline in coppicing and its effect on bird-life have already been discussed, but in short, arboreal species have benefited at the expense of scrub-loving ones. Arboreal species have also been favoured by the increase in the proportion of high forest, from around a half in 1947 to two-thirds by 1965. During this period, however, hardwoods probably declined slightly and the increase was due to afforestation or replanting with conifers, mainly Corsican and Scots pine, larch and spruce.

Initially these plantations, with their dense stands of alien species, were considered poor for birds. With maturity, felling and replanting, however, their age structure has become more varied and it is now recognised that they have enriched the regional avifauna by adding to its diversity. Woodcock, Goldcrest, Great and Coal Tits, Tree Pipit and Redpoll are typical breeding birds, but in their younger stages plantations may hold Nightjar or Grasshopper Warbler, whilst with maturity they attract Turtle Dove, Sparrrowhawk and rarely Long-eared Owl or even Goshawk. Mature larch and spruce are visited by Crossbills and after major irruptions a few pairs may breed, while Siskin are

5. Oakwoods with a formerly coppiced understorey are a feature of the Midlands countryside. *A. Winspear Cundall*

found in the same plantations in increasing numbers each spring.

Complementary to its woodlands, the West Midlands has a rich heritage of landscaped parklands, dating mostly from the eighteenth century. Many were laid out by Humphrey Repton or "Capability" Brown, who were the greatest exponents of their art, and as they have matured their subtle blend of small woods, copses, specimen trees, water and grassland has greatly enhanced the beauty of the countryside, especially in the Avon valley. They have also brought benefits to many birds. Owls, Kestrel, Stock Dove, Jackdaw and Starling all nest in the natural holes and crevices of the specimen trees and hunt or feed across the open grassland; ducks, Coot, Moorhen and warblers nest amongst the marginal or emergent vegetation of the pools; flocks of thrushes,

Woodpigeons and Lapwings feed on the grasslands in winter and the first two of these also resort to the woods for roosting. Nuthatch, Mistle Thrush and tits are also common in parkland, and the ornamental avenues of Wellingtonias provide roost sites for Treecreepers.

Today the country estate is such an integral part of the time-honoured Midland scene that the landscape would be unthinkable without it. Yet, with the escalating cost of upkeep and the burden of death duties, this is a very real possibility. Many have been donated to the National Trust, or opened by their owners to the public as a vital source of supplementary income, but others, like the Warwick Castle estate, which was "Capability" Brown's first independent commission, have been broken up and sold for agriculture or various other diverse purposes.

In the main their aquatic habitats are safe, though somewhat neglected in some cases, but the real evidence of decline is manifest in the trees. After nearly two centuries, many of these are well past their prime and in advanced stages of decay, with stag-headed oaks and derelict, wind-blown cedars an all too familiar sight. For many birds, such trees are now ideal, but soon they will have to be felled or they will topple and there has been insufficient replanting to ensure an adequate number of saplings to replace them. Neither is there much incentive to replant, with the result that there is now widespread concern over the consequent loss of our national heritage.

No review of the Region and its changes over the past half century would be complete without reference to the phenomenal growth in leisure and recreation that has occurred during the past three decades. Many factors have contributed to this, including a shortening of the working week, longer holidays and, until the end of the 1960s at any event, increasing affluence. More than anything, though, it can be attributed to the freedom arising from universal car ownership, which has brought all corners of the Region into easy reach of a steadily increasing proportion of the population. This has brought pressures to all parts of the countryside, but in a Region where the vast majority of the land is productively farmed, these have been most severe on a relatively few beauty spots and areas of general public access. Since 1968, when the Countryside Act set up the Countryside Commission and empowered local authorities to create country parks, the situation has been eased somewhat by the provision of proper car parks in suitable locations and more importantly countryside interpretation – a euphemism for environmental education. None the less, with over five million people searching for peace and solitude or a chance to "let off steam" in the countryside, the pressures are still acute. Hills, moors, heaths and water are most sought after and in these habitats birds are having to come to terms with increasing disturbance, ironically often from bird-watchers themselves. On water this varies in degree, from bankside picnickers or fishermen, to sailors, power-boaters or water-skiers. Despite the disruption, a healthy number of birds appear to survive in this habitat, partly because man's mobility on water is restricted and partly because many species visit the Region only in winter, when recreational pressures are least. However appearances are deceptive. Today, with so much more open water than ever before, numbers of wildfowl were bound to have increased. But the days of expensive new reservoirs may well be over and the Region's reserves of sand and gravel will one day be exhausted. Meanwhile sailing, power-boating, water ski-ing and most recently wind-surfing are insidiously, but steadily, increasing in regularity and spreading to more waters. Their effect on wildfowl is often dramatic, with an entire reservoir being emptied of birds at the appearance of the first boat. If the birds can drop in elsewhere this matters little, but alternative waters become fewer and fewer each year and the cumulative effect represents a very serious threat to both breeding and wintering wildfowl. The danger signs are there, often as subtle changes in species composition or behaviour rather than a decline in absolute numbers, but they are real none the less. For example, there appears to be a growing incidence of failed or non-breeding amongst wildfowl; and recently large numbers of Goosander have resorted to taking refuge on a tiny pool whilst the sailors are on their nearby reservoir. It is vital that the remaining undisturbed waters should be safeguarded now so that they can act as sanctuaries in the years ahead. If not, the future for wildfowl will be bleak indeed.

Upland, moorland and heathland birds are just as susceptible to disturbance, since in the main they share their habitat with man during the summer, when leisure activity is at its height and the birds are closely tied to their nest sites. Recreational activity may have had little influence on the disappearance of Woodlark and Red-backed Shrike, but it has certainly been a factor in the decline of the Nightjar. Tree Pipits too have declined, but Stonechats have made a welcome resurgence in recent years. On the moors the rugged terrain is less accessible to large numbers of people and this helps most species to hold their own, but even here there have been declines.

Overall, during the past fifty years the Region has gained in open water, but lost in heathland, marsh and general waste-land. Today there is more diversity of bird-life than half-a-century ago, but the populations of many common species have declined. Perhaps the most encouraging sign for the future is the growing aware-ness of, and support for, conservation, which is culminating in the creation of several permanent nature reserves and the management of other areas to meet the needs of wildlife. Welcome as this trend is, the events of the last fifty years surely emphasise that there is no room for com-placency in a world of rapid change and development. With the loss of inverte-brates and seeds from farmland and road-sides, the threat to trees and disturbance on moors, heaths and water, the bird-life of the Region is facing a bigger crisis than ever before.

So much for the broad changes of the past half-century. The next seven Chapters take a closer look at the habitats and bird-life of the Region as they are today. Many people have contributed to this review, particularly through the BTO Sites Register, but inevitably with an area of such size and diversity, not every site can be mentioned and it remains essentially a personal portrait, though hopefully none the worse for that. The chapters are based on the main physiographic sub-regions as defined on the accompanying map.

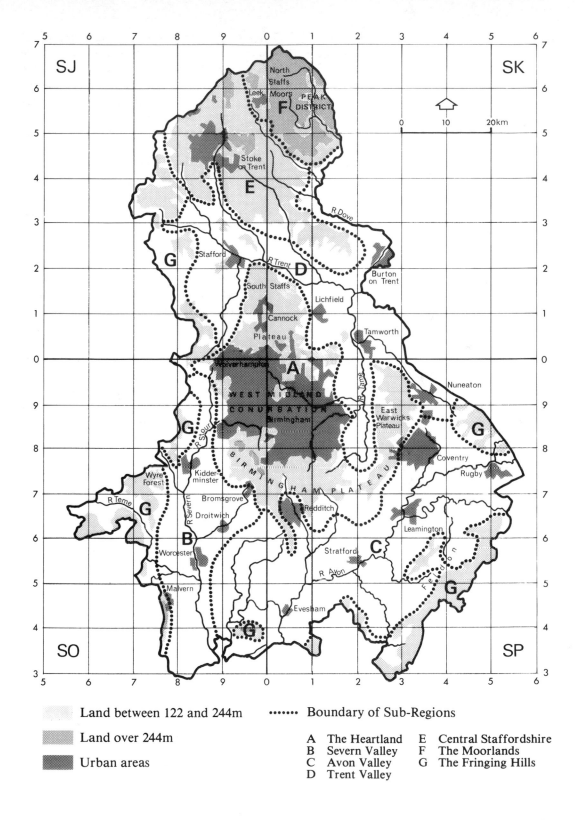

░░	Land between 122 and 244m	▪▪▪▪▪▪ Boundary of Sub-Regions
▒▒	Land over 244m	
▓▓	Urban areas	

A	The Heartland	E	Central Staffordshire
B	Severn Valley	F	The Moorlands
C	Avon Valley	G	The Fringing Hills
D	Trent Valley		

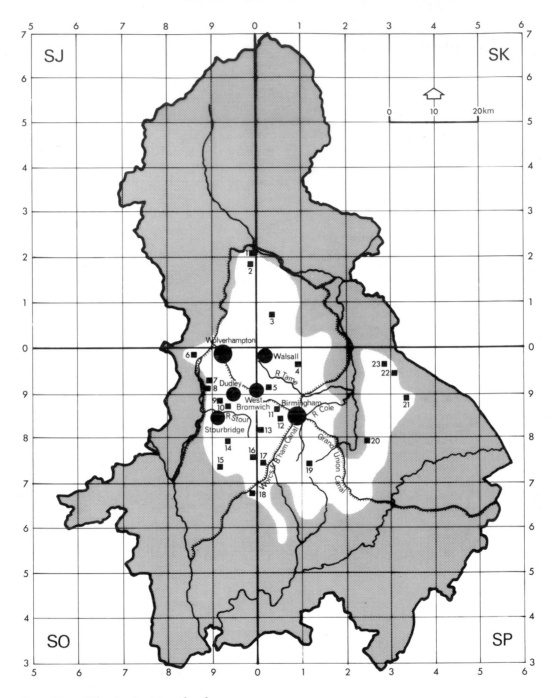

Selected localities in the Heartland

1 Shugborough Park	7 Baggeridge Wood	13 Bartley Reservoir	19 Earlswood Lakes
2 Sherbrook/Brocton	8 Himley Park	14 Clent Hills	20 Berkswell
3 Chasewater	9 Brierley Hill Pools	15 Chaddesley Wood	21 Arbury Park
4 Sutton Park	10 Saltwells Wood	16 Lickey Hills	22 Hartshill Hayes
5 Sandwell Valley	11 Edgbaston Reservoir	17 Bittell Reservoir	23 Bentley/Monks
6 Perton	12 Edgbaston Park	18 Tardebigge Reservoir	Park Wood

2 The Heartland

The heart of the Region is the Birmingham Plateau. This was first defined, though not mapped, by Lapworth (1913), since when it has been widely accepted as a physiographic unit despite varying opinion as to whether the 300 ft or 400 ft contour should delineate its boundary. In general terms though, it all lies above 100 m (328 ft) and rises to 243 m on Cannock Chase in the north and 315 m on the Clent Hills in the south-west. This is the second largest sub-region, with a horseshoe shape that covers 1,550 km², or 21 per cent of the total regional area, and measures 65 km from Cannock Chase in the north to Redditch in the south and 47 km from Stourbridge in the west to Nuneaton in the east. For the most part it is a level or gently undulating plateau between 100 and 200 m sloping generally to the east, where it merges with the surrounding landscape. To the north, west and south it falls away sharply into the valleys of various tributaries of the Trent, Severn and Avon respectively.

Apart from the higher areas being generally bleaker, climatic variations are small. Rainfall generally exceeds 650 mm per annum and rises to 760 mm on the high ground south-west of Birmingham, where summer storms are a feature of the micro-climate. With the heat emanating from factories and reflecting or radiating from buildings, Birmingham and the Black Country create a heat-island, which is most noticeable in winter when temperatures can be as much as 1°C above those in suburbs or the surrounding rural areas. On winter nights the warmer conditions attract substantial roosts of Starlings and during the bitter cold of the 1962/3 winter over thirty thousand were counted in the centre of Birmingham alone, huddled together along cornices, pediments and ledges. A normal winter, though, brings barely half this number. Similarly up to fifteen hundred Pied Wagtails have been recorded at factory roosts.

For most people this highly urbanised sub-region is their "back garden", where everyday bird-watching takes place. Teagle (1978) has provided a most fascinating account of the natural history of the Conurbation itself and much of the information quoted in this chapter has been gleaned from his work. Inevitably, the most popular habitats in the sub-region are the half-dozen or so canal feeder reservoirs, where a tremendous variety of birds can be seen during the course of a year. Of most importance, though, are the upland heaths of Sutton Park and Cannock Chase, which stand aloof and in sharp contrast to the industrial conurbation that adjoins them. Between them, they comprise the major part of the Region's heathland and provide a bridge between the lowland heaths of southern England and the northern moors. Cannock Chase is also of special significance as the regional stronghold of Nightjar. Finally there are some important woodlands, ranging from the conifer plantations of Cannock Chase to the sessile oakwoods of north Warwickshire, and one or two specialities, such as the winter flock of Twite at Chasewater and Black Redstarts in the industrial areas of the Conurbation. Reclamation of its feeding habitat may be placing the former species in jeopardy, but it is welcome to record that the latter has steadily increased over the past two decades and has become more widespread.

There are two distinct divisions to the sub-region, widely termed the South Staffordshire and East Warwickshire Plateaux, though neither accurately reflects their true geographical position. Between them lies the Tame basin. The South Staffordshire Plateau is much the larger part. Elliptically shaped, it extends from Cannock Chase in the north to Redditch in the south and from the Stour valley in the west to the Blythe in the east. The West Midlands Conurbation is dominant and only Cannock Chase in the north, the Clent Hills in the south-west and the open countryside south of Birmingham are really beyond urban influence. Within the heart of the Conurbation, though, the plateau is cleft from Walsall to Castle Bromwich by the upper reaches of the Tame. Prone to flooding, this area has never really invited settlement and even today provides an occasional open lung in a jungle of bricks and concrete.

The character of the plateau stems from its complex geology. There are small, but important outcrops of igneous and pre-Carboniferous rocks, but generally the plateau is underlain by Carboniferous coal measures or Permo-Triassic rocks. Folding and uplifting have brought each formation to, or sufficiently near, the surface for its mineral wealth to be exploited. The structure is completed by a patchy cover of glacial drift, mainly boulder clay.

In the very centre, the Middle Coal Measures are at or near the surface and it is these productive measures, with their coal, fireclay and limestone, that became the cradle of the Black Country, which is contained by Walsall to the north, Stourbridge to the south, Wolverhampton to the west and West Bromwich to the east. Mining of the thinner coal seams in the Cannock Chase field to the north came later.

Ironically, with three notable exceptions, such varied geology has produced a dull, dreary topography. These exceptions are the Clent Hills, a rounded outcrop capped with breccia and rising to 315 m; the Bunter beds of Cannock Chase, which reach 243 m; and the broken ridge of pre-Carboniferous rocks, which stretches from Sedgley in the north, through Dudley and Rowley Regis, to the Lickey Hills in the south, and reaches its highest points at Turners Hill (267 m) and the Lickeys (291 m). This ridge forms the watershed between the Severn and the Trent and marks a contrast to the landscape on either side. To the east the land falls gently to the flat basin of the Tame and its tributaries, with the only relief coming from drift-capped hillocks like those at Wednesbury, Solihull and Coleshill. To the west the whole drainage pattern has been disrupted by past glacial action, which caused the Severn to cut its great gorge at Ironbridge in Shropshire. As a result its rejuvenated tributaries have themselves cut deeper, steeper-sided valleys than the streams to the east.

In ancient times most of the plateau was Royal hunting forest and there are still remnants today on Cannock Chase and in Sutton Park. During the seventeenth and eighteenth centuries, however, there was a major transformation from a rural to an urban community, based on the supplies of coal and iron-ore, the growth of a skilled labour force and the creation and investment of capital.

The Black Country provided the raw materials and Birmingham attracted the business acumen. From the time the first canal was cut in 1769, Birmingham also became the centre of the canal network, with more waterways than Venice itself, and in 1826 a feeder reservoir was constructed to supply water to this system. Initially known as Rotton Park Reservoir, but now generally referred to as Edgbaston Reservoir, it covers 26 ha and is

6.   Large industrial complexes close to water are a favourite haunt of Black Redstarts.

Bill Smallshire

hemmed in on three sides by houses and on the fourth by old wharves and rolling mills. Nevertheless there are a few trees and shrubs around the perimeter that provide cover for passerines. Despite being used for boating, fishing and sailing, the water attracts some interesting birds. Over one hundred-and-twenty species have been recorded, ranging from regulars like Mallard, Tufted Duck, Pochard and Great Crested Grebe to rarities like Great Northern Diver, Red-necked Grebe, Ferruginous Duck and White-winged Black Tern. In fact almost anything might turn up and perhaps the most exotic sight was

that of a Flamingo, which once stayed for a few days and must have looked very incongruous in such a drab urban setting.

Before long Birmingham was firmly established as the undisputed capital of the Region, with an international reputation second to none as an industrial manufacturing centre. The advent of the railways strengthened this position and soon a vast assortment of trades could be found within the city, ranging from base to precious metals, screws to guns, and buttons to bedsteads. Finally, during this century, its versatile workers have switched to engineering, cycle, motor-cycle and vehicle manufacture, while its businessmen have further strengthened its importance as a commercial centre. Indeed the ability of its inhabitants to adapt to the needs of the time is perhaps the greatest asset of this fascinating city.

These fundamental changes in the industrial structure have also been reflected in the urban fabric and its bird life. As industries have declined, factories and land have become derelict and Goldfinches, Linnets, House Sparrows and even Corn Buntings have been seen in inner-city areas feeding on the weed seeds of disused land. Black Redstarts have shown a marked preference for large, derelict buildings and structures in power stations, gas works, railway stations and factories, especially those close to canals and derelict sites. For several years prior to its demolition, the derelict Snow Hill station was a favoured haunt, despite the young being vulnerable to predation by the local Kestrels.

Birmingham's physical form owes as much to the influence of large landowners as it does to natural features. As the city expanded so its wealthier residents moved upwind of the noise, smell and dirt of its factories. The refusal of the Calthorpe family to countenance workshops on their estate had a profound influence, with Edgbaston becoming an exclusive suburb of large, secluded houses in a delightfully sylvan setting. This character has even survived recent redevelopment and today, though no more than 3 km from the heart of Britain's second city, it can boast more trees than many a rural area.

Indeed Edgbaston Park is a unique urban sanctuary which, though close to Edgbaston Reservoir, is environmentally in an entirely different world. Part is occupied by a golf course, but the more interesting Site of Special Scientific Interest (SSSI) is managed as a nature reserve and its birdlife has been well studied by boys from the nearby King Edward's School. The central feature is the pool, lying at the foot of the slope from the hall. This in itself is magnificent, with islands and a mixture of submerged, floating and emergent vegetation. On the western shore is a reed-bed and marsh, backed by alders; whilst below the dam, on peaty, waterlogged ground, is a wet woodland of willow, alder, sycamore and downy birch. There is also a magnificent beech wood, which has oak, sycamore, holly and birch within it. The lake regularly holds breeding Great Crested and Little Grebes, Canada Goose, Mallard, Coot and Moorhen; Tufted Duck and Kingfisher are regular visitors that breed sporadically and Ruddy Duck is a recent colonist. Reed Buntings, Grey Wagtails and occasionally Reed Warblers nest around the margins, whilst the woods hold Stock Dove, Nuthatch, Great and Lesser Spotted Woodpeckers, Jay, Willow Tit, Blackcap, Willow Warbler, Chiffchaff, Tree Sparrow and a few pairs of Jackdaws, Magpies and Woodpigeons. Starlings and thrushes use the woods for roosting and rarities have included Merlin, Spotted Crake, Little Gull, Woodlark, Waxwing and Cetti's Warbler. A CBC has been undertaken on 16 ha here since 1966 and this has shown an average of 152 territories (943 per km^2), with Wren (26 pairs), Robin (20 pairs), Blue Tit and Blackbird (16 pairs each) and Dunnock and Willow Warbler (10 pairs each) the commonest species.

Not far from Edgbaston, in 1878 the Cadburys established the delightful garden village of Bournville around their new factory. With equal perception the Cadburys appreciated not only the need for good housing, but also the importance of recreation, and they gave considerable areas of open land to the residents of Birmingham, including the beautiful Lickey Hills. Elsewhere the city was less favoured and factories, with their attendant rows of terraced houses, spread in all directions, until they covered almost all available ground up to the rivers Tame and Cole. Gardens were small or absent and the few parks and cemeteries too isolated to create much wildlife habitat.

Beyond this, villages like Erdington, Handsworth, Harborne, King's Norton, Northfield and Selly Oak were expanding too, but here the houses had generous gardens and the environment was still sufficiently rural for species like Corn-crake to linger until 1933 and Hawfinch until 1948. Even in recent years, gardens in these areas have been visited by migrant Redstart, Pied Flycatcher, Woodcock, Wryneck, Ring Ouzel and Great Grey Shrike. During a four-year ringing study in a garden at Handsworth Wood (Edwards 1969) no fewer than 1,880 birds of twenty-three different species were ringed. Starling (54 per cent) was the most abundant, followed by Blackbird (21 per cent), Dunnock (5 per cent), Greenfinch (5 per cent), Blue Tit (3 per cent), Song Thrush (3 per cent), Great Tit (3 per cent) and Robin (2 per cent). The number of birds passing through and the small proportion of retraps (only 4 per cent of Starlings for example) was surprising, but ringing also showed an unexpected variety. Of the summer visitors, for example, visual observations had revealed only Willow Warblers, but in addition to the nineteen birds of this species that were trapped there were also three Garden Warblers, two Whitethroats and a Redstart.

For many years Bartley Reservoir was the mecca of Birmingham's birdwatchers. This is a drinking water supply reservoir of 46 ha made by damming a natural valley. Work was begun in 1925 and six years later the reservoir was almost full. In those days it was deep in open country-side, but today the houses and tower blocks of Birmingham have marched right up to its northern and eastern shores. The reservoir itself is harsh and unnatural, with concrete sides extending across the bottom for some thirty metres and a sur-

7. Tower blocks have now marched right up to Bartley Reservoir, which was once the mecca of Birmingham's bird-watchers. *J. V. & G. R. Harrison*

41

round of short grass and iron railings. At its deepest there is some twenty metres of water. Fearing pollution, Birmingham Corporation for many years resisted recreational use, but now this is being actively considered by its new owner, the Severn-Trent Water Authority. The old Corporation even resorted to a bird-scarer to dissuade gulls from roosting, but this proved largely unsuccessful as the records show. Nevertheless the bird population is erratic. Often little is present apart from a few duck like Goldeneye and a sizeable gull roost, but divers, the rarer grebes, most sea duck, Rock Pipit, Snow Bunting and a variety of passage waders and terns have been noted. For many years a Heligoland trap was operated in a nearby rick-yard and many finches and buntings were ringed. During 1950 over 400 birds were caught, of which Greenfinch (31 per cent), Chaffinch (18 per cent), House Sparrow (15 per cent) and Yellowhammer (12 per cent) were the commonest (Wolton 1951). At the western end of the reservoir, Bromwich Wood holds warblers, Spotted Flycatcher and Great Spotted Woodpecker, and Firecrests have been seen on more than one occasion.

For a time the Tame and Cole held the amorphous sprawl of Birmingham in check, but the city was bursting at its seams and during the last fifty years even these rivers have failed to halt the relentless tide of new housing estates which has engulfed the surrounding towns and villages and even spread to the very foothills of the Lickeys.

Three large, open lungs have survived within the city. The most remarkable of these is the Woodgate Valley, which, despite recent nibbling at its edges, still cleaves its way south of Quinton and Harborne to bring open countryside almost into Edgbaston itself. Here one can still find farmland species like Yellowhammer and Reed Bunting along with Tree Sparrow, Goldfinch and, in winter,

Fieldfare too, whilst chats are noted on passage and Stonechat has bred in recent years. To the north-west, separating Birmingham from the Black Country, is the urban oasis of Sandwell Valley – a legacy from the Earls of Dartmouth.

Sandwell Hall, which was built on the site of an earlier priory, has now been demolished along with an adjoining colliery, but Swan Pool remains, set in open farmland between two golf courses which are themselves entirely surrounded by urban development. The M5 motorway now bisects the area and during its construction the lake was drained and the opportunity taken to enlarge it for sailing. Sadly this led to the loss of some marshland and carr at the shallow end, but some 400 ha have been made into a country park and, whilst the part west of the M5 is used for formal recreation, that to the east is being left in a natural state and the farmland alongside the River Tame is being retained, apart from the construction of a balancing lake. New woods and hedges are being planted and pools created, with islands and natural shorelines, but even now the range of habitats is sufficient to attract a variety of birds. Out of 115 species recorded in 1979, for example, nearly 70 bred, including Swallow, Whitethroat and Yellow Wagtail. Priory Wood holds warblers; Tree Pipit used to breed; Swan Pool, though small, is visited by a variety of wildfowl; and rarities have included Stone-curlew and Caspian Tern. During winter, Lapwings flock on the fields and Snipe congregate in the damper hollows. The concentration of wildlife draws predators and amongst the owls reported was a Longeared in 1976. Whilst some regret the changes taking place, there is no doubt that Sandwell Valley is pointing to one way of reconciling the interests of people and wildlife.

It is perhaps fitting that the largest and most valuable of the three main urban lungs should be of royal descent. For such is Sutton Park, a surviving tract of old heathland covering a thousand hectares, which passed first from King Henry VIII to John Vesey, Bishop of Exeter, and hence to the Borough of Sutton Coldfield in 1529. The park's environment extends beyond its boundaries into the gardens of some of the biggest and most exclusive housing anywhere in Birmingham.

Apart from grazing, sometimes to excess, a little cultivation during the Second World War and a boy scout jamboree, the semi-natural heathland has suffered most from trampling feet, cars and accidental fire. The underlying Bunter beds support a typical heath vegetation in which birch and bracken cover much of the higher ground and heather, gorse or heath-grassland the slopes and open areas. Redpoll and Willow Warbler are typically found in the light birch woodland, though Tree Pipit has all but disappeared as a breeding species. Meadow Pipits, Skylarks and Linnets frequent the open heath and Stonechats sometimes appear amongst the heather and gorse, with burnt areas often attracting them in some numbers along with Whinchat and Wheatear. A Great Grey Shrike sometimes appears in winter, when the occasional Merlin or Hen Harrier quarters the heath. The older woods are dominated by oak – both pedunculate and sessile – with birch, mountain ash and holly in association, and bilberry occurs sparingly in the understorey. Great Spotted and Green Woodpeckers, Nuthatch and Redstart all nest in these older woods, though the last named has declined recently. Hawfinch, too, appears sporadically, though it is less regular than hitherto, and the woods are important winter roosts for thrushes and finches. In the wetter areas like Longmoor, the valley floor is completely different, with the waterlogged, peaty soils carpeted by purple moor-grass, *Sphagnum* moss and alder woods. Snipe can often be flushed from marshy ground, Redpoll and Siskin fidget through the alders in winter and Crossbills visit the conifers in good years. The park has not escaped entirely from industrial influence, as its valleys were dammed to provide water power, creating several mill ponds. Fortunately these have added to both its scenic beauty and its wildlife. Amongst the breeding species are Great Crested Grebe, Mute Swan and good numbers of Mallard, whilst in 1954 Black-necked Grebe also bred – an isolated occurrence but none the less remarkable for all that. At passage times the pools attract terns and waders and in winter they hold small numbers of wildfowl.

The last two decades have seen the arrival of the motorways, with the M5 and M6 somehow miraculously threading their way along the Tame Valley and through the Black Country and "Spaghetti Junction" becoming synonymous with the motorway age. They have also seen the city centre transformed into a concrete maze of roads, underpasses and tall office blocks, where pigeons, Starlings and sparrows rule supreme. In stark contrast, the lessons of high-rise living have been dearly learnt in the suburbs and the authorities have reverted to building houses. Pressure for land has become more acute and the city has spilt across its boundary to vast new estates like Chelmsley Wood, where it comes abruptly up against the M6 and the green belt. Nearby, the imaginative new National Exhibition Centre stands as a reminder that the pioneering spirit of this thriving city is still very much alive, whilst its ornamental lake affords a sanctuary for Canada Geese.

Although Birmingham and the Black Country now form a single Conurbation with a population of some two-and-a-quarter million, each retains its own peculiar identity. In fact the Black Country is unique. Even today it defies description,

8.   The extensive heaths, woods and lakes of Sutton Park form a unique urban oasis.

J. V. & G. R. Harrison

but in its heyday it must have presented an awe-inspiring prospect. Once a pleasant tract of ancient hunting forest and heath, with market towns at Wolverhampton, Walsall and Dudley, within the space of two centuries it was transfigured by man's quest for its wealth of coal, iron-ore, limestone, fireclay and road-stone. Exploitation began as early as the fourteenth century, based on coal-mining and iron production.

Glass manufacture began at Stourbridge late in the sixteenth century and Walsall specialised in the manufacture of ironmongery for harnesses and saddles. Specialisation of this kind is one of the fundamental differences between Birmingham and the Black Country. In Birmingham urbanisation spread outwards from a prosperous central core, engulfing smaller villages as it did so. In the Black Country

mines were sunk or works opened wherever raw materials were easiest to win and round them sprang up the mining villages and small towns.

Industrialisation was a gradual process spreading outwards from innumerable small centres and, despite their ultimate coalescence, most towns even today boast a separate identity and can point to tiny enclaves of undeveloped, often waste land, which separate them from their neighbours.

In 1769 the first canal was cut and soon a whole network of canals was opened to the major ports, enabling raw materials and goods to be easily transported. From hereon the pace of industrialisation quickened. The steam engine broke the reliance on water power and industry spread to new areas. Before long the canals were superseded by the railways and the whole industrial structure reached its zenith.

Amongst the canal legacy are three large pools known as Brierley Hill Pools, or Fenn Pools, which serve as feeders to the Stourbridge canal. The banks are natural and the surrounds of rough grassland and scrub are hemmed in by housing estates as well as overshadowed by the enormous Round Oak steelworks. Tipping from the steelworks still takes place, there is boating and fishing on two of the pools and the surrounding area is used for walking and horse-riding, but nevertheless there is a good bird population. Up to fifty Pochard and Tufted Duck occur in winter, both Little Grebe and Mute Swan breed, waders and terns occur on passage and the surrounding scrub holds a good variety of warblers including breeding Sedge Warbler. In recent years, too, the number of Reed Warblers has increased.

The depression brought change and everywhere mines, foundries and brickworks closed. In the twilight of an era, the oppressive pall of smoke that had permanently overhung the area slowly cleared and the glare from the blast furnaces gradually faded as the link between the natural resources and the industries of the area was finally severed. In the rush to a fortune, exploitation had often been haphazard, indiscriminate and wasteful. Valuable resources had been sterilised for short-term gains and the result was a legacy of pollution and dereliction. The labyrinth of workings below ground had caused subsidence and disrupted drainage, whilst above ground the land was littered with innumerable spoil heaps. Toxicity was often so great that it retarded natural regeneration, with pools remaining sterile, shrubs stunted and grass, if it grew at all, more likely to be yellow or brown than a refreshing green.

Even now such conditions are widespread, but amongst these unsightly, despoiled wastelands are some rewarding sites. Although the more toxic ones support only short, coarse grass holding a few Skylarks and Meadow Pipits, the better ones have a diverse flora including thorn scrub, gorse and other shrubs capable of holding warblers, finches and Reed Buntings. If disturbance is not too great Stonechats may exceptionally breed. Unpolluted pools can be a great attraction, with perhaps breeding Little Ringed Plover and a good variety of waders on passage, including Common and Green Sandpipers and Greenshank.

Surprisingly a few wilder sites survive. At Dudley, for example, the precipitous limestone ridge from Castle Hill to Wrens Nest was replanted a century and a half ago by the Earl of Dudley to compensate for the desecration wreaked by his limestone workings. Today there is a well-developed shrub layer beneath the mature beech and sycamore and, despite considerable disturbance in parts, colonies of Stock Doves flourish and Chaffinches, Wrens, Willow Warblers and Blackcaps are present in good numbers. In fact the Black Country contains quite a few bird-rich woods, though none is perhaps quite as good as Saltwells Wood at Netherton. Dating from medieval times, this is one of the oldest stands of coppiced oak woodland in the area. Amazingly even in 1836 this was described as a wooded district in advertising literature extolling the virtues of Cradley as a spa and even now skewbald and piebald ponies still graze the adjoining pastures. The wood contains beech, sycamore, willow and white poplar and an avifauna similar to that of Dudley's woodland, but with the addition of Green Woodpecker, Spotted Flycatcher and occasionally Wood Warbler. Teagle (1978) also records the nearby claypit as a breeding site for Red-backed Shrike until 1966. This wood and Rough Wood, near Walsall, both contain a stream and some marshy ground which add to their variety of habitat. The contrast between these woods, with their well-developed under-storey, and that of Warley Park, with its 40 ha of rolling beech and birch, is stark indeed. At Warley, Stock Doves and Magpies thrive, Nuthatch are occasionally seen and finch flocks, including Brambling, sometimes congregate in winter to feed on the mast. But with no understorey there are no warblers or ground-nesting species.

On the edge of the Black Country, but still within the shadows of pit mounds and mining subsidence, are Himley Park and Baggeridge Wood. Himley Hall was built by the Earls of Dudley, who also owned the nearby Baggeridge Colliery. With nationalisation the hall and the mine passed to the National Coal Board. Now the coal is exhausted and the hall is an educational centre, whilst the colliery has been demolished and its wastelands are being transformed into a country park. Four small lakes, created to supply power for an ironworks by damming a stream, are to be cleaned out to make an attractive link with the Great Pool and its backcloth of majestic beeches, but probably at the expense of wintering Water Rail. Little Grebe and Pochard frequent the pools in winter, whilst in summer Wood Warblers nest at Himley and Reed Warblers at Baggeridge. Woodcock rode at dusk, a good variety of woodland species occurs at all seasons and Bearded Tits once visited the pool.

Another peripheral site of note is the old airfield at Perton, near Wolverhampton. By the early 1950s this had become derelict and colonised by rank, coarse grass, thistles, willow-herb and meadow-sweet, with one or two marshy areas as well. Later, willow and hawthorn scrub invaded the western and northern flanks and there is also some mature oak and coniferous woodland on the latter flank, which holds Woodcock, Nuthatch, Stock and Turtle Doves, Blackcap and Garden Warbler. Amongst a good range of breeding birds, Red-legged Partridge, Lapwing, Skylark, Meadow Pipit, Yellow Wagtail, Whitethroat, Sedge Warbler, Reed and Corn Bunting and one or two pairs of Whinchat and Grasshopper Warbler are worth mentioning. Green Woodpeckers used to frequent the tarmac to feed on ants and the whole area became a rich hunting territory for raptors like Sparrowhawk and Barn Owl. In autumn and winter flocks of finches, especially Linnets and Goldfinches, fed on weed seeds and the occasional Great Grey Shrike would put in an appearance too. In 1974 Short-eared Owls appeared, to remain until 1976, and breeding occurred in 1974.

Since 1975, however, over half the airfield has been consumed by a major new housing development, with some interesting results. The disturbance created by earth-moving machinery exposed a rich feeding ground for insectivorous birds and up to thirty Stonechats were present during the hard weather of early 1976. The building operations also added two pools and a scrape to the habitat diversity, since when Canada Goose, Snipe and two pairs each of Redshank and Little Ringed Plover have bred; fifteen species of wader have occurred on passage; and Tufted Duck and Pochard have become winter visitors. To these can be added unexpected bonuses like Black Redstart, Spotted Crake, Hawfinch, Water Pipit and Black Tern – all of which indicate the riches that might be discovered by a diligent search of derelict land and construction sites.

Mention must also be made of Blackbrook Sewage Farm, which featured regularly in the Region's ornithological annals until its closure in 1961. The sewage farm itself covered 16 ha and was separated from a nearby stream by a narrow belt of marshy, mixed woodland. It was also bordered by rough pasture and arable land. Dabbling duck and Moorhens were regularly present, with the latter numbering 500 on more than one occasion in autumn, whilst in winter Snipe and Lapwing were numerous and a flock of Golden Plover was usually in the vicinity. Yellow Wagtail, Redshank and possibly Snipe bred, and Green Sandpipers were prominent amidst a good variety of waders that occurred on passage. With Buzzard seen several times in 1955, Stonecurlew in 1956, Peregrine in 1957 and Grey Phalarope in 1960, it is interesting to speculate on what might have turned up since if the closure had not intervened.

If the initial Black Country development depended on its raw materials, its sustained progress has been due to its labour-force. Traditional skills, passed down from one generation to the next, survived the depression to be harnessed by the lighter engineering trades which were now using steel as their raw material. Some of the traditional industries like glass and chain making also survived to be joined by newcomers like the steel and vehicle engineering works. Since the depression the centre of coal-mining has moved northwards to the Cannock area. Here both deep and shallow seams have been worked, with the galleries of modern pits being driven out like tentacles beneath Cannock Chase, whilst earth-moving equipment has lumbered across the countryside like giants, stripping the overburden and ripping out the shallow seams. Almost over-night, it seems, these opencast workings have been levelled and restored, with only the lower ground level and the lack of trees and hedges as reminders of the disruption that has taken place. The attendant mining settlements have not been swamped with industrial development, however, but have survived as distinctly separate villages in an area of curious transition between the hustle and bustle of the Black Country proper and the peace and tranquillity of Cannock Chase.

Just north of Brownhills, in the middle of this strange no-man's land, is Chasewater. Without doubt this is a most intriguing and unusual locality. Formerly known as Cannock, or Cannock Chase, Reservoir and locally as Norton Pool, it is a canal feeder reservoir of 85 ha, which was once in the middle of a lowland heath, but is now set in an industrial landscape of slag heaps and derelict land. Open heath has given way to a pleasure ground and go-kart track on the south shore and there is so much noise and disturbance from power-boating, water-skiing, sailing and unauthorised, indiscriminate shooting that it is surprising the water attracts any birds at all. Even the smaller, northern pool, known as Jeffrey's Swag, fails to escape disturbance. Being a canal feeder, water levels can vary by as much as five metres, though the decline in canal traffic has reduced the incidence of such a change, and at low levels the southern shore is still the best for waders notwithstanding the pleasure ground. The small fields and ponds that once lined the western shore between the embankment and the railway have now been filled with ash and topsoiled and during this restoration Little Ringed Plover bred on the bare ground. Along the eastern shore a mixture of heaths, marsh, ponds and spoil heaps, with some willow, hawthorn and a few larger trees still lingers to provide cover for a range of species, despite threatened reclamation as a golf course. To the north, spoil heaps have already been levelled in the interest of visual amenity and the old hawthorn hedges have virtually gone, but some heath, marsh and scrub have survived amidst pit heaps and slurry beds, whilst the willows fringing the pool and a feeder stream are good spots for migrants. To the north of Jeffrey's Swag is a flooded area, once known as Norton Bog, which the neighbouring pig farm has enhanced into a rich feeding area.

Over 180 species have been recorded at Chasewater, which is a place for quality

9. Set amidst an industrial landscape, Chasewater is noted for the quality rather than quantity of its birds.   *G. W. Ward*

rather than quantity of birds. Few species, except Coot, Tufted Duck and perhaps Goldeneye, appear in good numbers, but Common Scoter and Scaup are almost annual visitors and the water has attracted more than its share of rarities, notably sea-birds like Cory's Shearwater, Eider and Little Auk. In recent years, too, its gull roost has been blessed with a sprinkling of Glaucous and Iceland Gulls. At the water's edge a few Rock and Water Pipits might frequent the stone-pitched embankments and marshy shores in autumn and winter and exceptionally Snow or Lapland Buntings may appear. Great Grey Shrikes are quite regular here or on nearby Brownhills Common, but the speciality of the area is the wintering flock of Twite – the only locality in the Region apart from their moorland breeding sites

where this species can be seen with any regularity. Numbers fluctuate, but between 50 and 100 are usually present, though their future may be in jeopardy as their feeding area has been largely reclaimed. Waders and terns occur on both passages and have included rarities like Kentish Plover, Dotterel, Buff-breasted and Least Sandpipers and Caspian and White-winged Black Terns, whilst a few Jack Snipe and Water Rail usually winter in the nearby areas. Even the variety of raptors has proved exceptional, with records of Red Kite, Rough-legged Buzzard, Lesser Kestrel and Red-footed Falcon.

North of the Black Country and the Cannock coalfield the plateau rises to its summit on Cannock Chase – a magnificent expanse of woodland and semi-natural, largely unspoilt, heath. The Chase – or hunting ground of a cleric or lord – originated when Edward I sold his rights in the forest to the Bishop of Lichfield in 1290. After the dissolution it again changed hands and eventually passed to the Marquis of Anglesey, who sold the southern portion to the Forestry Commission. Much of the open heathland to the north, which formerly belonged to the Earl of Lichfield, is now in public ownership, whilst some of the woodland to the east belongs to private estates. Much of the extensive oak-birch forest has long since disappeared – felled to provide fuel for charcoal smelting and prevented from regenerating by over-grazing – but Brocton Coppice at least remains.

Over the years the Chase has also been subjected to military occupation, gravel extraction, afforestation with conifers and coal-mining beneath the ground, but today its 67 km² form the smallest of Britain's designated Areas of Outstanding Natural Beauty and it is to this uniquely varied and compact area that many of the Region's five million inhabitants come for recreation. At times up to a thousand cars and five thousand people can be present. Indeed the pressures are so enormous that Staffordshire County Council has already been forced to declare part a motorless zone and other management plans are being brought into being at the present time. Over 1000 ha have been scheduled as an SSSI and Brocton Pool and Oldacre Valley have been declared a Local Nature Reserve, but much of the same area is also a country park. It is to be hoped the conflict of interest between those anxious to conserve its ecologically rich, but fragile habitats, and those anxious to promote its recreational potential can be satisfactorily resolved.

The Chase is bounded by Stafford, Cannock and Rugeley and the high plateau ranges from 180 m in the north to 243 m in the south. It falls away abruptly to the Penk Valley in the west and the Trent Valley in the north, where some of the slopes are steep and dramatically picturesque with their mantle of woodland. It has also been deeply dissected by streams flowing northwards into the Trent, of which the Sherbrook is the best known. Bunter pebbles underlie the Chase, producing free-draining, pebbly soils that are agriculturally poor, with a tendency to be thin on the ridges and plateaux summits, and peaty in the valley bottoms. The same Bunter beds have considerable economic value, however, and are worked commercially for sand and gravel. Although pits are deep, altitude and the porous sub-strata prevent many from flooding and consequently they are less attractive to birds than river gravel pits. However, Sand Martins breed in small numbers, two or three pairs of Little Ringed Plover nest annually and Great Crested and Little Grebes, Kingfisher and Tufted Duck may breed where there is standing water.

Forestry is the principal profitable land use and the Forestry Commission has planted 2,700 ha. These are old-style plantations, up to sixty years old, with monotonously regimented ranks of larch, Sitka spruce, Corsican, Scots and lodgepole pines and a fringe of birch, beech and oak left as a token gesture to amenity along the roadsides. Public access is permitted, some forest walks and trails have been made, and to the ornithologist these plantations are beginning to offer new habitats. Though parts are still too dark and uniform in age structure to harbour many birds, with maturity more areas are being felled and the forest is beginning to acquire a varied age structure and consequently a more varied avifauna. Coal Tits find suitable nest holes in or near the ground, Redpolls, Goldcrests and Woodpigeons are numerous in the conifers and up to six pairs of Long-eared Owls breed. Chaffinches, Blue Tits, Robins and

Dunnocks frequent the woodland fringe, Reed Buntings nest alongside Yellowhammers in the bracken-covered rides, Nightjars nest in the young plantations, Sparrowhawks are frequently seen, and Siskin are often numerous in spring, have bred and could become established. Small parties of foraging tits or occasionally flocks of Crossbills may be encountered in winter, when there are also large roosts of Crows, Chaffinches and Bramblings. In 1979/80 a male Two-barred Crossbill delighted many bird-watchers during its lengthy stay. Following a severe attack from pine looper moths in 1954, nest boxes were erected in an experiment to see whether tits could be persuaded to nest in the denser plantations and if so whether this would control the pine looper caterpillars. The experiment was not continued to test this latter point, but it did show that Blue Tits, Great Tits and even Redstarts could be attracted into these otherwise forbidding plantations by artificial nest sites, with Great Tits out-numbering Blue Tits by nearly four to one. Great Spotted Woodpeckers also attacked the boxes to eat the young birds (Minton 1970).

Of much greater interest are the remnants of ancient hunting forest, with their scattering of old deciduous woodlands, extensive dry heaths on the plateaux tops and valley sides, and wet heaths along the valley bottoms. In 1957 some 800 ha of these fragile habitats were given to Staffordshire County Council by Lord Lichfield "to be used as a nature reserve or for public access" and therein lies the great management dilemma. Over 1000 ha centred on Brocton and the Sherbrook Valley and including most of the above, have been declared a country park and this area is almost identical to that scheduled as being of Special Scientific Interest. It is a rich and varied area, with notable examples of fen carr, eutrophic waters with emergent vegetation, wet heath,

freshwater marsh and peat bog. Its intermediate altitude and geographical location, however, combine to produce an almost unique environment which is a mixture of upland moor and lowland heath. This unusual transition is exemplified by bilberry and cowberry which are upland species, and the hybrid bilberry, which is not. Yet all three can be found on the Chase virtually at the national limits of their geographical and altitudinal range. This transition is also reflected in the bird life, with the Chase providing an important bridge between northern moors and southern heaths.

The summits and valley sides are dominated by heather communities and often the plateau appears flat, bleak and uninteresting, relieved only by patches of gorse, hawthorn or birch scrub. The more sheltered parts support acidic grasslands and bracken occurs on the deeper soils, where it is very invasive. Meadow Pipit is the commonest species, but with a little luck and a great deal of perseverance more interesting birds can be found. A hundred years ago Black Grouse were common, but they were shot out by 1923, whilst the small Red Grouse population, which was probably not indigenous anyway, lingered until the winter of 1962/3. Subsequent attempts at reintroduction have failed. In winter hordes of Fieldfare arrive, an occasional Hen Harrier, Rough-legged Buzzard, Merlin or Short-eared Owl quarters the heath and one or two Great Grey Shrikes regularly take up residence. Whinchats breed sparingly and in recent years Stonechats have returned to breed as well, whilst at dusk the eerie sounds of reeling Grasshopper Warblers, roding Woodcock and churring Nightjars punctuate the quiet of a summer's evening. With some 25 pairs, the Chase is the most important locality for Nightjar in the Region and it is this species and the deer which are vulnerable to human disturbance and which most concern conservationists.

10. The mixture of heath and woodland on Cannock Chase is important for species like the Nightjar.   *Bill Smallshire*

The wet heaths and mires of the Sherbrook valley are lined with alder coppice and a herb layer which includes greater tussock-sedge. Breeding birds include Willow Tit and Lesser Spotted Woodpecker, whilst in winter twittering flocks of Redpoll and Siskin, parties of foraging tits and a few Treecreepers and Goldcrests move through the trees. The Oldacre Valley is botanically different, with willow, birch and some oak, but its bird life is similar, except that its clumps of elder are regularly visited by Ring Ouzels and other thrushes in autumn. The relic oak or oak-birch woodlands like Brocton Coppice are most interesting. They are similar to, if less extensive than, their more famous counterparts in Sherwood Forest. Most of the gnarled oaks were planted some three centuries ago and their rich insect life indicates a long and continuous presence of primary woodland. The oaks are slow to regenerate, but attempts to replant are being made. Most of the woodland wildlife suffers from disturbance, but the range of breeding species is good, including all three woodpeckers, Woodcock, Tawny Owl, Nuthatch, Redstart, Tree Pipit and a variety of warblers, including Wood Warbler. In winter a few Brambling can usually be found in the small finch roosts. The deciduous or mixed woodlands at Shoal Hill and Brereton Hayes, though less often watched, can be equally rewarding, but Shugborough Park, with its fine mixture of mature hardwoods, has the greatest diversity of all, with roding Woodcock, all three Woodpeckers, Nuthatch, Hawfinch and perhaps even Pied Flycatcher, which has bred in the past. Here too, large numbers of finches roost in the rhododendrons in winter.

To the south of Birmingham, the South Staffordshire Plateau is completed by an arc of open countryside stretching from the Stour Valley in the west and passing north of Bromsgrove, Redditch and Henley-in-Arden to merge imperceptibly with the East Warwickshire Plateau at Kingswood gap. Northwards it abuts the Conurbation and to the south it is defined by that sudden drop into the Avon basin identified by features like Gorcott and Liveridge Hills, the Lickey incline and the lock flight at Tardebigge. The area changes considerably in character from west to east.

Although forming part of the major watershed between Severn and Trent, the only physical features of note are the Clent and Lickey Hills in the north-west. At Clent there are two rounded, breccia-capped hills of which Walton Hill is the highest. Most of the land is owned by the National Trust and the public enjoys free access, so the hills are extremely popular with trippers. Fortunately their steepness excludes cars, but horse riding and trampling feet still cause serious erosion of the short grass sward on the summits, where Meadow Pipits are the commonest breeding bird and quite large numbers pass through in autumn. The bracken-carpeted slopes, with their scattering of gorse, hawthorn scrub and isolated trees are more robust, however, and it is here that Linnets, Yellowhammers, Redstarts, Tree Pipits and Garden Warblers breed. Before their widespread decline, Wood-larks bred too.

Generally, bird life on the hills is not outstanding, although the altitude attracts some upland species on migration, with Ring Ouzels sometimes appearing in small numbers, together with Wheatears and an occasional Pied Flycatcher. There is also one record of a Dotterel. Woodland is quite extensive and together with the stream and pools just as rewarding. The stands of both broad-leaved and conifer-ous timber include a variety of species, but the oaks of Sling Common, with its willow-fringed pool, are among the more productive. This is good Green Wood-pecker country, Sparrowhawks are resident and Nuthatches occur in many of the woods along with Blackcaps and Garden Warblers. Dippers and Grey Wag-tails feature on most streams.

In contrast the up-faulted Cambrian rocks of the Lickeys are more wooded, with extensive plantations of larch and pine interspersed with broad-leaved woods in which oak, sycamore, beech and birch are all locally dominant. There are some interesting variations in ground flora, ranging from bare ground under the close canopy of beech or conifer, through hazel, holly and bramble to bilberry, gorse, heather and grass on the steeper, more open hillsides and summits. In summer day-trippers throng to the hills and woods, and horse-riding is commonplace, yet despite this the bird life is good. Gold-crests and Coal Tits are common in the conifer plantations and Crossbills often appear after irruptions. Redstart, all three woodpeckers, Nuthatch and Wood Warbler breed regularly and Firecrests did so in 1975 – the first record for the Region. Both Woodlark and Nightjar have been lost over the past fifty years, but probably not entirely through pressure on their habitat, although this may be why the shy Hawfinch has disappeared. In winter this is one of the more regular haunts of Brambling and in good years several hundred may flock with Chaffinches to feed on beech mast. At this time, too, mixed parties of tits are often joined by Treecreepers and large numbers of Redpoll and Siskin.

From the steeply undulating countryside around the Clent and Lickey Hills the ground falls away quickly south-west-wards into a level or gently undulating landscape of large fields and neat hedge-

11. Many woodland birds can be found amongst the variety of trees that clothe the steep slopes of the Lickey Hills. *P. Wakely*

rows. Apart from an occasional beech, hedgerow trees are few, but there are shelterbelts of larch or Lombardy poplar, some of which are managed as game coverts. Indeed, for an agricultural landscape conifers figure quite prominently. The extremely fertile, deep, sandy loams around Broome, Catshill and Tutnall are easily cultivated and are amongst the best agricultural soils in the Region. They are widely used for arable cropping or market gardening and the avifauna typically includes Grey Partridge, Quail in good years, Corn Bunting, Yellowhammer and mixed finch flocks in winter. East of the Lickeys the gently undulating topography gradually merges into the flat, featureless Keuper marl plateau. Superficial deposits of boulder clay are extensive and the resulting soils heavy, badly drained and agriculturally poor.

Indeed the Ministry of Agriculture, Fisheries and Food has classified an extensive area around Earlswood as amongst the poorest land in the Region. It is characterised by small fields, thick hedgerows and an abundance of young hedgerow oaks, and today, more than anywhere else, epitomises that bygone landscape which earned the reputation "leafy Warwickshire". Pastures and dairy herds predominate, there are some pigs and poultry, but often the poorest land is used only as paddocks by riding schools.

Everywhere mellowed red-brick buildings grace the countryside, occurring equally as isolated farmsteads or in charming villages and scattered hamlets dotted throughout a network of narrow lanes that have miraculously avoided improvement. Despite more than one envious glance at Wythall by the former City of Birmingham, only Redditch, with its world-wide reputation for needle manufacture, has blossomed into a town of any size. Today it is a fast-expanding new town with a population of almost 70,000.

To the ornithologist, though, this is a countryside of reservoirs, pools and woods. The impervious clays were ideal ground in which to construct the summit reservoirs that feed the canal system at Bittell, Tardebigge and Earlswood; and, though there are no major rivers, the fast-flowing streams that rise on the Clent Hills pass through interesting chains of small pools created by past generations to drive their watermills. These streams are a favourite haunt of Dippers and Grey Wagtails, whilst the pools themselves are visited by small numbers of wildfowl. With the occasional steep slope and so much heavy, unworkable land, numerous relics of former woodland have survived as well, especially around Earlswood, Redditch and north-west of Bromsgrove. Some are remnants of the former Feckenham Forest, like Chaddesley and Randan Woods. Here the Keuper marl is poorly drained and slightly acidic, whilst the glacial drift of the high ground is typically gravelly, excessively drained and strongly acidic. Both species of oak are found in the high forest, with sessile oak and its understorey of bracken mainly on the drift, and there is a richly varied shrub and coppice layer, which includes hazel, rowan, ash and birch. Alders are also abundant along the deeply incised valleys. Amongst the varied avifauna, Redstart, Woodcock, all three woodpeckers, Nuthatch and warblers are typical birds and a few Brambling are noted in most winters. In addition to the high oak forest, there are also extensive plantations containing a variety of conifers. Here again Woodcock rode; Coal Tit and Goldcrest are abundant; Crossbills have been noted fairly regularly in very small numbers, and Nightjar used to occur. Randan Wood is a Worcestershire Nature Conservation Trust reserve which has been very thoroughly studied by its owner, F. Fincher. Apart from a few public footpaths, though, it is private and less accessible than Chaddesley Wood, where 100 ha have been a National

Nature Reserve since 1973 and there are public footpaths and a circular walk, which was introduced to mark the Silver Jubilee in 1977. Furthermore, both the oak forest and the conifer plantations are being managed in the interest of wildlife, with sustained coppicing, erection of nest-boxes and the like, so all the typical birds can be found here.

Bittell Reservoirs nestle beneath the Lickeys, within the shadow of British Leyland's giant Longbridge car plant. Technically there are only two, Upper Bittell which covers 40 ha and Lower Bittell which covers 23 ha, but the latter is divided into two by a road across its northern extremity. Linking the reservoirs is a fast-flowing stream flanked by mixed woodland. Apart from the concrete dam at the southern end of Upper Bittell, both reservoirs are surrounded by rough pasture which provides grazing for wildfowl and exposes mud for waders whenever water levels are low. Upper Bittell is used for sailing and both waters are fished, but there is no public access other than from the surrounding roads and one or two footpaths, so disturbance is not too great. Upper Bittell is also the bleaker and birds often prefer the lower reservoir despite its smaller size.

The waters have been well watched for many years and hold a virtual monopoly of Worcestershire records for many species, notably sea-birds and terns. Outstanding amongst these are Velvet Scoter, Pomarine Skua and Sabine's Gull. There are good winter populations of Great Crested Grebe, Pochard and Tufted Duck, whilst over the past fifty years Wigeon have decreased and Shoveler increased. At this season, too, there is a large gull roost and the marshy shorelines hold good numbers of Teal and Snipe along with a few Jack Snipe. Herons are regularly present, whilst Great Crested Grebe, Sedge and Reed Warblers breed and Ringed Plover may have done so on one occasion.

Spring and autumn passage usually brings a good variety of birds, which has included a few Ospreys and a Pectoral Sandpiper. Away from the lakes a flock of Siskin regularly winters in the alders between the two reservoirs, all three woodpeckers occur, Dippers have appeared on the stream from time-to-time, Kingfishers are usually about the stream or canal and both Great Grey and Red-backed Shrikes have made sporadic appearances. Like Bittell, Tardebigge supplies water for the Worcester and Birmingham Canal, but it is generally too small to really attract wildfowl, though Goosander and Ruddy Duck have been recorded. Apart from fishing, though, there is little disturbance, and Great Crested and Little Grebes and Kingfisher breed. The surrounds include rough grassland, hawthorn scrub and willows backed by a narrow belt of oak woodland and these hold a good range of warblers, tits and finches. Reed Buntings and Yellow Wagtails also breed and Sparrowhawks are occasionally seen. After a summer of heavy boat traffic the water level is usually low and the exposed mud then attracts a variety of autumn waders in small numbers. A few terns are usually noted at this season as well.

Ten kilometres east of Bittell are more canal feeder reservoirs at Earlswood, where the three lakes cover some 26 ha. The public have access to much of the bankside and the area is very popular with fishermen and trippers. Sailing also takes place on the eastern lake, but the western-most lake is little disturbed. Willows surround and overhang all the lakes, providing some refuge for disturbed wildfowl, and the western-most pool has a small island and some marshy areas. Great Crested Grebes are regular breeders, with variable success, along with Mute Swans and the commoner waterfowl. Autumn and winter bring a sizeable flock of Canada Geese and the odd Water Rail skulks in the rushes and reeds. Amongst the small numbers of wintering duck an unexpected species like Goosander might turn up, a few terns regularly appear on passage and highlights over the years have included Ferruginous Duck and Purple Heron. Unfortunately, though, the lakes lack the quiet, muddy corners necessary to attract many waders.

The scrubby woodland fringing the reservoirs holds a variety of tits, including Willow Tit, but to the west New Fallings Coppice and the adjacent Clowes Wood together form a much more extensive area of oak-dominant woodland, with some beech and birch. Clowes Wood is a Warwickshire Nature Conservation Trust reserve in which bilberry features in a low under-storey that provides ideal conditions for Wood Warblers, a few pairs of which regularly breed. Although this is the speciality of the wood, most other woodland species are well represented, especially Jays and woodpeckers. In winter there is less bird life, but flocks of finches sometimes feed on beech mast and large numbers of Redwings and Fieldfares come into the wood from the surrounding fields to roost.

In many respects the East Warwickshire Plateau is a replica in miniature of the South Staffordshire Plateau. It has the same elliptical shape and measures 35 km from Tamworth in the north to Kenilworth in the south and 13 km across at its widest point.

For the most part exposures are of the barren Upper Coal Measures or Permian beds, but folding has brought the productive Middle Coal Measures within exploitable range. Indeed, between Bedworth and Polesworth the beds are almost vertical and there are parallel outcrops of Upper Coal Measures, Middle Coal Measures, shales and quartzite, interspersed with igneous intrusions of diorite. To the west the plateau falls away sharply into the Tame and Blythe valleys and to the north there is an abrupt drop into the Anker Valley. Eastwards, however, an extensive covering of boulder clay masks the boundary fault and here the plateau merges imperceptibly into the Avon valley.

A prominent ridge of sandstone and conglomerates extends from Coventry to Corley and south of this are extensive beds of Keuper marl with a patchy covering of glacial drift, chiefly boulder clay, which gives rise to rounded, gently undulating and often wooded countryside.

The southern edge of the plateau is again defined by that characteristically steep drop into the Avon Valley, identified at Hatton by a steep hill, railway incline and long lock-flight on the canal. The plateau ranges in height from 100 m in the south to 175 m in the north and again forms part of the main Severn-Trent watershed, being drained to the west by the Bourne and other small tributaries of the Blythe, and to the east by the Sowe and tributaries of the Avon. Between them, these streams have dissected the plateau surface, reducing it in parts to no more than a narrow neck between parallel valleys.

Coal, fireclay and roadstone at workable depths have resulted in a scarred landscape, but mining and quarrying have been mainly confined to an arc in the north and much of the associated wealth and economic growth was captured by the long-established valley towns of Tamworth, Nuneaton and Coventry. Consequently settlements have not coalesced as they have in the Black Country, although Nuneaton, Bedworth and Coventry have come close to doing so. The colliery villages themselves have remained small and, following pit and brick-work closures, some have even declined. Today there is a scatter of spoil mounds, but only two or three active pits, including Daw Mill, where the shaft was sunk as recently as 1957 and expansions are still taking place. Open-cast working ravaged the plateau above Atherstone, but the land has long since been restored to agriculture or planted with conifers and today it is hard to see where excavation took place.

On the other hand, stone quarrying is still very active and the vast quarries blasted out of the ridge at Nuneaton, Hartshill and Mancetter may ultimately have more impact on the landscape than coal mining itself. There is now, however, the prospect of a vast new coalfield being exploited to the south of Coventry.

Despite all this exploitation, the plateau is still essentially rural, even in the north, and the landscape is surprisingly attractive. Apart from suburban tentacles reaching out from Coventry and Nuneaton, the settlement pattern is one of larger mining villages in the north and smaller agricultural ones in the south, and the rich, red soils of the Upper Coal Measures are as productive as any in the Region. Here dairying and mixed farming are interspersed amidst a landscape of small fields and strong hedgerows dotted with sturdy young oaks. Amongst the more interesting birds, Corn Buntings breed in the arable-intensive areas and a few pairs of Curlew nest in some of the higher, damper meadows; whilst winter brings good flocks of Lapwing, Fieldfare and Redwing. Where drift occurs, soils are poorer and dairying predominates. Elm was once widespread and hosted a number of small rookeries, but now the elms have gone and the Rooks have moved into some of the many, widely-scattered, small oak-birch woods or conifer plantations that characterise the area. As well as being significant landscape features, some of these woodlands are of considerable ecological importance. The landscape is also enriched by parklands, with their copses, specimen trees and ornamental lakes. Apart from these lakes, however, water does not feature prominently on the plateau and it is really the woodlands that most interest bird-watchers.

Two woods, namely Many Lands and Bunsons, have both been well studied and provide a good insight into the bird life of typical oakwoods on the coal measures.

The former comprises five hectares of close-canopied oak, with an understorey of hazel that was once coppiced. Holly and bramble also occur in places and bracken or grass grow in the open glades. In 1965 there were 119 territories of all species in this wood, but by 1967 this had grown to 165 – an increase of 37 per cent. The density of Blue Tits, Wrens, Blackbirds, Song Thrushes and Robins was high and Woodpigeons were numerous.

Hole-nesting species such as tits, woodpeckers, Tawny Owl, Starling and Tree Sparrow were also well represented, but the absence of thick cover meant fewer Dunnocks and the close canopy precluded most finches except Chaffinch. The eleven hectares of Bunsons Wood contain a wider variety of trees, including oak, birch and rowan, and an understorey of holly, bramble, honeysuckle and bracken. With 220-246 pairs, the density of territories was only two-thirds that of Many Lands Wood, but the species variety was richer, with Blue Tit, Blackbird, Robin, Willow Warbler, Dunnock and Chaffinch the commonest breeding birds. Other breeding species included Blackcap, Garden Warbler, Wood Warbler, Chiffchaff, Tree Pipit, Tawny Owl, all three woodpeckers and most tits, whilst Woodcock, Little Owl and Redstart were also present during the breeding season.

Paradoxically the best sites occur in the north, in juxtaposition with pit-heads, spoil mounds and quarries. Here, on the steeply sloping, acid soils, the high oakwoods of Bentley Park, Monks Park and Hartshill Hayes are outstanding. These are all primary woods developed from coppice with standards and are generally acknowledged to be remnants of the old Forest of Arden. The first two are scheduled SSSIs in private ownership, but Hartshill Hayes is a Forestry Commission wood forming part of a country park. Floristically their interest lies in the combination of upland and lowland species, with both native oaks dominant and ash, beech and sycamore locally abundant, but more recent plantations of larch, pine and spruce have added to their diversity. Bracken dominates the under-storey on the drier slopes, but in the wetter areas ash, alder, hazel and elder are prominent. Hartshill Hayes also contains limes which were formerly coppiced to make blocks for the Atherstone hatmakers, whilst the range of habitats at Bentley Park Wood is enriched by a stream, ravines and wet areas. The breeding communities of these three woods embrace Woodcock, Turtle Dove, Jay, all three woodpeckers, Little and Tawny Owls, Redstart and a variety of warblers, including a few pairs of Wood Warbler; whilst passage brings warblers and finches; and winter, feeding and roosting flocks of tits, finches and thrushes.

Immediately north of Monks Park Wood lies Merevale Park. This is the ancestral home of the Dugdale family and its private grounds include a good variety of mixed woodland, parkland and ornamental pools bordered by rhododendrons, which thrive on the acid soils. A few duck and Coot frequent the lakes, where Great Crested Grebe and Ruddy Duck also breed, whilst the woods hold many warblers in summer, and roosts of finches and thrushes form in the rhododendrons in winter. In nearby George Eliot country lies Arbury Park. The home of the Newdigates, this charmingly picturesque oasis has survived in the centre of a private estate that has otherwise been ripped apart by mining and quarrying and all but engulfed by the housing estates of Nuneaton. Formerly a deer park, it has seen many changes over the years, being used as a prisoner of war camp during the last war and partly reclaimed for agriculture since. Amongst a wide variety of habitats are open parkland, farmland, mixed woodland, pools, a disused canal and subsidence flashes. Many alien trees have been introduced, but the natural woodland is mostly oak-birch, with a dense understorey of hazel, though in parts bracken is dominant. Most woodland birds occur, but it is the marshy, damp areas with their breeding Redshank, wintering Snipe and Lapwing and occasional Water Rail that are of greatest interest. A few duck frequent the pools and Ruddy Duck may have nested in recent years. Of the pools, Seeswood is usually worth a look, especially as it can mostly be seen from the causeway at the southern end. It owes its origin to the subsidence of a wood as a result of mining activity, and some of the old tree stumps can still be seen in the shallows. Breeding birds include Great Crested and Little Grebes, Mallard, Coot, Kingfisher, Lapwing and probably all three wagtails. In winter it is regularly visited by a flock of Canada Geese and small numbers of Teal, Wigeon, Shoveler, Tufted Duck and Pochard. Both wild swans have appeared too, though neither is regular, and each year brings a few passage waders or terns.

The woodlands and parklands further south are smaller and little watched. At Berkswell, though, there is a small heronry and Yellow Wagtails and Grasshopper Warblers can usually be found around the lake or damp meadows in summer. Woodcock rode over the tree-tops of a few conifer plantations, like the Forestry Commission's Hay Wood, and Tree Pipit and Grasshopper Warbler also bred here when the trees were younger. In their place Redpoll have become well established.

On the whole the East Warwickshire Plateau offers few surprises, though a wintering Short-eared Owl might be seen hunting across rough grassland or a Green Sandpiper flushed from a drained canal. None the less, its countryside sustains a rich and varied bird-life which is typical of much of the Region's heartland. Equally rich and varied avifaunas can be found along the Region's three main arteries, namely the rivers Severn, Avon and Trent, and these are explored in the next chapters.

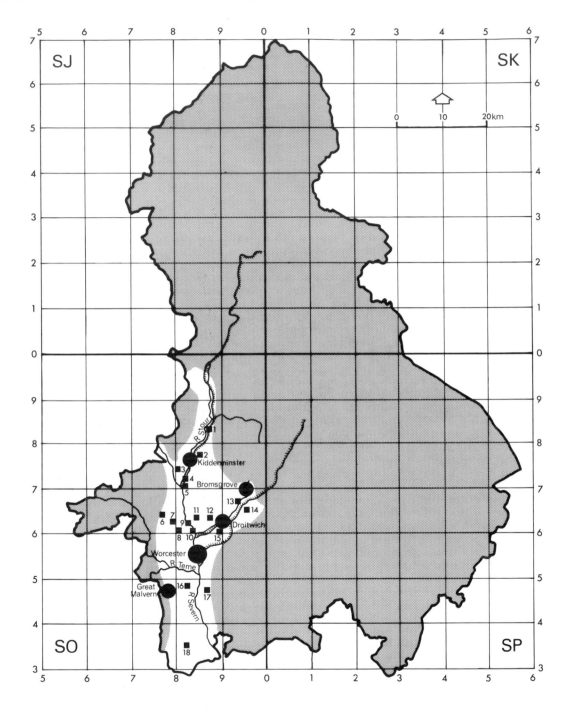

Selected localities in the Severn Valley

1 Whittington S.F.	6 Witley Court	11 Ombersley Park	16 Old Hills
2 Hurcott Pool	7 Ockeridge Wood	12 Westwood Park	17 Pirton Pool
3 Devils Spittleful	8 Monkwood	13 Upton Warren	18 Longdon Marsh
4 Wilden Marsh	9 Holt G.P.	14 Pipers Hill	
5 Hartlebury Common	10 Grimley	15 Oakley Pool	

3 The Severn Valley

The Severn is Britain's largest river. Rising in Wales, it follows a north-easterly course through the principality before swinging southwards through the English border counties and into the Bristol Channel. This circuitous course of some 354 km is directly attributable to glaciation, which caused the river to reverse its direction of flow and carve out its famous gorge at Ironbridge.

Once through this gorge the river pursues a purposeful, direct, southerly course in a deeply entrenched valley. It is in this state that it skirts the Wyre Forest to enter Worcestershire and the Region. It then continues in the same vein for the next 17 km as it successively traverses resistant beds of Carboniferous Coal Measures, Bunter sandstones and Keuper sandstones. Below Holt Fleet, however, the river bursts forth onto the Keuper marl and its valley gradually widens until, downstream of Worcester, the Severn is meandering gracefully between river terraces. Here, on the left bank, the low watershed with the Avon is scarcely discernible visually and is more readily identified by the transition from Keuper to Lower Lias beds and the effects this has on vegetation and farming patterns. On the right bank the Malverns rise dramatically out of the Keuper plain to give a very sharp, distinct definition to the valley. The close proximity of these western hills and the steep gradient of the river mean that nowhere is the valley more than twenty kilometres wide and mostly it is less than ten kilometres. This compact, linear area is thus the second smallest sub-region, covering a mere 700 km² or 10 per cent of the total land area.

The Severn is different from the Region's two other great rivers in a number of respects. Its volume of water is greater than that of either the Avon or the Trent and its quality is infinitely better too. To the West Midlands, therefore, the Severn is a valuable source of potable water, supplying not only the Conurbation but Coventry as well. Another difference is that whereas the Avon and Trent tend to flow bank high, the Severn and its tributaries are contained by high banks. This is the result of rejuvenation, which again has its origins in glaciation.

Climatically the Severn Valley benefits from the ameliorating influence of the Gulf Stream as it is funnelled up the Bristol Channel. The mean temperature in January is 4°C, and in July 16°C in the upper valley, rising in the lower valley to 17°C, which is higher than anywhere in Britain save the Solent and Thames Valley. In recent years, however, there has been a tendency for mean temperatures to fall, the incidence of south-westerly winds to decrease and summers to become wetter. Overall, rainfall averages around 675 mm per annum, but varies from 700 mm in the west and north to as little as 650 mm in the east. Heavy summer storms are frequent and the valley floors are prone to devastating late-spring frosts, but away from these frost pockets much of the area enjoys early, frost-free springs which are of considerable benefit to the horticultural industry.

The richly varied habitats and avifauna of this sub-region are closely related to its geology, topography and climate. Fast-flowing streams attract Dipper and Grey Wagtail; high river banks Sand Martin and Kingfisher and the mild climate Nightingale, though here this species is on the western fringe of its range and there has been some contraction following the recent incidence of wetter summers. Two other species that enjoy warm summers are Hobby and Cirl Bunting. Sightings of the former have increased in recent years and it seems this delightful raptor may be consolidating its position, whilst a very small, relict population of the latter seems amazingly to have survived until 1977, although none have been reported since. The gems of the area though are the lowland heaths on the Bunter sandstones around Kidderminster and at the foothills of the Malverns (see also Chapter 8). These have the richest flora of any in the West Midlands and a favourable climate, which is perhaps why Red-backed Shrike and Woodlark were able to survive longer here than elsewhere in the Region. Sadly both are now lost as breeding species, but Stonechat at least has returned in recent years as its numbers have recovered since the 1963 winter. Finally the water meadows of the lower Severn are as likely a place as any to find Corncrakes, though here again their appearances are becoming more and more erratic. For the ornithologist, though, the freshwater and saline pools of Upton Warren are perhaps the greatest attraction, being an important feeding ground for migrant waders.

Within the West Midlands there are three important tributaries to the Severn, namely the Stour, Salwarpe and Teme. The Stour, together with Smestow Brook, drains a triangular basin bounded by Wolverhampton, Kidderminster and the Clent Hills. It is a basin of marked contrasts, with the west and south predominantly rural, but the east heavily industrialised. However, the industrial area has been included in Chapter 2 as part of the Birmingham Plateau and it is only the rural part that is of concern here. The underlying sandstones yield well-drained, easily cultivated soils and this is part of a large area of intensive arable production in which sugar beet, potatoes, vegetables and some soft fruits are important crops. In places, though, the soils of the Bunter beds are sterile and prone to wind-blow and it is in such situations that the lowland heaths occur. The soils of the Keuper beds running up to the foot of the Clent Hills are generally the more productive.

To the west of the Conurbation, Smestow Brook flows through a landscape of large fields and neatly trimmed hedgerows, with few hedgerow trees. Corn Bunting is amongst its typical birds, whilst Little Ringed Plover has appeared at a small sand pit. Throughout its length the brook is accompanied by the Staffordshire and Worcestershire Canal and both Kingfisher and Grey Wagtail are to be found along this stretch. There is a small heronry nearby at Checkhill and Lesser Spotted Woodpecker frequents the willow and alder carr that lines an attractive, marshy stream backed by the poplar and larch plantations of Enville.

Having threaded its way through the Black Country, the Stour finally leaves Stourbridge for the open countryside and joins the Smestow at Stourton. Between here and Kidderminster it flows through some very attractive countryside, with the ground rising quite steeply on either side and Kinver Edge dominating the skyline to the west. On the slopes above the left bank is Whittington sewage farm, which for many years has been an attraction to Midland bird-watchers and which has an avifauna showing some interesting peculiarities. Originally an old-fashioned farm discharging effluent through valves onto the surrounding fields, this farm has sadly been modernised within the past decade. As a result wetland species have suffered from a gradual drying out of marshy areas, although some have survived in the few remaining damp spots. Passerines, however, appear to be increasing. In general this is a good roosting area for finches, thrushes, corvids and waders, but its main attraction is the wintering flocks of Curlew and Snipe, accompanied by a few Redshank and Jack Snipe. Indeed Snipe have reached 500 and the Curlew flock, which is the largest and most consistent in the Region, regularly exceeds 100 birds and reached a maximum of 340 in 1975. A Great Snipe was seen in 1954, Wheatears have bred and this area has also regularly produced records of scarce species like Corncrake and Quail.

The Stour seems unable to escape from urban influence for long, however, and soon it is engulfed on either side by the carpet factories, Victorian houses and new estates of Kidderminster. Cloth weaving was taking place in Kidderminster as early as the thirteenth century, but it was not until the eighteenth century that carpet weaving started. The fast-flowing streams to the east were quickly harnessed to provide water power to drive the looms and this has given the town some valuable wildlife habitats. There are a number of small mill pools to the north-east, around Caunsall, of which Islandpool is the best known, but the most important ones are the chain on the very outskirts of Kidderminster itself and particularly Hurcott Pool with its backcloth of mixed woodland. Here, on the very fringe of an urban setting, Great Crested and Little Grebes breed and Kingfishers can be seen regularly throughout the year. In winter small numbers of the commoner duck are present and occasionally something unexpected turns up.

Kidderminster lies at the heart of some very good bird country. In addition to its mill pools, it has a marsh just to the south, at Wilden, one or two lowland heaths and the Wyre Forest within easy reach. Wilden is an SSSI and one of the most interesting marshes in the Region, so it is fortunate that part is now owned by the Worcestershire Nature Conservation Trust and managed as a nature reserve. A good deal of the remainder belongs to the British Sugar Corporation and contains settling beds for the nearby sugar beet factory that enrich the general area by attracting wildfowl and migrant waders. The reserve itself consists of a very wet area of grazing meadow and marsh underlain by base-rich alluvial clay. Although alongside the Stour, it is actually kept wet by water flowing through the lower mottled sandstones of the adjacent Stour Hill into a series of springs that issue where the sandstones abut the clay. Poorly managed ditches, flanked by pollarded willows, permit some drainage, but the marsh remains wet even in summer and is consequently attractive to a wide range of breeding birds. Amongst these is a high density of Snipe as well as Redshank, Water Rail, Yellow Wagtail and Reed Bunting. This is a rich area for warblers too, with Sedge, Reed and Grasshopper Warbler, Whitethroat, Lesser Whitethroat, Garden Warbler and Blackcap all breeding. Green Woodpecker and Barn Owl also nest and both Whinchat and Nightingale may do so on occasions. Most autumns bring Rock and Water Pipits and the latter regularly stay well into winter, at which time the marsh is visited by mixed feeding flocks, which include Tree Sparrows, Chaffinches, Bramblings, Goldfinches, Linnets, Redpolls and Corn Buntings. Winter also brings dabbling duck, a few Pochard and Tufted Duck, Water Rail, Jack Snipe and Stonechat. Wader passage can be good both in spring and autumn, and Pectoral Sandpiper has been recorded along with most of the commoner species.

The Stour and the accompanying Staffordshire and Worcestershire Canal finally come to Stourport and their confluence with the mighty Severn. Stourport owes its very existence to the canal – in fact it is the only town in England to have been founded in consequence of a canal –

and in its heyday it was second only to Birmingham as the busiest port in the Midlands. Many of the fine Georgian buildings from this time have fortunately survived as a legacy of a bygone era to find a new lease of life with the growing popularity of the town as a canal leisure centre. Ornithologically speaking the town is less interesting, although a Red Kite once put in an appearance. It does have some claim to fame though, since it was here that Carrion Crows and Rooks were first noticed nesting on the switchgear of the giant power station that now dominates the landscape.

The Severn itself enters Worcestershire and the West Midlands from Shropshire at Arley. Here the rejuvenated river has carved a gorge through the Upper Coal Measures which, though less majestic than that at Ironbridge, is none the less beautiful with the hanging sessile oakwoods of Eymore and Seckley on either side. Sandwiched into a tiny piece of level ground between the river and the picturesque Severn Valley Railway is the water supply reservoir of Trimpley. Unfortunately the bare grass surrounds are treated too much like lawns and the water is too small for it to become really attractive to birds, especially as they have to contend with sailing and fishing as well. Even so it does hold a few duck plus the occasional passage wader, Rock Pipit has occurred and it has been graced by a visit from a migrant Osprey on at least one occasion.

The Severn's real gateway to the West Midlands though is Bewdley. This is a delightful Georgian town, where merchants' houses line the quaysides and Telford's bridge spans the river against a backcloth of beautifully wooded hillsides. Its wealthy citizens refused to have anything to do with Brindley's brash canal and consequently it has managed to avoid much of the unsavoury industrial development that has been inflicted elsewhere. It is immensely popular with day visitors, but still manages to sustain a varied bird population, with Mute Swans nesting on the river and Sand Martins and Kingfishers breeding in the river banks. Swifts nest beneath the eaves of the older houses and the garden conifers house several pairs of Collared Doves. On the left bank the land rises up to the house and gardens of Spring Grove, now the home of the West Midlands Safari Park. This inevitably has its collection of exotic birds and escapees can sometimes excite or confuse the unwary watcher. Of genuine origin, though, Storm Petrel, Red-necked Grebe and Osprey have all been unexpected visitors to the river.

Adjacent to the Safari Park is a fine tract of sandy heath which slopes gradually up to an outcrop of Bunter sandstone known as the Devil's Spittleful. A clump of Scots pine surmounts this outcrop, but otherwise the area is covered by heather and scattered birches. There are a few patches of gorse, hawthorn and bracken as well, whilst a birch wood and a developing oakwood add to the habitat diversity. The heath is a scheduled SSSI and also a Worcestershire Nature Conservation Trust reserve, but it is so close to Kidderminster that its footpaths and bridleways are very popular. Precisely what effect this has on its bird life is hard to judge. Certainly three species that have bred in the past – Woodlark, Nightjar and Red-backed Shrike – do so no longer, but then they have all declined everywhere. Otherwise there is still a good variety, with Green Woodpeckers in plenty, both Great and Lesser Spotted Woodpeckers, Tree and Meadow Pipits to host the Cuckoo's progeny, six species of tit, both leaf and scrub warblers, particularly Whitethroat and Willow Warbler, and good numbers of Chaffinch and Yellowhammer. Woodcock may even breed on the bracken covered slopes and winter regularly brings Siskin, Redpoll and Fieldfare and occasionally Great Grey Shrike.

12. Sandy heaths, such as Devil's Spittleful, are characteristic of parts of the Severn Valley. *A. Winspear Cundall*

In complete contrast the steep hillside on the opposite bank of the Severn is clothed by the conifers of Ribbesford Woods. Larch and spruce with a field layer of bracken is hardly a recipe for an exciting bird habitat, but in addition to the inevitable hordes of Woodpigeons and numerous Great Tits and Chaffinches, there are a few pairs of Green Woodpecker, Nuthatch, Chiffchaff and Willow and Wood Warblers, whilst Meadow Pipits nest on the open grassland of Stagborough Hill.

On the outskirts of Stourport is the Local Nature Reserve of Hartlebury Common. Like Devil's Spittleful, this is a dry, sandy heath, but whilst heather is still extensive and there is also some oak-birch woodland, slight differences in habitat are apparent. Here gorse is more abundant, often forming continuous patches, and there are areas of bare sand as well. More importantly there is a small pool, which, though heavily silted, provides somewhere for birds to drink and bathe in an otherwise dry habitat, and adjacent to this is a small, acidic bog overlying peat. In general the bird life is similar to that at Devil's Spittleful, but Stonechat has bred here in recent years.

By now the light, sandy soils of the Bunter beds have given way to the rich, brown soils derived from the Keuper sandstone. As the valley becomes less confined, so the low ridge on the east bank becomes a maze of narrow country lanes and tiny hamlets that first grew up when fruit growing and market gardening began to flourish. Today many of the old orchards have disappeared and the size of holdings, unlike in the Vale of Evesham, has increased, but the soils remain first class and this is still an area of intensive market gardening or horticultural production. Horticulture is less evident on the west bank, though a few orchards are dotted amongst oakwoods and small parklands like those at Shrawley and Witley Court. The former has interesting woods that include extensive areas of coppiced lime, while the latter is a National Trust property that is now in ruins. Its pools, parkland and woods are still intact, however, and they are visited by many people in summer. Great Crested and Little Grebes breed on the pools; Tufted Duck, Pochard and Teal are winter visitors in small numbers; Reed Warblers nest in the marginal and emergent vegetation and Water Rail are occasionally noted in winter. The parkland is a favourite haunt of both Barn and Little Owls and the woods hold breeding woodpeckers, Nuthatch, warblers and finches, with Fieldfare, Redwing, Brambling, Siskin and Redpoll as winter visitors. Similar birds also occur at Shrawley, though with greater emphasis on the woodland species and here Sparrowhawk or even Buzzard might be seen.

The weir and lock at Holt Fleet mark a significant change in the nature of the valley. From hereon the river reaches a more mature stage, with an ever-widening flood plain and the gradual development of terraces. Free from the resistant sandstones, it begins to meander and flooding is more prevalent and extensive. It is along these slower reaches, between this weir and the Gloucestershire boundary, that Worcestershire's only heronry is located. Recently established, this contains only a dozen pairs at present, but, if undisturbed, it will hopefully expand. Compared to the Avon and Trent, the Severn valley has few sand and gravel workings presumably because it is remote from the main markets, but below the lock at Holt Fleet alluvial gravels have been worked and there are also some derelict clay pits. All are similar and best dealt with together. They are typical wetlands, with transient areas of shallow, open water, exposed mud, reed-beds, and silt-beds which are regenerating into willow swamp. There are one or two groups of oak and a few alders, but everywhere willow is the dominant tree, forming a closed canopy in places. In winter, the riverside meadows are frequently inundated by flood water and this can be impounded by the spoil-banks, creating wet marshland and sometimes leaving willow and alder standing in open water. The most stable water is the largest of the pools at Grimley gravel pit and this is the most attractive to wildfowl, though they are also drawn to the flood waters in some numbers. Mallard, Tufted Duck, Coot and Mute Swan all breed and in winter they are joined by Shoveler, Teal, Canada Goose and a few Pochard and Wigeon. Small colonies of Reed Warbler, Sedge Warbler and perhaps an occasional pair of Water Rail breed in the reed-beds and at Holt the same habitat attracts up to ten thousand roosting hirundines and wagtails each autumn, with a Hobby or two sometimes in attendance. Colonies of Sand Martins nest in the faces of the sand quarries, with up to a hundred pairs at Holt; Kingfishers breed as well and Little Ringed Plovers nest on the bare quarry floor. Marshy areas are occupied by breeding Yellow Wagtails and Reed Buntings and in winter by Grey Herons, Lapwing, Snipe and perhaps the odd Jack Snipe, Water Rail or Green Sandpiper. The area is important for Lapwings, with over a thousand present from July onwards, and there is also a small, but varied wader passage, especially in autumn, which usually brings Green and Common Sandpipers, Redshank and Curlew. The surrounds to the pits hold a variety of passerines, especially in winter when thrushes and finches are present in good numbers. These often roost in the willow scrub, where they are joined by large numbers of corvids in spring, at which season Siskin and Redpoll may congregate in the alders. A good variety of warblers and other

13. Disused gravel and clay pits at Grimley have regenerated into alder and willow swamp.   *A. Winspear Cundall*

passerines breed in the willow scrub; Stock Dove, Tree Sparrow, tits and woodpeckers may be found amongst the decaying timber; and the adjoining mixed farmland supports Little and Tawny Owls, breeding Corn Buntings, Pheasants, both partridges and exceptionally Quail.

Small woodlands are a feature of this part of the valley and the two largest of these are Monkwood and Ockeridge Wood, which between them cover almost 200 ha. Both are mixed woods, with rich shrub and field layers beneath stands of oak, birch and alder. To the south of Monkwood is a small patch of gorse that attracts migrant Whinchat and Stonechat. Between them these woods provide a wide range of arboreal habitat and support a good list of breeding birds. All three woodpeckers, Stock Dove, Little and Tawny Owls, Jackdaw, Nuthatch and Tree

Sparrow nest in holes in the trees; Turtle Dove, Blackcap, Whitethroat, Lesser Whitethroat, Sedge Warbler and Nightingale in the shrub layer; Woodcock, Chiffchaff, Willow and Grasshopper Warblers in the field layer and Tree Pipit on the woodland fringe. A large Rook and Jackdaw roost forms in winter, when Siskin and Redpoll visit the alders and Barn Owl, Buzzard and Sparrowhawk might also be seen.

Although only small, parklands like those at Ombersley and Thorngrove are also of interest. Both are outstanding for their reed beds, which are amongst the largest in the Region, and both contain large colonies of Reed Warblers in excess of forty pairs. Thorngrove, in particular, is little disturbed and is used as an autumn roost by Swallows, Yellow Wagtails, Starlings and sparrows. One or two ducks,

Canada Geese and Mute Swans also visit the pool, though only Mallard is known to breed. The surrounding woods hold all three woodpeckers and a good range of breeding warblers whilst in winter thrush numbers are quite high.

At this point the Severn is joined by its second major tributary, the Salwarpe. The headwaters of the Salwarpe reach back to the Birmingham Plateau and the foothills of the Lickeys and for the most part its catchment is an undistinguished agricultural area, dominated by the towns of Bromsgrove and Droitwich and the M5 motorway. Farming is mixed, with an emphasis on stock rearing in the east and dairying in the west. The Salwarpe is unusual in two respects, however. Firstly it is out of character with the rest of the sub-region in that it flows through a broad, shallow basin rather than a deep valley. In this respect it is more akin to the Avon than the Severn. Secondly it has a natural salinity, which it derives from the salt deposits of its upper reaches. These deposits were known to the Romans and much later were to bring Droitwich fame as a spa town. Salt was mined in the area until 1972 and, for the ornithologist, interest is centred on the subsidence pools of Upton Warren, which are associated with the old Stoke Works. These pools are the central feature of an outstanding wetland and, together with the lake of nearby Westwood Park, form two of the most important aquatic habitats in this sub-region.

The watershed between the Salwarpe and the Avon basin is low and would be barely discernible on the ground if a scatter of small woodlands did not emphasise the low hills. These woods are remnants of the old Feckenham Forest. The highest point is the common land of Piper's Hill, where the mixture of mature oak, beech and sweet chestnut holds some interesting birds despite the pressures exerted by picnickers whose cars penetrate deep into the woods in defiance of the by-laws. Apart from a patchy cover of bramble, there is little or no shrub layer on the higher, level ground and it is here that Chiffchaff and Wood Warbler breed and flocks of tits, Chaffinches and sometimes Brambling forage amongst the fallen leaves for beech mast in winter. Westwards the ground falls away steeply to the plain below and here there is a better developed shrub layer that hosts Turtle Dove, Garden Warbler and Blackcap. Spotted Flycatchers can often be seen around the small pool, whilst amongst the holes and cavities of the older trees three species of woodpecker, Tawny and Little Owls, Nuthatch, Treecreeper, Redstart and Tree Sparrow all breed. Except for Redstart, similar birds are also found at nearby Hanbury Hall – a National Trust property in an attractive setting of arable fields, pastures and parkland with scattered oaks and exotic conifers. The small pool is visited by a few Mallard and Teal in winter, Lapwings flock to the fields and both Grey and Red-legged Partridge breed, but it is again the woodland species that hold most interest, with Stock Doves and Tree Sparrows present in good numbers.

Upton Warren lies either side of the Salwarpe, mid-way between Bromsgrove and Droitwich. This important SSSI, most of which is also a Worcestershire Nature Conservation Trust reserve, has one small and two large freshwater pools, one of which is a flooded gravel pit which the County Council uses for educational sail-training and fishing. Breeding birds on these freshwater pools include Great Crested and Little Grebes, Mallard, Tufted Duck, Canada Goose, Mute Swan and Coot. Garganey have been seen here as often as anywhere in the Region, usually on passage, though breeding has occurred. Shoveler and Shelduck have also bred and this is one of the principal breeding sites for Ruddy Duck. Passage often brings Little Gull and a variety of terns, which have included Roseate, White-winged Black and Caspian in the past; whilst Teal, Shoveler

14. The saline flash pools at Upton Warren provide an ideal site for passage waders.

J. V. & G. R. Harrison

and Pochard add to the variety of duck in winter. At this season too, parties of tits, Redpoll and Siskin visit the alders and willows alongside the river.

As previously mentioned, however, it is brine seepage into the three small flash pools that makes Upton Warren unique. The flora includes salt-marsh plants and fluctuations in the water level ensure plenty of exposed mud, so wader passage always promises to be exciting and over the years no fewer than thirty species have appeared. Greenshank, Spotted Redshank, Wood Sandpiper, Green Sandpiper and Ruff are regular, especially in autumn, and recently Temminck's Stint has been seen almost every spring, with up to six birds on one occasion. Other interesting waders that have occurred over the years include Avocet, Red-necked Phalarope and

Pectoral Sandpiper. Both the freshwater and saline pools have good margins of rush and sedge in which Water Rail, Lapwing, Redshank, Grasshopper Warbler and Yellow Wagtail breed, and where several Snipe and a few Jack Snipe congregate in winter. High hawthorn and blackthorn hedges, in which Lesser Whitethroat breed, enclose the surrounding pastures and both Kestrel and Little Owl nest in the vicinity. Amongst a varied population eight species of warbler breed, including several pairs of Reed and Sedge Warblers. Many other passerines occur on passage and there is a small winter roost of Corn Buntings. Hobbies are regular visitors during summer and early autumn; Marsh and Hen Harriers have appeared on passage and Gannet, Blue-winged Teal, Great Skua, Alpine Swift, Richard's Pipit and Marsh Warbler have been amongst the more unexpected visitors.

Just the other side of Droitwich is the other important SSSI wetland at Westwood Park. Hidden behind a high brick wall that formerly enclosed a large landscaped estate is a strictly private and delightfully secluded pool of 24 ha with extensive patches of reed. Open farmland flanks the northern and western shores and a plantation and coppice border the southern shore. There is disturbance from a limited amount of water ski-ing, sailing and an increasing amount of fishing. This affects some breeding species and may well account for the demise of the heronry which flourished here until 1963, but the reeds still provide safe nesting sites for Great Crested and Little Grebes, Mallard, Tufted and Ruddy Ducks and good numbers of Sedge and Reed Warblers. It was here in 1965 that Bearded Tits occurred for the first time ever in Worcestershire. Winter brings many more duck, including Gadwall, Shoveler and Pochard regularly and other species sporadically. Also at this season up to five thousand gulls, mostly Black-headed, roost on the water and Snipe, Jack Snipe and Water Rail can be seen around the margins. Passage regularly brings a few waders and terns and there is an autumn roost of hirundines. The damp woods are not without interest of their own, with two large rookeries and breeding Nightingale, Grasshopper and Wood Warblers, Blackcap, Garden Warbler and Tawny Owl. In winter parties of Redpoll and Siskin, and sometimes a few Brambling, are usually present.

Between Droitwich and its confluence with the Severn, the Salwarpe flows alongside the Droitwich Canal. For years this stretch of canal was one great reed-bed which held the largest Reed Warbler colony in Worcestershire. Considerable clearance has since taken place, particularly of scrub at the Droitwich end, but there are still many pairs of Reed Warbler; and Little Grebe, Mallard, Tufted Duck and Kingfisher breed as well. Overgrown hedges and copses along the canal harbour passerines in good variety and number, including breeding Whitethroat, Greenfinch, Bullfinch, Chaffinch and Yellowhammer. The land between the river and canal comes into its own in winter, when it is a rich feeding ground for flocks of Lapwing, Black-headed Gull, Fieldfare, Redwing and Starling. Some of the northern tributaries of the Salwarpe are bordered by wet meadows which seem ideal for Snipe, Redshank and Lapwing to breed, but there is no evidence that they do so.

More is known about Oakley, or Pulley as it is sometimes called. This is another pool that has been formed through subsidence as a result of salt extraction. Again it is an SSSI, important mainly for its marsh vegetation, especially *Phragmites*, which completely encircles the open water and forms the second largest reed-bed in Worcestershire. Areas of rank grass, sedge and meadowsweet fringe this reed-bed; there are scattered willows, patches of bramble and thick thorn hedges, notably at the southern end; a small partially submerged stand of conifers in the southwestern corner; and a belt of mature willows along the western boundary. There is a thriving Reed Warbler colony of between 60-80 pairs, with attendant Cuckoos, perhaps half as many Sedge Warblers, a few pairs of Grasshopper Warbler and many Reed Buntings. Blackcap, Whitethroat, Lesser Whitethroat, Chiffchaff and Willow Warbler also nest, whilst the water has attracted breeding Little Grebe, Pochard and Tufted and Ruddy Ducks. Several passerine species visit the hedgerows and in autumn many migrant warblers pass through the reed-bed, which at this season also hosts a large Starling roost. Less is known about the bird-life in winter, when there is some shooting, but Snipe and Water Rail frequent the marshy areas, Redwing and

15.   The fast-flowing River Teme and its tributaries are a stronghold of Dippers and Sand Martins.   *A. Winspear Cundall*

Fieldfare feed or roost in the hawthorn and in 1976 a party of Bearded Tits visited the reed-bed.

Below its confluence with the Salwarpe, the Severn enters Worcester. Famed for its sauce and porcelain, Worcester, like so many Midland towns, exhibits a curious blend of historic and contemporary architecture. The majestic cathedral stands above the alder-lined river, which is a favourite haunt of Siskin in winter. Here is yet another weir and by-pass lock and below this the last of the three main tributaries, the Teme, enters on the right bank.

For most of its length the Teme is closely hemmed in by hills and this must surely be amongst the more peaceful and picturesque parts of the Region, with its mosaic of small oakwoods, orchards and hop fields set amidst a maze of narrow lanes and tiny hamlets. All three woodpeckers are locally common in the woods, along with Nuthatch and Treecreeper, and most warblers and Redstart breed sparingly. Wherever the woods overhang one of the many tiny streams Spotted Flycatchers are common and in winter some of the woods hold sizeable Redwing and Fieldfare roosts. On the farmland Lapwing, Curlew and Yellow Wagtail nest in the damper cattle pastures and these are also visited by migrant Snipe and Redshank. Red-legged Partridge are fairly common and there are one or two pairs of Corn Bunting, though this is rather far west for them in the West Midlands. The intricate mixture of habitats in the Teme Valley also supports a good range of raptors. Kestrel is the commonest, but Sparrowhawks are fairly numerous,

71

Buzzards appear from time-to-time and, of the owls, Little is common and Barn quite widespread. Once again though it is the alder and willow-lined river and its tributaries that are the centre of attraction. In 1979 a comprehensive survey of the Teme was carried out for the Worcestershire Nature Conservation Trust (J. Day *in litt*) and this has provided much valuable information about the variety and density of its bird-life. It is very much a river of two stages. Downstream of Knightsford it meanders slowly across the Keuper marl between high earth banks. Here the deeper, more sluggish stretches support a good population of Kingfishers and a few pairs of Sedge Warbler nest where there is suitable bankside vegetation, though this species and Reed Bunting are generally scarce. Above Knightsford the flow is faster and the course across the Old Red Sandstone more direct. Rapids, shallows, rock faces and boulders are a feature of this stretch and these provide nesting and feeding sites for Dippers and Grey Wagtails. Common Sandpipers have also bred in the past, but there have been no positive records since 1943, although a pair was present in late May 1980. The river is especially important for Dippers, many pairs of which probably nest along the tributaries but come to feed on the boulder-strewn shallows of the parent river during late summer and autumn. Grey Herons also feed in the shallows and there are three areas of shingle banks backed by lush bankside vegetation that are used by moulting flocks of Mallard. Above all, though, the river is a major stronghold of the Sand Martin which, with colonies throughout its length, is easily the most numerous breeding species. Most of the earthbanks suitable for nesting occur downstream of Knightsford, yet surprisingly a count in 1976 showed most of the Sand Martins to be nesting above this point, where there are fewer suitable banks. Day (*op. cit.*) believes this to be

because the food supply is more abundant in the upper reaches. However, there appears to have been a substantial increase in numbers on the lower reaches since 1976, bringing the population of this stretch to the same level as that found by A. J. Tooby between 1942–4 (Harthan 1946).

Below Worcester and the Teme confluence, the valley of the Severn broadens, with the meandering river flanked on either side by terraces and the broad marl plain that sweeps eastwards to the low watershed with the Avon and westwards to the foothills of the Malverns. In places more resistant beds of sandstone bring some relief to an otherwise flat landscape of dairy farms, cattle pastures and willow-fringed streams. Astride the Avon watershed lies the "Capability" Brown parkland of Croome Court. Most of it lies within the Avon valley, but the largest area of water, Pirton Pool, belongs to the Severn catchment. This pool had established itself as a favoured haunt of Bewick's Swans, presumably moving to or from Slimbridge, and in 1977 a record herd of ninety was reported, but it remains to be seen whether drainage during 1978 will disrupt this habit. Certainly it caused a pair of Ruddy Duck to desert in that year.

Farming in this part of the valley is mixed and, where arable crops are grown Red-legged Partridge, a few pairs of Corn Bunting and sometimes Quail can be found. In the main though it is an area of grassland frequented in winter by large flocks of Lapwings, Starlings and thrushes and supporting breeding Lapwings, Skylarks and Rooks. A few Snipe winter in the ditches and Little Owls are widespread. Gravels brought down from the Malverns during the last Ice Age left pockets of poor soil around the foothills and these once supported rich lowland heaths similar to those around Kidderminster.

Sadly most have been reclaimed for agriculture or have disappeared under concrete during the past fifty years, although an outstanding exception is Castlemorton Common, but this is described later in Chapter 8 along with the hills themselves. Tooby (1945) has provided an interesting account of the changes in bird-life that resulted from the ploughing of Old Hills Common. Here from 1942–4 areas of gorse and bramble, with a few thorn thickets, were cleared and sown with wheat, but some grazing land, a few mature trees and a hillside of gorse and thorn scrub were left untouched. Linnets, Song Thrushes, Turtle Doves and warblers not surprisingly deserted the cultivated land for the surviving common, but Blackbirds, Chaffinches and Yellowhammers remained on the cultivated portion, where Tree Pipits actually increased in numbers. Overall, density increased by 18 per cent, from an estimated 2·95 pairs per ha in 1942 to 3·48 pairs per ha in 1944. However, the bulk of this increase occurred on the unimproved commonland, where density increased by 71 per cent, whereas on the newly-ploughed part it actually declined by 57 per cent. The commonest species, in descending order, were Linnet, Yellowhammer, Chaffinch, Whitethroat, Blackbird and Willow Warbler. Doubtless there have been many more changes since, but most of these species can still be found here and this is also a good place to listen for Nightingale. Around Madresfield the many small woods bring further diversity through hole-nesting species such as Stock Dove, Kestrel, Jackdaw and Green and Great Spotted Woodpeckers, whilst Buzzard has been recorded in the past.

Winter flooding along the lower reaches of the Severn is frequent, though in recent times steps have been taken to ameliorate its worst effects. Nevertheless, many of the meadows around Upton-on-Severn are permanently under grass, providing valuable summer grazing and it is here that an occasional Corncrake might appear, though their incidence has declined to the point where very few now occur. Not far from here is Longdon Marsh. Once an extensive wetland which was drained in 1870, Lees (1828) described it as "one of the most singular spots is this extensive flat, surrounded on all sides by deep ditches, and tenanted by flocks of geese. In autumn and winter it is covered with water, giving the landscape at that time the only natural lake-like appearance the county possesses. In winter the marsh is visited by various sea birds, as gulls, terns, Whimbrels, Curlews etc. Snipes and Lapwings breed in the marshy meadows. The Bittern is shot occasionally, and the Buzzard is seen lurking in thick coppices". Obviously most of this has been lost with drainage and today Longdon retains barely a vestige of its former glory. Even so, it is still low-lying and criss-crossed by many streams and ditches, though improvements carried out to the Severn in the late 1960s mean that it no longer floods regularly. Mallard, Lapwing and perhaps Curlew still breed, though the latter species is declining in the Severn Valley. The well-timbered hedgerows and undisturbed pastures are also ideal for many passerines. Redstarts nest in the pollarded willows, Reed Buntings and Tree Sparrows breed in good numbers and one or two Nightingales sing from hedgerows or thickets. Whinchat and Grasshopper Warbler are also present in summer, though evidence of breeding is lacking. In the final reckoning it is more than ironic, after years of progressive drainage, that this area should until quite recently have been under consideration for flooding to form a new water supply reservoir.

On this note the Severn finally leaves Worcestershire and the West Midlands for Gloucestershire and the Bristol Channel.

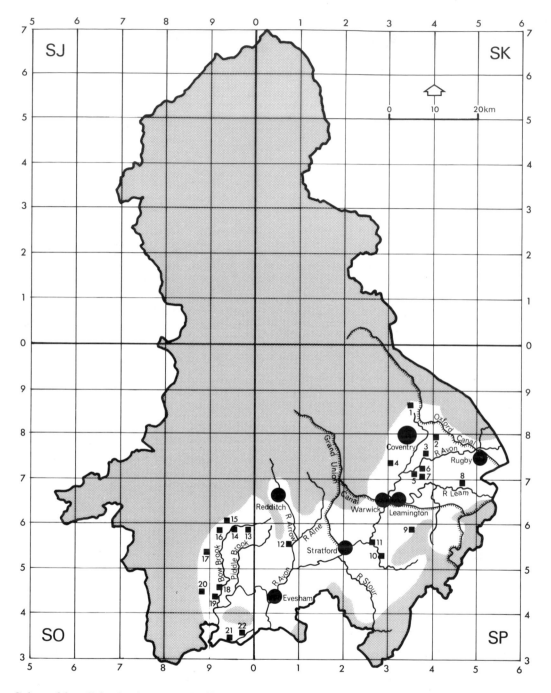

Selected localities in the Avon Valley

1	Bedworth Slough	7	Wappenbury Wood	13	Roundhill Wood	19	Defford
2	Coombe Abbey	8	Draycote Water	14	Hornhill Wood	20	Croome Park
3	Brandon Marsh	9	Chesterton	15	Goosehill Wood	21	Aston Mill
4	Crackley Wood	10	Walton Hall	16	Trench Wood	22	Beckford
5	Waverley Wood	11	Charlecote Park	17	Spetchley Park		
6	Ryton	12	Ragley Park	18	Tiddesley Wood		

4 The Avon Valley

The Avon and its tributaries drain most of Warwickshire and eastern Worcestershire. The basin is elongated about a north-east to south-west axis and is bounded to the north by the Birmingham Plateau and to the south by the Cotswold scarp. This is the largest of the seven sub-regions, covering 1,700 km², or 23 per cent of the land area. It measures 95 km by 32 km and mostly lies below 100 m, falling to 10 m where the river flows over the Gloucestershire boundary to join the Severn at Tewkesbury. Within this basin are one or two areas of higher ground, such as the Harbury Plateau and Bredon Hill, which are regarded as outliers of the Region's fringing hills and are dealt with in Chapter 8. The main tributaries are the Sowe, Leam, Stour, Alne, Arrow, Isbourne and Bow Brook.

By geological standards the Avon is a young river. It is also a river that has undergone immense change. It began life as part of an Avon-Soar drainage system flowing north-eastwards into the North Sea, but its valley was dammed by glacial deposits during the Ice Age and meltwaters from the highest lake then cut their way backwards through the watershed and into the Severn Valley.

This complete reversal of flow is why the tributaries of the Avon, except for the Leam, are short and flow at right-angles to the main river. Indeed some of the northern tributaries in the Sowe basin have even retained their original south-easterly orientation. In contrast to the Severn, the Avon flows across soft marls and clays and these have enabled it to adjust to rejuvenation swiftly and to develop a wide valley with extensive terraces. Because of their shortness, though, the tributaries bring only a small volume of water to their parent river so that its ability to recover from pollution is retarded.

The geological structure of the Avon basin is relatively simple. The oldest formation is the Carboniferous Coal Measures which dip down from the East Warwickshire Plateau to underlie Coventry and much of the Sowe basin. These are then down-faulted beneath beds of the Keuper series, which dip south-eastwards and are themselves covered unconformably by Jurassic deposits. In general the Keuper outcrops are north of the river and the Jurassic south, but some minor faulting results in local exceptions, notably around Stratford and west of Evesham, where the landscape is markedly undulating.

The Keuper formation is predominantly marl, but near the top is a thin band of Arden sandstone. Likewise the Jurassic formation is virtually all Lower Lias clay, but there are interbedded bands of White and Blue Lias, and a thin Rhaetic band separates the two formations. Because of the dip in the strata, these interbedded rocks usually outcrop on the south-eastern side of river valleys, where they form conspicuous, low escarpments which are often wooded. The best examples occur along the Avon Valley itself, between Warwick and Evesham, and in the Alne, Arrow and Stour Valleys. The beds of limestone are often associated with shallow soils and thorn scrub, though an interesting calcicole flora sometimes develops where scrub is cleared.

The picture is then completed by a covering of drift, which is extensive in the north-east, but gradually peters out towards the

Warwickshire-Worcestershire boundary, and by well-developed and locally extensive river terraces. The nature of the drift varies and this is reflected by the landscape and pattern of farming. The freely draining gravels yield good, productive arable land, but wherever drainage is excessive there are a few surviving remnants of lowland heath. These are most noticeable between Coventry and Rugby, but nowhere are they as rich or extensive as those of the Severn Valley. The boulder clay on the other hand yields medium-to-heavy loams suitable for mixed farming, though where drainage is impeded they often remain under woodland. If drift is absent, the differing structure and base content of the Keuper and Lias soils has most influence on farming and the landscape. Drainage of both is often poor, but the marl yields lighter soils capable of supporting both grass and arable crops. Like the drift they are neutral to acidic and sustain a wide variety of trees, especially oak and elm, but with ash prominent on the chalky drift north of Rugby. The Lias soils, on the other hand, are extremely heavy and were traditionally under permanent grass. However, with good drainage and proper management, more and more are going into arable production, particularly winter wheat. Most trees dislike their heavy, calcareous nature, with elm alone seeming to thrive on them, and it is here that the effects of Dutch elm disease have been most devastating.

Alluvial deposits occur throughout the length of the basin, which is narrow in its upper reaches, but widens out considerably downstream. The river has cut five identifiable terraces, although only the second and fourth are extensive, and the same terraces are also present on the tributaries. Apart from one or two small quarries, these extensive gravels have yet to be exploited commercially, largely no doubt because of their distance from the main markets and strong environmental objections. Meanwhile they continue to

nurture the intensive horticultural and market garden industries of the Vale of Evesham.

These industries also rely on the climatic advantages brought by the Gulf Stream, which are similar to those of the Severn. Rainfall seldom exceeds 700 mm and between Warwick and Evesham it is lower than anywhere in the Region, with around 600 mm. This means irrigation is often necessary, but this is offset by more sunlight, less wind and warm summers, with the July mean reaching 17°C. The big risk is radiation frosts in late spring, but these are usually confined to the valley floor, so that south-facing slopes above 75 m are excellent for orchards, with trees blossoming two-to-three weeks earlier than in less favoured districts.

The same advantages also attract birds. Hobbies are more plentiful here than anywhere else in the West Midlands, with at least five pairs, and Nightingales are widespread in the numerous small woods, most of which were formerly coppiced. Above all though, this is the stronghold of the Marsh Warbler, with the lower valley holding more than three-quarters of the entire British population (Sharrock 1976). Otherwise the land throughout is very thoroughly utilised and there are fewer natural or semi-natural habitats here than in the other sub-regions. With Draycote Water though, the Avon Valley does contain one of the two largest expanses of water in the Region and this has added considerably to its attraction for ornithologists in recent years.

The Avon is one of England's most romantic and peaceful rivers, flowing through a fertile landscape of pasture and arable land dotted with orchard and wood, farm and village, country house and landscaped parkland. It is a river of three phases. Upstream of Warwick it flows through open countryside, but is very much under urban influence, dominated by Rugby and Coventry and the effluent they discharge into it. From Warwick to the county boundary it becomes very much Shakespeare's Avon, whilst through Worcestershire it is flanked by large riverside meadows and the orchards and market gardens of the Vale of Evesham.

The Avon rises in Northamptonshire and its clean waters enter Warwickshire at the point where it is bridged by the old Roman Watling Street, now less glamorously called the A5. Here the river is 90 m above sea level and during the next 140 km, until it joins the Severn at Tewkesbury, it falls a mere 80 m. Shortly it is joined by another clean river, the Swift, before skirting to the north of Rugby. Like many a settlement Rugby can lay claim to ancient origins, but it really owes its reputation to its famous school, Rugby football and the legendary Tom Brown, and its existence to the advent of the railway and more recently the cement and electrical industries. Today it is a bustling little industrial town, unique in the Midlands for its yellow-brick buildings so characteristic of the London area. Inevitably the effluents from its industries and people find their way into the river and from hereon the Avon is polluted.

Westwards from Rugby the flood plain is small, but the willow-fringed river meanders through green meadows set in pleasantly undulating countryside. The varied geology ensures a varied scenery. There are old Scots pines, which serve as reminders of the former heaths that are embodied in many place names, a disused airfield and small woods on the more poorly drained soils. Generally though, the easily worked, productive soils are used for arable farming. Hedgerows are few and small, and trees sparse in this countryside of Skylarks, Corn Buntings and Yellowhammers.

At Bretford the Avon wends its way beneath the ancient stone bridge that carries the Fosse Way. Here there is some fish life and Kingfishers can be seen along the sandy banks, but the river's respite is short-lived and it soon finds itself under the threatening influence of Coventry, although it skirts round the city itself. Immediately downstream of Wolston is Brandon Marsh – an area where colliery subsidence originally created a flash and subsequent gravel working has added considerably to the extent of the aquatic habitat. With such a good variety of freshwater habitats, augmented by reed bed and willow scrub, it has developed into one of the most important ornithological sites in the Region. Scheduled as an SSSI, it has for several years been extensively studied and managed by local conservationists and, to their credit, although working is still in progress, the gravel operators have recognised its importance by establishing a permanent nature reserve in conjunction with the Warwickshire Nature Conservation Trust.

The original colliery flash lies in a meander of the river, to which it is connected at one end. As a result, water levels fluctuate markedly. Flooding occurs after heavy rain, but at dry times a good expanse of shallows or exposed mud attracts Snipe and dabbling duck in winter, plus a wide variety of passage waders, amongst which Ruff, Black-tailed Godwit, Greenshank and Green Sandpiper are fairly regular and Avocet has been seen. The river is flanked by sallow, which also grows vigorously on the adjacent marsh despite constant cutting back as part of a management programme, and there are areas of rush, sedge, reed-grass, reed-mace and *Phragmites*. Inevitably the habitat around the gravel workings is constantly changing and recently extraction has commenced on the south side of the river. Generally though there is always a good range of habitat. Within the workings themselves up to six pairs of Little Ringed Plover have bred regularly every year since 1962 along with a few pairs of Kingfishers and colonies of Sand Martins, providing there is a suitable quarry face, though the latter have declined in recent years. The open lagoons, with

16. Colliery subsidence and gravel extraction have created a rich and important bird habitat at Brandon Marsh. *C. Fuller*

their numerous small, vegetated islands, regularly hold breeding Great Crested and Little Grebes, Tufted Duck, Canada Goose and Mute Swan, and occasionally Teal, Garganey, Shoveler and Pochard as well; whilst winter brings good numbers of Tufted Duck, Teal and other dabbling duck. Lapwing, Redshank and Snipe breed amongst the extensive patches of rush and sedge, Sedge Warblers are common and some fifty pairs of Reed Warbler breed in the reeds and sallows along with the occasional pair of Water Rail. Despite their being very little *Phragmites*, Bearded Tits have become almost annual winter visitors and Spotted Crake has been recorded here more often than anywhere else and may even have bred. Those lagoons that are used as silt settling beds are quickly colonised by sallows and osiers. In past

autumns the ten-thousand roosting hirundines that have regularly used these and the reed-beds have frequently attracted Hobbies and in winter the same areas are used by roosting Reed and Corn Buntings. Two small coverts, one mixed and one coniferous, hold owls and woodpeckers, and Grasshopper Warbler and Whinchat also breed in the area. The number of gulls has declined since the closure of the nearby rubbish tip, but Great Grey Shrike and Short-eared Owl are still regular winter visitors and an intensive ringing programme has yielded two Great Reed Warblers and, most surprisingly, an adult male Barred Warbler in spring. Bittern, Little Bittern, Little Egret, Spoonbill, Marsh Harrier and Savi's Warbler all exemplify the quality of the marsh, but Ring-necked Duck, Rough-legged Buzzard and Arctic Skua have also been recorded – an impressive list indeed.

Above Brandon, on the high ground to the south, stands the Talbot car plant and behind this more sand and gravel quarries at Ryton. Although overshadowed in importance by Brandon, these quarries hold breeding Little Ringed Plovers and strong colonies of Sand Martins. Of greater interest, however, is their extensive backcloth of broad-leaved woodland. Ryton Wood, into which the quarries are making in-roads, the adjoining Wappenbury Wood and the nearby woods of Bubbenhall, Waverley and Weston, together cover over 300 ha and form one of the largest groups of woodland in Warwickshire. Oak is the dominant species in the first three, though birch is locally dominant on the sandier soils, and sallow and alder are widespread in the damper areas. All three have a rich understorey including hazel, which was formerly coppiced, hawthorn, blackthorn, buckthorn, bramble, dogwood and elder and sustain a good variety of breeding and wintering birds, though Wappenbury is perhaps the best. Here there is a stream flanked by boggy ground and heavy soils on which the thickets and dense coppice hold the largest Nightingale "colony" in the Region and many pairs of Willow and Long-tailed Tits, Blackcap, Bullfinch, Jay and Reed Bunting. The sandier, better drained ground to the south supports more mature oaks and birches, in which Stock Dove, Tawny Owl and all three woodpeckers breed; and a well-developed field layer of bracken amongst which several pairs of Woodcock nest. Garden Warbler and Tree Pipit breed along the rides or around the woodland fringe and in winter there are large roosts of finches, thrushes, Starlings and woodpigeons.

Waverley and Weston Woods are different. The latter, though broad-leaved, is birch dominant, but still holds breeding Stock Doves, woodpeckers and Jackdaws.

The former, however, is mainly a replanted stand of young pine, though still with a few stands of hardwoods. In the younger parts Turtle Dove, Grasshopper and Willow Warblers, Tree Pipit, Linnet, Redpoll and Yellowhammer all breed; Woodcock frequent the broad, bracken-covered rides and both Tawny Owl and Sparrowhawk occur.

Likewise, behind Brandon on the north side of the river, are more, if less extensive, woodlands at Brandon itself, New Close and Birchley. Regrettably these are too near to Coventry to escape disturbance and consequently are of less interest, though they do hold good populations of the commoner woodland species such as Robin, Blue Tit, Starling, Wren and Dunnock. Further downstream, at Baginton, is Coventry's airport, which in winter attracts Lapwings and occasionally Golden Plover, though no longer on a regular basis. Nearby is the site of an old sewage marsh, which was once the mecca of all Coventry bird-watchers. Like many such, it used to attract good numbers of Teal, Moorhen, Meadow Pipit and Snipe in winter; breeding Redshank and Yellow Wagtails and a variety of passage waders, including Wood Sandpiper and Black-tailed Godwit. Indeed, it was here in 1947 that the first Sanderling was recorded for Warwickshire. An enlightened Coventry City Council also set aside an area as a bird sanctuary, but sadly the works have been abandoned, drainage in the area improved and all that remains is a rough piece of open ground next to gypsy and industrial sites, with access from the somewhat euphemistically named Siskin Drive – though habitat suitable for Siskins would indeed be hard to find.

Coventry itself lies within the basin of the Sowe, which is the first of the Avon's major tributaries, although it also straddles another tributary, the Sherbourne. In the north it merges with Bedworth, but until recently its eastern expansion had been contained by the flood plain of the Sowe. During the past decade or so, though, this barrier has been breached and new estates have spread eastwards.

Coventry was an important town even in the Middle Ages and by the fourteenth century it was the fourth largest city in England, with a prosperous woollen industry and a thriving market. Later the Huguenots introduced silk ribbon manufacture and watch-making and the former eventually became the city's foremost industry. Both of course were primarily cottage industries, though some modernisation occurred when steam power was introduced, and both ultimately declined once import restrictions were lifted in the late nineteenth century. The workforce remained, however, and turned its skills to sewing machine and bicycle manufacture. The invention of pneumatic tyres greatly increased the comfort and popularity of the bicycle and Coventry remained the world centre for bicycle manufacture until the First World War.

Cottage industry survived in Coventry much longer than in most cities and factories did not appear until late last century or early this. The city consequently avoided the worst of the squalid overcrowding experienced elsewhere, only to be severely blitzed in 1940. Displaying characteristic resilience in the face of adversity, however, the last twenty-five years have seen a new city emerge, with an economy based on motor vehicles and textiles, and a brand new shopping centre and cathedral rising like a phoenix from the ashes of war. Indeed it is probably true that Coventry today owes more to Hitler than to Lady Godiva; and a contemporary Peeping Tom, with an eye for the right bird, might well locate a Black Redstart or two amongst the remaining preserved ruins.

To the south spacious housing estates and a new university lie behind the majestic tree-lined approach from Kenil-

worth and the Memorial Park is a valuable open lung for an urban area. Indeed such an area is more rural than urban and its bird life is rich and varied. Nuthatch and woodpeckers are not uncommon and Brambling sometimes join mixed finch flocks to forage through the leaves in winter. Commons, like that at Hearsall, with its oaks, hawthorn scrub, gorse and rank grass, harbour breeding Redpoll, Lesser Whitethroat and Willow and Long-tailed Tits as well as commoner species. Blackcaps occasionally winter and at this time there is also a large thrush roost.

Elsewhere, the city has other open lungs at Wyken Slough and the Coundon wedge, whilst to the east, just outside its boundary, lies Coombe Abbey. Here the City Council has established a country park, which incorporates the Abbey and its formal gardens, a large ornamental lake, and an extensive backcloth of broad-leaved woodland to the north. Southwards open farmland and arable fields within the park provide feeding grounds for Lapwing, Pheasant and Grey and Red-legged Partridges; winter grazing for wild-fowl; and a rich hunting ground for raptors like Kestrel and Barn, Little and Tawny Owls, all of which breed. Wood-cock, all three woodpeckers, Nuthatch and a good variety of corvids, tits, warblers and finches also nest in the woodlands and there is a large Blackbird roost in winter, together with a few thrushes, Siskin and Brambling. At this season Buzzard or Sparrowhawk might put in an appearance, whilst Hobby is occasionally seen in summer.

Inevitably the pool holds most interest. This is an SSSI with a good variety of wintering wildfowl and an important heronry – the largest in Warwickshire – on the main island. Surrounding this island is a reed-bed, with another at the western end of the lake, and these add to the habitat diversity. Amongst the breeding species on or around the lake are Great

Crested and Little Grebes, Canada Goose, Water Rail, Shoveler, Kingfisher, Sedge and Reed Warblers and all three wagtails. It is also an important feeding ground for hirundines and Swifts in summer and is visited by a few passage terns and waders. Winter brings a few Wigeon and good numbers of Tufted Duck, Pochard and gulls, whilst Shoveler have numbered as many as two hundred. Amongst the unexpected visitors have been Manx Shearwater, Storm Petrel, Bittern and Bearded Tit. North-east of Coombe, the Smite Brook flows through an extensive arable landscape of open vistas, shelterbelts, game coverts and windbreaks. There are no especially important habitats here, but Golden Plover regularly occur in some numbers each winter around Brinklow.

North of Coventry is Bedworth – a rather nondescript town of mining origin, which has diversified and expanded considerably since the last war. Although it straddles the Avon–Trent watershed, it has been included within this sub-region since its only site of ornithological importance drains into the Sowe. This is Bedworth Slough, which has recently been designated a Local Nature Reserve. Sandwiched between new housing estates and a busy by-pass, this is a small, unpretentious site of 5 ha. It comprises a shallow pool, which was formed through mining subsidence around the turn of the century, an extensive margin of reed-mace and a surround of rough, overgrown pasture. Great Crested and Little Grebes, Tufted Duck and Mute Swan breed and over seventy species have been recorded here, many on passage. Its real interest, however, lies in the autumn Swallow roost, which is one of the largest and most regular in the Midlands, with up to 25,000 birds on occasions though numbers have been lower of late. A ringing programme produced only forty-four retraps amongst seven thousand birds ringed over a period of forty-two nights, so quite clearly the passage is continuous and the roost constantly changing.

The rest of the Sowe basin lies south of Coventry and is drained by Finham Brook. Apart from Kenilworth – a quiet dormitory town of two distinct halves, with its old, red-brick cottages clustered around the historic castle at one end and new commuter estates at the other – it is a pleasantly undulating, rural area with small fields, thick hedgerows dotted with oaks, and several small woodlands of which Crackley Woods are best known. This is principally an oak-birch wood where Blackcap, Chiffchaff and Willow Warbler breed in reasonable numbers along with a few pairs of Garden Warbler, Great and Lesser Spotted Woodpeckers and Treecreeper. In winter flocks of Redpoll gather to feed on the birch seed. The neighbouring farmland is mixed and holds Whitethroat, Skylark and Yellowhammer in good numbers along with the occasional pair of Corn Bunting, Lapwing or Snipe.

Although pollution occurs upstream, it is not until near its mouth that the Sowe is finally destroyed by the effluent from Coventry's giant sewage works at Finham. The problem is not so much the quality of the effluent, which is not all that bad, but simply its sheer volume, which in dry weather equals the natural flow of the river. This excessive artificial load is then very quickly transferred to the main river, so within the first quarter of its catchment the Avon receives the effluent from more than half the population of its basin – a blow from which it never fully recovers.

At its confluence with the Sowe the Avon meanders round three sides of Stoneleigh Abbey and the National Agricultural Centre. Stoneleigh Park was landscaped by Repton, but since the war the expansive lawns and majestic trees have been gradually superseded by the permanent buildings of the Royal Show ground, and the weirs that were intended to widen the river and increase its impact now serve mainly to hold back toxic sludge.

Nevertheless, the setting of the river is picturesque enough as it wends past Ashow, then slithers like a snake around the sandstone bluffs on its way to Warwick and its confluence with the Leam. The bluffs are often wooded with oak, beech and sweet chestnut in which Nuthatch, all three woodpeckers and most species of tit can be found. On the adjacent fields Lapwings, Black-headed Gulls, Fieldfare and Redwing are numerous in winter and, notwithstanding the state of the river, Kingfishers, Grey Herons and Mute Swans can still be seen.

The Leam is one of the Region's cleanest rivers. It also has the largest basin of any of the Avon's tributaries, draining an almost circular area of some 20 km diameter known historically as the Feldon, or cleared lands. Despite its size, this is one of the most uniform areas, with a monotonously flat landscape that is relieved but rarely by a drift-capped hillock. It is underlain entirely by heavy Lias clay in which drainage is frequently poor. For centuries the land was regarded as too heavy to plough and remained permanent pasture, but ploughing was introduced as part of the war effort and with the help of increased mechanisation has continued ever since. Westmacott and Worthington (1974) described this area as "one of heavy clay soils within a flat or slightly rolling enclosure landscape with a large number of hedgerow trees, three-quarters of which are elms. There are very few roads and only an occasional small copse or covert, so that it is the receding layers of regularly-spaced elms that give the landscape much of its character, which would change entirely were they to be lost." Unfortunately this loss has subsequently occurred and views are no longer obscured by an intermediate horizon of trees, though serried ranks of dead elm still stand gaunt against the skyline as a melancholy reminder of a bygone age.

The same authors also noted that since 1920 arable farming had increased three-fold, but even so almost half the land is still under permanent pasture, and the mixed farming enterprises include stock rearing, dairying and cereal production. Hedgerows are often of poor quality, even where laying is still practised, but there is an abundance of nest sites for finches and buntings none the less. With the demise of elm, willow is perhaps the commonest tree – a sure indication that the ground is wet.

The high water table, however, brings invertebrates nearer the surface and this attracts large feeding flocks of Lapwings, Black-headed Gulls and Starlings. Even the villages of this remote area lack strong character and the deserted village of Wolfhamcote seems to exemplify an area which time has passed by.

Ironically this otherwise nondescript area contains a most important ornithological site, though perhaps it will come as no surprise to learn that it is man-made. Draycote Water was built to supply water to both Leamington and Rugby. It was first flooded in 1968 and covers almost 300 ha, which makes it the second largest water in the Region. Much of its shoreline is artificial, consisting of concrete dams and stone pitching, and it is intensively used by trout fishermen and sailors, ranking highly in the fishing league and having the largest inland sailing club in the country. Indeed the policy of the former Rugby Joint Water Board was to concentrate active recreation here and to keep Stanford Reservoir, in Leicestershire, as a quiet retreat for wildlife and more passive pursuits. As a consequence, first impressions for Draycote were not that favourable, but it was inevitable that a water of this size would attract birds. Even so, few could have imagined it rivalling Blithfield Reservoir in Staffordshire as the Region's leading water.

17. Despite intensive recreational use Draycote Water is still an excellent site for wildfowl and gulls.   *A. J. Richards*

The water is roughly triangular. To the north it has a natural, but steeply-shelving shoreline indented by one or two attractive bays, but on all other sides its banks are largely artificial. The immediate surrounds of rank grass are a rich feeding ground for flocks of finches, Tree Sparrows and Yellowhammers in winter and Yellow Wagtails in summer. Extensive tree planting has been undertaken and as this becomes established, so the passerine population is increasing. To the south the meadows stretching down to the Leam often attract grazing Wigeon, Grey Herons frequent the fields around the trout hatchery, and the sewage works to the east holds a few Teal, Snipe and the odd Jack Snipe.

Draycote's biggest disadvantage is unquestionably its restricted natural shoreline and the absence of any islands,

which limit its potential for breeding. As evidence of this, when only partly filled and islands were still showing, Little Ringed Plovers bred and Oystercatchers were suspected of doing so; whilst during the drought of 1976, when islands again emerged, Little Ringed Plovers bred once more and other species showed interest. The only waterfowl to nest regularly are several pairs of Mallard and a few pairs of Tufted Duck.

Winter though is different. Then there is no fishing and, whilst sailors spread themselves across the entire water, there is still sufficient refuge in the bays and corners for a considerable number of wildfowl. Counts of Teal and Wigeon reached their peak in the early days, when the reservoir was only partly full, and have since declined, but diving duck continue to increase and in 1977/8 Tufted Duck reached a regional record of 2,000 and Pochard exceeded 1,200. At the same time up to four Great Northern and two Black-throated Divers were present. The high numbers of Tufted Duck are interesting since this species has all but disappeared from Blithfield. Good numbers of Great Crested Grebe, Gadwall and Goldeneye are also attracted, along with up to 1,800 Coot, whilst sea-ducks, especially Scaup, appear quite regularly. Surprisingly it is favoured neither by Goosanders nor Cor-morants, though Shags appear almost annually. Divers, the rarer grebes, Smew and Bewick's Swans also occur most years and in the hard weather of February 1979 seven Red-necked Grebes were present. The gull roost is the largest in the Region, having been estimated on occasions to contain over 100,000 birds, about 80 per cent of which were Black-headed.

The reservoir is filled by pumping from the Leam during winter months only, so spring water levels are usually high. As a consequence wader passage is not that good, although a party of 54 Bar-tailed Godwits paused briefly in 1976 and a Spotted Sandpiper enlivened 1977, but with favourable weather conditions tern passage, especially of Arctic Terns, can be strong. Little and Sandwich Terns also appear along with a scattering of Little Gulls and Kittiwakes, whilst hirundine and passerine passage is also good. In autumn, with lower water levels, more waders appear. Greenshank and Common Sand-piper are most regular, both being equally at home on the artificial stone banks and the natural shoreline, but the absence of a gently-shelving shore limits the expanse of mud and hence the numbers of Dunlin, Curlew Sandpiper and Ringed Plover. Already, in just twelve years, Draycote has attracted a good selection of rarities, including Ring-necked Duck, Rough-legged Buzzard, Kentish Plover, Pectoral Sandpiper, Mediterranean Gull, Whis-kered, White-winged Black and Caspian

Terns and Red-rumped Swallow.

Like the Avon, the Leam also rises in Northamptonshire. It has one major tributary, the Itchen, which joins it at Eathorpe and it is here that water is abstracted to fill Draycote. Below Eathorpe the basin narrows into a proper valley and the landscape becomes more undulating and attractive. Farming changes too, with arable more in evidence, and though fields are larger, small woods break up the otherwise open vistas and one of these contains a recently discovered heronry. Corn Buntings, Yellowhammers and Skylarks are typical birds of the area and in good years Quail may be heard in one or two traditional places. Shortly before reaching its confluence with the Avon, the Leam passes through Leaming-ton – a fashionable spa town with tree-lined avenues and elegant Regency houses. Thanks to the generosity of its lord of the manor, Edward Willes, who donated the land, it has been able to make the most of its river by damming it to increase its width and by laying out parks and gardens along its entire length. Consequently, this town of pleasant prospect is cleft in two by a magnificent swathe of open space which brings the countryside into its very heart. Although concrete lined, the small reservoir alongside the Leam is often used by birds passing up and down the valley and a surprising variety has been recorded for such a small water. For many years this appeared to be a favoured haunt of Black-necked Grebe, though not so in recent years. The adjoining Welches Meadows contain reedmace, reedgrass, sedge and rush amongst which a few Snipe, Jack Snipe and Water Rail lurk most winters. In times of flood they also hold Canada Geese and sometimes Bewick's Swans en route to or from Slim-bridge. Flocks of finches come to feed on seeds, Kestrels and owls prey on small rodents and in summer the occasional Hobby hawks dragonflies above the trees.

On the opposite bank the scattered trees and scrub on Newbold Comyn are also of interest.

Right in the centre of the town a surprising variety of birds can be found in the formal Jephson Gardens. Nuthatch feed in the exotic specimen trees along with commoner species, whilst pinioned waterfowl on the river sometimes attract their wild cousins, with Tufted Duck, Pochard, Coot and Little Grebe occurring in winter and Mute Swans present throughout the year. Surprisingly, a Black-throated Diver spent five days here in 1979, affording excellent views at close quarters. A few pairs of Sand Martins nest in drainage pipes in a riverside wall and a Kingfisher has been seen sitting on the bridge of the main shopping street. The reeds and alders alongside Victoria Park afford seclusion in summer for Sedge and Willow Warblers, Blackcap and White-throat; and food for Siskin and Grey Wagtail in winter. Even a Common Sand-piper might appear on passage.

The Leam then joins its parent river shortly before the Avon enters Warwick and truly becomes Shakespeare's river. What more idyllic setting could commence this second phase than Warwick Castle, standing proudly on its sandstone bluff, whilst the willow-fringed river rolls leisurely over the weir beneath and swans dabble lazily amongst the weeds. For many this is the finest view in England and its sudden revelation is like unlocking the door into a bygone age. What a pity it is no longer graced by the Peregrines that reputedly nested on the castle last century. Warwick itself is a charming medieval town clustered around the elegantly slender tower of St. Mary's church. Despite its twentieth-century extensions, it remains one of the finest old towns in England, with so much more of interest than just its castle, and anyone exploring its narrow streets in summer will be greeted by the sound of screaming Swifts wheeling overhead.

From here unto the Worcestershire border the Avon meanders sleepily through countryside steeped in history. The historic houses of Charlecote, Compton Verney and Walton are all set in beautiful parklands that are the absolute epitome of everything so admired about the English countryside; the landscape is dotted with farmsteads full of character; and punctuated by picturesque villages with their olde-worlde, half-timbered, thatched cottages or beautifully-mellowed red-brick houses. Here the valley is wide, with extensive river terraces stretching southwards in a flat expanse to the low hills beyond Wellesbourne. Farms are large and the open landscape has been heavily denuded of trees in recent years through elm disease, though a few oak and ash remain.

For the most part, the valley is extremely fertile and heavily cropped and in consequence its bird communities are very typical of an arable area. For example a CBC on 95 ha at Wellesbourne in 1976 revealed 276 territories (or 290 per km²), with Skylark (65 prs.), Linnet (25 prs.), Blackbird (23 prs.), Dunnock (21 prs.), Corn Bunting (17 prs.) and Chaffinch (16 prs.) the commonest species. These figures make an interesting comparison with those from a nearby mixed farm astride the bordering hills. Here, at Moreton Morrell, CBC plots on 175 ha of farmland and 10 ha of woodland have been censused by Dr. E. R. Austin and the Coventry group of the RSPB since 1972 and, over eight years, they have held an average of 316 territories (170 per km²), with Blackbird (47 prs.), Wren (43), Robin (28), Song Thrush (24), Dunnock (22) and Blue Tit (20) the top six species.

There are one or two surprises in this intensely farmed district, particularly on the slopes of the low, bordering hills. An unusually ornate stone windmill, for

18.   Shakespeare's Avon meanders through a countryside dotted with beautiful parklands.

P. Wakely

example, is the only outward sign of Chesterton's existence, yet here, in a secluded, narrow valley well away from the beaten track, a small stream has been dammed by the local landowner to create a series of lakes set in gently rolling countryside. There is also a mill pool with an extensive reed bed and a backcloth of conifers. A surprising variety of birds can be found in this small area. Great Crested and Little Grebes, Mallard, Tufted Duck and Sedge Warbler breed and this is one of the few regular breeding sites in the Region for Pochard. Winter brings Canada Geese and Shoveler to the pools, Green Sandpiper sometimes winters on the stream and Bittern has been recorded in the reeds. Away from the water, Jackdaw and Kestrel nest around the church; Grey Partridge, Yellowhammer and Chaffinch

are common on the farmland at all seasons; Wheatear and Yellow Wagtail usually occur on passage and Corncrake bred in 1968 and 1969.

Some of the woodland also deserves mention. Wellesbourne Wood itself was once a dense, broad-leaved wood with considerable overgrown scrub much beloved by Nightingale. In 1973 it was mostly clear-felled and replanted with ash and sycamore, since when it has been largely neglected by bird-watchers. Nightingale and Grasshopper Warbler may still occur, however, and Nightjar has been heard on rare occasions, but all too little is known about the current status of most species.

Above all, though, it is the historic parklands that command attention. Being so private, little is known about the birds of Warwick Park, but the blend of water, arable fields, pastures and copses, together with a perimeter of mature, mixed woodland that includes both yew and larch, provides the right ingredients for many species. Both Woodcock and Wood Warbler are known to be present, Little Grebe, Mallard, Tufted Duck and Mute Swan breed and Wigeon, Shoveler and Teal occur in winter. Much of the New Waters has silted up and is now covered with *Phragmites* in which Water Rail are regularly present, a few pairs of Reed Warbler nest and, in winter, Redwings, Starlings, Reed and Corn Buntings roost. Summer brings Swallows and Swifts to hawk insects across the water. For many years this was also the only known breeding site for Hobby and in 1966 it was visited by a Honey Buzzard. It is to be hoped the recent sale of the park will not result in any major change to its mixture of habitats. Nearby the small oak woodland at Grey's Mallory has the second-largest rookery in the Region, with over 200 nests.

Scenically the mixed woodlands flanking Walton Hall on either side of the Dene Valley form one of the most attractive areas in Warwickshire. Although uniform in age and teeming with Rooks, Crows, Jackdaws and Woodpigeons, they also have less common species like woodpeckers, Nuthatch and Jay, whilst the river meadows are visited by flocks of Fieldfare and Redwing in winter. The final woodlands in this group are those at Compton Verney. Landscaped by "Capability" Brown, this estate has suffered more than most from the deprivations of elm disease, old age and decay and many of its trees require urgent replacement. For birds though, the mix of old native hardwoods, with their decaying timber, and conifers including yew, larch, cedar and Wellingtonia is ideal. Hole nesters like Barn and Tawny Owls, all three woodpeckers and Nuthatch are well represented and Jack-

daws are numerous. Treecreepers use the holes in the soft bark of the Wellingtonias for nesting and roosting and thrushes and Goldcrests feed on the yew berries in winter. Sparrowhawk and Nightingale also breed, Hobby is seen sometimes in summer and an occasional Hawfinch turns up in winter. There are two lakes here as well, both used for fishing, and these hold breeding Great Crested Grebe and a few Tufted Duck and Pochard in winter. Kingfishers are also regular visitors and the first British specimen of Upland Sandpiper was obtained here in 1851.

At Charlecote the Avon is joined by the Dene. By this time it has largely recovered from upstream pollution and has matured into a pleasant tranquil river with fish spawning in its shallows and broods of Mallard or Moorhen scuttling to the reeds for cover at the slightest sign of danger. At dusk a Barn Owl might emerge from one of the ancient oaks in Charlecote Park to hunt for its supper, and Jackdaw, Kestrel and Stock Dove also nest in the same places. Although the Avon downstream from Alveston is an ancient navigation which is now heavily used by pleasure boats, this upper part remains one of the few unspoilt stretches of river in the Midlands and it is a tragedy to think that it should now be threatened by a proposal to extend navigation above Alveston to link with the Grand Union Canal at Warwick. Such a proposal must inevitably harm the bird life of this quiet water, where Kingfishers and Sand Martins nest in the sandy banks, wagtails and passage Common Sandpipers feed along the muddy edges and even the secretive Water Rail may sometimes be seen. Of all the birds however, perhaps the Mute Swan would be most affected. Swans have declined dramatically on the Avon in recent years through lead poisoning and, whilst it is not known how much emanates from discarded anglers' shot and how much from boat hulls, the evidence points conclusively to higher mortality on the navigable stretches of the river, where dredging and turbulence disturb the bottom. At Stratford the disappearance of the swans has become a matter of civic concern and an active interest has even been shown by Government at ministerial level. In 1979 the only nest in the Stratford reach was fenced against vandals, but despite successful hatching the cygnets would have succumbed had they not been reared artificially. Thanks to a generous response to the mayor's financial appeal, a special swan reserve has now been created.

Swans are part of the time-honoured Avon scene and it would be hard to imagine the Royal Shakespeare Theatre or the riverside meadows and their pollarded willows without them. To most people the Avon is synonymous with Stratford, a small but thriving market town which derives both its prosperity and its problems from the millions of tourists that make their pilgrimage to Shakespeare's birthplace. Though its history probably goes back to Roman times, it is the Elizabethan buildings, with their twisted oak beams, that give the town its true character.

Yet swans are not all that Stratford has to offer. The pollarded willows, lush meadows and overhanging banks below the weir provide nesting sites for Little Owl, Yellow Wagtail, Kingfisher and Little Grebe; feeding and passage hirundines occur in good numbers and an occasional Common Sandpiper works its way along the muddy banks on passage. In winter a few Wigeon graze the meadows, which are also a favourite hunting ground for Short-eared Owls.

It is here the Avon is joined by the Stour, a small but pleasant river rising in the Cotswolds. The triangular basin of the Stour is 20 km long and its fast-flowing headwaters fall steeply from the scarp into the vale. Willows and alders then line the river as it meanders gently through a wide pastoral valley, flanked on the west by outliers of the Cotswolds and on the east by low,

glacial-capped hillocks. The valley is almost exclusively agricultural, with many delightful villages built of the attractive Cotswold stone, and just one small town at Shipston. Farming is mixed, but there is a bias towards livestock at the foot of the scarp and to arable further north. The park and woods at Ettington Hall add character to the landscape and provide a good habitat for birds. On the river itself, Dippers have bred at Stourton and Tredington and do so regularly at Burmington (J. A. Hardman *pers comm*), whilst Kingfishers occur in a few places. Corn Buntings, Yellowhammers and Lapwings characterise the farmland, one or two pairs of Curlew breed in the damper meadows, a small flock of Golden Plover sometimes visits the disused airfield at Long Marston in winter and this is also a good area for Little Owl.

Back on the Avon, the braided channel and osier-covered island at Welford attract a few Mallard, Moorhen and Sedge Warbler, but elsewhere the boat traffic causes too much disturbance for most birds. In winter, though, the river is quieter and during 1977 a family party of Cranes stayed along this stretch for several weeks. A feature of the valley between Stratford and Evesham are the low sandstone escarpments, with their mantle of woodland or thorn scrub. In these woods Great Spotted Woodpecker, Blackcap, Willow Warbler and Chiffchaff breed; winter flocks of tits and finches, especially Blue Tits and Redpoll, are numerous; and there are roosts of thrushes, Linnets and Greenfinches. By ringing in an alder, birch, sycamore and ash wood from 1968 to 1976, Hardman *et al* (1977) have provided a good insight into how bird-life can be affected by changes in management. Over 5,000 birds of 37 species were ringed during eight winters, with Redpoll (20 per cent) the commonest species, followed by Chaffinch (10 per cent), Blue Tit (10 per cent), Blackbird (9 per cent), Greenfinch (6 per cent) and Redwing (6 per cent). In 1972 the wood was thinned and much of the shrub layer was removed, since when tits have been more numerous, but roosting Redwing and Fieldfare fewer in number. Blackbird and Song Thrush numbers have changed little, however, and their roosting requirements seem to be less precise, whilst fluctuations in finch numbers appear to relate more to the abundance of seed crops than to any management regime. Another interesting study was the raptor census carried out over 23 km^2 of south-west Warwickshire between 1967 and 1972 (Cooper and Juckes 1977). This showed four species – namely Kestrel and Barn, Little and Tawny Owls – to be present, with approximately one territory per km^2. As well as population counts for these four species, tentative territory sizes were deduced for Tawny and Little Owls.

From the north come two more tributaries, the Arrow and Alne. Again these drain a predominantly rural basin, with its delightful villages of half-timbered, thatched cottages. The medium-sized farms are mostly dairy or stock-rearing units, set in a well-timbered landscape in which oaks are a notable feature. Indeed, most of this basin was formerly part of the old Forest of Arden and, whilst it is by no means certain that this was ever a true forest, it must have been well-timbered to earn its name.

The Arrow has its source near the meeting point of the Avon, Severn and Trent watersheds on the Lickey Hills. From here its swift-flowing headwaters have cut quite steeply into the plateau on their way to Redditch, where they divide the town into two, with the old standing proudly on its ridge whilst the new nestles snugly in the folds of the surrounding hills. Here, in the Arrow Valley Park, a small balancing reservoir regulates storm-water run-off to prevent flooding downstream. From time-to-time this attracts a few wildfowl, though in the main it holds little of interest. Further south the Arrow passes Studley, a

small town famed, like Redditch, for needle manufacture. It then receives the effluent from Redditch sewage works which, because of its high proportion of metal-finishing waste, causes some pollution.

The Alne and its tributaries rise on the southern edge of the Birmingham Plateau and flow generally south-westwards. The catchment area is almost entirely agricultural and consequently the water is of high quality. In their upper reaches, the streams are quite swift and Dippers have been recorded alongside Kingfishers on Preston Brook. This delightful, alder-lined brook is followed for some miles by the Stratford-on-Avon canal and Grasshopper Warblers and occasionally Redshank breed in the intervening sedge-covered meadows. These meadows are also a favourite haunt of Grey Heron, Snipe, Grey Wagtail and hunting Barn Owl in winter, with exceptionally a Green Sandpiper or Water Rail. Flocks of Siskin and Redpoll indulge in acrobatics as they feed on alder seeds; Jays, finches and all three woodpeckers also occur and thrushes gather to roost in the thick hawthorn hedges on winter nights. There is a heronry at Wootton Wawen and a pool that holds a few duck and the occasional passage wader like Common Sandpiper. Little Grebe can be seen on the river and the bankside osiers and poplar plantations are host to colonies of Reed and Sedge Warblers. The lower valley is wider, with a White Lias escarpment following the eastern bank. For most of its length this has a mantle of oak/ash woodland, which was coppiced before the war and exhibits typically varied shrub and field layers including hazel, elder, holly, primrose and bluebell, with some patches of hawthorn scrub. Most of the woodland is extremely private and managed as Pheasant coverts. Consequently it is quiet and undisturbed and sustains good populations of tits, all three woodpeckers, Nuthatch, Treecreeper, Kestrel and Barn,

Tawny and Little Owls. Sparrowhawks are increasing, Hobby appears occasionally and Long-eared Owls sometimes roost in winter.

The river Arrow and its tributary the Alne meet at Alcester, a market town of Roman origin which has retained its core of medieval, timber-framed buildings. From here southwards the river has been "improved" drastically and the landscape changes as the farming emphasis switches to arable production. Westwards, the ground rises to the ridge separating Worcestershire from Warwickshire and on this slope Ragley Hall nestles amongst its parkland trees and majestic mixed woodland. Ragley is the family seat of the Marquis of Hertford and its park is another of the many in the Midlands to have been laid out by "Capability" Brown. Many hedgerows have been removed, but some gnarled, old oaks remain standing in open fields as well as in the parkland itself. The holes in these are used for nesting by Kestrel; Barn, Little and Tawny Owls; woodpeckers and an occasional Redstart. Woodlands on the estate are mixed, with oak, birch, ash and chestnut prevalent amongst the native hardwoods; and larch, pine and Douglas fir amongst the planted softwoods. Many are kept as coverts for deer and Pheasants, which are reared on the estate, but they also harbour woodpeckers, Nuthatch, Nightingale and a few pairs of Wood Warbler. Sparrowhawk and Buzzard both breed, though the latter is irregular, and there is a heronry of some antiquity. In winter the woods hold good flocks of foraging tits, Redpoll and Siskin. The small lake is disturbed by anglers and sailors, but Great Crested Grebe and Canada Geese still usually manage to breed and colonies of Sedge and Reed Warblers inhabit the marginal vegetation both here and alongside the river Arrow. The reed-beds along the river are a traditional breeding site for Mute Swan. They are also used as a roost site by migrating Swallows and Yellow Wagtails and by Pied Wagtails and Reed Buntings in greater or lesser numbers throughout the year. Grey Wagtails can also be seen in the vicinity of the mill, as indeed they can at many locks and mills along the Avon itself. All these, together with Lapwings, gulls and the odd Redshank in the damper meadows, give the whole area a rich and varied bird-life.

Below the confluence with the Arrow, the Avon Valley is narrowed by a spur of high ground extending southwards from Redditch towards Evesham. The Avon descends Marlcliff weir and hugs the bluff of Cleeve Hill on its left bank as it leaves Warwickshire and enters Worcestershire and its third and final phase. This is the Vale of Evesham – a countryside once dotted with orchards, market gardens and straggling villages with their peculiar blend of traditional old cottages and contemporary houses. Each spring it would re-awaken to a vivid splash of colour from the blossom of apple, cherry, damson, pear and plum; but, although trippers still flock to the area on "Blossom Sunday", in truth this is now a flat, featureless and rather drab landscape, noticeably lacking mature trees and littered with glasshouses, polythene cloches and the paraphernalia of an intensive horticultural industry. Today exotic peppers and aubergines are grown alongside the traditional tomatoes and lettuce, whilst the current vogue to "pick your own" favours soft, rather than top, fruit and there has been a drastic reduction in the area of orchards since the last war. Of those remaining, many have been changed from plum to apple and replanting has been with new, low-growing and easily harvested trees that are less attractive to birds.

The expansion of market gardening was said a hundred years ago to have resulted in a decrease in all birds, except finches, and from our present knowledge this would seem to be correct. Today though, there are even fewer tits, Starlings, Jackdaws, wood-

19.   The greenhouses and polythene cloches of an intensive horticultural industry are now as common in the Vale of Evesham as orchards used to be.   *A. Winspear Cundall*

peckers, Tree Sparrows and Little Owls because of the removal of so many of the older trees. "Pershore" plums and cherries, in particular, have declined and, with the latter, the Hawfinch; but Chaffinches are still widespread, though thinly distributed; Goldfinches appear to have increased and are now commonplace and Redpoll have joined them as recent colonists. Unfortunately not all birds are welcome because of the damage they can inflict in such an intensively farmed district. Woodpigeons attack brassicas, for example, particularly during hard winters and Bullfinches systematically destroy fruit blossom. In 1963 no fewer than 111,000 Woodpigeons were shot on eight farms in the Vale and 163 Bullfinches were trapped on one single fruit farm.

At the heart of the Vale, surrounded on three sides by a tight meander of the Avon, is the pleasant old market town of Evesham. The focus of the town is its bustling busy market place, but behind this, in a splendid setting overlooking the river, the ruined abbey stands as a reminder of a history that can be traced back for twelve-hundred years. Pied and Grey Wagtails can be seen along the riverside and its flood meadows and this was once another traditional haunt of Mute Swans, but, as at Stratford, they have declined in recent years. More surprisingly, a pair of Common Sandpipers bred nearby in 1944. In this, its final phase, the Avon is a typical eutrophic clay river fringed with bur-reed, bulrush and *Phragmites* growing on a shelf at the foot of the steep banks. This emergent vegetation is an important habitat for breeding Moorhens and also

91

shelters the broods of Mallard that are hatched in the crowns of pollarded willows. Coot are occasionally encountered as well and Grey Wagtails frequent the locks and mills. On the debit side, most of the river meadows have been improved with a consequent decline in breeding Redshank, Curlew, Snipe and Whinchat, whilst Corncrakes, which were common in the early 1950s, have disappeared altogether.

North of Evesham, the higher, pleasantly undulating ground around the Lenches is more interesting. A mixture of orchards and woods in an agricultural landscape provides a rich variety of habitat, though notes on the birds of this area some thirty years ago (Harthan 1953) clearly indicated the extent of change that had already taken place, with the regrettable loss of Hawfinch, Woodlark, Red-backed Shrike and Nightjar and declines in many other species, including Lapwings and wintering larks. As everywhere the only beneficiaries of agricultural change were said to have been those birds, like the corvids, that had adapted to it. In the 1940s the late Anthony Harthan also used to record large skeins of White-fronted Geese as they moved to and from their Severn feeding grounds. Presumably such movements still occur throughout the valley, but go unrecorded as they mostly take place at night.

West of the Lenches lies a large rectangular basin drained by the Piddle and Bow Brooks and bounded roughly by Redditch, Evesham, Pershore and the M5 motorway. Underlain by Lias clay, this extensive tract of rural countryside is very much a no-man's land, with small, isolated villages served only by a maze of narrow lanes. Much of the land is agriculturally poor, especially that between Feckenham and Inkberrow, and the pattern is one of small dairy or stock-rearing farms set in a flat landscape relieved somewhat in the north by occasional domed hills. Ridge and furrow is still very evident and the small fields are enclosed by well-timbered hedgerows, though again the tree cover has been decimated by elm disease. Yellowhammer, Dunnock, Robin, Blackbird and tits, including Long-tailed, are the typical hedgerow birds; whilst Grey Partridge, Lapwing and Skylark breed in the fields; Swallows around the farms and Little and Tawny Owls in the trees. Winter brings large flocks of Lapwings, Rooks, Fieldfares, Redwings, Starlings, finches and Yellowhammers. A CBC on 73 ha of farmland in 1978 showed the commonest species to be Yellowhammer, Wren, Blackbird, Chaffinch, Skylark and Dunnock in descending order.

The real interest of this area, however, lies in its woodlands. Feckenham was once the centre of an ancient Royal hunting forest and even now this delightful, sleepy little village of red-brick Georgian cottages is encircled by small woodlands that have survived on the steeper or more badly drained ground. Today, though, Himbleton rather than Feckenham is the centre of these relict woodlands, which for the most part are typical Midland coppice-with-standard woods, containing pedunculate oaks with an understorey of hazel, bramble and honeysuckle. Nowadays they are used mainly for Pheasant rearing and, with the cessation of coppicing, they have become almost impenetrable, except where rides are maintained for shooting. Nightingales used to revel in these woods when they were coppiced, but they have declined as conditions have become more overgrown.

None the less this is still their stronghold, with perhaps half-a-dozen pairs in most woods and even more at Grafton. Indeed if the West Midlands has any true Nightingale country this is it. Other typical breeding birds are Little and Tawny Owls, all three woodpeckers, and a variety of corvids, tits, warblers and finches. Winter invariably brings roosts of thrushes and enormous numbers of Starlings, and many of the woods have a few birch or alder

20.   Young oaks on the edge of dense woodland in a typical Nightingale habitat.

A. Winspear Cundall

which attract feeding flocks of Redpoll.

Some of the richest areas have been clear-felled and the remaining woodland tends to be uniform in age and structure. Nevertheless, there are some small differences that help to improve the variety of bird-life to be found. Hornhill Wood, for example, is a Worcestershire Nature Conservation Trust reserve where coppicing is being reintroduced along with selective thinning of the standard oaks. Woodcock are present throughout the year and breeding Nuthatch, Redstart and Tree Sparrow are indicative of more mature trees. In Bow Wood, pines were planted some fifteen years ago and are now well-established. Grasshopper Warbler and Tree Pipit both bred when the plantations were young and a few pairs may still do so. Redstart, Yellowhammer and Reed Bunting also breed and eight species of warbler used to

do so, but most have declined as the conifers have grown. Woodcock also occur here in winter. Several of the Worcestershire and Warwickshire woodlands are owned by Messrs. L. G. Harris & Co. Ltd., the brush manufacturers, and their management regimes are of interest. One such is Roundhill Wood, which was formerly oak-hazel coppice, but where all except the perimeter oaks have been replaced by alder and birch. Another is Goosehill Wood, where just a few mature birch survived the early 1960s felling and replanting – again with alder and birch. In both cases the scrub-like conditions created by these sub-climax species attract many pairs of Willow Warbler and Blackcap as well as a few pairs of Grasshopper Warbler, Nightingale and Tree Pipit. Not surprisingly, good numbers of Redpoll, Siskin and tits, including Long-tailed, visit the woods in winter. At Goosehill shooting creates some disturbance, but a few Woodcock are usually present and a pair or two breed. It is also a regular haunt of Turtle Dove, Little and Tawny Owls, all three woodpeckers and for many years a pair of Hobbies bred here too. Finally there is Trench Wood, with its floristically rich rides and plantations of oak, birch and alder standing alongside the old, traditional coppice-with-standards woodland. Here the typical breeding birds are augmented by many pairs of Tree Pipit and a few of Sparrowhawk, Woodcock, Whinchat and Wood Warbler, whilst winter again brings large numbers of roosting Redwing and Fieldfare and feeding flocks of Redpoll and Siskin. Although strictly private, Messrs. Harris have generously given permission for members of the Worcestershire Nature Conservation Trust to visit their woodlands.

To the south, around Pershore, is another group of woodlands largely related to the Croome Estate. At Drakes Broughton there are both oak woods and mature conifer stands, but Tiddesley is perhaps the most varied. This wood was almost clear-felled in the last war, but a few old oaks remain and there are stands of pine to the south, conifers and broadleaved scrub to the north, a few areas of old birch and one or two marshy spots. In consequence it exhibits a rich scrub avifauna, including many pairs of Turtle Dove and Bullfinch as well as the expected Woodpigeons, corvids, tits and Redpolls. A few Grasshopper Warblers and Nightingales also nest, but truly arboreal species are scarce. Winter brings a large Starling roost; good numbers of thrushes, Bullfinch, Chaffinch and Linnet; a few Redpoll and Siskin and the occasional Woodcock or Sparrowhawk.

Apart from the main brooks, along which Kingfisher and Sedge Warbler breed, there is little open water, but a small pool at Dormston does hold a few duck, Canada Goose, Little Grebe and Kingfisher in winter. Of the parklands, two are worth mentioning, namely Croome Court, which is another of the many landscaped by ''Capability'' Brown, and the smaller estate of Spetchley, where the gardens and deerpark are set in a landscape heavily denuded of elms. Croome is becoming rather neglected and fragmented as its trees die or are removed, but the lake is mildly interesting, with breeding Mute Swan, and the range of habitats supports a good variety of the commoner passerines. Though probably less diverse than Croome, Spetchley does have the advantage of being open to the public, so it affords a better opportunity to see the birds. Mallard, Canada Geese, Moorhen and Coot visit the lake and lawns in the gardens and the same species, along with Great Crested and Little Grebes, breed on the large lake in the park, where there is a wooded island. The mixed trees of the gardens and the old oaks in the parkland are used by breeding Barn and Little Owls, Green and Great Spotted Woodpeckers, Stock Dove and Nuthatch. Treecreeper and Spotted Flycatcher are also well represented and there is a good variety of tits and corvids. Lapwing and Grey Partridge breed in the fields, a few pairs of Reed Warblers nest in the small reed-bed and there are roosts of Pied Wagtails, Reed Buntings and finches in winter.

Part of the parkland at Croome was commandeered in the Second World War for Defford Airfield, but some trees were left around the perimeter and there is a small plantation of oak and ash. The airfield has since reverted to agriculture and much is now ploughed, but there has been some replanting with oak, alder, willow and pine.

Red-legged Partridge and Lapwing breed, Snipe and Meadow Pipits are winter visitors and Golden Plover, Curlew and Redshank occur on passage. Dense areas of bramble and hawthorn provide nest sites

21. The lower reaches of the Avon are of national importance for Marsh Warblers.

A. Winspear Cundall

for scrub warblers, Turtle Dove, Grass-hopper Warbler and Nightingale; and winter roosts for large numbers of thrushes and finches, especially Chaffinch and Linnet. Barn and Little Owls, Kestrel and Buzzard are regularly present, Whinchat breed and Woodcock, Short-eared and Long-eared Owls are winter visitors, though the latter two species are not regular.

From Evesham onwards the Avon meanders in wide sweeps through a frequently flooded plain. Nevertheless the terraces make excellent arable land, where Corn Buntings are present throughout the year and Quail appear from time-to-time in summer. Near Fladbury, a small irrigation reservoir has attracted a Canada Goose flock of about fifty, and in 1979 it was visited by an Osprey (G. H. Green *pers comm*). To the ornithologist, though, this

stretch of river has a special importance, since it is along these lower reaches that the Marsh Warbler breeds. With the bulk of its national population, the West Midlands has a special responsibility to conserve this rare warbler and its habitats. Ironically it is subjected to excessive disturbance by in-considerate bird-watchers, so precise locali-ties cannot be disclosed, but fortunately reserve agreements for several of its breed-ing sites have been secured by the Worcestershire Nature Conservation Trust. But for these the species would have been severely diminished by now.

The attractive market town of Pershore, with its Georgian red-brick buildings has, like Evesham, a history going back twelve hundred years or more and an Abbey dating from the eleventh century. Beyond it the valley is dominated by the dome of Bredon Hill, lying in the landscape like

95

some giant, dormant animal. The villages around Bredon are as picturesque as any in Britain and often in sharp contrast to one another, with half-timbered and thatched cottages on the north side and clusters of Cotswold stone houses to the south. The hill itself is considered along with the Cotswolds in Chapter 8, but is mentioned here because of its significant influence on the microclimate of the whole valley, most of which lies within its rain shadow. High summer temperatures, especially during the "climatic optimum" of the 1930s, convection currents and unstable air conditions often produce storms and these lead to humid conditions, which are very conducive to high insect populations and the kind of luxuriant plant growth which Marsh Warblers relish. Indeed climatic factors are probably more important to Marsh Warblers than the type of habitat, since within the valley birds have occurred in bean fields, orchards and a disused quarry as well as in the more traditional areas of nettles and meadowsweet. Osiers, when they were cut to make baskets for the fruit growers of the Vale of Evesham, were also much frequented, but now that cutting has ceased all the beds except Strensham Lock island have been cleared. A few derelict remnants and the occasional river bank tree are all that remain of this once extensive habitat.

To the north of Bredon Hill, the Avon slithers like a serpent through entrenched meanders to pass Nafford, with its lock, weir and delightful little island, which is a Worcestershire Nature Conservation Trust reserve; and Eckington with its ancient sandstone bridge. This is another of the more attractive stretches, with overhanging willows, small reed beds, and occasional plantations of poplar. Sedge Warbler and good numbers of Reed Warbler breed along the river wherever there are reeds or suitable emergent vegetation; and the occasional pair of Redshank may be found in some of the damper meadows, although most have now gone as pastures have been improved. Unfortunately the river is prone to flood after heavy summer storms and nests are often lost. In winter, though, the same flooded meadows may attract dabbling duck like Teal and occasionally flocks of Wigeon or herds of Bewick's Swans.

Away from the river, there is a small lake at Strensham where Great Crested Grebe breeds, whilst on the opposite side of Bredon Hill, alongside Carrant Brook, black currants figure prominently in another area of intensive horticulture and the bushes are used by a variety of birds for feeding and roosting. Throughout the lower Avon Valley quarries are unusual, so the small sand pits at Aston Mill and Beckford assume more than usual importance. Sand Martins breed in the quarry face, Little Ringed Plover nest in the pit bottoms and the flooded areas hold breeding Mallard and Tufted Duck along with small numbers of the commoner wildfowl in winter. During winter this general area might also be frequented by a Hen Harrier.

Still hugging the foot of Bredon Hill, the Avon finally flows beneath the M5 motorway and leaves Worcestershire and the Region for its confluence with the Severn at Tewkesbury.

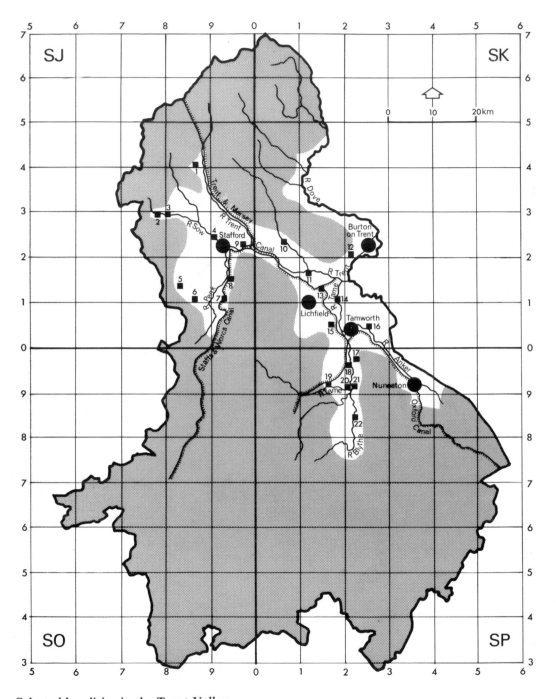

Selected localities in the Trent Valley

1 Trentham Gardens	7 Gailey Reservoir	13 Fradley Wood	19 Minworth S.F.
2 Bishops Offley	8 Teddesley Park	14 Elford G.P.	20 Ladywalk
3 Copmere	9 Tixall	15 Hopwas Hayes	21 Shustoke Reservoir
4 Tillington	10 Blithfield Reservoir	16 Alvecote Pools	22 Packington Park
5 Mottey Meadows	11 Kings Bromley	17 Kingsbury Wood	
6 Belvide Reservoir	12 Branston G.P.	18 Kingsbury W.P.	

5   The Trent Valley

This sub-region has four main divisions, based on the rivers Trent, Sow, Tame and Dove together with their respective tributaries. North of the Trent and Sow, drainage is from north-west to south-east, through parallel valleys which penetrate like fingers into the high ground of the Central Staffordshire Plateau. In this area tributaries are few. In contrast the drainage pattern south of these rivers is dominated by two basins, through which the Penk and Tame flow northwards into the Sow and Trent respectively. Between these basins lies the high ground of the Birmingham Plateau, from which streams radiate in all directions.

This is the fourth largest sub-region, embracing some 1,150 km² of low-lying land, or 15 per cent of the regional area, mostly below 100 m and falling to 43 m where the Trent enters Derbyshire just below Burton. The underlying rock is entirely Trias and principally Keuper marl, though sandstones do occur, notably to the east of the Penk and west of the Tame. Quaternary deposits have most influence on the landscape, however, with a patchy covering of glacial drift across most of the marl, and alluvial deposits throughout the lower reaches of the rivers. These are most extensive in the Trent, Tame and Dove valleys, downstream of Rugeley, Coleshill and Rocester respectively. The flood plain alluvium usually consists of fine clay, but the associated terraces frequently contain fluvio-glacial gravels of considerable importance to the building industry and these have been extensively quarried.

Within the flood plain, the clay soils are usually heavy and the high water table impedes drainage and causes waterlogging.

In the narrower valleys north of the Sow and Trent, permanent cattle pastures and small dairy farms of less than 150 ha predominate. On the terraces and higher ground soils are better drained, though sometimes stony and acidic, and they are widely cultivated for arable and root crops, especially in the Sow and Tame basins and around Lichfield.

For convenience, the bird-life of the West Midlands Conurbation and the Potteries is dealt with in Chapters 2 and 6 respectively, so, although parts of the Trent Valley and more especially the Tame basin are highly urbanised, this chapter is not primarily concerned with these. Rather it deals with an extensive area of open countryside containing a string of small towns like Stafford, Lichfield, Nuneaton, Tamworth and Burton-on-Trent. Climatically everywhere is sheltered from the extremes of weather; temperatures are about average for the Region and rainfall varies from 650 to 750 mm.

The Trent is an enigmatic river. Its source is almost within sight of the Irish Sea, yet its horseshoe course of 274 km embraces the backbone of England before reaching the Humber estuary and the North Sea. Indeed it is the great divide of England, separating uplands from lowlands and northerners from southerners. It has a similar impact on the avifauna, with more breeding records of species like Teal, Oystercatcher, Snipe, Curlew, Redshank, Black-headed Gull and Common Tern – all of which have a generally northern distribution – than similar habitats in the Severn and Avon Valleys. It also holds the major share of wintering Goosander and Ruddy Duck. Within this sub-region are a host of valuable wetlands, both natural and man-made, and it is to these that most ornithologists are attracted. The unimproved washland meadows of the Sow and Penk catchments are good examples of a fast disappearing habitat; the canal-feeder and water-supply reservoirs of Belvide, Gailey and Blithfield are important to wildfowl, passage waders and roosting gulls; whilst the industrial land and extensive gravel workings of the Tame and lower Trent Valleys have created a chain of wetlands which is of national importance to migrating birds in particular.

Rising as a series of small streams high on Biddulph Moor, the Trent begins life as a fast-flowing, clear stream beloved by Dippers and Grey Wagtails. Its headwaters have been dammed at Knypersley to form a canal feeder reservoir for the Caldon Canal, but, if this intervention of man portends what lies ahead, there is every reason for failing to heed its warning in such pleasant surroundings. It is an abrupt shock, therefore, to be suddenly confronted with the forbidding bricks-and-mortar of Stoke and the river seems to falter perceptibly, as if unprepared to face such a hostile environment. But there is no turning back and the Trent is subjected to a spate of effluent discharges as it threads its way midst factories, houses, canals, roads and railways. As a result it leaves Stoke almost shamefully, a poor apology of its former self.

The transformation on leaving Stoke is equally abrupt, for suddenly, on the western bank, the seemingly endless terraced houses give way to the floral splendour of Trentham Gardens. The inspiration of the Duchess of Sutherland, these magnificent gardens have long been the playground for people not only from the Potteries, but from much further afield. The lake is used for angling, rowing and water-skiing, a miniature railway skirts one shore and vegetation has been cleared to make way for a caravan site.

Nevertheless, away from the popular parts many birds still manage to find a quiet corner. Small, wooded islands offer some sanctuary from power boats to a few wildfowl, and Mallard and Great Crested and Little Grebes breed beneath the overhanging rhododendrons. Mallard, Tufted Duck and Pochard occur in winter along with a few Teal, Shoveler, Wigeon, Mute Swan and Canada Geese. Spring, and more especially autumn, usually bring Common Sandpipers, one or two Redshanks and a few terns and recently the lake was graced by two Ospreys. Kingfishers sometimes nest in the river banks and one of the bridges is a favourite haunt of Grey Wagtails. Grey Herons have attempted to nest in the past, but without success. The lakeside is wooded and on the western shore this merges into a steep hillside of larch plantations, backed by mature oak and beech woods. These woods, which are scheduled as an SSSI and are private, offer undisturbed breeding sites to a variety of woodland species, such as all three woodpeckers, warblers including Wood Warbler, tits, Nuthatch, Tree Pipit, Redstart and, in the conifers, Redpoll and Goldcrest. Many of these frequently stray into the public areas on feeding forays.

Siskin and Brambling are winter visitors and over-wintering Blackcap have been regularly noted in recent years. At this season, too, the rhododendrons are used by roosting finches, Redwings, Song Thrushes and Blackbirds. Mistle Thrushes devour the yew berries in autumn, Treecreepers roost in the Wellingtonias and an occasional Dipper appears in spring to give credence to the rumour that they formerly bred.

The pattern of sharp contrasts continues as the river squeezes between the high ground of the wooded Tittensor Common to the right and the open Downs Banks to the left. Somehow, sandwiched in between, are Meaford Power Station, the

Trent and Mersey Canal, which accompanies the river for the next 30 km, and the railway. Black Redstarts have bred at the power station.

Between Stoke and Sandon, across the drift covered marl, the river continues its south-easterly course between tree-lined meadows and wide, rich pastures which provide a foretaste of its lower reaches. At Sandon, however, it again encounters the Bunter beds and the high ground of Sandon Hall and Hopton Heath close in on either flank just before the confluence with the Sow.

The Sow rises in the west, near the Shropshire border just south of Ashley. Its early course is southwards, through a valley cut deep into the sandstone, but at Bishops Offley it turns eastwards through two small millpools. Similar millpools also occur on a tributary at Gerrards Bromley. For their size, these secluded, willow-fringed pools hold a surprising variety of breeding and wintering wildfowl and most also have small reed-beds, marshy areas, patches of alder and some woodland to add to their species diversity. Great Crested and Little Grebes, Tufted Duck, Ruddy Duck, Mute Swan and Canada Goose breed and other duck appear in winter, along with Water Rail and Kingfisher. A few pairs of Reed Warbler nest in the reed-beds, Grasshopper Warbler is also present and other warblers breed in the scrub and woodland.

To the east is Copmere. This is a much larger, natural lake of glacial origin, with a fringe of reeds and a backcloth of broad-leaved woodland on three sides, including many elms which have succumbed to Dutch elm disease. Mallard, Mute Swan, Ruddy Duck and both Little and Great Crested Grebes nest on the mere and a good colony of Reed Warblers and perhaps one or two Water Rail do likewise in the reeds. Grasshopper Warbler, Sedge Warbler and all three wagtails are also present and may sometimes breed. Winter

brings good numbers of Mallard, Tufted Duck, Pochard, Shoveler and Goosander, along with a few Wigeon and Water Rail. Sparrowhawk frequents the woodland, which also has a good variety of breeding species including all three woodpeckers, many corvids and tits, Nuthatch, Redstart and woodland warblers, including Wood Warbler.

Below Eccleshall the valley opens out into wide pastures. The Sow is joined from the north by Meece Brook, then flows south-eastwards through Stafford and skirts the northern edge of Cannock Chase to its confluence with the Trent. On either side of Stafford are flood meadows of immense value and it is to be hoped that current efforts to save them from impending river improvements and a consequent lowering of the water-table will be successful. North-west of the town, around the confluence of the Sow and the Darling and Doxey Brooks, are the Tillington-Doxey marshes. Within an area hemmed in on three sides by housing estates, on the fourth by the M6 motorway and dissected by the main electrified railway, lies a mosaic of small pools, reed-beds, marshes and wet pastures flanked by drainage ditches. Conditions are ideal for breeding Redshank, Snipe and duck, including Ruddy Duck, Pochard and perhaps even Garganey. It is also a rich feeding ground visited by large numbers of hirundines. Sedge Warblers and Yellow Wagtails breed in good numbers and there is a large autumn roost of the latter species, accompanied by Pied Wagtails and Swallows. At this season as well a few waders occur, whilst in winter, thrushes and Lapwing are numerous, up to thirty Jack Snipe may be present and Snipe may reach a peak of 600 in October and November. South-east of Stafford the water meadows along the Penk, just before its confluence with the Sow, are similar, but less exciting, though even here Snipe may reach 400.

22.   Wetlands such as Tillington are important for marsh-loving species.   *D. Smallshire*

The Penk basin contains much more of interest, however. It is roughly circular, with a diameter of 7 km, and extends back to the broken high ground near the Shropshire border, which forms the watershed between the Trent and Severn. To the west of the Penk, the drift-covered marl lies in a wide, flat expanse of heavy clay soils on which dairy farming and general arable cropping are widespread. The fields are quite large and enclosed as often by fences as by hedgerows. Hedgerow trees are not numerous, but marlpits are quite common. East of the Penk the underlying rocks belong to the Bunter series, but the farming pattern is much the same, though holdings are often larger and there is a greater emphasis on arable production. The whole basin is served by a network of little lanes, which connect small, compact villages to the main roads.

Unimproved water meadows are a feature of some of the western streams, where there is very little gradient. These provide summer grazing for cattle and the occasional hay crop, and as such are often important in the farming economy. Nevertheless, a high water-table and frequent flooding limits their agricultural value and there is constant pressure for river improvements. Unfortunately such improvements might benefit the farmer, but they would have a disastrous effect on wildlife and they are consequently seen by ornithologists as a perpetual threat. The meadows of the Penk Valley, in particular, are of some importance as the stronghold of Shoveler in the Region and the Mottey Meadows have been scheduled as an SSSI of national importance. Because they seldom dry out completely, conditions are perfect for ducks and waders. As well as Shoveler, other breeding birds include Mallard, Tufted Duck, Redshank and good numbers of Snipe, Lapwing, Curlew and Yellow Wagtail. One site even held a Black-headed Gull colony until river improvements and the 1976 drought caused it to dry out and become deserted. Alders and willows frequently line the streams, providing perches for Kingfishers and food for tits, Redpoll and Siskin in winter.

These water meadows attract a wide range of wildlife and provide an abundant food source for predators. Sparrowhawk, Kestrel and Barn, Little and Tawny Owls are all seen regularly, whilst Short-eared Owl may also hunt the grasslands in winter. The surrounding arable land holds Grey and Red-legged Partridge and sometimes Quail, whilst the hedges, thickets and corners of scrub are the haunt of tits, Whitethroat, Lesser Whitethroat, Willow Warbler, Chaffinch and Yellowhammer.

In addition to these wetlands, the very south of the Penk basin contains two canal-feeder reservoirs, for which the WMBC operates a permit scheme for visitors. The first of these is Gailey Pools, which are bisected from north to south by the M6 motorway and from east to west by the A5 trunk road. Despite the continuous background noise from traffic and disturbance from anglers and sailors, the bird life is good, though the upper pool is less interesting because of the considerable pressure from trout fishermen during March-October. The lower pool contains an island with several mature willow and Scots pine, which for many years have held a large heronry and a winter roost of Cormorants. The shores are artificial, but have some emergent vegetation; the surrounds are wooded with a mixture of mature oak, beech, pine, alder and silver birch and there are also scrub areas of bramble, gorse and hawthorn. Wintering wildfowl are the main attraction, with good numbers of Tufted Duck, Pochard, Goldeneye and Canada Geese and an occasional scarcer species like Scaup or Smew. Along with nearby Belvide, Gailey can lay claim to fame – or notoriety – as the first breeding site of Ruddy Duck in the Region (in 1961), but its other breeding birds include Great Crested and Little Grebes, Tufted Duck, Canada Goose, Mute Swan, Sedge Warbler and a few Reed Warblers. Good numbers of wagtails, hirundines and Swifts occur on passage, along with a few terns and waders.

The second is Belvide Reservoir. Constructed in 1834 and referred to in the 1920s as "Bellfields" by A. W. Boyd and H. G. Alexander, Belvide will always have a special place in the affections of Midland bird-watchers and members of the WMBC in particular. After years of protracted negotiations, it was with great delight and not a little pride that the Club succeeded in obtaining a lease of the reservoir from the British Waterways Board in 1977. This would not have been achieved without the efforts of Cecil Lambourne, who prior to his retirement in 1978 had held many offices with the WMBC including that of Conservation Officer, and it is fitting to pay tribute to him here. The effect of the lease is to secure for the future the reservoir and its surrounds as a bird reserve, free from other activities save fishing and grazing. Lack of disturbance makes Belvide very important as a breeding and moulting water and, with up to 5,000 waterfowl over its 75 ha at any one time, it carries a higher density of birds than any other water in the Region. It has natural banks around three sides and a brick-faced dam to the east. The margins provide rough grazing and contain rushes, *Glyceria* and a bed of bulrushes, with a more extensive marshy area around the inlet at the west end. The gently-shelving shores also expose mud at low water level and, since obtaining their lease, the WMBC have carried out considerable work to improve and manage the habitat, notably through the construction of

23.   The WMBC reserve at Belvide holds the highest density of wildfowl in the Region.

D. Smallshire

islands. Smallshire and Richards (1976) have provided a full account of Belvide and its birds. There are very good numbers of breeding waterfowl, including Shoveler, Ruddy Duck, Little and Great Crested Grebes and occasionally Gadwall, Garganey, Teal and Pochard. Curlew, Snipe and Redshank also breed around the margins and Little Ringed Plover bred whilst the reservoir was drained for repairs in 1968. Winter wildfowl numbers can be spectacular, with over 3,000 Mallard, 400 Shoveler, 120 Goldeneye and 60 Goosander on record. Slavonian and Black-necked Grebes, White-fronted Geese and Smew have also appeared with some regularity over the years. There is a large gull roost, but perhaps the most outstanding feature is the early spring and autumn concentrations of moulting Ruddy Duck, which have steadily increased in recent years and now exceed 400. There are also large moulting flocks of up to 2,000 Coot and 700 Tufted Duck. In summer the reservoir is an important feeding area for hirundines and more especially Swifts, which regularly reach 3,000, whilst spring and autumn passage bring good numbers of waders, terns and Yellow Wagtails. Surrounding the reservoir are several deciduous copses which also hold a good variety of birds, including the occasional Lesser Spotted Woodpecker and Haw-finch. Almost anything might turn up at Belvide and amongst the two hundred plus species recorded to date have been Crane, Marsh Sandpiper, Black-winged Stilt, Whiskered, Caspian and White-winged Black Terns, Pomarine Skua, Wryneck and Golden Oriole to name but a few.

Woodlands are scarce in the Penk Valley, apart from a few small ones of which those around Belvide are typical. This makes the small copses and open parkland of Teddesley Park even more important than would be the case in a more heavily-timbered landscape. Covering 300 ha and sloping gently down from the foot of Cannock Chase to the Staffordshire and Worcestershire Canal and the Penk itself, there is less woodland than open parkland, but the latter contains many scattered oaks which are past maturity and in advanced stages of decay. Their holes and cavities provide homes for woodpeckers, Nuthatch, Jackdaw and owls; the farmland holds good numbers of Grey Partridge and the woods both Pheasant and Woodcock. Even here one cannot escape the influence of water, as there are streams through the park, a small pool in its south-west corner and others just outside it on which grebes, ducks, Mute Swan and Canada Geese breed and an occasional wader appears on passage.

The Trent and Mersey, Staffordshire and Worcestershire, and Shropshire Union canals also contribute to the richness of bird life, especially where they run alongside the rivers. Here birds like the Kingfisher enjoy the best of both worlds, nesting in the natural banks of a river free from boat disturbance, but turning to the less polluted canal for food. Bushes, hedges and thickets abound alongside the canals, where they harbour breeding Sedge Warbler, Whitethroat, Willow Warbler, Chiffchaff, Chaffinch, Bullfinch, Yellowhammer and Reed Bunting. Wherever hedges are overgrown, Lesser Whitethroat is not infrequently encountered as well. Mallard, Moorhen, Coot and even the occasional Tufted Duck breed along the canals and if, as often happens, the adjacent pastures are wet, Snipe and Lapwing may breed and will almost certainly flock in winter together with a few duck. At Tixall Wide – a boat turning point on the Staffordshire and Worcester-

shire canal – there is a substantial area of marsh and *Phragmites*, where Great Crested and Little Grebes breed together with Mute Swan, Canada Goose and Reed and Sedge Warblers. In winter Teal and Wigeon join commoner species to feed in the flooded meadows.

Convention insists that the Sow joins the Trent, but logic dictates otherwise, as the main river arrives at a higher level, pouring its murky waters over a weir to descend on its cleaner tributary like a raptor on its prey. It is an unhappy merger, welcome to the Trent, but degrading to the Sow and in dry weather the respective flows remain clearly distinguishable for some distance downstream. If the merger is unhappy, though, its setting could hardly be more spectacular, with the majestic wooded slopes of Shugborough Park and Cannock Chase to the right and the ancient Essex bridge spanning the river. At a more mundane level, the Sow provides sufficient dilution for the Trent to support fish and from hereon Grey Herons may be seen standing in the shallows as the river hugs the steep, northern scarp of Cannock Chase, with its mantle of conifers, birch and bracken.

From now on the character of the Trent changes. Free at last from the constraining Bunter beds that have confined it to a narrow valley, it meanders back and forth across the drift-covered Keuper marl, which is the true Midlands plain. This is a countryside of riverside meadows and scattered farmsteads; of grazing cattle and lofty hedgerow elms. At Rugeley, power station and coal mine stand side-by-side against the hills of Cannock Chase as the valley slowly widens and the river, almost with a gesture of defiance, leaves the railway and canal that have accompanied it from Stoke and swings first eastwards and then north-eastwards, beyond Alrewas, towards Burton and Derbyshire.

Southwards, the rolling plain, which stretches across the countryside to the delightful city of Lichfield, is dominated by the three lofty spires of the majestic cathedral, which are known locally as the "Ladies of the Vale". Close to the cathedral the twin Minster and Stowe pools attract a few wildfowl, whilst to the north of the city the tiny Hanch reservoir, despite its steep banks and small size, sometimes holds interesting species like Goldeneye and Goosander. Within the vale, the rich, fertile soils are used for a variety of general arable crops and even horticulture as well as dairy farming. There is also a scatter of small, mixed woodlands, mainly kept as coverts for fox or Pheasant. Fradley Wood, which stands alongside the Trent and Mersey Canal, is perhaps typical of most. Here the original oaks were largely felled between 1966 and 1971 and replaced with stands of spruce, larch, pine and some beech and sycamore. Some of the older hardwoods survived, however, and birch has managed to regenerate through self-setting to the extent that it now covers over half the wood. Bracken, bramble and elder feature prominently in a generally varied under-storey that provides plenty of suitable nesting habitats. Changes in silviculture have naturally affected the bird life, which has been the subject of a CBC since 1966, (M. J. Austin *in litt.*). Over fourteen years the 40 ha of Fradley have held an annual average of 287 territories (718 per km^2), with Willow Warbler (39 prs), Wren and Blackbird (31 prs each), Robin (27 prs), Dunnock (21 prs) and Blue Tit (20 prs) the commonest species. Rooks deserted the wood in 1969 following the felling of the larger trees, Nuthatch disappeared about the same time and Treecreepers, Blue Tits and Great Tits all declined. As the conifers have grown, however, Goldcrests and Redpolls have increased. Woodcock breed sparingly and Green and Great Spotted Woodpeckers erratically, whilst a Nightjar churred in 1971 and Redstart was present in 1966.

By Kings Bromley the flood plain is over a mile wide and the first of many fluvio-glacial gravel workings are encountered. The unusual feature of this quarry is that it is worked wet, with the gravel being extracted by dredger. Throughout the extractive operation, therefore, it has been attractive to birds, despite competition from fishing and sailing. Fortunately there are one or two islands and plenty of willow scrub and rank vegetation on these and around the shoreline to provide a safe refuge for birds. Indeed some of the birds from Blithfield Reservoir come here to roost, including in recent years Cormorants, which have numbered as many as 240 (S. C. Brown *pers comm.*), though less than fifty is more normal. Nevertheless the main attraction is again winter wildfowl, especially Mallard, Pochard, Tufted Duck, Goosander, many Canada Geese and the occasional Bewick's Swan.

At Kings Bromley the Trent also collects the clean waters of the Blithe and by Yoxall Bridge dilution and natural self-cleansing enable it to support chub, dace, roach and gudgeon. Here Goosander and Cormorant might be seen on the river itself, whilst the nearby village has some claim to notoriety as one of the places where Collared Doves were first recorded in the West Midlands, back in 1961.

The Blithe drains an agricultural valley of little ornithological interest save for Blithfield Reservoir. Here the river has been dammed and a 320 ha drinking water reservoir created. Ever since it was first flooded in 1952, Blithfield has figured prominently in the Region's ornithological annals and by 1963 it was rated the third most important reservoir in Britain for wildfowl. It is crossed by a causeway which carries the B5013 from Rugeley to Uttoxeter and good views can be obtained from this, though to see the best of the birds it is necessary to walk round much of the perimeter. Again the WMBC administers a permit scheme for bird-watch-

24.   Within a few years of flooding Blithfield Reservoir became the third most important reservoir in Britain for wildfowl.   *T. W. W. Jones*

ing on behalf of the South Staffordshire Waterworks Co. The deepest part is south of the causeway, where some 40 ha have been used for sailing since March 1972. Trout fishing takes place on the whole of the water, both from bank and boat. Apart from the causeway, southern dam and some reinforcing of the eastern and south-western shores with stone-filled gabions, the entire shore-line is natural and the grassy surrounds provide grazing for over a thousand each of Mallard and Wigeon in winter. North of the causeway the reservoir divides into two arms fed respectively by the Blithe and Tad. At their northern end these arms are shallow and, with low water, a large expanse of mud is exposed providing the best and most extensive area for passage waders in the entire Region. Up to 260 Dunlin and 160 Ringed Plover have been recorded in autumn, when Little

Stint, Curlew Sandpiper, Greenshank, Spotted Redshank and Ruff are also regular and a more unusual species like Kentish Plover or Pectoral Sandpiper might turn up. In recent years Dunlin have tended to winter as has an occasional Common Sandpiper. Rush and sedge are common in these shallow bays, which are favoured by wintering wildfowl, and the willows and alders along the inlet streams are often visited by Siskin at the same season. In winter, too, up to 20,000 gulls use the reservoir for roosting and these usually contain one or two Glaucous Gulls and less often an Iceland Gull. Cormorants now regularly number up to 50, Goosander 150, Teal 500 and Ruddy Duck over 200. In 1976 Pochard reached a total of 1,200, which was then a record for the Region. Lord (1976) analysed the wildfowl counts from 1955–74 and showed between two and three thousand duck to be present on average from October to February inclusive. Except for Great Crested Grebes, of which forty or more are reared most years, Blithfield is less important as a breeding area, but Little Grebe, Mallard, Tufted Duck and, less often, Teal and Shoveler do nest. Passage often brings good numbers of terns and passerines and over the years an impressive variety of raptors has appeared, with Osprey an annual visitor. There is a mixed plantation of pine, spruce, larch and sycamore on the northern shore, whilst the peninsula between the two arms contains an old oakwood with a field layer of bracken.

Great and Lesser Spotted Woodpeckers, Redstart and Sparrowhawk all breed in these woods along with other, commoner woodland species, and the rhododendrons in the plantations are used in winter by roosting thrushes and finches. Barn, Little and Tawny Owls and Kingfisher also nest in the area. All-in-all over two hundred species have been recorded at Blithfield and half of these have bred. Notable amongst a host of rarities have been Little Egret, Lesser Yellowlegs, White-rumped Sandpiper, Sabine's Gull, Whiskered, Caspian and White-winged Black Terns, Richard's Pipit, Shore Lark, Golden Oriole and Snow Bunting.

Reverting to the Trent, the river is again joined by the Trent and Mersey canal at Alrewas, an unspoilt, picturesque village of thatched and half-timbered houses. To be more correct the canal crosses the river on the level – an unusual feature forced upon the engineer, James Brindley, by the prohibitive cost of constructing an aqueduct across such a wide, marshy flood plain. Below this crossing, the Trent descends another weir and wends its way between rushy banks set in a flat landscape of tree-lined thorn hedges. At this point its water is probably as clean as anywhere throughout its entire course, but sadly it is about to succumb to its second disaster. This time,

however, the tranquil, rural scene gives not the slightest hint of the devastation about to be wrought. The effect is dramatic. Polluted by a welter of artificial discharges, the more powerful Tame injects its turgid waters into the Trent, instantly reducing it to a state where fish can barely survive. From here to its confluence with the Dove the Trent is again a largely sterile river.

The Tame drains a large basin which lies mostly to the north-east of Birmingham and separates the West Midlands Conurbation from Coventry. It is the most complex and probably the most important division of the Trent sub-region. The Tame itself is 85 km long and, with its tributaries, drains almost 1,500 km^2, though this includes the built-up area already considered in Chapter 2. Most of its basin lies between 100 and 200 m above sea level and comprises drift-covered marl. It is extremely varied country, ranging from the built-up areas of the Black Country and Birmingham to wide expanses of open farmland and it embraces large industrial complexes, collieries, gravel workings and landscaped parklands alike.

Ornithologically the Tame is one of the most important areas in the Region. Indeed it is arguably of national importance to a wide range of migrants. Yet few sites are individually outstanding. Rather, numerous small habitats link together to form a complex chain of wetlands, the true worth of which far exceeds the sum of the individual parts. In many respects the Tame Valley is to the West Midlands what the Lea Valley is to London, even to the extent of sharing the same geographical situation to the east of the conurbation and the same north-south orientation. The best habitats are wetlands and most have been created by man, usually in his quest for sand and gravel. Like the wetlands of the Penk basin, they are attractive to wildfowl, and Teal, Garganey, Shoveler, Pochard and Shelduck are amongst the breeding species. However they attract more passage waders than does the Penk, especially during spring when water levels are usually high at the reservoirs. Terns, too, are regularly noted and Common Terns have recently re-established themselves as a breeding species. Migrant passerines always appear in good numbers and in recent years occasional reports of Savi's Warbler have been received, although breeding has not been substantiated. The valley has also become a stronghold of Little Ringed Plover, with some 15 pairs in 1973, and both Ruddy Duck and Black Redstart have increased as breeding species. It is also an important wintering area for Green Sandpiper and Redshank, whilst Merlin has been noted more frequently of late.

The Tame rises in the heart of the Black Country and the Wolverhampton and Oldbury arms meet at Bescot before wending their way south-eastwards through Birmingham's northern suburbs. For much of its upper reaches the river weaves its way in a blue-brick or concrete culvert through factories, houses, roads and railways, collecting on its way a spate of industrial waste, sewage effluent and general detritus. It has little direct influence on bird-life whilst in this state and the birds of this area have therefore been dealt with in Chapter 2, which deals with the Conurbation as a whole.

Rapid storm-water run-off further impairs the water quality and for many years the Tame had the unenviable reputation of being the dirtiest river in Britain. In recent years, though, its condition has improved. Its problems stem from the sheer volume of water added to the river, which is so great that there is simply insufficient natural flow for adequate dilution. Downstream of the giant sewage works at Minworth, for example, two-thirds of the river's flow has been artificially introduced. By now the valley is broadening out and beginning to lose its urban dominance and this is as good as anywhere to begin our examination of its bird-life. The character of the valley is unique. Frequently it is an incongruous mixture of large industrial buildings and highly productive arable land; of extensive areas blighted by sand and gravel extraction yet surrounded by little villages and narrow lanes that might be a million miles from Birmingham. To those ornithologists whose senses demand neither aesthetic nor salubrious surroundings, it can be a rich and rewarding territory indeed.

Like most modern, hygienic sewage-works, Minworth no longer holds the same interest as hitherto. Even so, a little-disturbed installation of 350 ha inevitably attracts birds, and whilst the river may be too polluted to be of interest, there is a small pool on the opposite bank to enrich the habitat. A few old sludge lagoons survive to attract dabbling duck, mainly Mallard, Teal and Shoveler, along with the occasional wader and good numbers of wagtails.

There is sufficient marshy vegetation and sallow to hold a few Sedge and Reed Warblers, whilst between and around the beds are areas of rough wasteland and scrub with a variety of trees. These provide homes for breeding Whitethroat and Whinchat and a rich feeding ground for autumn finch flocks.

A mere 4 km to the east lies another giant installation – Hams Hall power station. Between its cooling towers and the river lies an area of some 50 ha, which, up to 1963, was little more than wet, rough grassland where two or three pairs of Curlew bred and small flocks of Teal and Wigeon wintered. Between 1963–5, however, sand and gravel was extracted and, though half of the pits were filled with flyash, the remainder were retained for circulation of the cooling water prior to discharge back into the river. This influx of warm water from the power station inhibits freezing and consequently areas of water here remain open long after surrounding ones have frozen. In 1970, thanks largely to the efforts of H. T. Lees, who later became honorary head warden, the Central Electricity Generating Board established this area as the Ladywalk Nature Reserve. Botanical colonisation of the ash infill has followed and the colony of southern marsh orchids is of special interest, but the extensive reed-beds, areas of willow scrub and vegetated islands all combine to provide cover for a wide range of breeding species in an undisturbed habitat. This was one of the first areas in the West Midlands to be colonised by Little Ringed Plover and amongst a good range of breeding duck are Mallard, Teal, Shoveler, Tufted Duck, Shelduck and sometimes Garganey. In winter up to two thousand waterfowl frequent the marsh, including good numbers of Teal, Tufted Duck and Pochard and a few Wigeon. Cormorants, too, are often present outside the breeding season and unusual visitors have included Red-necked Grebe, Temminck's Stint, Pectoral Sandpiper and Lapland Bunting. The rough grassland and scrub hold breeding Meadow Pipit, Whinchat and occasionally Stonechat, whilst two or three pairs of Black Redstart nest on the power station itself, with others in the nearby Coleshill gas works. There is also an interesting woodland, though somewhat depleted by Dutch elm disease, but still with a good range of commoner species and within it a small pool that is visited by Kingfisher, Moorhen and Water Rail. Both Spotted and Little Crakes have been seen here as well. Because of the many inherent dangers, the reserve is strictly private, but access can be obtained through a permit scheme operated by the WMBC and there are hides affording good views over the more interesting parts. A good indication of the variety of birds to be seen at Ladywalk is given in the reserve's annual report for 1978 (H. T. Lees *in litt*), which records 140 species for the year, of which half bred.

Near to Hams Hall the Tame is joined by the Blythe – a river which has two major tributaries, namely the Cole and Bourne. The Cole rises in open countryside south of Birmingham, near Wythall, but for most of its course passes through the south-eastern suburbs of that city before joining the Blythe north of Coleshill. The water is of dubious quality, but it does support some small fish and Kingfishers are seen quite regularly in the urban area around Hall Green and Sarehole Mill. It also feeds the small, ornamental lake at Trittiford Park, where the resident Mallard are often joined by Canada Geese and sometimes by other, less usual species.

In complete contrast the Blythe rises within one kilometre of the Cole, but flows throughout in open countryside as it meanders across the Green Belt between Solihull and Knowle before finally turning northwards to join the Tame. This beautifully clean and tranquil river is a source of potable water and supports thriving coarse and game fisheries. Flanked on either side by overhanging alders and willows, its sandy banks provide nest sites for Kingfishers, whilst Curlew and Redshank breed in its lush water meadows. Apart from commuter villages like Hampton-in-Arden and Coleshill, it is hard to imagine that at this point Birmingham and Coventry are no more than 10 km apart. Above the flood plain the farmland is excellent and since the last war more and more land has been brought into arable cultivation. This has brought about a reduction in the number of hedges and trees and an idea of the effect this has had on the bird-life can be gathered from a survey of 330 ha of mixed farmland near Maxstoke which was surveyed in 1959 and then revisited in 1979 (J. Robbins *in litt* via Warwickshire Biological Records Centre).

Originally there were thick hedgerows dotted with oaks – the type of landscape that characterised much of the Tame and Blythe valleys – and commonest amongst the fifty or so breeding species in 1959 were Chaffinch and Whitethroat, with around

75 pairs per km^2, and Blackbird, Wood-pigeon, House Sparrow, Yellowhammer and Dunnock, with over 40 pairs. Of the scarcer species, six pairs of Curlew bred and Corncrake was amongst the unexpected. Twenty years later there had been an overall reduction in bird life, especially amongst those species which depend on trees and hedgerows for nest sites, or those, like Moorhen, Curlew and Lapwing, which rely on permanent damp pastures. The change to arable farming had benefited Red-legged Partridge and Skylark, however, and brought Corn Bunting into the area for the first time. Despite the loss of hedgerow trees, Lesser Spotted Woodpecker also increased – probably reflecting its high population following Dutch elm disease – but more surprisingly Lesser Whitethroat had increased too.

The gem of the Blythe Valley is the Earl of Aylesford's seat at Packington. Here the unpretentious hall is surrounded by an extensive park containing many very ancient, stag-headed oaks reminiscent of those in Sherwood Forest. The rotting timber in these provides nest sites for a variety of hole-nesting species including Kestrel, Barn, Tawny and Little Owls, all three woodpeckers, Redstart and Tree Sparrow. Beneath the oaks are extensive areas of bracken over which Woodcock rode. The estate also includes farmland, broad-leaved and mixed woodland, gravel-pits and mature lakes, the largest of which is fringed with *Phragmites*. Great Crested and Little Grebes, Ruddy Duck and Pochard are amongst the breeding species, large numbers of Mallard congregate in autumn and in winter up to a thousand duck are present. Some of the old gravel pits have been developed as a commercial fishery, but nevertheless they hold many birds, including a resident flock of 150 feral Greylag Geese and a good population of Canada Geese. Sedge and Reed Warblers also breed.

Nearby the Little Packington pits, with their open banks and shores, became the focus of ornithological interest in 1975 with the appearance of a White-tailed Plover – the first ever recorded in Britain. However, they were well known to local bird-watchers as a good spot for passage waders long before this and at least twenty-six species were recorded between 1973–8. In addition, a small colony of Sand Martins and a few pairs of Tufted Duck and Little Ringed Plover used to breed and Bearded Tits occasionally appeared in the reed-beds in autumn.

The drought in 1976 though, showed these pits to be too high above the water table to reliably hold water and consequently too risky to be part of the commercial fishery. They are now being filled with household refuse and the land restored, but even this has had its rewards as vast numbers of gulls, mainly Lesser Black-backs, have been attracted and diligent searches have revealed a few Glaucous or Iceland amongst them most winters. Gorse and bramble grow in profusion on the dry, sandy soils around the pits and these support a variety of passerines. The whole estate is strictly private and is used primarily as a trout fishery, but it is still managed very much with an eye to conservation of its wildlife.

The dry, sandy soils around Packington contain a few remnants of old heathland, including an interesting shallow pool and peaty bog surrounded by oak–birch woodland at Coleshill Pool. This is now hemmed in on three sides by motorways and dual carriageways and during their construction the drainage pattern was much disrupted. In 1976 the pool dried out completely, but its water-level is now back to normal. A few duck breed; Sedge, Reed and Grasshopper Warblers nest in the marginal reeds and sedges; and reasonable numbers of Tufted Duck are present in winter.

The surrounding woodland also holds a good range of breeding species including warblers, woodpeckers, both Marsh and

Willow Tit and sparingly Tree Pipit.

The Bourne is a small stream which rises on the high ground south of Atherstone and flows south-eastwards at first, then turns north-westwards near Fillongley. Kingfishers nest in its banks and Snipe breed in some of the damper meadows. Despite receiving some sewage effluent, it is still sufficiently pure for water to be abstracted to supply Shustoke Reservoir. This 40 ha reservoir, which is administered by the Severn-Trent Water Authority, has steeply shelving, concrete lined banks. Both sailing and trout fishing occur, yet it still attracts some wildfowl, whilst the grassy banks and mixture of mature alders and conifers that flank the river along its northern side hold breeding and wintering passerines. All three wagtails breed, but it is winter wildfowl that are the principal interest, with Wigeon, Pochard, Tufted Duck, Goldeneye and a few Goosander amongst the usual visitors. Migrant sea-duck, especially scoter, are seen fairly regularly and all three divers have occurred in the past decade. There is also a winter gull roost, particularly in hard weather.

Rejoining the Tame, the river turns northwards beyond Hams Hall to meander through a wide flood plain where sand and gravel extraction has been taking place for over forty years at places such as Whitacre Heath, Coton, Marston, Bodymoor Heath and Middleton. This extensive network of active and exhausted pits has resulted in a chain of pools, marshes and scrubland in all stages of natural succession from pioneer colonisation through to virtual maturity. Many of the habitats are transient, but the birds can move from site-to-site according to the prevailing conditions and so this chain is without doubt the richest area of the valley. With tipping space at a premium and water in great demand for recreation, gravel operators have found little difficulty in finding profitable uses for their spent quarries. Many have been filled with pulverised fuel ash from nearby power stations and then restored to agriculture, but some 250 ha have been rescued by Warwickshire County Council for Kingsbury Water Park. In addition the water authority has a proposal for a major river purification scheme, of which the first in a string of seven large lakes has recently been completed, so a major wetland habitat seems assured for the future, though its bird-life will be under severe pressure from competing uses.

Already nearly twenty pools in the Bodymoor Heath area have been incorporated into Kingsbury Water Park and there is a variety of other habitats as well. The public has access to the whole area and to two hides overlooking a nature reserve. Despite a wide range of sports including fishing, sailing, wind-surfing, water-skiing and even power-boat racing, there is usually sufficient quiet water for birds, especially in winter, and the area is managed with conservation of the wildlife very much in mind. Garganey has bred in the past and Gadwall and Shelduck are both recent colonists, whilst the regular breeding wildfowl include Great Crested and Little Grebes, Tufted Duck (25 pairs) Pochard, Shoveler, Mute Swan and Canada Goose. A flock of the latter species approaches 500 each autumn and this sometimes draws in other geese, some of which are probably genuinely wild. Bewick's Swans appear most years, though Whoopers are less regular. Winter also brings many Coot and up to 900 duck, including a concentration of around 60 Goldeneye in March. Although the river is polluted, the duck take readily to it whenever the pools freeze over and Sand Martins nest in the river banks, though Kingfishers seemingly prefer the smaller streams and lakes. There are many gravel islands to attract migrant waders and now that water levels can be partially controlled a good variety may be seen. A few Wood Sandpipers appear most autumns, but

25. Kingsbury Water Park, with its Common Tern colony and nine species of warbler, is outstanding amongst the Tame Valley's chain of wetlands.   *J. V. & G. R. Harrison*

spring passage is often best for waders in general and as many as nineteen species were noted in the spring of 1979. Tern passage, too, can be good if weather conditions are favourable. Wherever there is sufficient bare gravel Little Ringed Plover breed and two pairs of Ringed Plover also nested in 1980. After an absence of five years, an incipient colony of Common Terns re-established itself in 1980 and Oystercatchers have stayed well into summer and may yet try nesting as well.

Away from the open water many of the old silt beds have been colonised by reed, reedmace and willow or alder scrub. These hold good breeding populations of Reed and Sedge Warblers, together with a few Snipe and Redpoll and an occasional pair of Water Rail. In winter they are visited by

Snipe, Jack Snipe, Water Rail and roosting finches, thrushes and buntings. Bittern has been recorded, Great Grey Shrike is a not infrequent visitor and a Marsh Harrier sometimes passes through in spring. The extensive areas of rough grassland, bramble and hawthorn hold breeding Yellow Wagtails, Reed Buntings, Grasshopper Warblers and many Whitethroats and are a favoured haunt of finch and thrush flocks and hunting Short-eared Owls in winter. In recent years also one or two Long-eared Owls have roosted. Lesser Whitethroats nest in the overgrown hedgerows, and the mature, rotting willows, ash and oak harbour Tawny Owl and Great Spotted Woodpecker. Blackcap and Garden Warbler also breed and, with nine nesting species, it is the warblers as much as anything that are the great attraction of the area. Overall, no fewer than 190 species have been recorded, including such rarities as Little Bittern, Ring-necked Duck, Kentish Plover, Caspian and Whiskered Terns and Lapland Bunting.

The nearby Middleton Pool, although an SSSI, has to a large extent been eclipsed by Kingsbury Water Park and today its main interest is the breeding Ruddy Duck, although a good range of wildfowl still occurs in small numbers and Lesser Spotted Woodpecker and Spotted Fly-catcher inhabit the surrounding woods. This part of the valley is also well known for its wintering Golden Plover, with a flock of up to 1,000 which is usually centred on Drayton Bassett. By now the valley is beginning to shed its industrial mantle and assume a more rural one, with extensive arable and potato fields. Hedge-rows are few, fields large and trees scarce and this is a countryside very much to the liking of Corn Bunting.

The Tame skirts to the south-west of the historic market town of Tamworth, which is dominated by its castle. Like many Midland towns, Tamworth has grown con-siderably in recent years to accommodate overspill from Birmingham and there are now extensive housing estates to the east. In the shadow of the castle is the confluence between the Tame and Anker. In its upper reaches, the Anker is no more than a small stream flowing through open farmland frequented by Skylarks and Lap-wings, with a few Golden Plover and the occasional Short-eared Owl to enliven the winter. For much of its length the river marks a sharp contrast. On the left bank the towns of Nuneaton, Atherstone and Polesworth, together with their associated colliery spoil heaps and road-stone quarries, rise up the scarp face onto the plateau above; whilst on the right bank is a wide pastoral plain of rich, fertile farmland and small villages. Scattered copses or coverts hold all three woodpeckers, Barn and Tawny Owls, a few warblers and the

usual corvids, tits and finches. The industrial towns of Bedworth, Nuneaton, Hinckley and Atherstone, with their coal mines, engineering, hosiery and hat-making factories, contribute to the polluted state of the river, but below Poles-worth the Anker passes through the subsidence pools at Alvecote and these enable it to effect some self-purification. These pools have arisen through colliery subsidence since 1940 and they form one of the oldest nature reserves in the Region. There are several shallow pools, which total 125 ha in area. Their depth varies be-tween one and three metres and the surrounding pastures are liable to flood. Among a wide range of habitats are extensive reed beds, fen, marsh, alder and willow carr, scrub, wasteland, farmland, colliery spoil-heaps and a canal, as well as the river and pools. It is a generally open area, but there are a few oak, birch, alder and willow and several old tree stumps in the pools which are much appreciated by passage terns. At low water levels, the exposed mud attracts a variety of waders and up to thirty species have been recorded over the years, including Black-tailed Godwit and Wood and Pectoral Sand-pipers. Almost anything might occur and over 175 species have been recorded since the last war, of which around 90 have bred. Breeding birds include up to 25 pairs of Great Crested Grebe and Tufted Duck, 10 pairs of Pochard and six pairs of Mute Swan, plus Shelduck, Little Ringed Plover, Snipe, Redshank, Kingfisher, Grass-hopper, Reed and Sedge Warblers and Yellow Wagtail.

The moulting herd of Mute Swans may number as many as 200 and autumn brings a mixed Swallow and Sand Martin roost up to 25,000 strong. In winter, duck numbers reach around 800, with a good range of species; flocks of Lapwing may be 2,000 strong; Golden Plover are often present in the general locality; and finches and thrushes can be numerous. A few Water

26.   Alvecote Pools are one of the oldest nature reserves in the Region and a major moulting ground for Mute Swans.   *J. V. & G. R. Harrison*

Rail secrete themselves in the marshy areas, but can usually be spotted by the patient observer.

The pools, which are owned by the National Coal Board, are scheduled as an SSSI and managed as a nature reserve by the Warwickshire Nature Conservation Trust, so they are little disturbed, except for some fishing. Two threats hang over them however. With the ever worsening energy crisis, the NCB may at any time resurrect proposals which it abandoned some years ago for open-cast mining to considerable depths. If this were to happen, then the area as it is today would totally disappear. The other threat will have a less profound effect, but a corner of the reserve lies within the proposed route for the M42 Birmingham–Nottingham motorway.

Below Tamworth the Tame is finally free of industrial influence – except of course for the legacy still reflected by its water quality – and it presents an attractive prospect as it meanders amidst pastures, grasslands, cereal and potato fields, past Hopwas and Elford, to its confluence with the Trent. This is an area for open grass-land species such as Skylark, Grey Partridge and Pheasant, with Reed Bunting in the hedgerows and thickets. In winter the washlands are visited by Mute Swan, Mallard, Teal, Lapwing, Snipe and perhaps an odd Jack Snipe. Finches and thrushes flock to the pastures, a few Corn Buntings occur and hunting owls are regular, with an occasional Short-eared joining the commoner species.

At Elford yet another gravel pit provides a momentary reminder of the scene upstream, but this is only a transient scar on the landscape as it is scheduled to be filled with pulverised fuel ash and returned to agriculture. This is a pity from the ornithological viewpoint, since this pit has proved to be one of the most interesting in the entire valley. Why it should be so good is not clear, but it may be quite simply that it is more remote and less disturbed than most. Whatever the reason, it has provided breeding records for Black-headed Gull, Common Tern and Oystercatcher.

Throughout the Tame Valley woodland is scarce, but the woods at Hopwas Hayes and Kingsbury are worthy of mention. Strangely enough, both are partially owned by the Ministry of Defence and, as such, are strictly private, being used as firing ranges. No doubt the birds quickly realise that they are not the gunners' target, since they seem quite prepared to tolerate the noise in exchange for freedom from human disturbance and the bird-life in both woods is quite good. Hopwas Hayes contains a mixture of oak with hazel, which was formerly coppiced, a variety of other hardwoods, plantations of pine and larch, and much bracken throughout the field layer. The felling and replanting cycle promises a good age diversity and is helping to perpetuate a few areas of open heath, scrub or bog. Breeding species here include Woodcock, Sparrowhawk, all three woodpeckers, Redstart and a variety of warblers including, sparingly, Wood Warbler. In winter the rhododendron bushes are used by thrushes and finches for roosting. Kingsbury Wood is similar, but less diverse, being basically an oak-hazel wood in a damp situation. It is an SSSI, but more so for its flora than its bird-life, since there are bands of limestone present which give rise to a luxuriant, calcicole ground flora. Amongst its breeding birds are Kestrel, all three woodpeckers, Tawny Owl, Woodcock, Marsh Tit, Blackcap, Garden Warbler and Chiffchaff; whilst in winter it is host to roosts of finches,

thrushes and Starlings.

The Tame joins the Trent near Alrewas and from here to Burton, the Trent, with its willow-fringed banks and expansive flood plain, is picturesque, if sterile. Apart from willow, poplar plantations are evident in a flat, pastoral landscape that is dominated by the cooling towers of Drakelow power station. Here, on the right-hand bank in Derbyshire, the CEGB has established another reserve, with an impressive list of bird records (Frost 1978). On the opposite bank, in Staffordshire, are the less known, but equally important gravel pits at Branston. These pits are of longer standing than the Drakelow reserve and their breeding records over the years have included Common Tern more regularly than anywhere else in the Region, Black-headed Gull, Little Ringed Plover and, in 1979, Ringed Plover – a welcome first record for the Region. Water Rail has bred and Shelduck has attempted to do so, but without success. In some respects this is an unusual gravel pit, with industrial development encroaching on one side, but open countryside on the other. Some parts have been filled with ash from the power station and then restored to agriculture, but others are being used by the gravel company for silt disposal and so there is the usual mixture of open water, shallows and muddy areas, with a substantial reed-bed and plenty of sedge and rush. Add to this small islands, which provide nest sites for Canada Geese and Tufted Duck, and it is readily apparent that the habitat is a good one indeed. Most years bring a good variety of passage waders, with Dunlin usually peaking in late autumn and some staying into winter. Among the more exciting species to have been recorded are Pomarine Skua and Peregrine. In freezing weather the warm water discharged into the river from the power station attracts ducks and in 1979 six Smew were present, including four drakes. These drew many bird-watchers to the area, with the result that two Lapland Buntings were discovered in a nearby meadow about the same time.

Beyond Branston the Trent wends its way past the brewery chimneys and Victorian buildings of Burton-on-Trent. Downstream of the town its course becomes braided and its waters eddy and swirl as they tumble over weirs to the confluence with the Dove. Thankfully, oxygenation from these weirs and dilution by the clean Dove water do much to improve the Trent's quality before it finally leaves Staffordshire and passes entirely into Derbyshire.

The final division of this sub-region is the Dove Valley. Rising high on the moors, the upper reaches of the Dove are more appropriately considered in Chapter 7, but here we are concerned with the lower reaches below Rocester. This section of the Dove featured regularly in historical records and the Tutbury area has some claim to fame as the place where, in 1852, Sooty Tern was first recorded in Britain – even more amazingly killed by a stone thrown by a small boy! A hundred-and-one years later a Tawny Pipit appeared in the same area. Today the Dove is a beautiful river in a valley that epitomises the tranquil life of a bygone age. Only the small market towns of Tutbury and Uttoxeter are reminders of the twentieth century. For the rest, the river meanders through a wide pastoral flood plain flanked to the south by the wooded slopes of Needwood Forest. Small dairy farms are dotted along narrow little lanes and each one has its resident Swallows perched outside. In winter the fields are full of Lapwing, Fieldfare and Redwing and in summer the wetter ones hold nesting Curlew. On one occasion recently the rasping call of a Corncrake was heard on a still, summer's evening. Kingfishers can be seen on the river and its tributaries and even upland birds, like Dipper and Common Sandpiper, may come as far south as Tutbury.

The northern bank is entirely within Derbyshire, so the heronry at Norbury is not strictly within the scope of this account, except that the Grey Herons themselves can often be seen standing motionless in the shallows. There are few sites of particular note within the Dove valley, but one such is on the Churnet shortly before it meets the main river. Here, within a private sports field, is a small fishing lake which contains two islands on which Great Crested Grebe and Mute Swan breed and where a few duck congregate in winter.

This portrait of the Trent Valley concludes the review of the Region's three major river systems. For the richness and variety of their bird-life they could hardly be bettered and the West Midlands is indeed fortunate to possess three such distinctly different lowland sub-regions. As we shall see in the next three chapters, it is equally fortunate to be able to complement these with a wide variety of uplands as well.

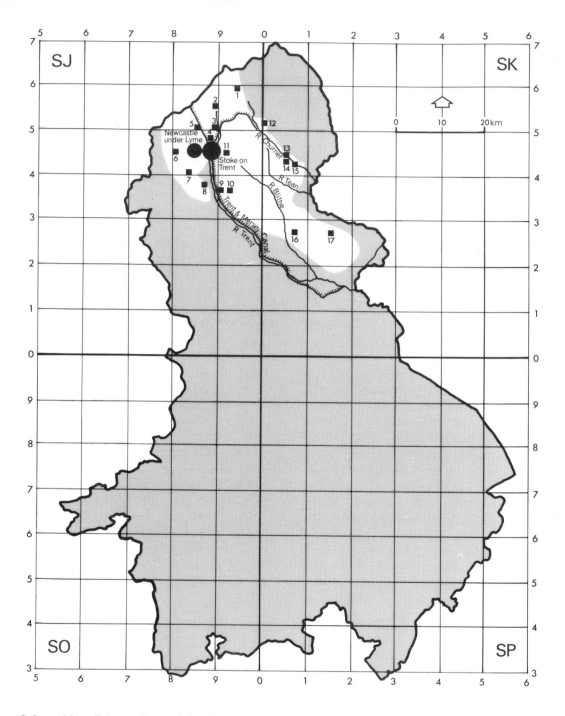

Selected localities in Central Staffordshire

1 Rudyard Reservoir	6 Keele	11 Park Hall	16 Bagots Wood
2 Knypersley	7 Hanchurch Hills	12 Coombes Valley	17 Greaves Wood
3 Ford Green	8 Tittensor Common	13 Oakamoor	
4 Central Forest Park	9 Downs Banks	14 Dimmingsdale	
5 Westport Lake	10 Moddershall	15 Alton Towers	

6 Central Staffordshire

A large, low plateau covers that part of Staffordshire lying between the rivers Sow, Trent and Churnet, but excluding the river valleys themselves. This forms the third smallest sub-region, with an area of 800 km^2 or 11 per cent of the regional area. It is roughly rectangular, 20 km wide and stretches 50 km from the Cheshire border in the north-west to Needwood Forest in the south-east. Altitude varies from 336 m in the extreme north to 100 m over much of the southern part and this influences climate to some extent, with rainfall varying from 900 mm to 700 mm respectively. It is indeed an area of transition, forming a bridge between the lush lowlands of the Trent Valley to the south and the bleak moorlands to the north and between the urban complex of the Potteries to the west and the sylvan surroundings of the Churnet Valley and Needwood Forest to the east.

The plateau has two main divisions, stemming largely from its geological structure. The higher, northern half is underlain by Carboniferous and older Triassic rocks and is dominated by the Potteries. The lower, southern half is comprised mainly of Keuper marl and is predominantly rural.

The geology of the Potteries is complex. The Carboniferous rocks strike north to south and have been folded into two south-pointing anticlines and a north-pointing syncline. Subsequent denudation has exposed the oldest beds of Millstone Grit in the eastern anticline, which extends southwards from the Cheshire border to pass between Stoke and Leek, whilst the newest beds – the Keele group – are exposed in the trough of the syncline west of Newcastle.

Between these two is a crescent-shaped exposure of Middle Coal Measures encircling Stoke to the north and then extending eastwards as far as Cheadle.

These steeply dipping, productive measures were of course the cradle of industrialisation in the Potteries and even today faulting and subsidence from the labyrinth of underground workings causes earth tremors. In general, the coal measures and more especially the Millstone Grit form higher ground, which slopes away gently to the Triassic plain of the west and south.

West of the Potteries, the junction between the Carboniferous and the Trias is largely a faulted one, with the resistant bands of grit and sandstone resulting in the rugged ridge running north-eastwards from Mow Cop to the Cloud. Elsewhere the Triassic strata lie unconformably on the Carboniferous, with the Bunter beds successively transgressing all the strata in the coal measure series until they meet the Millstone Grit east of Longton. Radiating outwards from the Potteries, successively newer beds are encountered, indicating that the Trias has been elevated into a dome and subsequently denuded. Further structural complications arise from some important faults, running in the direction of both dip and strike, with the main faulting to the east and west of Newcastle.

In contrast the geology of the lower, southern half of the plateau is much simpler. It is underlain almost exclusively by Keuper marl, which extends in a wide band virtually from Shropshire in the west to Derbyshire in the east. In the Needwood Forest area, however, the Keuper series is overlain in places by Rhaetic shales and

Tea Green marl. The whole area exhibits a thin and discontinuous covering of glacial drift, mainly boulder clay, which occurs especially on the plateaux tops and valley floors, but has often been eroded from the valley sides.

Such diverse geology has led to a varied landscape, which has been further manipulated by man's extraction of the earth's riches. More than anything, though, it is a landscape of woodlands, many of which have considerable ecological importance. There must be more Redstarts in this sub-region than in all the others put together and here Pied Flycatcher also has one of its two strongholds. Amongst the raptors, Sparrowhawk is widespread, Buzzard appears irregularly and Goshawk occurs very sparingly. Warblers are numerous throughout, especially Wood Warbler, and most of the woodland holds woodpeckers and a full range of the commoner species. It is somewhat surprising, therefore, to find this area largely neglected by ornithologists, with only the few sites on the doorstep of the Potteries receiving due attention. It is perhaps fitting to begin by examining these before moving on to the lesser known localities.

Pottery has been made in north Staffordshire since Roman times, but it was not until the discovery of coal in the same seams as the clay that prosperity really arrived. Prior to then, pottery manufacture was largely a cottage industry, supplementing the meagre income of farmers in isolated moorland settlements. The triangular saucer of the North Staffordshire coalfield, however, contained over twenty workable coal seams to provide energy, as well as the clays and marls necessary for the manufacture of earthenware, pottery, bricks and tiles. Furthermore the salt and lead used for glazing were available nearby in Cheshire and Derbyshire respectively, whilst gypsum from the Burton-on-Trent – Tutbury area

was used to make moulds for shaping the pottery. Thus all the necessary raw materials were available for tile and pottery manufacture and the coal mining and iron smelting industries also flourished.

During the eighteenth century the demand for better quality pottery inspired Josiah Wedgwood to establish his Etruria works and subsequently the names of Spode, Wedgwood and Stoke-on-Trent became renowned throughout the world for quality china. This reputation was not won easily though. The highest quality china required the highest quality materials, which meant importing china clay from Cornwall and ball clay and flints from Dorset. In the early days the transport costs were so prohibitive that, through Wedgwood's initiative, the Trent and Mersey canal was opened in 1777. So once again the West Midlands exhibited an ability to overcome its inherent disadvantages and in so doing added yet more variety to its habitats and birdlife.

The Potteries continued to flourish through the nineteenth century, with coal and ironstone mining second only in importance to pottery manufacture itself, and this basic pattern has changed very little during the twentieth century, though there has been some diversification into brick and tile manufacture in association with the extraction of clay.

Like the Black Country, the Potteries developed as several separate settlements clustered around a mine or quarry. In particular the six main towns of Tunstall, Burslem, Hanley, Stoke, Fenton and Longton all grew up on or very near to the Black Band marls which, with their clay, coal and ironstone, formed the richest seam in the coalfield. As they grew from villages into towns they coalesced into a single, linear urban area, which today is one of the most compact industrial regions in England. Although now amalgamated politically into the city of Stoke-on-Trent, each town still jealously guards its individual identity.

Thus the 'five towns' so vividly described by Arnold Bennett in his novels are really six, whilst a seventh, Newcastle-under-Lyme, lies just to the west beyond Fowlea Brook and really forms part of the same urban complex, distinguished more by social and historical factors than by physical ones. The Potteries of Bennett's day was a sombre assortment of belching chimneys and innumerable bottle kilns dotted indiscriminately amongst rows of tiny, Indian-red terraced houses, each one topped by an amber chimney pot and fronted by a blue-brick pavement. Today Stoke is trying to break away from this popular, if unenviable, image. It is now a city of over a quarter-of-a-million people, but it retains its strong bias towards manufacturing industry, with coal mining and pottery still prominent. The bottle ovens have largely disappeared – victims of modern technology – but still a quarter of the houses are over sixty years old; and its slate-grey colliery slag-heaps, deep marl-hole scars, mounds of pottery waste and, more recently, abandoned railways have left a legacy of dereliction which probably surpasses that of any other British city.

Paradoxically this industrial scene is not devoid of bird life and Smith (1938) records Grasshopper Warblers on Rough Close Common, Longton, as well as small parties of migrating Stonechats. Swifts breed beneath the eaves of the Victorian buildings and Wheatear, Yellow Wagtail, Meadow and Tree Pipits are all recorded as having nested on the slag heaps of Longton and Etruria. Smith also records Twites, Linnets and Goldfinches being caught in birdcatchers' snares.

Although the urban form is very compact, simultaneous growth from several nucleii has led to pockets of undeveloped land hemmed in by colliery spoil-heaps, disused railways, canals and polluted streams. Frequently such areas have escaped the bulldozers because they occur along watercourses like the Fowlea and Ford Green brooks, where the ground has subsided to become marshy and prone to flooding. More often than not, they are much abused wastes, used for fly-tipping, playgrounds and the exercise of dogs. Notwithstanding this, they harbour a surprising variety of birds. Kestrels and even owls occupy any derelict buildings; good numbers of Meadow Pipits breed amongst the willowherb, nettle and rank grass, along with the occasional pair of Grey Partridge, Wheatear and Whinchat; and in autumn flocks of Goldfinches and other finches feed on the thistle heads and weed seeds. The wetter areas of reed-grass and reed-mace shelter thousands of roosting Swallows each autumn, along with smaller numbers of Yellow and Pied Wagtails and even an occasional wader. In winter, the same areas provide cover for good numbers of Snipe and Jack Snipe and even a few Mallard and Teal can be found in the wettest spots. At Holden Lane, mining subsidence has left two areas of open water and here Mute Swan, Great Crested Grebe and Little Grebe all manage to breed and Kingfishers appear from time-to-time.

Clearing away derelict land like this is an enormous, daunting prospect, but, like so many of the Region's problems, it is being tackled with enthusiasm, vigour and foresight. Already conical pit mounds have assumed more natural shapes, trees and grass are growing on slag heaps, flooded marl pits have been transformed into natural lakes and disused railway lines into pleasant, tree-lined walks. Of course the naturalist may well wish the land had been left derelict, but reclamation is the new image of Stoke-on-Trent and it need not necessarily mean a reduction in wildlife. Even now these new habitats are beginning to attract birds and bird-watchers.

Right in the very centre, the former Hanley Deep Pit colliery is being reclaimed into the 50 ha Central Forest Park. Already

27.   Westport Lake typifies the new image of Stoke-on-Trent and promises to become a
real oasis in an urban wilderness.   *D. Smallshire*

the old pitmounds have been reshaped to more natural contours, seeded with wild grasses and planted with forty-thousand forest transplants in a project that is unique for an urban area in this country. There is a small pool and once the trees become established the bird life should improve considerably. Even now Skylarks, Meadow Pipits, Linnets, Goldfinches and Kestrels are breeding and Stonechats have appeared on passage just as they used to on the old slag heaps.

Not far away is Westport Lake. This was once the site of Port Vale's football ground before the pressure of water pumped from nearby marl workings caused it to subside and flood. Now some 40 ha, sandwiched between the main Manchester railway line and the Trent and Mersey Canal, have been reclaimed as a water park, including a

10 ha lake with an average depth of two metres. Three smaller pools and a small marshy area of reed-grass and water horsetail, flanked by willow and alder, have been declared a nature reserve. Mounds have been constructed to screen an adjoining pottery-waste tip and once the newly-planted young trees have matured, this should become a real oasis in an urban wilderness despite its being the main outlet for water recreation in the city, with sailing, fishing, swimming and picnic areas. Fortunately agreement has been reached to keep boats off the water from November to March, inclusive, and as a result a good winter bird population is developing, although the prospects for breeding are obviously much poorer.

Wildfowl are the main feature, with better numbers of Tufted Duck, Pochard,

Coot and Mute Swan than anywhere in the Potteries. A few Great Crested Grebe and Goldeneye are usually present, but dabbling duck are scarcer and more erratic, although Wigeon have occurred and Mallard have exceeded fifty on occasions. Amongst the more unusual records are ones of Slavonian Grebe, Long-tailed Duck and Shag. The water also attracts good numbers of Lesser Black-backed and Herring Gulls in winter, though gulls from the Potteries generally seem to resort to Cheshire for roosting, and both Kittiwake and an Iceland Gull have been spotted amongst the regular species. A few waders and terns appear on passage, Pied and Yellow Wagtails frequent the shoreline, Rock Pipit has been noted once or twice and Stonechat has occasionally wintered. The nature reserve attracts a variety of passage warblers, whilst winter brings Goldfinch, Redpoll, perhaps Siskin and once a Great Grey Shrike.

On the south-eastern side of the city, at Park Hall, is the largest and most ambitious of the reclamation projects, covering 176 ha of hilly countryside, sandwiched between two housing estates, and reaching a height of 250 m. Originally a deer park, which was reclaimed for agriculture and then exploited for coal, sand and gravel, these hills have had nothing if not a chequered history. With the cessation of sand and gravel extraction, Blue Circle Cement dedicated them to the City and County Councils. Widely regarded as an unofficial tip, they were at that time a danger and an eyesore, with pit-shafts, sludge tips, settling lagoons and unfenced canyons scattered amongst more natural areas of moorland, grass, gorse, woodland and water.

As is often the case with rough areas, birds readily moved in wherever they were free from disturbance. A surprising variety of unexpected species found its way to the small pools, Sand Martins bred in the quarry face and Skylark, Meadow Pipit,

Wheatear, Whinchat, Linnet and on occasions Stonechat and Short-eared Owl amongst the grasslands and gorse. The wetter pastures held nesting Redshank, Snipe and Curlew and winter feeding flocks of Lapwing and Golden Plover. Reclamation of the area for a country park began in 1975, with the object of creating a mosaic of heath, grass, meadow, scrub, wood and wetland between the existing conifer plantations of the rural east and the broad-leaved woodland of the more urban west. As yet, little is known about the effects of this scheme on the bird life of the area, but it will be interesting to monitor this over the next year or two.

Beyond its urban fringe, the Potteries is surrounded by much pleasant countryside. To the north and east the elevated dome of Millstone Grit and Middle Coal Measures has produced an area of high, open countryside rising to 336 m at Biddulph Moor and extending northwards to the river Dane. The hard grit bands have formed pronounced, west-facing scarps at Brown Edge, Baddeley Edge and more especially Mow Cop, where the sham eighteenth-century castle ruin affords a fine panorama across the Cheshire plain in one direction and the headwaters of the Trent in the other. Soils are poorly drained, wet, strongly acidic and often stony throughout and this is an area of small dairy farms. The landscape is reminiscent of the moorlands in its bleakness, though it is less wild. The small fields are nearly all permanent pastures, much frequented by Lapwings and Skylarks and divided by neglected, overgrown thorn hedges that host a mixture of breeding finches and corvids, especially Magpies. There is a scatter of very small oak or birch woods, mostly on the steeper slopes, and Green Woodpeckers are common in these, but otherwise trees are few. Numerous tiny settlements are dotted amongst a maze of little lanes, but only the small mining town of Biddulph is of any importance.

Within this area two canal-feeder reservoirs are worthy of mention. Firstly, nestling at the foot of the moors, is the long, narrow Rudyard Reservoir, which stretches for 3 km and covers 71 ha. Constructed in 1797, it is surrounded by pastures and a margin of mixed woodland along the southern shore that holds Sparrowhawk and breeding Redstart and Wood Warbler. Around the inlet at the northern end the lake is shallow and marshy, with a vegetative cover of water horsetail, yellow flag, reed-grass, alder and willow and patches of exposed mud when water levels are low. It is here that dabbling duck, grebes, Coot, passage waders and migrating passerines are likely to be found. In summer the reservoir is a popular resort of fishermen, sailors and the general public, which limits its breeding potential, although Great Crested Grebe, Mallard and Mute Swan usually manage to raise broods. A few terns occur on passage and in winter small numbers of Goldeneye and Goosander are usually present. In recent winters, too, a small mixed herd of Bewick's and Whooper Swans has frequented the area, moving between here, Tittesworth and Endon. It is to be hoped this becomes an established feature, since this is the only area in the Region where Whooper Swans are regular.

Secondly, on the lower ground near Stoke, is the delightful Knypersley Reservoir and the 45 ha of Greenway Bank country park, with its own serpentine lake and extensive mantle of oak woodland. Here, where the swift-flowing Trent passes through a miniature gorge, is perhaps as good as anywhere to savour the typical bird-life of the area. Dippers and Grey Wagtails breed along the river, whilst the oakwoods hold breeding Stock Dove, Great Spotted and Green Woodpeckers, Nuthatch, Redstart and woodland warblers, including Wood Warbler. The reservoir itself is not outstanding, but Mallard and Great Crested and Little Grebes breed, there are small numbers of Tufted Duck and Pochard in winter and a few waders and terns occur on passage.

West and south of the Potteries, the high plateau extends southwards from the Cheshire border, between the Trent and Meece Valleys as far as Swynnerton. Underlain by Permo-Triassic red marls and sandstones, this is an area of well-drained, gravelly soils which are thirsty, acidic, prone to leaching and frequently under woodland. In places there is a thin covering of coarse or medium textured drift. This is a rolling landscape of wooded valleys and open farmland. Mature oak and beech woods are interspersed with pine and larch plantations and areas of natural birch woodland. Understoreys of rhododendron and open areas of heather are indicative of acidic soils and occasionally there are outcrops of bare sandstone. Farms are larger than on the higher ground to the east, but they still concentrate on dairying or cattle rearing. It is a pleasant, undulating open countryside in which fields are often enclosed by fences rather than hedges. There are numerous small streams, where Grey Wagtails breed, and these have often been dammed to create tiny lakes. Wherever there is a reedy margin, as at Black Lake, Little Grebe might nest as well, and an occasional Water Rail might be seen in winter.

Most of the woodlands are strictly private and best dealt with in two broad groups – broad-leaved and coniferous – since the bird-life of the woods within each group tends to be very similar. The older broad-leaved woods are usually dominated by beech or oak, the latter sometimes stag-headed and decaying. They contain a good variety of breeding woodland species, including all three woodpeckers, Nuthatch and Redstart. Wood Warblers also breed wherever the shrub layer is sparse, whilst in winter Brambling are regularly noted in mixed finch flocks feeding on beech mast. The small woods around Keele have a well-developed understorey of rhododendron, holly and yew which provides food and shelter for large numbers of Redwing and Fieldfare in winter; and breeding sites for Mistle and Song Thrushes, Blackcap and Garden Warbler. Otherwise the understorey often consists of a little bramble amongst a blanket of bracken which may conceal a Woodcock or two. In winter the birch woodland, which is sparser and more like scrub, contains good populations of feeding tits and finches, especially Redpoll and Siskin. The conifer plantations are mostly larch and pine, often with a perimeter planting of birch or oak for amenity purposes. Their breeding birds include Sparrowhawk, Woodcock, Coal and Long-tailed Tits, Goldcrest and owls, including the occasional pair of Long-eared. Many of the typical birds from all these woods can be seen with patience from the public areas around Trentham Park or in the Forestry Commission's woodlands, two of which are worth describing in more detail.

The largest of these is Swynnerton Old Park, which straddles the sandstone ridge of Hanchurch Hills and rises to over 200 m. This was once an extensive area of mixed woodland and heath and some of the oak-birch woodland and odd patches of heather and gorse still remain amidst the large Scots pine and larch plantations of the early 1950s. There are also stands of beech, sycamore and sweet chestnut. In former days churring Nightjars were a familiar sound on summer evenings, but now the trees have grown too tall and this species has deserted the area, though it might return when felling takes place a few years hence. An unusual feature is the large House Martin colony on a water tower situated in the middle of the heath. The other Forestry Commission woodland is Tittensor Common. Here the fir plantations are about ten years old and have a typical avifauna, but there are also stands of mature oak, beech and chestnut where

Redstart and Wood Warbler breed. This is also a good place for winter tit and finch flocks, including Siskin, Redpoll, Chaffinch and Brambling, whilst Crossbill has occurred in the past.

Overlooking the Trent, on the opposite bank, are Downs Banks – a National Trust property covered in birch scrub and bracken, with scattered hawthorns and a small conifer plantation. Tree Pipit, Whitethroat and Lesser Whitethroat breed. The countryside from here across to Cheadle and the Churnet Valley has deeply-cut, parallel valleys like those of the Blithe and Tean, small streams and numerous woodlands. Dairy farms still predominate, permanent pastures are common and the small fields are enclosed by thick thorn hedges. Large numbers of Fieldfare and Redwing feed on the pastures in winter, whilst in summer Lapwing, Yellowhammer and Bullfinch are notable amongst the breeding species.

Always, though, it is the woodlands that are the focus of interest. Most of them are private and were probably planted two centuries ago for charcoal burning. Nevertheless, there are places where birds can be seen by the public. At Moddershall, for example, a road follows the valley northwards and is flanked on either side by old oak woodland that was formerly coppiced and a few pine and larch plantations. There are good breeding populations of Willow Warbler, Great, Blue and Coal Tits and a large rookery. Sparrowhawk, Woodcock, woodpeckers, Marsh and Willow Tits, Whinchat, Redstart, Wood Warbler, Tree Pipit and Tree Sparrow also breed and in winter the woods host large numbers of roosting Fieldfare, Redwing and Blackbird. A pool and one or two marshy areas within the wood add to the diversity, with breeding Kingfisher, Snipe, Lapwing and Sedge Warbler.

Further east is the picturesque Churnet Valley, which, with a Germanic flavour to its architecture and a wooded gorge, is known as the "Staffordshire Rhineland". Cut into the sandstones and shales by glacial melt-waters, the valley is 100 m deep and at times almost sheer. Industry came early and on a very small scale. It also declined quickly, so the valley has escaped the worst of the usual deprivations and dereliction. Instead it has inherited a rich industrial archaeology set in beautiful, wooded countryside that is every bit as charming as that of the nearby Peak District National Park, although it has earned no higher designation than that of an Area of Outstanding Natural Beauty. For much of the length from Longsdon to Alton there is an almost continuous mantle of hanging woodland on either bank and it is convenient to deal with this as an entity, although physiographically the woods to the north fall within the moorland sub-region.

The Caldon Canal, which has been largely restored for pleasure cruising, and a disused railway follow the river along its narrow valley, and the severed land between is often marshy and liable to flood. Although slightly polluted, Dipper, Kingfisher, Grey Wagtail and, very sparingly, Common Sandpiper all find homes on the river or its purer tributaries, whilst Reed Bunting and Grasshopper and Sedge Warblers breed in the wetter meadows. But again it is the woodlands that command attention, being the best and most extensive upland woods in the Region.

There is great variety too. Some of the old, natural woodland has been reclaimed for agriculture or felled and replanted with alien larches, pines, spruce or cypress. Often there is a fringe of oak, ash or sycamore, however, and an understorey of rhododendron which holds winter finch and thrush roosts. There are also some areas of sessile oak and ash, with sycamore and wych elm as secondary species; and understoreys of hazel, bramble, holly and elder, most of which are important to birds as a source of food. Former management of these woods was frequently coppice-with-standards, though coppicing has long since ceased. A Sparrowhawk might be seen soaring overhead almost anywhere and both Willow Warbler and Redstart breed in good numbers, the latter particularly where there are old trees with rotting timber. Other breeding species include Woodcock, all three woodpeckers, Marsh and Willow Tits, Nuthatch, Blackcap, Garden and Wood Warblers, Tree Pipit, Spotted Flycatcher and, in a few places, Pied Flycatcher as well. Goldcrest and Redpoll are well established as breeding species in the conifer plantations and Long-eared Owls nest sparingly. In winter the woods are used by thousands of roosting Woodpigeons, Jackdaws, thrushes and finches. Near Froghall the steep hillside is covered with well-grown birches, which give way to Scots pines on the summit, whilst the National Trust's nature reserve at Hawksmoor contains stunted oaks, extensive areas of birch and bracken-clad hillsides beloved by Woodcock, Tree Pipit and formerly Nightjar. It is probably from the ancestors of these old oaks that the nearby village of Oakamoor derived its name. This is the centre of the Churnet woods and its small picnic area forms an ideal base from which to explore a mosaic which includes oak, ash, birch, beech, alder, willow and poplar amongst the hardwoods and pines, larches, hemlock and spruce amongst the conifers.

Most of the typical birds of the area can be seen in Dimmingsdale, where there is a delightful walk through a sheltered valley of mixed woodland. Nor should Alton Towers, with its mock-Gothic ruins and pagoda fountain, be ignored. Despite the many thousands that come to enjoy it as a pleasure ground, the influence of "Capability" Brown is still apparent in the beautiful blend of parkland, lakes and woods, many of which are little disturbed especially in winter. Several of the typical breeding species can be found in the quieter

28. Coombes Valley RSPB reserve offers the best chance to savour the bird-life of the Churnet woodlands, which includes Redstarts and Pied Flycatchers. *Bill Smallshire*

parts and the rhododendron cover is widely used as a winter roost.

Undoubtedly the best opportunity to savour the bird-life of the Churnet woodlands, however, is to visit the RSPB's reserve at Coombes Valley. This is a superb setting where steep, wooded hillsides drop down to a fast-flowing stream whose crystal waters bubble across rocks and boulders on their way to join the Churnet. Dipper, Grey Wagtail and Kingfisher all breed here and a few pairs of Pied Flycatcher occupy the nest-boxes dotted amongst the overhanging sessile oaks. Beneath these old oaks is an understorey of holly, but there are also open glades, which are the result of felling or fire. These are initially colonised by bracken and eventually by birch and rowan, but meanwhile

they host Tree Pipits, Whitethroats and, in 1976 at any rate, Nightjar too. The oak woodland is ecologically very rich on its own, but greater variety is provided by conifer plantations. All this adds up to a diverse avifauna. The clamouring of Rooks is a familiar sound in spring and high summer brings Redstart (up to 35 pairs), Blackcap, Garden Warbler, Willow Warbler, Chiffchaff and Wood Warbler. All three woodpeckers and Nuthatch are resident, Sparrowhawk are often present and four species of owl, including the scarce Long-eared, nest most years. In winter, roosting flights can be spectacular as huge numbers of Fieldfare and Redwing, together with hordes of Woodpigeons and Jackdaws, flock into the woods. For many years this has been the only RSPB reserve in the Region, but now the Society has acquired a similar woodland at nearby Rough Knipe, but this is not open to visitors.

Away from the woodlands and streams, there are one or two old quarries and sand pits of interest. Often these are surrounded by birch scrub, heather, bracken and gorse. Some have Sand Martin colonies and, though basically dry, there may be sufficient water for Little Ringed Plover to nest. Wheatear and Whinchat are sometimes seen around these areas, more often than not on passage, but possibly breeding in one or two places.

Further south and east, the sandstone gives way to drift-covered marl and a more agrarian landscape of fewer roads and settlements, where dairying remains the commonest form of farming. There are no sites of particular ornithological significance, but passing mention must be made of Chartley Moss, where round-leaved sundew and cranberry feature in an interesting flora.

West of the Blithe, the landscape is once more characterised by fragmented woodlands. These are remnants of yet another ancient Royal hunting forest – Needwood Forest – which once stretched from Abbots Bromley to Burton-on-Trent. Clearance has been extensive, since much of Needwood lay on more fertile ground than forests elsewhere. Today the largest remnants are Bagots Wood and the Marchington Woodlands, which extend for eight kilometres along a north-facing scarp dropping 40 m into the valley of the Dove. If anything, this area has been more neglected than any by ornithologists, but recent sightings of Hawfinch and the discovery of breeding Pied Flycatchers (S. C. Brown *pers comm*) suggest that it would repay closer investigation.

The woodland of today is very mixed, though in earlier times it was mainly oak with coppiced hazel. Quite a few large, old oaks are still standing, ash and sycamore are dominant in places and birch scrub regularly occurs along the margins, but in other parts the old broad-leaved woods have been cleared and replanted with larch, Corsican and Scots pine. Alder, willow and poplar also occur where the ground is moist. The understorey varies considerably as well. Sometimes it is well-developed, with holly, bramble, elder and hazel in the shrub layer and a variety of ground plants; sometimes it is sparse, with little other than bracken. Clearance for forestry and the occasional fire has left open, grassy glades and these attract birds like Redstart and Tree Pipit which require room for their display flights. Many of the woods are used for Pheasant-rearing and are consequently private and well-keepered. Their breeding birds are much the same as those of the Churnet woods, with Sparrowhawk, Woodcock, all three woodpeckers, Nuthatch and many Blackcaps as well as those already mentioned. The surviving old oak woodland at the southern end of Bagots Wood holds the second-largest heronry in the Region, with some 50 pairs.

Otherwise the Forestry Commission replanted some 400 ha of Bagots Wood with Corsican pine some twenty-five years ago, since when the variety of its bird-life has diminished considerably. Even so there is a roadside margin of birch, beech and oak together with a few stag-headed oaks and large numbers of yew to serve as reminders of its former glory. Open rides, bracken and heather augment the woodland fringe and Woodcock and Tree Pipit are plentiful here, Lesser Spotted Woodpecker and Sparrowhawk both nest, and two or three pairs of Whinchat are usually present. Sadly, though, the wood is no longer suitable for Grasshopper Warbler and Nightjar. Up to ten pairs of the latter species used to breed, but there have been no records now for at least five years. At the opposite end of the forest to Bagots Wood, Greaves Wood compresses into a few hectares a surprising species diversity and age structure that consequently supports an excellent range of birds.

Breaking up the woodlands are arable farms on which Little Owl is common and Barn Owl not infrequent. This is pleasant countryside, noted for its ancient hedgerow hollies, oaks and ash. Marlpits are numerous and the hawthorn, hazel and ash that grow up around them provide nest sites for many finches and warblers, including Lesser Whitethroats. Drainage is through tiny streams flowing southwards, on which Kingfishers occur. Parkland features on the eastern side of the Forest, with scattered beech and oak and shelter-belts of larch, pine and other conifers that also house breeding Sparrowhawk, Lesser Spotted Woodpecker, Nuthatch and, less frequently, Redstart and Wood Warbler.

A distinct upland flavour emerges throughout this account of Central Staffordshire and its woodlands. This immediately evokes a strong association with hill-country and leads naturally into the next chapter, which looks at the Moorlands.

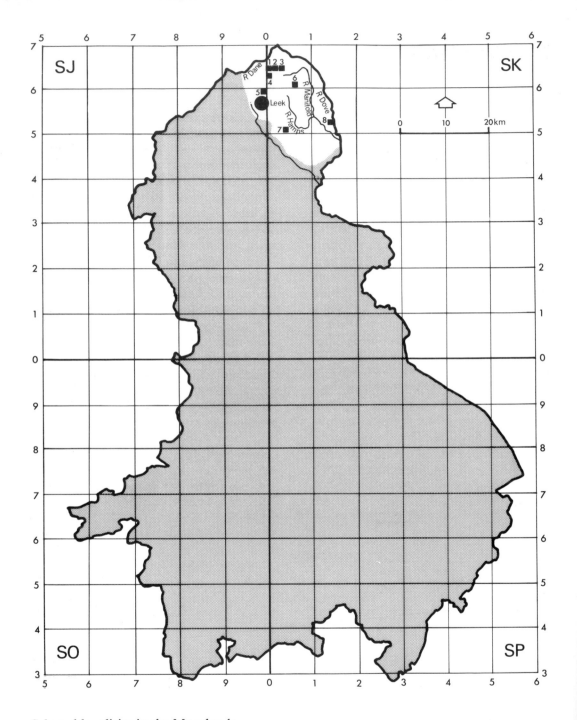

Selected localities in the Moorlands

1 Blackbrook Valley   3 Gib Torr   5 Tittesworth Reservoir   7 Swineholes Wood
2 Goldsitch Moss   4 The Roaches   6 Swallow Moss   8 Dovedale

7 The Moorlands

The southernmost tip of the Pennines protrudes into north-east Staffordshire, forming the smallest, yet most distinctive, of the seven sub-regions and one which supports an avifauna quite different from that of any other part of the West Midlands. It is bounded by the River Dane and Cheshire to the north-west, the River Dove and Derbyshire to the east and the River Churnet to the south and west, and is roughly rectangular, measuring approximately 25 km by 15 km. It covers some 300 km^2, or a mere 4 per cent of the total area of the Region, and mostly falls within the Peak District National Park.

Again there are two distinct divisions, arising from geological and drainage differences. Most of the area is underlain by impervious Millstone Grit, but the extreme south-eastern corner, adjacent to Derbyshire, forms part of the porous Carboniferous Limestone dome. Together these rocks have produced a bleak, upland plateau, mostly over 300 m, with higher ridges of resistant sandstones and quartzites. The highest point of all is Oliver Hill, standing at 513 m above sea level. The climate on this hilly ground is much more hostile than elsewhere in the Region. At Buxton, which is just within Derbyshire, rainfall is as high as 1,230 mm, January's mean temperature is as low as 2°C and July's no higher than 14°C. On average rain falls on 211 days each year, snowfall is heavier and more persistent than elsewhere, ground frost is recorded on 111 days and high winds are frequent. This is the only part of the Region that does not experience a moisture deficiency in summer.

The top of the Carboniferous Limestone plateau is covered with thin deposits of Trias or Loess drift, which have produced shallow brown-earths over hard limestone. As the porous, heavily fissured limestone drains very freely, lime is leached from the soil. In the main the short, but nutritious, grass sward is grazed by dairy cattle or sheep, though an occasional field is cut for hay. Fields are enclosed by drystone walls of white limestone and trees are few, apart from the occasional shelter-belt of sycamore. Small, attractive villages of stone cottages are dotted across the plateau and along the river valleys and are linked by a network of narrow lanes. Skylarks, Meadow Pipits and Lapwings are typical birds of the plateau top, but in spring both Whinchats and Wheatears use the drystone walls as song posts and the evocative calls of Curlew fill the air.

Typically for an area of high precipitation, the Carboniferous Limestone plateau has been dissected by numerous streams, which have cut deep, steep-sided and narrow valleys known as dales. Ravines, potholes and falls have all been sculptured by running water as it makes its way into the rivers Dove, Manifold and Hamps. These rivers, which flow south-eastwards to eventually join the Trent, have been famous amongst anglers since the days of Izaak Walton and Charles Cotton. As the limestone is so porous, many valleys are characteristically dry and even the Manifold and Hamps flow underground in dry weather.

The steep-sided dales are often clothed with woodland, in which ash is dominant, and they are renowned for their scenic beauty. Unquestionably Dovedale, the western side of which is in Staffordshire,

29. Tourism in the dales puts pressure on the birds, especially in beauty spots like
Dovedale. *Bill Smallshire*

is the most famous, but parts of the Manifold are no less magnificent and the
National Trust owns substantial areas in
both the Dove and Hamps valleys. On
summer weekends especially, the dales are
thronged with visitors, yet Dipper, Kingfisher and Pied and Grey Wagtails still frequent their crystal streams and breed
wherever visitor pressure is not too great.
Even Common Sandpipers nest sparingly
on the remoter stretches. In the woodlands
that clothe the valley sides, Chaffinch and
Willow Warbler are the commonest birds
and a count of the latter in 1977 revealed

300 singing males along a 14 km stretch of
the Manifold between Hulme End and
Waterhouses. Woodpeckers, Redstart and
Wood Warbler also nest in good numbers.
Occasionally outcrops of reef limestone
form sheer, rocky cliffs like those at
Wetton Mill, or hills such as Bunster Hill
and these are frequented by Kestrels, Jackdaws and Stock Doves, which nest in the
caves and crevices.

The Millstone Grit plateau is quite
different and instantly recognisable by the
much drabber, greyer stone of its walls.
Here the dominant shales are interbedded

with hard sandstones. High rainfall, aggravated by windborne industrial pollution and impeded drainage, has produced wet, strongly acidic soils in which peaty layers are commonplace. Even on the moderately steep slopes of the valleys, moisture penetration is insufficient to prevent superficial wetness and the whole area is consequently clothed with moorland vegetation. There are few villages, but isolated farmsteads are liberally scattered and a maze of minor roads gives unusually good access for such an area. Although less popular than the dales, many people take advantage of this easy access to visit the moorlands on summer weekends. Sheep graze widely, but pastures are often poor and only the presence of a few dilapidated drystone walls distinguishes them from the open moorland. Wherever the pastures are most nutritious, cattle are found.

The main road from Leek to Buxton marks a major watershed and the higher, flatter parts of the plateau on either side have been moulded into a rounded topography, with poor soils dominated by cotton-grass and tufted hair-grass. In the hollows peat has accumulated and blanket bog, with *Sphagnum* moss, has developed. Such areas can be readily identified by names like Goldsitch Moss and are characterised by the breeding of species such as Mallard, Teal, Curlew, Snipe, Lapwing and occasionally Golden Plover. To the west of the watershed, the upland streams flow into the Dane and hence to the Mersey, whilst to the east they flow via the Hamps and Manifold to the Trent. Save for a few sheep, this is a deserted, bleak and often barren landscape, with scarcely a tree to be seen, although within the past decade some 25 ha have been planted with young conifers at Gib Torr. To the ornithologist, though, this bleak, exposed area can be very exciting, although the bird-watcher has to work hard to find the more interesting species. It is particularly desolate in winter, when little

except Red Grouse, corvids or an occasional Hen Harrier or Short-eared Owl can be seen, but bursts into life again each spring and summer and Yalden (1979) has provided a good account of the bird community at these seasons. In general migrants arrive late, but by mid-April Meadow Pipits are returning to their territories and within a short while they are the commonest birds across the moor, with many pairs parasitised by Cuckoos. Skylark, Curlew and Snipe are also well distributed, Short-eared Owls are regularly noted and nest most years and there are a few pairs of Wheatear, Whinchat and Stonechat. Upland species like Dunlin and Whimbrel also pass through on migration, but most bird-watchers come to the moors hoping for raptors. Both Merlin and Hen Harrier breed intermittently and may appear at almost any time of year, but Peregrines are few and far between these days.

Inter-bedded in the Millstone Grit are sandstones and quartzites which are resistant to weathering and form ridges, like the Weaver Hills, Ipstones, Morridge, the Roaches and Ramshaw Rocks, which stand above the general plateau level. In places rugged, sculptured rock outcrops resemble the well-known granite tors of Dartmoor, springs issue everywhere and the coarse-textured, stony soils are strewn with boulders. Wherever streams occur, the Millstone Grit is quickly eroded into deep, narrow valleys known locally as cloughs. Soils are still very damp and peaty, but surface water does run off the slopes more readily. Heather, bilberry and cowberry are usually dominant on the shallow-soiled, but better-drained slopes; with bracken on the deeper soils of the footslopes and purple moor-grass in the wetter areas. Enclosed fields frequently peter out partway up the valley sides, as in the Blackbrook Valley, and, although a few sheep graze the heather, much of it is rotationally burnt and managed as grouse moor. The

30. The North Staffordshire Moors have a distinct avifauna, which includes Ring Ouzel, Red Grouse and most of the Peak District's Black Grouse population.   *Bill Smallshire*

agricultural policy of the European Economic Community – much criticised in many respects – provides grants to farmers for stocking marginal lands and 460 ha of heather moorland around the Roaches was recently enclosed and turned over to sheep at a very high stocking rate. In the long term this would have inevitably led to moorland grasses replacing the heather to the detriment of habitat diversity in general and the grouse in particular, but fortunately the Peak Park Joint Planning Board has now interceded and bought the area. The transition between enclosed fields and open hillside is often the favoured haunt of Ring Ouzels, but birds can prove elusive and are easily overlooked. Outcrops of bare rock and streams often feature in their territories, whilst in late summer post-breeding flocks can sometimes be seen feeding on bilberries or mountain ash. Patches of willow scrub on the moors host Willow Warblers and Redpolls, along with a few Blackbirds.

Although the higher slopes are virtually devoid of trees, save for an occasional stunted hawthorn or mountain ash, the steep valley downslopes are often well wooded, usually with sessile oak or birch scrub intermixed with old Scots pines, rowan, willow or larch plantations. Bilberry, cowberry, crowberry and heather occur in and around the woodland fringe and Black Grouse can sometimes be flushed from these areas, though they are much more likely to be seen in early spring when they form leks in the rough pastures. Indeed the Staffordshire section of the Peak District appears to be a stronghold for Black Grouse, which depend on its unique patchwork quilt of varied habitats and especially its small birch-scrub woodlands and luxuriant crop of bilberries. Meadow Pipits are again abundant; Red Grouse numerous; Linnets and Twite breed sparingly on the heather moor; and Whinchats can be found amongst the drystone walls and bracken. Tree Pipits occur around the woodland fringe, where they prefer the moorland sides, whilst within the woods themselves Willow Warbler is probably the commonest species, favouring especially the birch scrub. Redstarts also occur in reasonable numbers, Chaffinches are widespread, and one or two pairs of Great Spotted Woodpecker, Spotted Flycatcher, Woodpigeon, Tawny Owl and, sparingly, Long-eared Owl may be encountered. Outside the breeding season the woods are more deserted, with even the Robin apparently forsaking them in winter, but foraging parties of tits may be seen and small flocks of Fieldfare roost on the open heather or in the willow scrub. Indeed the latter species actually bred on the moors between 1974 and 1977 – an unusual event,

but one in line with the recent expansion of its range.

Streams, gullies and wet ground abound, but large areas of standing water are scarce on the moors and Tittesworth Reservoir, at 190 m, is the nearest the Region has to an upland reservoir. Set amidst upland grazing, but with a few conifer plantations around its shores, it can be easily viewed from the road which crosses it at the northern end or from the small picnic area nearby. The water itself is well-stocked with trout and heavily fished. Amongst the breeding species are Little Ringed Plover, Snipe, Common Sandpiper, Kingfisher, Sedge Warbler and probably all three wagtails; whilst winter brings Mallard, Teal, Wigeon and Pochard. Ospreys – no doubt attracted to the trout – have occurred more than once, with at least two in 1976, and in late summer Grey Herons congregate around the ample food supply. Curlew are numerous, both on passage and in summer, when they leave their moorland breeding territories to feed around the reservoir; and a good variety of other waders occurs in small numbers on passage. Away from the water, both Sparrowhawk and Woodcock breed in the conifer plantations.

The moorlands of the West Midlands may only be small, but they are very varied. Often they support a richer bird community than their more extensive counterparts elsewhere and for some species they are extremely valuable. Unquestionably they are the most distinctive of the Region's uplands.

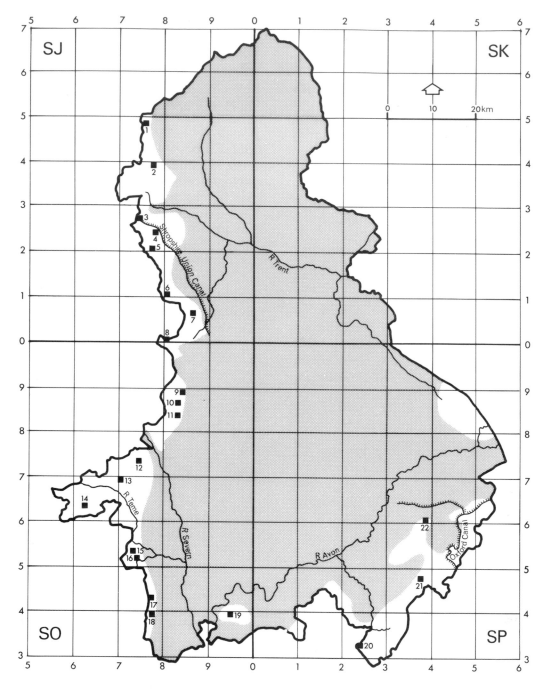

Selected localities in the Fringing Hills

1	Betley Mere	7	Chillington	13	Dumbleton Dingle
2	Maer Hills	8	Patshull	14	Kyre Pool
3	Knighton Reservoir	9	Highgate Common	15	Ravenshill Wood
4	Loynton Moss	10	Enville	16	The Knapp
5	Aqualate Mere	11	Kinver	17	Malvern Hills
6	Weston Park	12	Wyre Forest	18	Castlemorton Common

19	Bredon Hill
20	Wolford Wood
21	Edge Hill
22	Ufton Fields

8 The Fringing Hills

Inevitably from its definition, the fringe of the Region contains several separate areas of countryside, which at first appear to have little in common. Nevertheless, they have two unifying characteristics. Firstly, reverting to the analogy with the lemon-squeezer, they all combine to make the outer rim, though in truth this is chipped or cracked in one or two places, and secondly they often exhibit considerable scenic beauty. In all, these discrete areas extend to some 1,150 km^2, or 16 per cent of the total regional area, making this the third largest sub-region.

Strictly speaking the moorlands of north Staffordshire are a very significant part of this outer rim, but they are sufficiently extensive for geographers to widely regard them as a separate physiographic unit and since their bird-life is so very different as to demand separate consideration, they have been accorded a chapter of their own. It is convenient, therefore, to examine the rest of the Region's fringe by working round in an anticlockwise direction from the moors. Doing so reveals six distinct upland divisions. Firstly there are the meres, mosses and parklands along the Cheshire and Shropshire borders; then in quick succession come the heaths of Enville, Highgate and Kinver; the extensive Wyre Forest; and the beautiful wooded hills along the old Worcestershire–Hereford-shire border, which culminate in the rugged Malverns. Then follows the first major gap, or crack in the rim, through which both the Severn and Avon escape, before the wooded scarp slopes and wide arable fields of the Cotswolds form a complete barrier to the south-eastern edge. Finally there is a small, low and rather nondescript plateau which separates Avon from Soar and Warwickshire from Leicestershire, after which comes the second major gap through which the Trent and Dove invade the East Midlands. There are a few unrelated lowland areas too, such as those draining to the Tern and Thames, and for completeness these are dealt with in sequence with the uplands.

The periphery of the Region contains so much variety that it perhaps holds the key to why the West Midlands lacks a real identity of its own, since here it impinges unequivocally on that of surrounding counties. Derbyshire's dales, Yorkshire's moors, Shropshire's meres, the Welsh hills and the Cotswolds of Gloucestershire are all transgressed, albeit merely as a microcosm of their whole. Their representation may be small, but the breath-taking beauty of Dovedale, the rugged grandeur of the Malverns and the picturesque charm of Broadway all testify to their being second-to-none in beauty.

The countryside of the meres and mosses along the Cheshire and Shropshire borders, west of Stoke-on-Trent and the Meece and Sow Valleys, is extremely pleasant. In the north-west, around Betley, is a small area which drains into the River Weaver and which is really part of the great Cheshire plain. Betley Mere covers 12 ha and is an SSSI. Like Copmere, it is set in a pastoral landscape and is fringed on three sides by reed-beds and on two by broad-leaved woodland. There is also an area of wet meadow land to the north-west. Anglers and the general public cause some disturbance, but notwithstanding this, it can boast Great Crested and Little Grebes, Ruddy Duck, Mute Swan, Snipe, Curlew,

Kingfisher, Grasshopper, Reed and Sedge Warblers and Yellow Wagtail among the species that have bred. In winter it regularly holds Snipe, Water Rail, Coot and Tufted Duck, along with the occasional Goldeneye, whilst the woodlands also harbour a good range of breeding and wintering species.

South of Betley is the watershed between the Mersey and Severn, beyond which the boundary of the Region is defined by the River Tern and the dissected plateau that extends from the Maer Hills, through Ashley and southwards towards Adbaston.

Maer Hills reach a height of 200 m and command extensive views across the Cheshire plain. As such they are a potentially good vantage point to watch visible migration, and Meadow Pipits, Skylarks, hirundines and, once, six Crossbills have all been noted. Stag-headed oaks, clumps of beech, or old Scots pines surmount the hills; oak woodland has survived along the western edge; mixed woodland with birch scrub and an understorey of bracken or rhododendron along the northern edge; and some fine old beeches to the east. Otherwise the hills are covered with pine and larch plantations and a few sycamore and ash. Apart from one public footpath, they are private property. A wide variety of arboreal and scrub species breeds, including Goldcrest, Coal Tit, Sparrowhawk, Woodcock, Tawny Owl, Tree Pipit and a good range of warblers and other tits. Nightjars have bred in the past and could do so again once the forestry operations re-establish suitable areas. In winter the dense cover of rhododendrons and conifers is used by roosting thrushes and finches and amongst other things ringing studies have shown that male Chaffinches outnumber females at the start of the winter, but by March this relationship has been reversed. Brambling too are most numerous in spring. The nearby Maer Hall Pool has a stand of *Phragmites* and harbours breeding Ruddy Duck and Reed

and Sedge Warblers, whilst the surrounding woodland holds Sparrowhawk, woodpeckers, Nuthatch and Treecreeper. Even Pied Flycatchers once bred in a nest-box. Bishops Wood, though less interesting, is more accessible. Owned by the Forestry Commission, this is an extensive coniferous woodland with a mixed broad-leaved edge, mainly of sycamore, birch and maple. It is dissected by rides, some of which are also public footpaths, and its breeding birds include Tree Pipit, Jay and a good range of warblers, tits and finches; whilst thrushes are again regular in good numbers during winter.

The Shropshire Union canal takes advantage of one of the minor depressions in the outer rim to enter the Region and here, straddling the county boundary, is the small feeder reservoir of Knighton. On the Shropshire side are gently shelving banks and on the Staffordshire shore are hawthorn, gorse and a few oaks that hold winter thrushes and finches, with occasional passage migrants. Winter wildfowl include good numbers of Wigeon and a few Goldeneye, there is a small gull roost and one or two waders occur on passage.

The mosses and meres reappear again around Norbury. Of glacial origin, they were all once open water, but over the centuries many of the meres have silted up as drainage has been improved. Vegetation has then steadily encroached upon the water and dead plant material has accumulated to form peat. On the Staffordshire Nature Conservation Trust's reserve at Loynton Moss, the entire lake surface is covered by floating reed, which houses colonies of Reed Warbler and the occasional Water Rail. Surrounding the meres is a raised peat bog, which for a couple of centuries at least has been dry enough to support scrub and ultimately alder and willow carr. At Loynton many of the alders are dead and provide a valuable ecological niche for hole-nesting species such as woodpeckers, Redstart and Tree

31. Aqualate, with its extensive reed-bed backed by mixed woodland, is the best of the Staffordshire meres and holds the Region's largest heronry.   *T. W. W. Jones*

Sparrow. Behind the carr, where the ground has dried out even more, is oak-birch woodland in which Sparrowhawk and scrub warblers nest, whilst beneath is a herb layer of bracken that is host to breeding Woodcock. In winter these woodlands are also used by woodpeckers, corvids, thrushes and finches for roosting.

In general the mosses have more botanical than ornithological interest and this is especially so at Loynton, where the inflowing water is basic, whilst the raised bog is acidic, supporting *Sphagnum* moss and bog myrtle. The meres, with their greater expanse of open water, attract more birds and the largest of them, Aqualate, has been scheduled as an SSSI of national importance. This highly eutrophic water covers 60 ha, but is now barely a metre deep. Again there is a good margin of reed

and reedmace, backed by alder and willow carr, and mixed woodland which includes oak, black poplar and pine. Further habitat diversity is provided by wet grazing meadows traversed by drainage ditches and the undisturbed nature of this strictly private site makes it superb for birds. Teal, Shoveler, Tufted and Ruddy Duck all breed around the pool and Great Crested and Little Grebes, Water Rail and Reed Warbler in the reed-beds. Snipe, Curlew, Redshank and Lapwing nest in the wet meadows, whilst the adjoining woodland holds a long-established heronry which, with around 75 pairs, is the largest in the Region. Both Bittern and Bearded Tit can be counted amongst interesting passage birds, whilst in winter the mere holds up to 2,000 duck, with the shallow water and margins of reed favouring surface-feeders.

Canada Geese and Wigeon regularly number around 400, Shoveler have exceeded 500 and Goosander 100, and there is usually a small Cormorant roost at this season.

Superb as the meres and mosses are, it is the parklands of Aqualate, Chillington, Patshull and Weston that tend to steal the ornithological limelight. Here, in the quiet seclusion of these private estates, are a number of unusual breeding species, including several pairs of the shy Hawfinch and on occasions Buzzard and Pied Flycatcher. To avoid disturbance to birds and property, however, it is prudent not to disclose their precise whereabouts. All four estates contain a combination of lakes, mixed woodland and open parkland. At Aqualate the mere and its surrounding woodland are integral to the park, but there is also open grassland with scattered beech and oak and several Wellingtonias. Water is a central feature of Chillington's 400 ha too, but here the pool is an artificial one contrived by "Capability" Brown as part of his improvements to the estate two centuries ago. By now the trees on the island have matured and there are extensive reed-beds around the shoreline. West of this pool is a large, mixed woodland, known as Big Wood, and to the east is a canalised extension of the lake, fringed with alder, and set in a parkland of small woods, arable fields and scattered trees. Oak is the dominant tree in the woods, with beech, chestnut and sycamore as secondary species. There are also a few scattered yews, whilst the conifer plantations are largely larch and spruce. Unfortunately the southern extremity of the park is affected by the proposed M54 motorway, but how seriously this will disturb the bird life remains to be seen.

Patshull also has a large lake with areas of reed and marsh. There are many fine beeches amongst the small woods and scattered parkland trees and rhododendron and bramble are common in the shrub layer. Amongst the conifer plantations on the eastern part of the estate are several stands of spruce. The lake is used for fishing and the western part of the park is being developed as a leisure complex with golf course and holiday chalets. This is likely to have an adverse effect on bird life, especially the shier species. At Weston Park, which is the ancestral home of the Earl of Bradford, there is already public access to the house in summer and a country park in the grounds. Here the main habitat is open parkland, with trees in clumps or as isolated specimens. Woodland is less extensive, but what there is is used for Pheasant rearing. Water is less significant as well although there are two small pools within the park and a larger one, fringed with woodland, outside, but still within the estate, at White Sitch.

An extremely varied avifauna includes breeding Great Crested and Little Grebes, Shoveler, Teal, Ruddy Duck and Canada Goose on the lakes; Grasshopper, Reed and Sedge Warblers in the reeds or marginal vegetation; Snipe in the marshy areas and Sparrowhawk, owls, Woodcock, all three woodpeckers, Redstart, warblers, including Wood Warbler and finches in the woods. Nightjar has occurred in the past, but does so no longer, while Chillington has a small heronry.

In winter the alders and birches attract feeding flocks of tits and finches, especially Redpoll and Siskin, whilst Chaffinch and Brambling gather to feed on mast beneath the beeches. Large thrush and finch roosts, especially Chaffinches and Greenfinches, also form in winter, when the lakes hold many duck, notably Tufted, Shoveler and a few Goldeneye. Small numbers of Cormorants use Chillington as a roost. Although there is little opportunity for the casual watcher to explore these parks, many of the typical woodland species can be seen from the footpath along the Lower Avenue leading from Chillington towards the Shropshire Union canal. There are a few hornbeams here and this is as good as anywhere to look for Hawfinches in the winter or early spring.

Further south the landscape changes yet again as the Triassic sandstones of western Staffordshire give way to an outcrop of Bunter pebbles and Permian beds centred on Enville. Characteristically this outcrop was once covered in heathland, but since the last war much of it has been cleared and afforested with conifers. Some good tracts of heather with scattered birches have survived at Highgate Common, but unfortunately this area of 115 ha is now an intensively used country park and human pressure and disturbance have squeezed out the more sensitive species. Tree Pipits still breed, however, and have been observed nonchalantly feeding young within a few feet of a car park. Surprisingly a Nightjar has been heard within the last year or two and a migrating Hobby made a brief appearance, but the prospect of Woodlark ever returning to this once-favoured haunt seems slim indeed. South of the Common, on the golf course, are some mature pines, a small oakwood and an alder and poplar lined valley, which attract a variety of passerines.

Similar habitat was found at Enville Common until the early 1950s, when the Forestry Commission undertook extensive afforestation with Scots and Corsican pines and both larches. Known as "The Million", this area augments the many established, mature Scots pines and larches that are so characteristic of the district. Fortunately some old oaks, birches and small patches of heather have survived and there is an adjoining oak-birch wood with a field layer of bracken that adds more diversity. When young, the plantations were a Nightjar stronghold, with as many as fifteen pairs in 1966, but now the trees have grown too large and the species has deserted the area. It is indeed a pity that the whole forest is of uniform age, since with a varied age structure the Nightjar might still

be present. Nevertheless, for an area dominated by conifers, there are some interesting birds. Several Woodcock rode, Redpoll and Goldcrest breed in good numbers, Long-eared Owls have bred and the older pines are a likely place to find Crossbills after irruptions. Indeed breeding occurred here three times between 1959 and 1964, which is quite exceptional for the West Midlands. Sparrowhawk also breed regularly whilst winter brings good numbers of tits and Goldcrests. Amongst the unexpected was a pair of Montagu's Harriers in the spring of 1951. Nearby, the secluded parkland of Enville Hall, with its oaks and beech woods, provides a good, quiet sanctuary where up to thirty pairs of Grey Herons nest and Hawfinch, Buzzard, Raven and Redstart might be seen.

Another sandstone ridge extends southwards from Kinver into Worcestershire. This too is clothed with birch woods of all ages, sometimes in association with rowan or thorn; or stands of pine, larch, spruce or sycamore with a field layer of bracken. Some old oaks have survived as well. The National Trust, Staffordshire County Council and Hereford–Worcester County Council all own parts of Kinver Edge southwards as far as Kingsford Country Park and they provide the public with access to large areas. Meadow Pipits breed on the open grasslands and Wheatear have done so in the past, Linnets frequent the scrub areas, especially patches of gorse, and Woodcock, Redstart and Tree Sparrow inhabit the areas of scattered birches. The ridge was once *the* place for Woodlark, but it has not occurred regularly since the early 1960s, though Nightjar still manages to maintain a tenuous hold.

West of the Severn the Old Red Sandstones give rise to some of the most beautiful countryside in the Region. This is an area of distinct character, reminiscent of the Welsh border country, with fast-flowing rivers, steep-sided valleys clothed

32. The WMBC and the Worcestershire Trust have a joint reserve in the Wyre Forest, where Redstarts, Wood Warblers and Pied Flycatchers are amongst the breeding species.

A. Winspear Cundall

in deciduous woodland, orchards and hop-yards on the gentler slopes and open hillsides culminating in the Malverns. There is a characteristic avifauna, which is distributed throughout the whole area and not just restricted to a few sites.

It is here, amongst the sheep walks and scattered deciduous woods, for example, that Raven and Buzzard might be seen soaring overhead; where Pied Flycatchers, Redstarts and Wood Warblers add colour and song to the woods each summer; and Dippers and Grey Wagtails haunt the bubbling streams.

There are some sites of particular importance, notably the Wyre Forest, which is the largest expanse of woodland in the Region. The Forest lies immediately west of the Severn, near Bewdley, on the sandstones, marls and conglomerates of the coal measures. Needless to say, it is yet another remnant of ancient hunting forest, dominated by oak, but with some beech, birch, holly and yew; and a varied under-storey which includes hazel, hawthorn, elder, blackthorn, bramble, rose, heather and bracken. Inevitably much clearance has occurred in the past and since the last war the Forestry Commission and private interests have replanted large areas with stands of spruce, pine, larch, hemlock and redwoods. Nevertheless the large tracts of broad-leaved woodland support a sufficiently interesting and varied flora and fauna to have warranted designation as a National Nature Reserve. In addition a further tract of oak woodland has just been secured as a nature reserve by the

Worcestershire Nature Conservation Trust and the WMBC, thanks largely to a legacy from the late Dr. L. F. Dale.

The Forest is divided into two by Dowles Brook, which marks the county boundary. Only the area to the south is really in Worcestershire and the Region, though this is a distinction which is seldom observed by bird-watchers. Strictly speaking the bottom of this steep-sided valley belongs to the Severn sub-region, though this distinction is even more arbitrary and has been ignored in favour of treating the Forest as an entity. The soils are mainly free-draining and acidic, so the sessile oaks that dominate the old coppiced woodlands have a calcifuge field layer of bracken, bilberry and heather, which is favoured by the Wood Warbler. Valleys like the Dowles exhibit more variety, however, with ash, wych elm, small-leaved lime, alder and wild service; and a better-developed shrub layer which holds birds like Garden Warbler and Blackcap. The field layer varies as well. In the wetter areas there are acidic communities which include *Sphagnum* moss and *Molinia*, whilst at the other extreme some of the clay soils support base-rich communities including dog's mercury and primrose under ash, hazel and dogwood. Ecologically it is indeed a fascinating habitat, where the sessile oaks show characteristics intermediate with pedunculate oaks, and features of Welsh and East Anglian woodland exist side by side. For the ornithologist it is further enriched by the streams, with their gravel and boulder-strewn beds and high, overhanging banks where Dipper, Kingfisher and Grey Wagtail breed; and by the small pastures and apple, plum or damson orchards along the valley floors, which provide the ideal blend for Pied Flycatchers and Redstarts. The Pied Flycatcher in particular has benefited from a major nest-box scheme, which was initiated some years ago and currently has nearly two-hundred boxes producing five-hundred nestlings of a variety of species each year. The more mature woodland, and especially the standard oaks, provide sites for many hole-nesting species like Tawny Owl, all three woodpeckers and many tits. Sparrowhawk, Woodcock, corvids, warblers and finches all breed in very good numbers and there are a few pairs of Hawfinch. Golden Oriole and Black Stork are amongst the unexpected species that have occurred. In winter the woodland holds many thousands of roosting corvids, thrushes and finches. Even the conifers are not without interest. Good numbers of Turtle Doves, Coal Tits, Goldcrests and Redpolls breed, Siskin occur in winter and are increasingly staying into spring and Crossbills are recorded as often here as anywhere in the Region.

Some of the Forest is open to the public and its popularity for recreation is rapidly increasing. Despite the growing human disturbance, most of the species characteristic of the area can still be seen from these public places and especially from the footpath alongside Dowles Brook. Fortunately there are detached woods like Eymore Wood, Far Forest and Rock Coppice, which are adjacent to the Forest proper and possess similar habitats, but which are private and secluded and offer better sanctuary for many breeding birds.

Further south and west the steep-sided valleys of the Teme and its tributaries are also clothed in woods and orchards, especially around Stanford and Alfrick. Some are conifer plantations, but many are sessile oakwoods with an avifauna similar to that of the Wyre Forest, though obviously their diversity is restricted by their smaller size. Few have been studied in any detail, but their breeding birds typically include Sparrowhawk, Woodcock, Tawny Owl, all three woodpeckers, Nuthatch, Redstart, many warblers, Tree Pipit and an occasional pair of Pied Flycatchers. Buzzards and Ravens are also noted during the breeding season and may occasionally nest in one or two places. At Menith

Wood, which is part of the Dumbleton Dingle SSSI, a nest-box scheme has been in operation since 1972 and boxes have been occupied by Pied Flycatcher, Redstart, Nuthatch and good numbers of tits. Here too, Snipe breed in the wetter meadows and Dipper and Grey Wagtail along the stream.

Within the same SSSI, the Worcestershire Nature Conservation Trust has a reserve of 25 ha at Hunthouse Wood. This is a remarkable old, deciduous woodland clothing several deep, steep-sided valleys. Once there was a coal mine in the centre of the wood, but there is little remaining evidence of it today. Oak and ash are foremost amongst a good variety of broad-leaved trees, though birch dominates where the larger trees were felled during two world wars. Several species of warbler breed, all three woodpeckers are present and Tawny Owl, Sparrowhawk and Buzzard have all been recorded.

The woodland bird-life of the two Worcestershire Nature Conservation Trust reserves at Ravenshill and Knapp-Papermill is very similar. Ravenshill is a mixed wood with stands of oak, ash, birch, alder, poplar and softwoods on a down-slope to a small stream and tiny pool. The shrub layer is well developed with hazel, dogwood, guelder rose, spindle and bramble and is sufficiently dense to hold Nightingale. Grasshopper Warblers also breed here and in winter good numbers of Redpoll and thrushes occur along with a few Siskin. Deciduous woodland covers the steeper slopes at the Knapp-Papermill as well, but rough pastures, orchards, and a delightful stream with fast-flowing shallows and deeper, slower reaches, add to the habitat variety. These provide the right conditions for Barn and Little Owls, all three wagtails and both Dipper and Kingfisher, together with the usual woodland birds. Winter again brings thrushes, Redpoll and Siskin, plus an occasional Hawfinch; whilst a young Cormorant once appeared on the stream.

In this rolling countryside standing water is scarce, but there are small lakes at Stanford Court and Kyre Pool. The latter is a secluded, alder and willow-fringed pool, with a backcloth of mature beech and stands of conifers along the northern shore. Mallard, Coot and Great Crested Grebe all breed around the pool and Kingfishers nest in its steep banks, whilst winter brings a few Tufted Duck. The surrounding woods hold Green and Great Spotted Woodpeckers, Nuthatch, Marsh Tit, Treecreeper, warblers and Goldcrest.

Pleasant though these waters are, the hills quickly recapture attention. Not all are clothed in woodland, especially the main range which begins at Abberley and passes southwards through Woodbury and Berrow Hills to culminate in the bare outcrop of the Malverns. In all, this range is some 34 km long, with the Malverns themselves stretching for 13 km. There is some woodland at the eastern end of Abberley, where Tree Pipits may be found, and on the lower slopes of the Malverns, where Wood and Garden Warblers and Blackcap are amongst the breeding species.

The Malverns are composed of Archaean or pre-Cambrian rocks, with Silurian deposits along their western flank. They reach a height of 425 m and are devoid of trees on their summits, having a short grass sward or blanket of bracken. Erosion from the feet of countless visitors is a serious problem in certain areas, but despite having been a popular haunt of day-trippers for years, a fascinating, and in some respects unexpected, avifauna has lingered on around the hills and this has been well documented by Palmer-Smith (1978). Skylarks and Meadow Pipits are still both common on the short summit grasslands and the latter sometimes falls victim to the Cuckoo. Up to six pairs of Wheatears used to nest around British Camp and there may still be a pair or two on the hills. Woodlark used to frequent those areas with short grass and scattered

33. The Malvern Hills rise dramatically out of the Severn plain, whilst in the foreground Castlemorton is one of the district's few surviving commons.   *P. Wakely*

trees, but they have certainly now disappeared, as have Nightjars from the bracken-covered slopes. Even so Tree Pipits and a few Redstarts and Grasshopper Warblers still survive in such places. Linnets and Yellowhammers are common in areas of scrub and the Green Woodpecker is widespread, feeding amongst the plethora of anthills in the short turf and often nesting in orchards around the foothills – a niche occupied by the Wryneck before the last war. On the credit side, Palmer-Smith said Stonechats had once been common, but had since deserted the hills. However, following the recent succession of mild winters, one or two pairs have again nested and it seems this species recolonises the hills at times of high population. It is of course regularly seen on passage. Autumn also brings a

strong southerly passage of thrushes, especially Song Thrush, and amongst these Ring Ouzels are regular in small numbers. Kestrel is the common predator of the hills, but Buzzards have appeared from time-to-time and Ravens are seen more often here than anywhere else in the Region. One was once observed in aerial combat with a Peregrine. A few rarities like Snow Bunting and Shore Lark have been reported, attracted perhaps by the altitudinal habitat which is unique to this area. For many years, though, the real speciality of the hills were Cirl Buntings, a pair of which often nested in or around the old Wyche Quarry, with the cock singing regularly from the car park. Sadly there have been no records since 1972 and it now seems that this species, too, has disappeared. A similar fate has also befallen the Red-backed

Shrike, which used to nest regularly on the hills or adjoining commons, but does so no longer. However, its loss has been partially offset by the increasing incidence of Great Grey Shrike in winter, especially on Castlemorton Common. Commons were once typical of the foothills around Malvern, until, like Old Hills, they were ploughed up during the last war. Castlemorton somehow survived and here, notwithstanding attempts at control, the grazing land is constantly being encroached upon by dense patches of gorse and bramble, which hold Linnet, Yellowhammer, Whitethroat, Whinchat, Stonechat and Grasshopper Warbler. A small stream traverses the common and this is flanked by a marshy area of rush and sedge, which dries out in summer, but is wet in winter, when it is visited by Snipe and the odd Jack Snipe and Water Rail. Other waders occasionally drop in on passage.

From the Malverns the long line of the Cotswolds can be seen across the Severn and Avon Valleys. Comprising beds of Middle and Upper Lias with cappings of Inferior Oolite, this scarp runs from southwest to north-east and provides a strong physical definition to the Region. Although it dominates the Region's landscape for over 50 km, much of its length, notably the higher, south-western end, actually lies in neighbouring Gloucestershire and only in the north-east, when it reaches Warwickshire, does it strictly become part of the Region. Nevertheless there are some significant outliers within both Warwickshire and Worcestershire.

The most dominant of these is Bredon Hill, which is an enormous dome rising to a height of 293 m and measuring 6 km across. To any historian Bredon is full of interest. Even today, surveying the panorama from its summit, it is not hard to understand why Housman was so inspired by this view. All the main communications by-passed the hill and time seems to have done the same to the ring of neat little villages around its footslopes. The contrast between the south side, with its charming Cotswold stone houses and walled gardens, and the north side, with its half-timbered and thatched cottages and gardens enclosed by hedges, could hardly be sharper. Most of the hill is farmed, being either under arable cultivation or grazed by sheep. Many of the shallower, southern slopes are under new grass leys, but the steeper slopes have rough pastures, scrubland or small mixed woods and there are several small parklands and one or two quarries. For a predominantly agricultural area the birdlife of Bredon is good. Parklands, like Overbury, with their specimen oaks, ash and chestnuts, hold a variety of hole nesting species including Stock Dove, Tawny and Little Owls, Nuthatch and woodpeckers. Pheasants are found in the woods and both Grey and Red-legged Partridges in the open fields. Skylarks, Meadow Pipits and Lapwings breed on the summits and Redstarts in the scrub and woods, but Woodlark is no longer present and it is well over a hundred years since Stone-curlew nested on the hill. Brambling are often present in winter and amongst the unusual, both Dotterel and Black Redstart have occurred on passage.

For the West Midlands it is a fortuitous quirk that Broadway, the gem of the Cotswolds, is in Worcestershire not Gloucestershire. Known as the "Painted lady of the Cotswolds" – alluding surely to the butterfly of that name – this must be one of the best known and most beautiful villages in Britain, where scarcely a building is misplaced. Behind the village, Broadway Hill, at 319 m, is one of the highest points of the Cotswolds and from the hunting tower on its summit there are vast panoramic views across the Vale of Evesham. Somewhat surprisingly few records are ever received from this area, but large flocks of Rooks are a familiar sight on the arable fields of the dip slope, finches are numerous at all seasons and

Corn Buntings breed in one or two places. Again Stone-curlew bred here in the early nineteenth century.

Further to the north-east, in Warwickshire, is another prominent outlier of the scarp, Meon Hill. Rising to a height of 194 m, the hill is surmounted by a group of beeches and its steep slopes are also dotted with oaks, limes and chestnuts. Simms (1949) described autumn migration across the nearby Stour Valley and showed that birds were heading for this hill and were orientated along the scarp. During September, Skylarks, Meadow Pipits, Pied Wagtails, Yellowhammers and finches were all noted moving south-westwards, with daily numbers often exceeding a thousand. Peak passage occurred in mid-October, however, by which time Yellow Wagtails, Mistle Thrushes, Goldfinches and a Merlin had joined the larger numbers of Chaffinches, Greenfinches and Yellowhammers. These movements then fell away rapidly towards the end of October. Although we now know that migration takes place on a much broader front than was realised thirty years ago, there can be little doubt that, under certain climatic conditions, it is advantageous for birds to follow the scarp rather than cross it and, with the exception of Merlin, the same species could just as easily be seen today.

Beyond Meon Hill the main scarp is displaced south-eastwards by some 15 km and is so heavily dissected by the headwaters of the Stour and its tributaries that its strong physical form is lost in a landscape of isolated, rolling hills like those above Ilmington and Brailes. On the higher ground, where hedgerows give way to drystone walls, Corn Buntings occur, whilst on the permanent pastures of the steeper slopes Lapwing breed and Grey Partridge are widespread. Across this rolling countryside are a few scattered woodlands, like those at Whichford and Wolford. None is outstanding, but most hold a good range of woodland birds. For example,

Wolford, which is principally an oak-birch wood with some ash and an understorey of bramble and old hazel coppice, holds breeding Barn, Little and Tawny Owls, all three woodpeckers, Nightingale, most warblers including Grasshopper Warbler, Tree Sparrow and a good range of tits and buntings.

The main scarp then continues north-eastwards into Northamptonshire, becoming progressively more broken by cols towards the north-east. Here the hills never reach the heights of those further south and they are capped by marlstones rather than limestones, which yield deep, rich brown ferritic soils and a similarly coloured building stone. Even so, the long, straight scarp of Edge Hill, with its hanging beech woods overlooking the battlefield below, is still a significant feature. That it has remained under woodland is probably due to its being too steep even for grazing. The bird-life of these woods has not been thoroughly studied, but it seems less exciting than might have been expected, perhaps because of disturbance from people and forestry operations. Nuthatch, Treecreeper and all three woodpeckers are certainly present, along with good numbers of tits, but some anticipated species like Wood Warbler appear to be absent. Behind the scarp, on the dip-slope, the rich brown soils are widely used for growing cereals and in places ironstone is quarried. This is an open landscape of large fields and an occasional hedgerow tree, where Woodpigeons, Crows and Rooks rule supreme and where Quail may sometimes be heard calling on summer evenings.

On such high ground, standing water is scarce, but there are small, ornamental pools at Upton House and Farnborough Hall and a canal-feeder reservoir at Wormleighton. The latter feeds the Oxford canal, which is heavily trafficked in summer and one of the first canals to suffer water shortage. By late summer the reservoir water-level is regularly low, exposing good areas of mud. These conditions attract small numbers of passage waders and both Great Crested Grebe and Tufted Duck breed. All three waters hold small numbers of duck in winter, with Mallard in particular congregating at Farnborough.

Once again the outliers of the scarp are more interesting, especially the low plateau centred on Gaydon. This consists primarily of Blue Lias shales and limestones with a capping of boulder clay. The dip in the strata produces scarps along the northern and western faces and at Ufton this is clothed with superb oak-hazel woodland that is still coppiced. The mature oak standards house a number of hole-nesting species like Stock Dove, Nuthatch and woodpeckers. The shrub layer is dominated by hazel, which is coppiced in rotation, so that all age classes are present, but it also includes hawthorn, wild rose, wayfaring tree and dogwood. Bramble dominates the ground flora on the higher and drier parts and dog's mercury, bluebell and primrose carpet the lower slopes. The vigorous growth of hazel from its coppice stools favours Nightingale and up to ten pairs are present, whilst the varied shrub layer in general is attractive to warblers like Blackcap, Garden Warbler and Chiffchaff. Tawny Owl and Woodcock also breed in this wood, which is scheduled as an SSSI of national importance and which is extremely private.

The soils of this plateau are often thin, with bands of White-Lias limestone only a few inches beneath the surface. In places the limestone has been quarried for the cement industry, leaving a variety of derelict sites ranging from shallow, dry excavations to extremely deep and wet ones. Where the ground has been disturbed the soils are highly calcareous and their flora, especially their orchids, is of considerable interest. Fortunately for wildlife there has been little attempt at restoration, partly because the cost far exceeds the value of the land afterwards, and many areas have been allowed to revert to nature, some of them for the best part of twenty-five years. Consequently there is now a unique mixture of downland, woodland and aquatic species. Hawthorn is invasive and quickly forms a dense, impenetrable scrub which is used by feeding and roosting thrushes. Large numbers of Starlings also roost along with a few Reed Buntings, wagtails and finches and as many as fourteen Long-eared Owls have been recorded in one hawthorn thicket. Shallow depressions lie wet in the winter, providing feeding areas for Snipe, Lapwing and Woodcock, and some even attract passage waders. In summer, though, they are mostly dry and Yellow Wagtails breed along with a good range of warblers, which includes Reed and Sedge Warblers where there is suitable vegetation. The manner of extraction has sometimes left deeper, parallel furrows, which have filled with standing water, and intervening ridges, which with their high rodent population are a rich feeding ground for Barn and Short-eared Owls in winter. Elsewhere quarrying has been so deep as to leave pools backed by sheer rock faces 20 m high, with holes and ledges which provide "natural" nest sites for Jackdaw, Stock Dove and Kestrel. The open water brings a few breeding waterfowl like grebes, Mallard, Canada Goose and Coot. On the Local Nature Reserve at Ufton Fields, which is managed by the Warwickshire Nature Conservation Trust, there are good numbers of Grasshopper Warbler and Turtle Dove, whilst Nightingale has appeared recently – a sure indication of the succession into scrub and developing woodland. Winter brings good numbers of thrushes to the reserve, a small roost of Corn Buntings and the occasional Great Grey Shrike.

The final part of the perimeter rim is the low plateau north of Rugby, which marks the watershed between the rivers Avon and Soar and separates Warwickshire from Leicestershire. This is a nondescript plateau with gentle slopes from its summits into broad, shallow valleys. Glacial action has left a thick covering of chalky boulder clay, with flints, and patches of glacial gravels. This supports mixed farming in a countryside where ash is the dominant tree – the more so since elm disease struck – and fox coverts and brakes are noticeable landscape features.

Most of these coverts and brakes comprise scrub oak and ash, with an understorey of hazel, elder and bramble; or dense thickets of hawthorn and blackthorn. Between them they house breeding Woodpigeon, Turtle Dove, Blackcap, Willow Warbler, Chiffchaff, Treecreeper, Jay, Magpie, Rook and a variety of tits and finches, whilst in winter they are thronged with roosting flocks of Starling, Redwing and Fieldfare. At this season a few Reed and Corn Buntings can usually be found too, along with an occasional hunting Sparrowhawk. Yellowhammer, Skylark and Lapwing breed on the neighbouring farmland and the latter is also present in large numbers in winter along with small numbers of Golden Plover at a few favoured localities and the occasional Short-eared Owl. Amongst the forty-two species recorded on a CBC plot at Willey in 1978 the commonest breeding birds were Dunnock, Blackbird, Wren, Robin, Chaffinch, Skylark, Yellowhammer, Blue Tit and Willow Warbler in descending order, and it is perhaps fitting to conclude this review of the Region with a community which is typical of so much of it.

9   The Systematic List

The systematic list follows the sequence in the *List of Recent Holarctic Bird Species* (Voous 1977). Each species account begins with a brief statement of its current numerical status and seasonal and geographical distribution. The numerical status is based on that established for breeding species in the *Breeding Birds of Britain and Ireland* (Parslow 1973) and adapted for the *Atlas of Breeding Birds of the West Midlands* by Lord and Munns (1970), who reduced each of Parslow's ranges to 2 per cent which is the percentage of the British and Irish land area that the West Midlands covers. The numerical status, based on the period 1969–78, has now been extended to cover both winter visitors and passage migrants. The definitions used are given below.

Most of the historical records have been abstracted from *The Birds of Staffordshire* (Smith 1938), *The Birds of Worcestershire* (Harthan 1946) and *Notes on the Birds of Warwickshire* (Norris 1947), which together with the *Atlas of Breeding Birds*

of the West Midlands (Lord and Munns 1970) and *The Atlas of Breeding Birds in Britain and Ireland* (Sharrock 1976) were the standard works of reference. To save space, in-text references to these works have been abbreviated to just the authors' name, WMBC *Atlas* or BTO *Atlas* respectively, or simply the *Atlas* surveys.

The main analysis covers the fifty years 1929–78, referred to as the fifty-year period or period under review, but has been kept up-to-date by the inclusion of relevant data up to December 31st 1980. The species' accounts summarise all available information on preferred habitats or localities, distribution, population, density, flocks, roosts, migration and ringing recoveries. *West Midland Bird Reports* Nos 1–46 (1934–79) have provided most of the data and records from this source have not been specifically referenced. Acknowledgements are also due to the Wildfowl Trust and the British Trust for Ornithology for supplementary data on wildfowl counts and on Common Bird Census and ringing

	Annual Number of Breeding Pairs	Annual Winter or Passage Population
Very rare	More than 1 but less than 6 records ever.	
Rare	More than 5 records but less than annual.	
Very scarce	1 or 2	Up to 5
Scarce	3–20	5–50
Not scarce	20–200	50–500
Fairly numerous	200–2,000	500–5,000
Numerous	2,000–20,000	5,000–50,000
Abundant	20,000 +	50,000 +

recoveries respectively. Mean and extreme arrival and departure dates are given for all migrants and these have been computed from the first and last dates given in the WM *Bird Reports*. For less common species, histograms show the five-yearly and monthly distribution of records, generally based on dates of arrival rather than presence, whilst for species that have occurred less than 20 times records are listed in full. Since 1958 records of nationally rare species have been assessed by the *British Birds* Rarities Committee, whilst these and local rarities have always been subjected to scrutiny by the WMBC. A few records are still pending and these are noted as being "subject to acceptance by the *British Birds* Rarities Committee". Otherwise all the records that are included have been accepted by these bodies. Observers' names have been omitted throughout.

Wherever possible, breeding populations have been obtained from specific surveys. Otherwise estimates have been made using data from the WMBC and BTO *Atlases* and the Common Bird Census. These have been done in three ways:—

i) by multiplying the number of 10-km squares in which a species was proved or suspected of breeding (1966–72) by the density per square assumed in the BTO *Atlas*.

ii) by multiplying the mean 1978 national CBC densities for farmland and wood-land (*Bird Study* 27:35–40) by the area of land devoted to each within the Region (5,657 km^2 and 337 km^2 respectively) and then adding an allowance for the population in other habitats. (Farmland densities were in fact halved to allow for most territories being around the perimeter.)

iii) as in (ii), but using the mean 1978 regional CBC densities from nine farmland and six woodland plots.

None of these provides more than a rough estimate and the use of the regional CBC data is dubious because of the small sample size. Nevertheless it could reflect any real regional differences.

These three estimates were then used to define a probable breeding range, though any that were markedly different from the other two were ignored. Inevitably the range is wider for some species than others, and in some cases estimates by all three methods were not possible because of lack of data.

Plant names throughout this book follow those in the *Flora of the British Isles* (Clapham, Tutin and Warburg 1962), except in the case of cultivated or exotic species not included in that work. Botanical plant names and their source are given in Appendix IV. Finally, all place names are generally based on those that appear on Ordnance Survey 1:50,000 maps and Appendix VI contains a gazetteer with four-figure national grid references.

Red-throated Diver
Gavia stellata

Very scarce and irregular winter visitor and passage migrant.

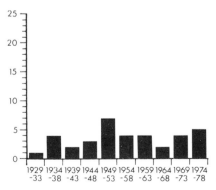

Smith, Harthan and Norris all described the Red-throated Diver as an intermittent visitor to the West Midlands during the nineteenth century and earlier years, with a majority of records from the rivers Avon, Severn, Tame and Trent. There were a few records during the early years of this century, including one at Bittell in November 1926, while between 1929 and 1978 the species appeared on 33 occasions, involving 36 birds. The Red-throated was thus the scarcest of the three divers visiting the Region over the fifty-year period as a whole, though during the last decade the Black-throated Diver became the least frequent.

Five-yearly Totals of Red-throated Divers, 1929-78

A marked increase in the visits of divers occurred during the 1950s, followed by a decline during the subsequent decade. This trend was dramatic in the cases of the two larger divers, but less pronounced in the present species, for which the records display a comparatively well-dispersed temporal distribution. Three birds appeared in each of the years 1951, 1952, 1972 and 1976, but the species occurred in successive years on only four occasions (1951-2, 1955-6-7 and 1961-2).

The Red-throated Diver has appeared in all months except August and September. Over 80 per cent, however, occurred between October and March and within this period the records are remarkably evenly spread, with no one month accounting for more than 16 per cent of birds.

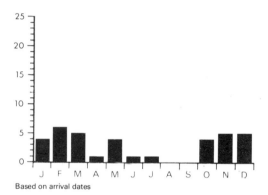

Based on arrival dates

Monthly Distribution of Red-throated Divers, 1929-78

There is only one April record but four during May, perhaps indicating overland spring passage, although the even spread of records during the autumn and winter months suggests that most occurrences are related to random and hard-weather movements rather than to regular seasonal movements. Several records have involved oiled or sickly birds.

Most individuals have stayed for less than a week, and many for only one or two days, though a bird remained at Kingsbury from February 10th to May 2nd 1951.

Not unexpectedly, the most frequently visited localities during the fifty years were the larger, well-established Staffordshire reservoirs with five records (six birds) at

153

Table 3: Duration of visits of Divers, 1929–78

Species	Number of days			
	1–9	10–19	20–29	30 and over
Red-throated Diver	29	3	3	1
Black-throated Diver	23	7	7	4
Great Northern Diver	20	8	2	15

Note: An injured Black-throated Diver held in captivity during early 1963 is excluded from this table.

Blithfield, five at Belvide and six at Chasewater, but elsewhere Bittell produced four birds and Bartley three in the same period.

Although falling outside the period of detailed analysis, it is worth mentioning here the remarkable influx of divers (and Red-necked Grebes) inland into Britain during the very cold weather of February 1979. In the West Midlands five Red-throated Divers were recorded between mid-February and mid-March, including two at both Bartley and Draycote.

Black-throated Diver
Gavia arctica

Very scarce and irregular winter visitor. One record of summering.

The earliest records of Black-throated Diver came from Staffordshire in 1896 and 1902 (Smith). There were then 37 records between 1929 and 1978, involving at least 42 birds. Nominally, therefore, this species was the second-most numerous of the divers during the fifty-year period, but this impression results entirely from a remarkable incursion during the 1950s.

Prior to 1950 and from 1964 to 1978, it was in fact the scarcest of the three. Of the 42 birds, no less than 32 (76 per cent) occurred between 1950 and 1965 and, during this period, only in 1953 and 1961 were there no fresh arrivals. By contrast,

there was only one record during the 1940s and one (of two birds) between 1966 and 1973 inclusive. During the last few years there have been signs of a revival, though this has been less marked than in the case of the Great Northern Diver. The reasons for these marked fluctuations are not readily apparent, although there is a partial but not consistent correlation with cold winters. A comparable though somewhat less pronounced situation is apparent, however, in records from the London area (Chandler and Osborne 1977).

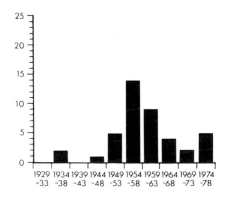

Five-yearly Totals of Black-throated Divers, 1929–78

During 1960 a Black-throated Diver remained at Chasewater from May to September and the species has in consequence been present in all months. All arrivals, however, have been between October and May, with a peak in December and January. 60 per cent of birds appeared in these two months compared with 47 per cent of Great Northern Divers and only 25 per cent of Red-throated Divers.

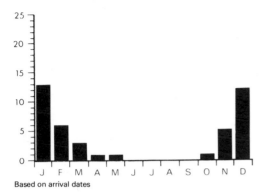

Based on arrival dates

Monthly Distribution of Black-throated Divers, 1929–78

Belvide and Chasewater have been the most favoured localities, with ten and nine birds respectively, while there have been five at Blithfield since 1955. Outside Staffordshire, Bittell with three records, and Bartley, Draycote and Earlswood with two, were the only localities visited more than once during the fifty years. Both occurrences at Draycote, a large and relatively new water, involved two birds. (This reservoir produced three further records during the cold weather early in 1979, while in mid-February two were present at Bartley and single birds at at least three other localities.)

Great Northern Diver
Gavia immer

Very scarce and irregular winter visitor and passage migrant. One record of summering.

Smith, Harthan and Norris documented about 15 records prior to 1929, including one as far back as 1678 in Warwickshire.

With 42 records involving 45 birds, the Great Northern Diver was the most numerous of the divers visiting the Region between 1929 and 1978, though it was outnumbered during the late 1950s by the Black-throated Diver. Records increased markedly during the 1950s, declined precipitously, and then resurged strongly in the mid-1970s. The species appeared annually from 1954 to 1960, and again from 1974 to 1978.

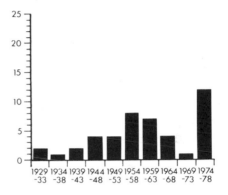

Five-yearly Totals of Great Northern Divers, 1929–78

Arrivals of this species tend to be concentrated at the end of the year and decline through the winter months; 31 per cent of birds appeared during November. Rather than birds being driven inland by adverse conditions, this pattern suggests a deliberate penetration inland at the time of the immigration of winter visitors into British coastal waters and this species, in fact, displays a much greater tendency to make protracted stays in the Midlands than the other two divers (see Red-throated Diver). There have been no arrivals between April and September inclusive, although a bird which appeared at Shustoke in February 1948 remained until the following December.

During the fifty years, nine birds occurred at Chasewater, seven at Blithfield and Draycote, five at Bartley and Belvide, and three at Bittell and Shustoke. In February 1956 two Great Northern Divers

155

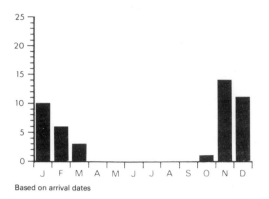

Based on arrival dates

Monthly Distribution of Great Northern Divers, 1929–78

were present at Blithfield at the same time as two (possibly three) Black-throated Divers, while in 1977/8 four Great Northern Divers wintered at Draycote.

Little Grebe
Tachybaptus ruficollis

Widespread and fairly numerous breeding species. Autumn and winter concentrations on many waters.

The Little Grebe was described by both Smith and Norris as widely distributed whilst Harthan remarked on an autumn influx of passage or wintering birds. It is still a widespread breeding species, absent only from the highest moors, the dry heaths around Enville, parts of the Teme Valley and the Stour Valley in Warwickshire. During the WMBC *Atlas* project (1966–68) breeding was confirmed or suspected in fifty-one 10-km squares (66 per cent of those surveyed) and this distribution was described by Lord and Munns as indicating a decrease, probably dating from the severe winter of 1962/3. During the BTO *Atlas* project (1968–72) the number of 10-km squares for which breeding was probable or confirmed increased by 22 per cent to sixty-two, or 81 per cent of those surveyed.

This probably indicates a real increase, but may to some extent reflect the especially thorough coverage resulting from a second period of survey work. Favoured breeding sites are usually relatively small, well-vegetated pools and lakes. Several pairs breed annually at such localities as Alvecote, Kingsbury and Upton Warren. Where suitable conditions exist, the Little Grebe will breed quite close to urban centres, with Brierley Hill Pools and Edgbaston Park being good examples of regularly used sites. Reflecting its ability to occupy smaller waters, the Little Grebe has a wider regional and national distribution than the Great Crested Grebe and the British breeding population has been estimated to be approximately double that of the Great Crested (Sharrock). In the West Midlands, however, although breeding in fewer squares, the larger species is probably the more abundant, as much higher breeding concentrations are to be found at several waters. Overall there are probably now 200–300 pairs of Little Grebe in the Region, compared to the 20–200 estimated by Lord and Munns.

During the autumn Little Grebes desert many of the smaller pools, and the larger waters, which are not especially favoured as nesting sites, receive a considerable post-breeding influx. Among noteworthy concentrations may be mentioned 40 at Edgbaston Reservoir in August 1953, 90 at Blithfield in August 1955, 65 at Alvecote in December 1957, 40 at Bittell in September 1961 and 70 at Kingsbury in September 1978.

In the autumn of 1976, following an exceptionally dry summer during which many smaller pools dried out, remarkable totals were reported at the larger reservoirs. At Belvide, 68 on August 18th was a record for the reservoir, while 100 at Draycote on October 31st constituted the highest number ever recorded in the Region at a single locality.

34. Staffordshire and Warwickshire are among the best inland counties for breeding Great Crested Grebes.   *S. C. Brown*

Great Crested Grebe
Podiceps cristatus

Fairly numerous and widespread breeding species and winter visitor, with recent signs of an increase in the wintering population.

The Great Crested Grebe is a fairly numerous breeding species, occupying nearly all suitable waters within the area. The regional and national *Atlas* projects revealed confirmed or probable breeding in fifty-five 10-km squares, that is 71 per cent of all squares. The species has been the subject of three national BTO enquiries and the total number of birds for the West Midlands Region has shown an increase on each occasion, as Table 4 shows.

Table 4: Censuses of Great Crested Grebe

County (1973 boundaries)	1931 Counted	1965 Counted	1965 Estimated	1975 Estimated
Warwickshire	115	228	244	
Worcestershire	74	43	45	
Staffordshire	144	205	224	
Total	333	476	513	598

The national increase in this species dates from the early years of the century, when it was saved from virtual extinction by legal protection. During the last thirty years or so the increase in suitable breeding waters, resulting from the excavation of gravel pits and new reservoirs, has benefited the species considerably, though more recently this has to some extent been offset by increasing disturbance from recreational activities. Nevertheless, Warwickshire and Staffordshire now rate among the best inland counties in terms of breeding numbers, but in Worcestershire, where such habitats are fewer, the species remains relatively scarce. Certain localities support especially noteworthy colonies. At Alvecote 21 broods were reared in 1963 while at Blithfield up to 30 pairs have nested, and at least 40 young were reared in 1978. At all localities the breeding success depends greatly on the suitability of the water-level, as marked fluctuations can either flood nests or leave them stranded and open to predation.

In the early 1970s a change in the wintering habits of this species became evident. Previously, many birds left the area during the winter months though a few were usually to be found at the larger reservoirs (Boyd 1929–39, Smallshire and Richards 1976). During the winter of 1973/4, however, around 100 birds were present at Blithfield, a water which had normally held only about 20 at this season. The increase has been maintained in subsequent winters with 230 birds counted in February 1976. At the relatively new Draycote Water – of comparable size to Blithfield – winter numbers have normally remained low, and frequently the species has been out-numbered by the Little Grebe, but in January 1978 up to 114 were present.

In general, peak numbers tend to be in spring or autumn and the following counts are among the highest reported from the localities concerned: 335 at Blithfield on August 25th 1954; 83 at Alvecote in March 1966; 60 at Kingsbury on May 23rd 1968; 50 at Bittell on March 6th 1969; and 60 at Belvide on August 13th 1971.

Red-necked Grebe
Podiceps grisegena

Very scarce and irregular winter visitor and passage migrant.

Tomes (1901) mentioned several nine-teenth-century records from the River Avon; Smith quoted three early occurrences in Staffordshire and Harthan referred to a single bird at Bittell in December 1921. The Red-necked Grebe was then recorded on 34 occasions between 1929 and 1978, with three records involving two birds. Prior to the mid-1950s the species was very rare, with only seven records from 1931 to 1953, but it has appeared in all but six of the last twenty-five years.

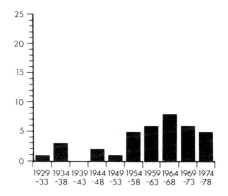

Five-yearly Totals of Red-necked Grebes, 1929–78

Birds have appeared in all months except June and August, with the winter months December to February accounting for 61 per cent. A further 25 per cent, however, appeared during the passage months of March, April and October. On July 2nd 1964 a bird in full summer plumage occurred at Belvide, while several spring records have also involved birds in breeding dress.

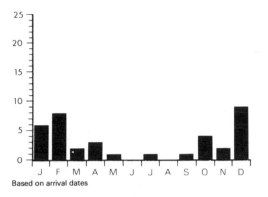

Monthly Distribution of Red-necked Grebes, 1929–78

Of the 37 birds, 22 remained for only a single day, while seven stayed for three weeks or more. During the fifty-year period Belvide (13 birds) and Blithfield (11 birds) were the most regular haunts, while Bittell produced four birds. One or two individuals appeared at Chasewater, Draycote, Gailey, Kingsbury, Shustoke and Trentham.

During the severe weather in February 1979 (and therefore just outside the period of detailed analysis) unprecedented numbers of Red-necked Grebes were recorded inland in Britain. In the West Midlands Region about 24 were reported between late January and the end of March. At Draycote in mid-February no less than seven were to be seen, in company with a Black-throated Diver and two Red-throated Divers, while at Bartley two Red-necked Grebes, two Black-throated Divers and two Red-throated Divers were observed on February 17th.

Slavonian Grebe
Podiceps auritus

Very scarce winter visitor and passage migrant.

Previous authors described the Slavonian Grebe as an occasional or irregular winter visitor, but the actual number of records is not precisely documented. In the fifty years 1929–78, 94 Slavonian Grebes were recorded in the area, compared with 184 Black-necked Grebes. The species occurred in thirty of the forty-three years 1936–78, and in every year since 1969. During this last decade it has gained ground on the Black-necked, with 35 and 45 birds respectively. Thus the Slavonian Grebe is apparently increasing as a visitor to the Region, whereas the Black-necked Grebe has recently shown a decline (Hume 1977). A similar increase has been noted in the London area (Chandler and Osborne 1977). Most records are of single birds, but there are five records of two birds, two records of three birds and one record of four (Belvide, February 1937).

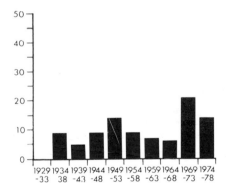

Five-yearly Totals of Slavonian Grebes, 1929–78

Slavonian Grebes have arrived in all months except May and June, while in 1937 and 1970 birds which appeared in February, at Belvide and Westwood respectively, stayed through into May.

Traditionally, the Slavonian Grebe has been principally a winter visitor, with 72 per cent of records between November and February, and a peak in December. During recent years, however, signs of a limited spring passage have emerged, with eight

March–April records (nine birds) in the seven years since 1972 from a total of only eleven (13 birds) for these months during the period 1936 to 1978.

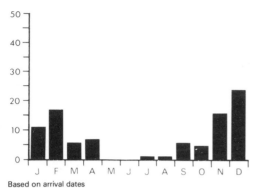

Based on arrival dates

Monthly Distribution of Slavonian Grebes, 1929–78

Overall, the most productive waters have been Belvide (30 birds), Blithfield (17), Draycote (9), Bartley (6), Bittell (5) and Chasewater (5). During the last ten years, however, the most favoured localities have been Draycote and Blithfield, each with nine birds, and at both waters this species has outnumbered the Black-necked Grebe over the same period.

Of the 94 birds, 47 remained for only a day and most stayed less than a fortnight. There are three records of over ninety days, however, the longest involving a bird which remained at Blithfield from November 21st 1971 to at least March 5th 1972.

Black-necked Grebe
Podiceps nigricollis

Regular, but scarce, passage migrant and winter visitor. One breeding record.

Previous authors described this species as an occasional or irregular winter visitor without precisely specifying the number of records. From 1929–78, however, 184 Black-necked Grebes were recorded in the Region, with occurrences in forty-two of

the fifty years, and in every year from 1948 to 1976 inclusive. Between 1944 and 1968 the records displayed a relatively even spread. Numbers then increased somewhat during the late 1960s and early 1970s, but the last five years have witnessed a substantial decline and in 1977 the species went unrecorded for the first time since 1948. This decline has been remarked upon elsewhere (Hume 1977), but no immediate explanation has been suggested, although the small British breeding population has also declined in recent years (Sharrock *et al*. 1978).

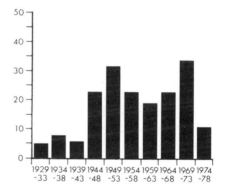

Five-yearly Totals of Black-necked Grebes, 1929–78

The species has appeared in all months, with a substantial peak in September, which accounts for 27 per cent of birds. The autumn and early winter months, from August to December, account for 76 per cent.

The monthly distribution of records over the period as a whole mirrors that for 1934 to 1962 described by Lord and Richards (1964), who commented on the absence in the West Midlands of a real spring peak as mentioned by Witherby *et al*. (1940). Between 1973 and 1977 inclusive there were no records at all for the first seven months of the year (cf. Slavonian Grebe).

Of the 184 individuals, 88 were recorded on only a single date, while 23 birds have remained for over a month. There are 20

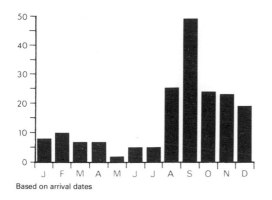

Based on arrival dates

Monthly Distribution of Black-necked Grebes, 1929–78

records of two birds, two records of three birds, one of four, and one of seven (Bittell, November 16th–23rd 1946).

Over the period as a whole, Belvide produced 52 birds, followed by Bittell with 27, Blithfield with 15 and Bartley with 13. The remaining 77 birds were distributed among 18 other localities. During the past decade Blithfield has produced seven birds, Belvide six and Draycote six.

In 1954 a pair bred in Sutton Park, being watched feeding a single juvenile during July and August. This remains the only breeding record.

Fulmar
Fulmarus glacialis

Rare vagrant. Eleven twentieth-century records.

Hastings (1834) referred to a bird caught near Fladbury in 1820, while Smith mentioned one during December 1862 at Perry Barr, which was then in Staffordshire but is now part of the West Midlands County. Records this century have come from Evesham on March 6th 1938, Bromsgrove on March 8th 1943, Lickey on June 23rd 1952, Astley on May 28th or 29th 1959, Sutton Coldfield on April 20th 1965, Swallow Moss on June 5th 1966, Malvern

on April 28th 1968, Bittell on June 8th 1971, Welland on September 7th 1974, Draycote on June 3rd 1977 and Chasewater on May 1st 1978.

Of the grand total of 13 records, eight have come from Worcestershire, perhaps a reflection of that county's rather more westerly position. That nine of the occurrences have been since 1952 may be no more than a reflection of the recent increase in the number of observers. It is noteworthy, however, that Fisher (1952) found British inland records to be predominantly in the months September to November, and January to March, but since 1952 only one individual in the Region has conformed to that pattern (a bird in September), while the remainder have been in spring and early summer, with June the most frequent month.

Cory's Shearwater
Calonectris diomedea

One record.

An exhausted bird, believed to be of the North Atlantic race *borealis*, was picked up at Chasewater on October 2nd 1971 (*Brit. Birds* 65:324). It subsequently died when released on the coast. This remarkable occurrence appears to be the first record from an inland county, but came in a year which produced good numbers on the coasts of Britain and Ireland (Sharrock and Sharrock 1976).

Manx Shearwater
Puffinus puffinus

Rare storm-driven visitor.

Smith and Norris documented 21 Staffordshire and Warwickshire records of the Manx Shearwater between 1872 and 1923. In addition there are a number of early, but undated, records from Burton

(Smith) and several from Worcestershire (Harthan).

Subsequently there were no records until 1943, but then the thirty-six years to 1978 produced 33 birds, including one record of two individuals at Leamington Spa on September 8th 1974. These 33 birds were distributed as follows:

Warwickshire	: 9	July	: 5
Worcestershire	: 7	August	: 2
Staffordshire	: 11	September	: 24
West Midlands	: 6	October	: 2

Nearly all birds appeared after autumn gales and this is reflected by the monthly and yearly pattern of records. Three birds occurred in September 1954, three in September/October 1958, five in the autumn of 1966, six during September 1974 and four in the autumn of 1978. One or two birds appeared in nine other years.

Surprisingly there have been two West Midland recoveries of Manx Shearwaters, both ringed on Skokholm (Dyfed). In 1946 one, which was ringed on August 26th, was at Ipstones three days later, whilst in 1974 one was at Leamington Spa on September 9th, just nine days after ringing. In the 1930s several birds were released experimentally in the Region and four of these later returned to Skokholm, the quickest taking just five days.

Storm Petrel
Hydrobates pelagicus

Rare vagrant. Eight records this century.

Smith referred to several early records around Burton, Harthan quoted an early report from Worcester and Norris listed ten occurrences in Warwickshire between 1820 and 1890.

The following eight records, all between 1936 and 1978, are seemingly the only fully documented records of Storm Petrel this century and it is thus a much scarcer visitor than the Leach's Petrel : Evesham on November 16th 1936, Wickhamford on November 2nd 1952, Perry Barr on November 16th 1966, Bewdley on July 13th 1968, Kings Norton on July 16th 1970, Coombe Abbey on September 13th 1970, Stafford on November 6th 1977 (kept in captivity until 11th, when died) and Draycote on September 17th 1978.

Six of these records have involved birds presumably driven inland by coastal gales. The Evesham and Wickhamford birds were, in fact, found dead, while the individual at Coombe Abbey was killed and eaten by a Lesser Black-backed Gull, suggesting it was in poor condition. The two July records are more surprising and apparently involved birds in good condition.

Leach's Petrel
Oceanodroma leucorhoa

Very scarce storm-driven visitor.

Harthan and Norris listed a total of nine nineteenth-century records of the Leach's Petrel in Warwickshire and Worcestershire, while both authors referred to Tomes' (1901) statement that he had details of a dozen more along the River Avon. Smith noted that the species had appeared in Staffordshire on several occasions and listed a number of occurrences up to 1910.

Between 1935 and 1978 a further 71 Leach's Petrels were recorded in the Region. Of this total, nearly 80 per cent arrived during two weeks in the autumn of 1952, when a nationwide wreck occurred in late October. Between October 25th and November 9th, 56 birds were reported, from Evesham in the south to Leek in the north, but unfortunately 47 of these were found dead.

Of the remaining 15 records, two were in 1950, three in 1953, three in 1978 and singles in seven other years. Eight were in September, three in October, two in November and two in January. There were five records from Blithfield between 1966 and 1978.

Gannet
Sula bassana

Rare vagrant.

Smith, Harthan and Norris provided details of the following early records of the Gannet: one at Coleshill in 1678, eleven nineteenth-century records, two on the Staffordshire border by Clifton (Derbys) in 1900, and single birds at Crowle in 1901, Oakamoor in 1914 and Cheadle in 1922.

Between 1944 and 1976 there were 26 records of this maritime species, with one record involving two birds.

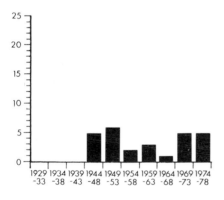

Five-yearly Totals of Gannets, 1929–78

As might be expected, the autumn months have produced most occurrences, with 56 per cent in the period August to October, but there is a surprising secondary peak in June (cf. Fulmar). The decline in records during the mid-1950s and 1960s, and the prominence of June records are both features of note, but, with so few records, it is doubtful whether these patterns have any genuine signifi-

cance in a species which is essentially a vagrant to the Region. Clearly, however, not all birds have been storm-driven.

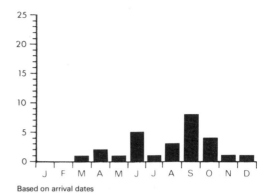

Based on arrival dates

Monthly Distribution of Gannets, 1929–78

Cormorant
Phalacrocorax carbo

Principally a winter visitor, not scarce but localised. One or two remain all year.

Prior to the mid-1940s the Cormorant was a regular, but scarce, visitor to the Region. Smith considered it uncommon, except in south Staffordshire, and Harthan that it was a regular visitor in small numbers between September and the end of May, yet Norris could quote only seven Warwickshire records.

Numbers wintering in Staffordshire then began to increase markedly, however, with the tree-covered islands at Aqualate,

Chillington and Gailey providing attractive roosting sites for the species. At Gailey 14 birds were noted in October 1946 and by the winter of 1956 numbers up to 40 were occurring, with occasional birds present even during the summer. Initially, these birds visited nearby waters only infrequently, but Belvide soon began to attract feeding parties, with 27 birds recorded in December 1947 and 42 in December 1950. Since 1971 the numbers visiting Belvide have again declined, and this has been attributed to increased disturbance at Gailey (Smallshire and Richards 1976).

Blithfield Reservoir, which filled in 1952, at first attracted only small numbers of Cormorants, but a progressive increase occurred over the years, no doubt related to the stocking of the reservoir with trout. During the past decade this reservoir has become the stronghold of the species, with up to 65 birds in recent winters and one or two birds throughout the summer. The birds do not roost at Blithfield, but collect together at dusk before flying off in formation. A number of different roosting sites has apparently been used over the years and there is probably a varying pattern of interchange between feeding and roosting sites within Staffordshire. In November 1979 an exceptional count of 240 was made at a roost at Kings Bromley (S. C. Brown *pers. comm.*).

Although numbers elsewhere in the Region are still comparatively small, there are clear signs of an increase, both in numbers and geographical spread. Atypically high locality totals in recent years have included 12 at Bittell in April 1973, 14 in the Tame Valley during October 1976 and November 1978, and 16 at Draycote on April 13th 1978. This expansion has created some concern among fishing interests, as this species is reputed to consume unacceptably large quantities of game-fish. The Cormorant also takes coarse-fish, however, and there is still considerable argument as to whether the resulting benefit to fisheries is outweighed by the damage to trout. At Blithfield, organised culling was introduced during 1977 and in that year about 40 birds were shot. In the winter of 1977/8 numbers again rose to 60, however, and it seems at least possible that culling merely results in the immigration of other individuals to make up the natural holding capacity of the reservoir. At Draycote, which also supports a flourishing trout fishery, the number of Cormorants has not yet proved a problem. The relative lack of perches and disturbance-free loafing areas may perhaps make this water less attractive than Blithfield, but it will be instructive to follow the development of the Cormorant population at Draycote. In view of the strong emotions which the subject raises and the clear indication that Cormorants will increase still further as more waters are stocked with fish, it is important that the true effect of the species on game fish stocks be properly determined, so that the necessity and extent of control measures can be objectively assessed. Certainly it seems birds might come from far and wide, since six Cormorants, all ringed as nestlings, have been recovered in the

Table 5: Maximum Counts of Cormorant at Principal Waters since 1964

	1964	1965	1966	1967	1968	1969	1970	1971	1972	1973	1974	1975	1976	1977	1978
Aqualate	—	32	—	—	—	20	—	23	—	16	—	13	37	40	26
Belvide	22	—	17	18	11	17	19	6	15	12	3	10	14	3	18
Blithfield	11	26	24	30	22	35	30	21	35	42	42	47	65	65	47
Gailey	22	—	30	24	13	18	20	24	28	—	—	25	33	45	—

Region – three from the Farne Islands (Northumberland) and one each from Orkney, Wigtown and Puffin Island (Gwynedd).

Shag
Phalacrocorax aristotelis

Scarce, but increasingly regular, visitor.

Until the early 1970s the Shag was an extremely scarce visitor to the Region. Smith referred to nine late-nineteenth century and early-twentieth century Staffordshire records, while Norris listed seven Warwickshire records between 1869 and 1923. By 1962 Lord and Blake were still able to describe the species as only a very irregular visitor to Staffordshire. During the 1970s, however, the Shag underwent a remarkable change in status, with occurrences in every year between 1973 and 1978 inclusive and a total of 16 birds during the latter year. There have been 59 records, involving 89 birds, since 1938. Of this total 51 per cent of records and 60 per cent of birds have been since 1974.

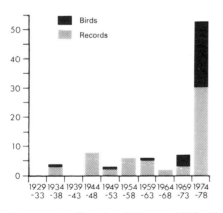

Five-yearly Totals of Shags, 1929–78

The species has appeared in all months except June, but most records have been in the autumn and winter months. Eighty-two per cent of birds arrived in the period August to January, with September and November the peak months.

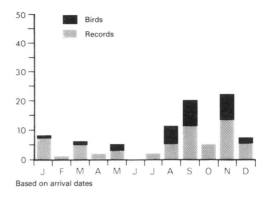

Based on arrival dates

Monthly Distribution of Shags, 1929–78

Most occurrences have involved only one or two birds, but there are two records of three birds, two records of five, one record of six (Himley, November 20th–26th 1978) and one record of seven (Draycote, September 1974). The occurrence of parties of Shags inland (and in low-lying coastal districts) has been related to eruptive movements from the breeding areas, which result from food shortages following periods of onshore winds (Potts 1969).

The most productive localities have been Belvide (nine records totalling 18 birds), Draycote (nine records totalling 17 birds), Blithfield (twelve records totalling 14 birds) and Chasewater (five records totalling five birds). One, ringed as a nestling on the Farne Islands (Northumberland) on June 29th 1961 was recovered at Redditch on March 11th 1962.

Bittern
Botaurus stellaris

Very scarce visitor, most often during winter months.

Harthan and Norris quoted nineteenth-century authors to the effect that the Bittern was not infrequent at that time, while Smith stated that it bred commonly

165

in Staffordshire at one time and was seen in most winters up until the early twentieth century.

Between 1933 and 1977 it was recorded on 46 occasions, but it has become more frequent since the mid-1960s, with five occurrences in 1971 and seven in 1976.

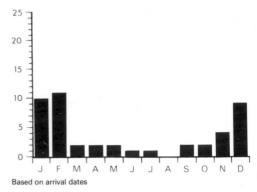

Based on arrival dates

Monthly Distribution of Bitterns, 1929–78

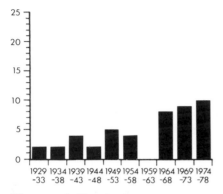

Five-yearly Totals of Bitterns, 1929–78

A similar increase in records from the London area was attributed by Chandler and Osborne (1977) to a recovery of the East Anglian breeding population following the cold winter losses of 1962/3, though Day and Wilson (1978) have shown that, away from the Minsmere–Walberswick area, numbers in East Anglia are in serious decline.

Bitterns have appeared in the Region during all months except August, but two-thirds of arrivals have been between December and February and have tended to be associated with periods of severe weather.

Thirty-five records have involved only a single date, but there are five records of more than a week. In 1976 a bird remained at Kingsbury from February 12th to March 31st and was heard booming towards the end of its stay. There are six records from both Alvecote and Brandon; three from Aqualate, Kingsbury and Sutton Coldfield; two from Baginton, Blackdown, Blithfield, Chasewater and Coombe; and single records from fifteen other localities.

Little Bittern
Ixobrychus minutus

Rare vagrant. Seven records.

Smith, Harthan and Norris listed the following early records: two shot at Sutton Coldfield prior to 1836, one at Kings Bromley about 1838, one shot near Badsey in 1865, one at Hanley in May 1901 and one shot between Warwick and Stratford prior to 1904.

The two recent occurrences involved a male at Brandon Marsh on May 29th and 30th 1976 (*Brit. Birds* 70:415) and a bird at Kingsbury from June 24th to July 2nd 1980. Nationally, 1976 was a good year for the species, with a total of eleven birds (O'Sullivan *et al.* 1977).

Night Heron
Nycticorax nycticorax

Very rare vagrant. Four records.

There are one early and three recent records of the Night Heron, namely an adult at Bradley Green in the 1870s (Willis-Bund 1891), an immature at Moneymore Gravel Pit on June 13th 1971 (*Brit. Birds* 65:326), an adult at a gravel pit near Worcester from June 17th to 22nd 1978 (*Brit. Birds* 72:509) and an adult at another site near Worcester on July 27th 1979.

Records of Night Herons are frequently suspected of relating to escaped birds, as there is a free-flying colony at Edinburgh Zoo. Sharrock and Sharrock (1976), however, demonstrated a distinct April-June peak in the pattern of occurrences and concluded that most records in fact involved wild birds.

Squacco Heron
Ardeola ralloides

One record.

The only record is of a bird at Coton-in-the-Clay, Staffordshire, on May 17th 1874 (Smith).

Little Egret
Egretta garzetta

Rare vagrant. Seven records.

The Little Egret has been recorded in the Region on seven occasions. Four of these were pre-1836 records from Sutton Park (Norris), while the other three have all occurred since 1967, namely two birds at Blithfield on May 11th and 12th 1967 (*Brit. Birds* 61:333), one at Belvide on May 30th 1974 (*Brit. Birds* 68:310) and one at Brandon on May 29th 1978 (*Brit. Birds* 72:510).

This species has appeared in Britain far more regularly since the early 1960s and May has become the classic month for its arrival (Sharrock and Sharrock 1976).

Grey Heron
Ardea cinerea

Fairly numerous resident.

The *Atlas* projects (1966–72) revealed that Herons were nesting in eighteen 10-km squares within the Region, while breeding was considered probable in a further two squares. Apart from a temporary decline following the severe winter of 1962/3, the number of pairs in Staffordshire has remained fairly stable over the past twenty-five years, but in both Warwickshire and Worcestershire there has been a decline, as Table 6 shows.

In Staffordshire there are well-established heronries at Aqualate, Bagots Wood and Gailey, though in 1979 the latter moved temporarily from its traditional island to nearby Fullmoor Wood. There are also others at Chillington and Enville. Those at Aqualate and Bagots Wood are of great antiquity, the former dating from before 1686. In recent years over 50 pairs have nested at each site.

Traditional Warwickshire heronries include Berkswell (now West Midlands), Coombe Abbey, Ragley Park and Wootton Wawen, while another site has recently been discovered in the centre of the county

Table 6: Censuses of Grey Heron

	1952	1960	1965	1978
Warwickshire	71	39	33	36+
Worcestershire	61	33	6	14
Staffordshire	123	163	127	180
Total	255	235	166	230

35. Although there are only a dozen heronries in the Region, Grey Herons may be seen at most waters outside the breeding season.   *S. C. Brown*

and in 1978 it supported 14 nests. The early history of Warwickshire heronries was described in detail by Norris, including the now extinct site at Warwick Park which in 1850 contained as many as 80 pairs. Currently, the county total is only half that of 1952, but it has been stable for the past twenty years.

In the Region as a whole the species suffered a decline following the severe winter of 1962/3, but total numbers subsequently recovered. In Worcestershire, however, the species did not recover, and none apparently nested between 1966 and 1975. The traditional sites, at Croome and Westwood, remain unoccupied, but, encouragingly, a small new heronry was founded or discovered near Worcester during 1976 and by 1978 it held 14 pairs, which produced 31 young. Unlike most

heronries, which are in tall trees, this one is in relatively low scrub.

Post-breeding concentrations are regularly noted at reservoirs and gravel pits and the following are among the more noteworthy totals: 32 at Bittell on September 4th 1955, 60 at Blithfield on August 2nd 1961, 43 at Tittesworth on July 29th 1970, 27 at Kingsbury on September 1st 1974, 25 at Pirton on August 24th 1976 and 25 at Draycote on August 15th 1977.

The vast majority of ringing recoveries relate to the Gailey heronry and were summarised by Minton (1970). Subsequent recoveries reinforce the conclusions made at that time, namely that Grey Herons disperse in random directions, with less than a quarter of the birds travelling more than 80 km. A high proportion is recovered in the first year of life, but birds ten and

168

eleven years old have been recovered. There have now been three foreign recoveries of Gailey-ringed Grey Herons. All were during the winter following ringing and came from Coruna (Spain) in 1965; Loire Atlantique (France) in 1969 and Pontevedra (Spain) in 1974. As would be expected from the random dispersal of Gailey birds, Grey Herons ringed in areas adjacent to the Region have been recovered within it. There have also been winter recoveries of two birds ringed in Sweden and one ringed in Norway.

Purple Heron
Ardea purpurea

Very rare vagrant. Three records.

There is one early record of a Purple Heron, at Wetmore on July 1st 1856 (Smith); and two recent ones, of an immature along a canal near Earlswood from August 31st to September 15th 1956 and an adult at Wormleighton Reservoir on April 18th 1970 (*Brit. Birds* 64:343).
An exceptional influx of Purple Herons into Britain occurred during the spring of 1970 (Sharrock and Sharrock 1976).

Black Stork
Ciconia nigra

One record.

A Black Stork was flushed along Dowles Brook in the Wyre Forest on May 31st 1956. Records of this species are open to the suspicion of being escapes from captivity, but enquiries at the time failed to reveal such a source and May has produced a number of British records of Black Storks in recent years (Sharrock and Sharrock 1976).

White Stork
Ciconia ciconia

Very rare vagrant. No recent records.

An immature White Stork was captured at Beacon Hill, near Coleshill, on September 26th 1896 (Norris). According to Garner (1844) several had occurred along the River Dove on the Staffordshire–Derbyshire border, while Smith quoted an undated record from Abbots Bromley.

Glossy Ibis
Plegadis falcinellus

One record.

Smith quoted one genuine Staffordshire record of a bird shot at Fradley in 1840 and also mentioned a bird shot at Walton-on-Trent (Derbys.) in 1847 or 1848.

Spoonbill
Platalea leucorodia

Rare visitor in early summer and autumn.

Smith and Harthan listed five records of the Spoonbill between about 1830 and 1900. There were no further records until 1955, but since then the species has occurred on the following 14 occasions – a dramatic increase which is in line with the national trend: one at Blithfield on June 25th 1955; one at Brandon from June 16th to 20th 1957; one at Ladywalk and Kingsbury from October 20th to 22nd 1957; one at Brandon from July 4th to August 4th 1965; one at Aqualate from August 19th to 26th 1965; one at Blithfield from June 19th to July 26th 1966, which also visited Gailey on June 26th; four moved between Draycote and Brandon on October 19th and 20th 1972, and were seen later on the 20th in flight over Bartley and then at a pool at Claverley (Shropshire); one at Blithfield from October 26th to November 9th 1972; one at Brandon on May 14th 1973; one at Alvecote on June 13th 1973; one at Blithfield from July 13th to August 16th 1974 (which briefly visited Belvide on July 14th), with two between July 27th and August 10th; two at Blithfield from July 2nd to 4th 1976; one at Belvide intermittently from late-June to late-July 1976; and one at Draycote from May 25th to June 2nd 1979.

Of these 14 records, which involved 19 birds, 62 per cent were in June or July and 23 per cent in October. Blithfield with five records (seven birds) and Brandon with four records (seven birds) have been the most favoured localities.

Mute Swan
Cygnus olor

Widespread and not scarce as a breeding species, but signs of a significant decline. Fairly numerous in winter.

The Mute Swan is a familiar sight on lowland waters, including reservoirs, suburban parks, pools, gravel pits, rivers and canals, but unless action is taken, there are signs that in the future this may no longer be the case.

Previous authors all agreed that the species was common and widely distributed, and the *Atlas* surveys (1966–72) showed that Mute Swans are still widespread, with breeding confirmed in sixty-six 10-km squares (86 per cent). The only gaps in its distribution were the extreme south of Warwickshire and the Carboniferous limestone district of north Staffordshire, where in both cases there is little suitable habitat.

This widespread distribution belies its numerical strength, however. Sharrock quoted studies showing that the species was increasing nationally until 1959–60, but then declined rapidly until about 1965, after which numbers stabilised. The decline particularly affected non-breeding flocks. Continuing studies suggest that nationally the Mute Swan population is not seriously threatened, but that in certain regions, including the Midlands, substantial decreases have occurred. During a national census in 1955–6 (Campbell 1960) a total of 198 breeding pairs and about 270 non-breeding birds was located in the West Midlands, but this was probably an underestimate. Since 1961 an intensive study has

Table 7: Numbers of Breeding and Non-breeding Mute Swans in the West Midlands in 1955/6 and 1978
The latter are extrapolated figures based on a sample census.

County (pre-1974 boundaries)	Year of Census	Breeding Adults	Non-breeding Birds	Total
Warwickshire	1955/6	130	99+	229+
	1978	114	127	241
Worcestershire	1955/6	78+	61+	139+
	1978	28	26	54
Staffordshire	1955/6	188	111	299
	1978	98	130	228

been made of the Mute Swan population in an area of 1,440 km^2 in south Staffordshire (Minton 1968, Coleman and Minton 1980) and up to December 1978 a total of 4,060 swans had been ringed within the study area. On the basis of densities revealed by this study, the total breeding and non-breeding population for the West Midlands in the early 1960s was estimated to have been around 1,200 birds (Lord and Munns), though this was probably optimistic. A sample census in 1978 as part of the national BTO survey, and long-term counts of moulting flocks, both reveal a continuing decline.

The results of the 1955/6 census are generally regarded as an under-estimate while the 1978 totals are extrapolated from the results of a sample census (Ogilvie 1981). Direct comparisons are therefore difficult, but an overall decrease of at least 20 per cent is implied with a decrease of more than 60 per cent in Worcestershire. In Warwickshire numbers have apparently changed relatively little. These figures compare with a national decline of 8–15 per cent.

Mortality of Mute Swan progeny has formed an important part of the south Staffordshire study. Nearly 50 per cent of all clutches laid in the study area between 1961 and 1978 were stolen by humans. Of the cygnets hatched, 76 per cent survived to fledging. The highest mortality after fledging occurred during the first year of life with peak periods in October and March. Overhead wires were the commonest known cause of death. Birds in the second

and later years showed mortality peaks between January and April, with overhead wires again the commonest known cause.

Other increasing threats to Mute Swans include poisoning from lead-shot discarded by anglers and oil pollution. Post-mortems conducted by the Ministry of Agriculture, Fisheries and Food on 206 birds found in central and southern England between 1973 and 1977 revealed that 107 had died from lead-poisoning. The Trent and the Warwickshire Avon were the worst affected areas (Hunt 1977). At Stratford-upon-Avon, a site once famed for its Mute Swans, numbers had declined to only six by 1975 and, in an attempt to avert the complete elimination of the species, a special swan reserve has been established through public subscription. Here, again, lead poisoning has been implicated in the decline (Hardman and Cooper 1980). In 1966, 85 birds died in an oiling incident at Burton-on-Trent. At Tamworth, oiling killed 75 birds in October 1974, while about 45 birds were oiled in the Tamworth–Kingsbury area in February 1978. Oiling incidents of this kind tend to involve flocks of non-breeding immatures and young adults and this can have disastrous effects on the nucleus of future breeding populations.

Ringing studies have shown (Coleman and Minton *in litt.*) that movements are mainly governed by a reluctance to traverse ground above 150 m and a strong preference to travel along river valleys and low ground below 80 m. Significant

Table 8: Selected Annual Maxima of Mute Swans at Four Traditional Moulting Sites

	1950	1956	1961	1968	1972	1978	1980
Alvecote	33	60	76	129	146	125	85
Belvide	—	67	—	8	6	24	48
Blithfield	—	135	147	30	30	25	45
Chasewater	128	99	57	53	56	42	47

interchange takes place between flocks up to 25 km apart, but only about 5 per cent of the population moves further than 50 km and only 1–2 per cent more than 80 km. The majority of long distance movements takes place in the first two years of life, particularly the second year. Regular movements to special moulting sites occur annually, with Alvecote currently the most important site. Although young birds may move well away from their natal area prior to pairing, they tend to return to it, especially females. In the south Staffordshire study area 85 per cent of paired breeding birds retained the same mate from year-to-year, but non-breeding pairs were less stable. Divorce rates were 3 per cent for breeding birds and 9 per cent for non-breeding ones.

Bewick's Swan
Cygnus columbianus

Regular winter visitor and passage migrant, not scarce.

The Bewick's Swan only established itself as a regular wintering species in England during the 1930s, but increased dramatically from the late 1950s onwards (Cramp and Simmons 1977). This national pattern is reflected by the development of records in the West Midlands Region. Harthan knew of no records prior to 1933, whilst Smith and Norris could list only eight records, involving 54 birds, between 1904 and 1926, including a party of 40 on February 27th 1904, flying along the River Dove near Clifton, on the Staffordshire–Derbyshire border. Only five further parties, totalling 33 birds, occurred between 1929 and 1941, but, apart from 1949, the species has been annual in the Region since 1945. In all a total of 78 records, involving 642 birds, was reported between 1904 and 1959. Since 1960 Bewick's Swans have appeared annually in each of the three major counties and precise details of every occurrence have not been documented.

Maximum counts generally involve transient parties, many doubtless on their way to or from Slimbridge (Glos.), and only rarely do large numbers remain for any length of time in the Region. At Pirton, which has become an increasingly regular haunt, however, birds were present from November 5th onwards in 1977 and reached a total of 90 on December 24th. At Draycote on February 23rd 1980, there were 60 during the day and these were joined by a further 35 just before dusk. Only two of these were immatures.

Typically, birds appear in the area between late October and late March, extreme dates for healthy birds being October 11th (Belvide in 1975) and May 22nd (Brandon in 1955). Injured birds,

Table 9: Annual Maxima of Bewick's Swans at more Regular Haunts

	1960	1961	1962	1963	1964	1965	1966	1967	1968	1969	1970	1971	1972	1973	1974	1975	1976	1977	1978
Alvecote	17	—	18	—	10	11	6	20	15	16	30	5	14	38	16	—	14	3	1
Belvide	—	—	1	—	17	6	17	3	—	28	32	11	2	2	16	—	25	10	4
Bittell	—	18	9	—	5	4	—	13	2	—	—	4	—	32	13	19	7	1	1
Blithfield	15	43	23	37	57	43	42	22	23	27	13	20	16	24	34	32	37	22	19
Chasewater	—	3	1	5	14	11	2	9	12	5	1	4	20	5	6	10	9	2	4
Draycote	—	—	—	—	—	—	—	—	3	—	14	24	3	13	24	7	16	34	16

36.　During the past fifty years Bewick's Swans have become increasingly regular winter visitors, often no doubt on their way to or from Slimbridge.　　　*S. C. Brown*

however, remained at Rednal from December 10th 1945 to September 5th 1946 and at Alvecote from April 21st to May 28th 1958.

during recent years and has been recorded in all but five years since 1949. It has accumulated totals of 96 records and about 347 birds since 1929.

Whooper Swan
Cygnus cygnus

Scarce winter visitor and passage migrant.

During the nineteenth and early twentieth centuries the Whooper Swan was an irregular winter visitor to the Region, but nevertheless occurred more frequently than the Bewick's Swan. Since that time the Bewick's Swan has increased greatly and is now by far the more numerous visitor. Yet the Whooper Swan has also shown a significant, if less spectacular, increase

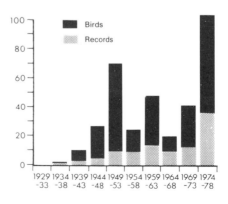

Five-yearly Totals of Whooper Swans, 1929–78

173

The monthly distribution of records suggests that birds visiting the West Midlands fall principally into two categories: those moving spontaneously through, and more recently into, the area during the November–December passage period, and those displaced into the Region during midwinter, hard-weather movements.

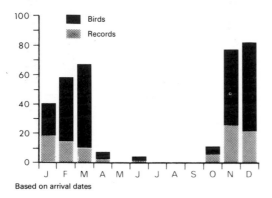

Monthly Distribution of Whooper Swans, 1929–78

In areas where the former element is absent, such as the London region, a pronounced mid-winter peak is apparent (Chandler and Osborne 1977).

Most occurrences have involved under five birds, while the largest party during the last fifty years was of 37 birds, at Bittell on March 7th 1953.

During the period of analysis there have been ten records at Kingsbury, nine at Bittell, eight at Belvide, six at Alvecote and five at Upton Warren. The most noteworthy development has been in the Rudyard Lake–Longsdon Mill area, where Whooper Swans have wintered each year since 1976/7. These birds move about between two or three different sites and it has proved difficult to determine the exact number of individuals concerned. Currently, it appears two or three parties of about half-a-dozen birds are involved, and it is to be hoped that they form the nucleus for a regular and expanding wintering population.

Bean Goose
Anser fabalis

Very rare vagrant. Two records during the last fifty years.

The early history of the Bean Goose in the West Midlands Region is rather confused. Tomes (1904) stated that it was the most frequent grey goose to drop out of migratory flocks, but Norris considered that the birds concerned were probably Pink-feet. There are, however, specimens collected in the nineteenth century at Barford (1841) and Welford, while Smith refers to a bird in a collection at Pelsall. Harthan quoted Tomes as having seen specimens that had been shot, but noted that he gave no dates.

There are only two recent records, both involving single birds, at Bittell on March 27th 1941 and at Alvecote from January 30th to February 27th 1972.

Pink-footed Goose
Anser brachyrhynchus

Scarce or very scarce autumn and winter visitor.

Tomes (1901 and 1904) said the Pink-footed Goose was formerly more common

Table 10: Flock-size of Whooper Swans, 1929–78

Size of Flock	1	2–5	6–10	11–20	21–40
No. of Records	27	52	14	2	1

along the Avon and Severn Valleys, but it appears to have been very scarce during the late-nineteenth and early-twentieth centuries, with subsequent authors able to quote very few positive records. The species appeared in the Region during twenty of the fifty years 1929–78, however, with notable increases in frequency during the mid-1950s and again during the 1970s. The 43 records involved a total of 267 birds.

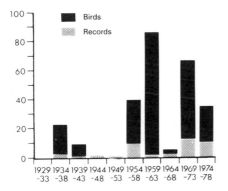

Five-yearly Totals of Pink-footed Geese, 1929–78

A number of the more recent records of single birds have undoubtedly concerned feral birds, but the monthly distribution of records indicates that the majority of occurrences involved wild individuals. The main arrival of Pink-feet in Britain takes place during October (Atkinson–Willes 1963, Cramp and Simmons 1977) and this month has produced seven records. The months of December and January, when cold-weather movements might be expected, have produced a further 19 occur-

rences. These three months together thus account for over 60 per cent of all records, while all the larger flocks were also during these months.

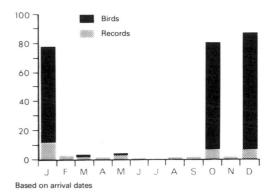

Based on arrival dates

Monthly Distribution of Pink-footed Geese, 1929–78

The largest party up to 1978 was of 80 birds at Rugeley on December 10th 1960, while there were three records involving between 20 and 30 birds.

Most occurrences have involved only a single date. All eight records spanning more than one day concerned single geese, and at least five are strongly suspected of referring to feral birds.

Blithfield has produced five records, Alvecote four, and Kingsbury, Belvide and Upton Warren each three records during the period under review.

Although just after the period of detailed analysis, it should be mentioned here that during severe weather at the end of January 1979 flocks totalling several hundred Pink-feet were noted moving across Staffordshire.

Table 11: Flock-size of Pink-footed Geese, 1929–78

Size of Flock	1	2–5	6–10	11–20	21–30	80
No. of Records	29	4	4	2	3	1

White-fronted Goose
Anser albifrons

Annual winter visitor and passage migrant, not scarce, but variable in number.

The White-fronted Goose is by far the most numerous of the grey geese visiting the Region, with 206 occurrences totalling approximately 5,080 birds having been reported between 1935 and 1978. There are in addition five reports of skeins of unknown size heard flying over the area at night.

Smith could quote only eight positive records prior to 1935, however, whilst Norris knew of only one. Both Harthan and Norris, though, considered that flights of unidentified geese were probably of this species and some impressive flocks were recorded in the early 1940s, when Harthan observed flights of up to 500 birds over the Avon Valley in January and March, presumably from the wintering population on the Severn estuary. Although these were the largest flocks to be recorded, the

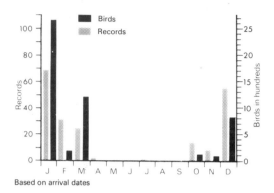

Based on arrival dates

Monthly Distribution of White-fronted Geese, 1929–78

White-fronted Goose has in fact been noted rather more frequently since the 1950s.

December, January and March have proved the most productive months, accounting for 71 per cent of records and 92 per cent of birds. There have, in fact, been more records in February than March, but the latter month has produced several large parties of geese, with birds presumably departing from their wintering grounds in the south-west, whereas the largest flock to appear in February held only 31 birds.

Somewhat over half the records have involved between one and five birds, but in general flock size has been very variable, as illustrated by Table 12.

Apart from odd individuals suspected of being feral, most reports have concerned transient parties, with birds staying for only a day or two or merely over-flying the area. In the Belvide–Wheaton Aston area, however, flocks have wintered on a number

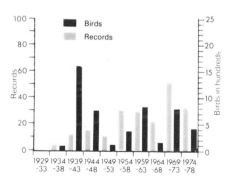

Five-yearly Totals of White-fronted Geese, 1929–78

Table 12: Flock-size of White-fronted Geese, 1929–78

Flock Size	1–5	6–10	11–20	21–50	51–100	101–200	201–500	Total
No. of Flocks	109	29	26	22	10	5	5	206
Total Birds	189	217	368	795	732	689	2,092	5,082

of occasions. In 1939/40 about 40 birds were observed, while in several winters since 1968/69 a similar number of birds has occurred, with 80 on January 10th 1970. A family party of five birds of the Greenland race *flavirostris* stayed from October 30th 1949 to February 27th 1950 and four adults joined a flock of Siberian birds on February 14th 1970 (Smallshire and Richards 1976).

The species has occurred at Belvide in 28 years since 1935, at Blithfield in 17 years since 1956 (with the largest flock 150 on January 18th 1977), at Bittell in nine years since 1945 and at Kingsbury in nine years since 1962. Systematic observations would no doubt reveal that Whitefronts appear annually over the Severn and Avon Valleys.

Lesser White-fronted Goose
Anser erythropus

One record.

A Lesser White-fronted Goose observed at Packington Park on several occasions between January 28th and June 18th 1978 was clearly an escape.

Greylag Goose
Anser anser

Small resident feral population.

The Greylag bred in England until about 1831 (Rooth 1971) and there is some evidence that before this period wild birds occurred not infrequently in the Region (Mosley 1863; Hastings 1834; Tomes 1904). Norris, however, did not admit the species to the Warwickshire list; but Smith gave four dated records, including 43 flying west over Oakamoor on March 11th 1923 and two at Belvide in May 1936; and Harthan one, of six birds at Bittell in March 1929.

There were no further records until 1952 since when the species has been reported regularly, and annually since 1965. Nearly all the more recent records refer to feral birds, which have bred in the area since at least 1967. The principal breeding site is at Packington Park, where birds were first introduced in the mid-1960s. By the early 1970s this flock had increased to 50. In 1973, 70 young were raised and a total of 124 birds was counted on 28th July. One or two pairs have also bred at Brandon, Enville, Ryton and West Park in recent years and the current feral breeding population of the Region is probably between 10 and 20 pairs.

Since 1970, birds showing the characteristics of Greylag × Canada Goose hybrids have been noted at a number of localities.

Birds from Packington frequently visit other localities in the nearby Tame Valley, especially Kingsbury, where up to 90 have been recorded in autumn. Small numbers of feral Greylags are intermittently observed at all the principal pools, lakes and reservoirs in the Region, often in association with large flocks of Canada Geese.

During late November and December 1965 between 30 and 40 birds were reported from water meadows at Ripple. This site has been visited by White-fronts on a number of occasions and it seems possible that this flock consisted of wild birds.

Snow Goose
Anser caerulescens

Feral birds recently recorded as scarce visitors.

Records of escaped Snow Geese have been published in the WM *Bird Reports* since 1970, and up to 1978 there were 40 reports, including two involving a pair. The monthly distribution of arrival dates is as

Table 13: Monthly Distribution of Records of Snow Geese, 1929–78

	Jan	Feb	Mar	Apr	May	Jun	Jul	Aug	Sep	Oct	Nov	Dec
No. of records	7	4	4	3	5	3	—	3	2	4	2	3

expected of a feral species, with no well-defined seasonal pattern.

Birds have appeared in widely scattered localities, but Kingsbury has been visited in four years and Belvide and Bittell in three years since 1970. One or two hybrids have begun to appear in the Region in recent years.

Canada Goose
Branta canadensis

Widespread and fairly numerous resident.

The Canada Goose was first introduced into England during the seventeenth century, but for many years its population was centred on a relatively limited number of sites, principally private ornamental lakes and pools. During the 1950s and 1960s, however, a rapid increase took place, largely as a result of the artificial redistribution of birds and the consequent establishment of new and vigorous breeding groups (Sharrock). The British population increased from about 3,500 in 1953 to 10,500 around 1968 and 19,500 in 1976 (Blurton-Jones 1956, Ogilvie 1969). Because of its threat to agricultural interests, the Ministry of Agriculture, Fisheries, and Food undertook an investigation of population trends in the West Midlands, where numbers have risen from 104 in 1953 to 860 in 1968 and 2,230 in 1976 – a rate of increase in excess of that of the British population as a whole (B.E. Jones *in litt*).

Smith made no mention of the species in Staffordshire, whilst Harthan and Norris could list only eight breeding sites. By 1968, however, the WMBC *Atlas* had confirmed breeding in thirty-one 10-km squares (40 per cent) and by the time the BTO *Atlas* survey had been completed in 1972 breeding had been confirmed or suspected in forty-nine 10-km squares (64 per cent). About 300 young were reared in Staffordshire during 1975, while in the Lower Tame–Blythe Valley area, which is the Warwickshire breeding stronghold, about 170 young were reared in 1977. The species is now established in nearly all areas except the Carboniferous limestone district of north Staffordshire, the higher, drier ground of central Staffordshire and the Lower Lias beds south of the Avon, where suitable waters are scarce. The current breeding population is probably around 500 pairs compared to the 20–200 pairs estimated by Lord and Munns. Much of this used to be concentrated in the Aqua-late–Chillington–Patshull area. Indeed Patshull was the original centre of population growth in the late 1950s (Atkinson-Willes 1963).

The greatest concentrations of birds generally occur in autumn at the larger lakes and reservoirs. Annual maxima reflect an increasing population, though the effects of disturbance and considerable interchange between sites makes interpretation of unco-ordinated counts difficult.

The annual average mortality of West Midland Canada Geese ringed under the BTO scheme between 1967 and 1978 has been calculated at 23 per cent ± 2 per cent and this may be compared with the average for all British Canada Geese over the same period of 20 per cent ± 1 per cent (Jones *in*

37. The regional Canada Goose population has expanded rapidly and, because of the threat to agricultural interests, it has been the subject of special studies.   *M. C. Wilkes*

prep.). The most important cause of death is shooting, which accounted for over 63 per cent of deaths reported.

The Region's Canada Geese are relatively sedentary, with 93 per cent of all known movements being less than 50 km, and they live largely within a home range of between 2,000 and 8,000 km^2. Individual breeding colonies exist on a number of waters within the home range and regular interchange of birds between these waters occurs at all times of the year except the moulting period (Jones and Minton *in prep.*).

Movements further than 50 km are almost always associated with the moult migration to the Beauly Firth. This phenomenon principally involves non-breeding Yorkshire birds (Dennis 1964, Walker 1970), but now seems to be drawing birds from further afield. The first two records

of West Midland birds undertaking this migration came in 1976 and by 1978 54 birds had made the movement.

Between December 9th 1976 and January 2nd 1977 a bird of one of the dark-breasted races, believed to be *occidentalis*, was seen at Belvide. During this period the same individual also visited Gailey. At Kingsbury, a bird showing the characteristics of the race *interior* was observed on January 12th 1978; another individual with the characters of the race *hutchinsii* was recorded between March 11th and 31st and again on August 28th of the same year; and one with the characteristics of the race *minima* was present during the summer and in October of 1979. One or two birds with the characteristics of the race *minima* also occurred at Grimley and Holt in mid-September 1979. All are probable escapes.

179

Table 14: Peak concentrations of Canada Geese at Selected Sites in six years between 1954 and 1978

	1954	1959	1965	1970	1974	1978
Aqualate	27	85	115	—	120	487
Belvide	31	—	79	350	270	201
Bittell	13	62	87	94	167	330
Blithfield	40	65	53	106	350	422
Gailey	21	97	70	120	139	97
Kingsbury	—	—	162	450	300	400

Barnacle Goose
Branta leucopsis

Small resident feral population. Otherwise a very rare vagrant, with three recent records possibly involving wild birds.

Smith gave a single Staffordshire record of the Barnacle Goose, a bird at Tutbury in 1859, whilst Tomes (1901 and 1904) referred to the species as a rare visitor to Warwickshire and Worcestershire, but Norris considered the evidence inconclusive.

There were no further published records until the early 1960s, but, as with several other species of geese, records of feral birds then became commonplace. Not all occurrences have been fully documented but there are reports for all months and from all the principal lakes and reservoirs.

Only the following three records suggest the possibility of wild birds: 11 birds at Brandon on January 25th and 27th 1970, five birds at Belvide on December 26th 1976 and a family party of seven at Blithfield on September 30th 1978.

Hybrids between Barnacle Goose and Canada Goose have been reported on a number of occasions, while in February 1975 a Barnacle Goose × Whitefront hybrid occurred at Gailey and in 1980 a Barnacle × Snow Goose hybrid was noted at Belvide.

Brent Goose
Branta bernicla

Rare and irregular winter vagrant.

Smith and Norris listed five nineteenth-century Staffordshire and Warwickshire records of the Brent Goose, while Tomes (1901) described the species as of erratic occurrence in Worcestershire. This century has produced a further 25 records (55 birds) of which 19 were during the past fifty years.

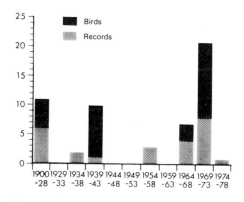

Brent Goose: 1900–28 and Five-yearly Totals, 1929–78

The species thus decreased during the second quarter of the century, but subsequently recovered and appeared annually from 1969 to 1973. This regional pattern is no doubt allied to fluctuations in the British wintering population of dark-bellied Brent Geese, which declined dramatically in the early part of the century, but then increased nearly seven-fold between the mid-1950s and the mid-1970s (Ogilvie and St. Joseph 1976).

All occurrences have been in the passage and winter months, a distribution strongly suggesting that all records have involved wild birds.

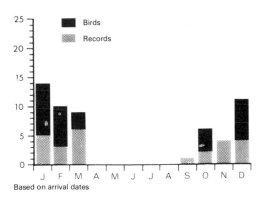

Based on arrival dates

Monthly Distribution of Brent Geese, 1900–78

The largest parties were ten at Alveston on January 25th 1941, five at Chasewater on February 4th 1910 and seven at Brandon on December 21st 1969. There have been five records from Blithfield, three from Chasewater and two each from Bittell, Draycote, Kingsbury and Oakamoor.

Subsequent to the main analysis, a dark-bellied bird occurred at Belvide on December 22nd 1979.

Red-breasted Goose
Branta ruficollis

One record.

A bird of this species, undoubtedly an escape, was observed at Kingsbury between September 29th and October 14th 1973.

Egyptian Goose
Alopochen aegyptiacus

Very scarce visitor and resident.

Records of feral Egyptian Geese have been documented in the WM *Bird Report* only since 1971, following the addition of the species to the British list, though the species was mentioned incidentally by Smith. There have been 18 records in the nine years to 1978, though several records have undoubtedly involved the same individual. There have been six records from Blithfield, two each from Edgbaston Reservoir, Kingsbury, Trittiford Park and Upton Warren, and singles at five other localities. Most occurrences involved one or two birds, often in association with flocks of Canada Geese, but at Upton Warren there were nine on November 2nd 1971 and four on April 5th and 6th 1975. The monthly distribution of arrival dates was as shown in Table 15.

Five records have involved birds remaining for a month or more, and one individual has been resident in the Edgbaston Park area since early 1978.

Table 15: Monthly Distribution of Records of Egyptian Geese, 1970–78

	Jan	Feb	Mar	Apr	May	Jun	Jul	Aug	Sep	Oct	Nov	Dec
No. of records	2	1	2	2	—	—	—	3	2	—	2	4

Ruddy Shelduck

Tadorna ferruginea

Recently recorded as a very scarce feral species.

Two early records of the Ruddy Shelduck from the Birmingham area were referred to by Chase (1886). There have been thirteen records in recent times, all considered to involve escapes from captivity, though the May-June tendency is noteworthy. In 1974 birds were at Belvide from May 7th to 9th, Ladywalk on September 8th and Alvecote on September 9th. Two birds were at Ladywalk on May 4th and at Kingsbury on May 6th 1976; a ringed duck was at Ladywalk from May 8th to 12th 1977, visiting Kingsbury on May 9th; one was at Knighton during May 1977; a drake was at Brandon on June 22nd 1977; another drake was at Brandon on June 11th 1978; one flew over Branston on June 13th 1978 and remained at Drakelow (Derbys.) from June 12th to 18th; two, possibly three, were at Alvecote on November 12th 1978; eight circled Wilden on July 22nd 1979; up to five frequented Holt Gravel Pit between August 8th and 19th 1979; and one was at Kingsbury on August 15th 1979.

Shelduck

Tadorna tadorna

Very scarce as a breeding species, but regular and not scarce as a passage migrant.

Smith said the Shelduck occurred regularly on certain pools in Staffordshire, but Harthan described it as only an occasional visitor to Worcestershire in spring and autumn and Norris, too, called it a rare or occasional visitor to Warwickshire.

The Shelduck declined in many parts of Britain during the nineteenth century, but subsequently recovered as a result of protection. With the increase in numbers came a tendency to nest further inland (Sharrock). Reflecting these national trends, migrant Shelduck increased substantially in the West Midlands during the middle decades of this century.

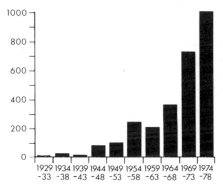

Five-yearly Totals of Shelduck, 1929–78

More dramatically, birds were noted with increasing regularity in Warwickshire along the Tame Valley and in 1970 a pair nested at Ladywalk, rearing six young. The BTO *Atlas* (1976) showed Shelduck breeding in only four non-maritime English counties and, at more than 100 km from the coast, the Tame Valley constituted easily the furthest inland breeding site. One or two pairs have since nested annually in the Middleton–Lea Marston–Ladywalk area, while breeding has also taken place at Alvecote in at least three years since 1973. Between 1970 and 1978 a minimum of eleven broods totalling 80 young were hatched in Warwickshire. In the latter year breeding extended to Worcestershire, a pair rearing three young at Upton Warren, and in 1979 breeding occurred for the first time in Staffordshire, when a pair with eight young was seen at Stretton Hall. None the less the total regional population is still probably only one or two pairs, since in 1979 there was no confirmed breeding from Warwickshire, although it resumed again in 1980. In the Tame Valley the breeding sites have centred on the gravel workings and the adjacent river. Major changes in land usage are currently in progress in this area and it is to be hoped that landfill does not interfere with the colonisation of the area by Shelduck.

As a migrant, numbers continue to increase. The seasonal pattern remains much as described by Hawker (1967), though the proportion of November records has increased. The most productive months are April, when birds are migrating back to their breeding grounds, and September and December, which months encompass the varyingly-timed return phase of the moult migration (Cramp and Simmons 1977).

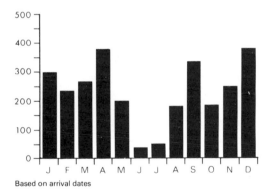

Based on arrival dates

Monthly Distribution of Shelduck, 1929–78
(Young reared in the area are excluded from the figures)

Away from breeding sites, the most frequently visited localities are Blithfield, Belvide, Chasewater and, latterly, Draycote, though many areas are now visited annually.

The majority of occurrences have involved between one and ten birds, but there are 23 records of more than ten individuals (excluding broods), including the following maxima: 27 at Chasewater on November 29th 1969, 29 at Blithfield on January 3rd 1976, 27 at Blithfield on November 26th 1978, and a regional record of 43 at Kingsbury on December 14th 1979.

Mandarin
Aix galericulata

Scarce visitor. Has bred.

The Mandarin is an East Asiatic species introduced into Britain during the mid-eighteenth century. In the twentieth century a feral population became established and the species was admitted to the British list in 1971 (Sharrock). Since that year there have been 19 records involving 39 birds from a total of 14 localities in the West Midlands, including a report of 20 birds, mostly juveniles, at Aston Mill on July 14th 1978. The monthly distribution of the remaining birds was as shown in Table 16, records clearly involving the same individual being included only under the month of first appearance.

There have been three records from Blithfield, and two each from Belvide, Compton Verney and Ladywalk.

Sixteen records have involved only a

Table 16: Monthly Distribution of Records of Mandarin, 1971–78

	Jan	Feb	Mar	Apr	May	Jun	Jul	Aug	Sep	Oct	Nov	Dec
No. of birds	1	1	—	3	2	—	—	—	—	6	3	3

single date. Individuals more or less resident at Brocton and Mary Steven's Park are clearly associated with local wildfowl collections.

Wigeon
Anas penelope

Fairly numerous, but local winter visitor. One or two birds regularly summer.

The Wigeon is a regular winter visitor to all the major lakes and reservoirs, especially those with adjacent grassland for grazing, though significant numbers are restricted to Aqualate, Belvide, Draycote and, in particular, Blithfield.

Smith said it figured greatly among wildfowl arriving at pools and reservoirs in autumn and both Harthan and Norris described it as a regular winter visitor, though the latter added that it was local. The numbers at specific waters have varied over the years, but the total population has probably not altered significantly in the last fifty years. At Belvide, 2,500 were reported on February 5th 1949 and over 1,000 birds were noted in several years prior to 1950. Numbers then declined significantly, an event probably associated with the filling in 1952–3 of Blithfield Reservoir, where up to 1,000 Wigeon were soon wintering and 1,842 were counted on

December 12th 1965. Draycote Water, which filled in 1968–9, attracted 1,000 birds in both 1969 and 1970, but numbers have since been considerably fewer. At smaller waters such as Alvecote and Bittell numbers have also declined recently.

At more favoured sites, notably Blithfield, this species is the second-most numerous duck during much of the winter, while in February and March, when Mallard are dispersing and numbers of Wigeon reach a maximum, it becomes the most abundant species. Peak numbers at Blithfield are among the highest at any non-coastal English locality.

The average first arrival date over forty-three years has been August 25th and the average last departure date over forty years May 1st. Extremes have been recorded on July 3rd (1975) and June 4th (1978), but these may be confusing as one or two birds are regularly noted in summer at Blithfield and occasional summer records have also come from such localities as Alvecote, Belvide and Draycote.

Many Wigeon have been ringed in the West Midlands. In the 1910s some were ringed at Middleton Hall, but the vast majority have been ringed since 1968 at Blithfield. The country and month of recovery of regionally ringed Wigeon are as shown in Table 19.

It will be seen that there is a strong indication that the birds which winter in

Table 17: Five-yearly Means of the Annual Maxima of Wigeon at Selected Localities, 1954–78

	1954–58	1959–63	1964–68	1969–73	1974–78
Alvecote	178	124	64	65	68
Aqualate	260	204	197	442	389
Belvide	286	122	200	250	129
Bittell	154	148	126	80	27
Blithfield	1,060	1,120	1,320	1,212	904
Draycote	—	—	—	720	270

Table 18: Mean monthly concentrations of Wigeon at Principal Localities, based on all available counts, 1953/4–1977/8

	Oct	Nov	Dec	Jan	Feb	Mar
Alvecote	3	15	44	60	60	56
Aqualate	9	54	148	216	225	127
Belvide	18	38	86	86	102	57
Bittell	16	50	71	53	69	38
Blithfield	119	466	676	678	884	631
Shustoke	15	21	32	60	84	64

the West Midlands breed east of 53°E and all May and June recoveries are also north of 61°N. However, one Blithfield bird has been recovered in Iceland and a juvenile ringed in Iceland on June 18th 1927 was found in Rugby in March 1929, which indicates that some Icelandic breeding birds also winter here.

Autumn passage is well shown by recoveries of birds in the more westerly parts of the Soviet Union and in Denmark. There is evidence of birds spending subsequent winters on the continent, but many return to Britain and have been recovered at varying distances and directions from their ringing site.

Gadwall
Anas strepera

Not scarce and increasing, both as a passage migrant and winter visitor. Very scarce as a breeding species.

The English breeding population of Gadwall dates from introduction in about 1850. Numbers increased substantially during this century, with feral birds no doubt being joined by wild birds of continental origin (Sharrock). The birds nesting in England are considered rather sedentary, but are supplemented in winter by immigrants from Scotland, Iceland and Western Europe (Cramp and Simmons 1977).

Table 19: Summary of Recoveries of Regionally-ringed Wigeon

	Jan	Feb	Mar	Apr	May	Jun	Jul	Aug	Sep	Oct	Nov	Dec	Total
Iceland	—	—	—	—	—	2	—	—	—	—	—	—	2
Finland	—	—	—	—	—	—	—	—	1	—	—	—	1
Sweden	—	—	—	—	1	—	—	—	—	—	—	—	1
Soviet Union													
East of 53°E	—	—	—	—	11	2	—	—	2	—	—	—	15
West of 53°E	—	—	—	—	1	—	—	—	5	3	—	2	11
Denmark	—	—	—	—	—	—	—	2	7	1	1	—	11
Germany	—	—	—	—	—	—	—	—	—	—	2	—	2
Holland	2	1	—	—	—	—	—	—	—	—	1	1	5
France	1	—	—	—	—	—	—	—	—	—	1	1	3
British Isles													
0–100 km	—	1	—	—	—	—	—	—	—	1	—	3	5
101–400 km	3	3	—	—	—	—	—	—	1	—	—	2	9

38.   Gadwall have appeared with increasing regularity in recent years and have established a small breeding population.   *T. C. Leach*

In the West Midlands, the Gadwall was a rather scarce migrant and winter visitor up to the late 1960s. Smith described it as uncommon in Staffordshire, whilst Harthan listed just four Worcestershire records and Norris only two from Warwickshire. However, there was circumstantial evidence of breeding at Belvide in 1923 and 1924 and a bird raised by pinioned parents at Burton-on-Trent in 1933 was recovered at Meriden before the end of the year.

By 1962 Lord and Blake were able to describe the species as a regular passage migrant in small numbers, most often in April, November and December. This remained an adequate summary of the status until the late 1960s, at which time the number and geographical spread of transient and winter visitors began to increase, while in 1970 breeding was con-

firmed for the first time at Belvide. The BTO *Atlas* showed breeding confirmed in one 10-km square in the Region, probable in another and possible in a further five during 1968–72. The first Warwickshire breeding records came in 1979, from Brandon and Kingsbury.

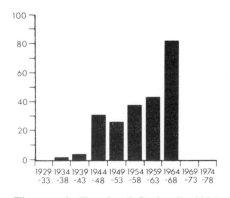

Five-yearly Totals of Gadwall, 1934–68

186

	1969	1970	1971	1972	1973	1974	1975	1976	1977	1978
Belvide	2	6	8	7	8	4	7	49	4	20
Draycote	5	0	0	1	5	5	7	29	23	44
Kingsbury	2	3	2	4	5	13	22	16	23	22
Westwood	2	2	0	5	9	10	14	26	12	10

Based on arrival dates

Monthly Distribution of Gadwall, 1934–68

With only one or two pairs throughout the Region, the breeding status of Gadwall remains tenuous, but during the last few years wintering populations have been established, principally at Draycote, Kingsbury and Westwood. The changing status is well-illustrated by the increase in annual maxima since 1969 at such principal localities.

The reasons for this increase and the origins of the birds involved remain open to speculation. Birds can now be found somewhere in the Region during all months, but there are very few in June and July. The main arrival takes place from late August onwards and numbers are at their highest between November and January. Most have departed by early April, but odd pairs linger into the early summer, raising hopes of more widespread breeding in the future.

Teal
Anas crecca

Widespread and fairly numerous winter visitor. Scarce breeding species.

The Teal is a fairly numerous winter visitor to the Region, favouring waters with emergent vegetation and extensive shallows. Previous authors all considered it common in winter and, though numbers vary considerably from year to year, there has been no significant long-term change.

The annual fluctuations probably reflect the severity of the winter, both here and on the continent, and the temporary fluctuations in suitable habitats within the Region itself. For example, especially favourable conditions existed at Draycote immediately after the reservoir was filled and 1,500 birds were estimated there in January 1969, though subsequently numbers have been much smaller. At Ladywalk, where warm water from Hams Hall Power Station inhibits freezing, up to 450 have been recorded during cold spells.

At many localities, numbers increase from August onwards to reach a peak in December or January. At Blithfield, however, low water levels regularly produce extensive muddy creeks and flats each autumn, so the maximum is characteristically earlier, though the highest ever concentration was 1,400 in December 1961 – a season which produced exceptionally high numbers nationally (Atkinson–Willes 1963).

The breeding status of the Teal is less

Table 21: Five-yearly Means of the Annual Maxima of Teal at Selected Localities, 1954–78

	1954–58	1959–63	1964–68	1969–73	1974–78
Alvecote	303	260	121	185	211
Aqualate	270	259	151	66	154
Belvide	216	109	249	288	207
Bittell	67	176	88	59	174
Blithfield	688	1,012	630	691	614
Brandon	173	188	145	420	182
Upton Warren	—	—	83	130	72

clearly defined, though the species is certainly scarce. Smith described it as nesting sparingly in lowland Staffordshire, while Harthan (1961) could cite only two Worcestershire records, at Bittell in 1892 and 1953. Norris considered that Teal possibly bred with some regularity in Warwickshire, but positive evidence was slight. Lord and Blake (1962) stated that breeding occurred sparingly in the northern moorlands, north-west and west Staffordshire and Trent Valley areas. Between 1947 and 1978, 21 reports of breeding appeared in the WM *Bird Reports* including records in three years at Belvide and Ladywalk and a report of four broods at Brandon in 1971.

During systematic studies for the WMBC *Atlas* breeding was confirmed or suspected in four and six 10-km squares respectively, almost all of which were in the Trent Valley sub-region. This distribution was considered by Lord and Munns to show an extension into north-east Warwickshire, but a disappearance from north Staffordshire. Sustained census work for the BTO *Atlas* increased the number of 10-km squares with confirmed or suspected breeding records to combined totals of nine and seven respectively, with further records in the Trent Valley, breeding again confirmed on the moorlands and isolated occurrences in the Severn and Avon valleys. Overall, the breeding population is probably between 5 and 10 pairs, which is at the lower end of the range 3–20 postulated by Lord and Munns.

The earliest recorded recoveries of Teal concerned Staffordshire ringed birds, which were found in Sweden (July 1911) and Holland (December 1913). Since then the only other recovery of a West

Table 22: Mean monthly concentrations of Teal at Principal Localities, based on all available counts from 1953/4 to 1977/8.

	Sept	Oct	Nov	Dec	Jan	Feb	Mar
Alvecote	49	50	75	100	71	59	52
Aqualate	32	74	87	140	103	79	47
Belvide	33	58	66	92	89	61	28
Bittell	17	28	56	71	60	66	37
Blithfield	370	482	494	441	273	246	117
Brandon	38	36	58	106	62	41	47

Midlands ringed bird has been in Belorussiya SSR in October 1976. However, two birds ringed in Denmark were found in Staffordshire (1933 and 1962) and a Dutch bird was shot in Worcestershire in January 1976. In 1927 a Cumbrian ringed Teal was found in Staffordshire and in the late 1930s three birds ringed in Dyfed were recovered in the Region.

A bird of the North American race *A.c.carolinensis* was reported from Baginton between January 15th and 17th 1953 and another at Kingsbury on April 16th 1980, is subject to acceptance by the *British Birds* Rarities Committee.

Mallard

Anas platyrhynchos

Widespread and numerous, both as a breeding species and winter visitor.

The Mallard is a widespread breeding species and, at most waters, is the most numerous of the wintering wildfowl. Norris, Harthan (1961) and Lord and Blake (1962) all described the species as a common resident and winter visitor. During the BTO *Atlas* project (1968–72)

breeding was confirmed in every 10-km square. The species nests on all types of water including small pools at various altitudes, suburban parks, rivers and canals, but the greatest concentrations are at the larger lakes and reservoirs. At Belvide 44 broods totalling 214 young were reared in 1970 and 39 broods totalling 230 young in 1971, while about 40 pairs bred at Packington Park in 1976. Up to a dozen pairs frequently breed at Alvecote, Kingsbury and Ladywalk. Sharrock assumed a density of 20 pairs per 10-km square based on 1972 CBC data, but the CBC showed an increase of 40 per cent in the farmland population between 1972–8. This would imply a density of 28 pairs per 10-km square, or a regional population of at least 2,000 pairs. However, the national farmland CBC indicates an even higher density of $3 \cdot 34$ pairs per km^2, which would imply 9,500 pairs on farmland alone. Whilst this may be too high, it seems likely that the total breeding population of the Region is between 2,000–10,000 pairs, which accords well with Lord and Munns' range of 2,000–20,000 pairs.

Waters such as Belvide and Blithfield act as moulting grounds, and at such

Table 23: Mean monthly concentrations of Mallard at Principal Localities, based on all available counts between 1953/4 and 1977/8

	Sept	Oct	Nov	Dec	Jan	Feb	Mar
Aqualate	955	1,114	1,146	1,410	1,079	712	326
Belvide	281	404	567	549	528	286	122
Bittell	114	156	198	198	166	140	70
Blithfield	951	1,046	1,178	1,189	992	666	218
Chillington	244	348	359	338	263	165	89
Coombe	311	528	490	479	437	375	143
Packington	1,241	867	573	397	367	295	230
Patshull	249	300	278	390	297	249	129
Sutton Park	125	130	147	184	184	120	90
Warwick Park	300	157	172	136	110	42	37
Westwood	114	112	116	133	111	88	44

Table 24: Five-yearly Means of Annual Maxima of Mallard at Selected Localities, 1954–78

	1954–58	1959–63	1964–68	1969–73	1974–78
Aqualate	1,410	1,940	1,484	1,720	1,640
Belvide	496	599	998	2,100	1,224
Bittell	283	346	232	333	310
Blithfield	1,260	1,326	1,926	1,543	1,592
Draycote	—	—	—	1,100	983

localities the breeding populations are supplemented by a post-breeding influx around June, which often involves several hundred birds and includes a high proportion of drakes. The autumn immigration gains momentum during September and at most localities numbers reach their peak at the end of the year. At Belvide, Smallshire and Richards (1976) recognised three principal influxes, in late October/early November, mid-December and late January.

At the request of the Wildfowl Trust, a study of the proportion of the Region's wintering wildfowl frequenting minor waters, rivers and canals was carried out during 1954/5. In the case of the Mallard, this proved to be a significant proportion, reaching approximately 20 per cent in mid-winter and as much as 35 per cent in August and from February to April (A. R. M. Blake *in litt.*).

The spring exodus begins in February and, if the weather remains mild, numbers may decline rapidly at the main wintering sites, so that relatively few birds remain by early March.

The total wintering population does not appear to have altered significantly in the last twenty-five years, though at Belvide there has been an increase. There have been local decreases resulting from increased disturbance (e.g. sailing activities at Gailey and Shustoke) but these have been more than offset by the advent of Draycote.

The largest concentrations recorded were 2,549 at Blithfield in January 1967, 3,100 at Belvide in January 1969 and a remarkable report of 4,200 at Packington Park in September 1974.

The only foreign recovery of a Mallard concerns a duckling ringed at Blithfield in June 1965 and found in Jutland (Denmark) in November 1966. Within Britain, 20 birds have been recovered in random directions from the point of ringing and at distances up to 200 km.

Pintail
Anas acuta

Regular, but scarce, winter visitor and passage migrant.

The Pintail is a regular, but scarce, winter visitor to the Region. Over much of the area its current status is little different from that described by Smith, Harthan and Norris, with one or two birds appearing regularly at Belvide and intermittently during most winters at Alvecote, Aqualate, Bittell, Brandon, Kingsbury, Ladywalk and Upton Warren. There are occasional reports from all other waters of significant size. The creation of large reservoirs at Blithfield (1952) and Draycote (1969) led to a substantial increase in both frequency and numbers, but even these two sites seemed to be most attractive immediately after filling, as numbers

39. The Pintail is a regular, but scarce, winter visitor to a few favoured waters.

T. C. Leach

have subsequently declined.

Over thirty-four years, the average first arrival date for winter visitors has been September 10th and over thirty-nine years the average last departure date has been April 15th. The extreme arrival and departure dates are July 8th (1967) and May 24th (1975), though these may be mis-leading because birds have twice appeared in June. Most birds pass through the Region and only at Blithfield and, to a lesser extent, Belvide, Draycote and the Tame Valley, is the species at all consistently present through the winter.

Notable locality maxima include the following, most involving transient parties

Table 25: Five-yearly Means of the Annual Maxima of Pintail at Selected Localities, 1954–78

	1954–58	1959–63	1964–68	1969–73	1974–78
Alvecote	3	2	2	4	3
Belvide	9	6	5	9	7
Bittell	2	2	2	5	2
Blithfield	21	8	15	12	8
Draycote	—	—	—	16	5

Table 26: Mean monthly concentrations of Pintail at Belvide and Blithfield, based on all available counts between 1953/4 and 1977/8

	Oct	Nov	Dec	Jan	Feb	Mar
Belvide	0	1	1	0	1	0
Blithfield	1	2	4	4	4	3

recorded on only a single date: 30 at Belvide on February 14th 1943; 24 at Bartley on January 31st 1948; 36 at Bittell on January 8th 1949; 32 at Chasewater on December 20th 1950; 23 at Gailey on December 23rd 1954; 36 at Blithfield on December 30th 1954; 40 at Draycote on February 23rd and March 2nd 1969; and 21 at Shustoke on February 17th 1974.

Pintail occasionally appear during summer, with June or July records at Blithfield in five years between 1957 and 1977 and odd records in these months from Belvide, Ladywalk and Upton Warren. A drake which was resident at Draycote during 1976 and 1977 had perhaps escaped from a collection.

Garganey
Anas querquedula

Regular, but scarce, passage migrant. Irregular and very scarce breeding species.

In the West Midlands the Garganey is principally a passage migrant to shallow pools fringed by rank vegetation, though single pairs have bred in at least ten years since 1945. Nesting has occurred twice at Alvecote, Baginton and Belvide, and was proved or strongly suspected in the Tame Valley on four occasions between 1967 and 1972. During the *Atlas* projects (1966–72) breeding was confirmed in two 10-km squares within the Region and considered probable in a further three. However, even allowing for the secretive behaviour of breeding birds, it seems un-

likely that the regional population in any one year exceeds one or two pairs.

As a migrant the species was regarded by Smith, Harthan and Norris as occasional, but it has subsequently shown a steady increase and was described as a regular visitor by Lord and Blake (1962).

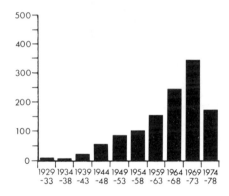

Five-Yearly Totals of Garganey, 1929–78

The species is now more or less annual at Alvecote, Belvide, Bittell, Blithfield, Brandon, Draycote, the Tame Valley and Upton Warren, but is considerably more numerous in some years than in others. The late 1960s and early 1970s were particularly productive. At specific localities numbers rarely exceed a dozen, and are typically much less, but during August 1968 up to 26 were present at Blithfield and up to 16 at Upton Warren, while there were 15 at Belvide on August 16th, 1975.

The average dates of first and last birds over thirty-five and thirty-four years respectively have been March 31st and September 26th, whilst the extreme dates were March 2nd (1969) and October 21st

(1971). The spring movement reaches a peak in late April and early May. In the autumn it frequently returns at the end of July, but August is the peak month, though in good years birds are often present throughout the autumn at certain localities and September arrivals are perhaps not always distinguished from birds lingering from earlier months.

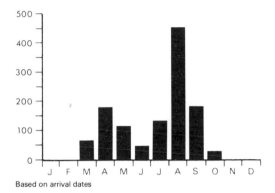

Based on arrival dates

Monthly Distribution of Garganey, 1929–78

Blue-winged Teal

Anas discors

Very rare vagrant. Two records.

A drake Blue-winged Teal was present at Upton Warren from March 6th to 27th 1968 (*Brit. Birds* 62:464). British records of this North American duck prior to 1970 are considered unlikely to involve escapes (Sharrock and Sharrock 1976). More recently, a male and an immature were present at Upton Warren again, from September 27th to October 12th 1980.

Shoveler

Anas clypeata

Fairly numerous winter visitor and passage migrant. Scarce as a breeding species.

Previous authors all regarded the Shoveler as a regular winter visitor and passage migrant and this is still the case, though there are considerable local variations in abundance. In the first half of this century numbers at any one locality rarely exceeded fifty, and in Warwickshire and Worcestershire flocks of a dozen or so birds were considered notable. On many waters numbers remain small, but at certain localities a large increase in passage and winter concentrations has occurred during the last twenty-five years. In the early 1950s the annual peak at Belvide was usually around 50, and a concentration of 100 birds on February 20th 1949 was described as exceptional. By the late 1960s, however, the yearly maximum usually exceeded 100 and reached 385 in 1974, 300 in 1976 and at least 460 on October 4th 1980. Similarly, 200 were recorded at another important site, Coombe Abbey, in 1969 and over 100 birds have also occurred at Alvecote, Brandon and Upton Warren in recent years. Exceptional numbers occasionally appear at Aqualate, including the regional maximum of 520 on October 16th 1977. By contrast, numbers at Blithfield have decreased during the present decade. In the 1950s and 1960s the peak was regularly 100 or more, with as many as 230 as late as 1972, but since then numbers have only once reached treble figures. Draycote has not yet proved particularly attractive to the species, with a maximum of only 40 in December 1975.

Prior to 1950 peak numbers were frequently during the spring, typically March (Atkinson–Willes 1963). More recently, along with a general increase in numbers, the maximum concentrations have usually occurred during October or November. During the last decade, the mean October maximum at Belvide has risen to over 180, compared with 61 during the period 1954 to 1978. In several recent years there have probably been in excess of 800 birds in the Region at this season,

Table 27: Five-yearly Means of the Annual Maxima of Shoveler at Selected Localities, 1954–78

	1954–58	1959–63	1964–68	1969–73	1974–78
Alvecote	6	2	2	90	62
Aqualate	73	279	176	100	288
Belvide	69	116	81	151	272
Blithfield	106	117	135	93	47
Upton Warren	—	—	18	56	108

though in the winter months numbers decline significantly.

As a breeding species the Shoveler is scarce, nesting regularly only in the Penk, Sow and Tame basins, though it has also bred more than once in recent years at Brandon and Coombe. Before 1940 there are few published records of nesting, although Tomes (1901) referred to nesting at Bittell in 1884 and 1898, Norris mentioned apparently fairly regular nesting at Middleton up to 1891 and Smith described it as regular in Staffordshire. Following the national trend (Cramp and Simmons 1977), records increased during the next few years and by 1970 Lord and Munns were able to describe the species as not scarce. During the *Atlas* projects (1966–72) breeding was confirmed in seven 10-km squares (9 per cent) and considered probable in a further four (5 per cent). In the 1970s one or two pairs have bred almost annually in the Tame Valley (mainly Kingsbury and Ladywalk), but during the same period there is evidence of a decline at Belvide, which is historically the most consis-

tent breeding site. Here the species bred regularly until 1948, erratically to 1956 and then annually until 1972, with eight broods in 1970 (Smallshire and Richards 1976), but in the following six years no young were seen. Similarly, breeding occurred at Blithfield in at least six years between 1955 and 1966, but there has been only one subsequent record, in 1972. Lord and Blake (1962) described breeding as regular at Aqualate, where there were four or five pairs in 1953. This may still be a regular site, but recent information is lacking. Breeding has always been – and remains – very irregular in Worcestershire and the only published record this century appears to be that of a pair with eight young at Upton Warren in 1947. In 1970 Lord and Munns estimated the Shoveler's breeding strength as 20–200 pairs, but this seems to have been an over-estimate and it is doubtful whether more than five pairs nest in most years.

Shoveler are rarely ringed nowadays, but were apparently caught with some success at Middleton Hall early this century. At

Table 28: Mean monthly concentrations of Shoveler at Aqualate and Belvide, based on all available counts from 1953/4 to 1977/8

	Sept	Oct	Nov	Dec	Jan	Feb	Mar
Aqualate	75	124	88	61	56	33	37
Belvide	34	61	70	44	39	36	35

40.   The Region's waters, particularly those in Staffordshire, appear to hold a significant proportion of British wintering Shoveler.   *M. C. Wilkes*

least two recoveries resulted. One ringed on January 10th 1915 was reported from Gwynedd in January 1917, whilst the other survived five years from ringing on January 16th 1915 to recovery in Jylland (Denmark) on August 7th 1920.

Red-crested Pochard
Netta rufina

Very scarce, but increasingly regular visitor, though most records are considered to involve escapes.

During the twentieth century the west European breeding range of the Red-crested Pochard extended significantly, reaching Denmark about 1940. In the next few years continental birds visited Britain not infrequently, and the species was considered a rare, but regular, autumn visitor to south-east England from about 1952 to 1962 (Ryman 1959). Following a decline in the Danish breeding population, however, the species returned to vagrant status after the mid-1960s. Conversely, small numbers of escaped birds have bred regularly in one or two English counties since 1968 (Cramp and Simmons 1977).

In the West Midlands, Red-crested Pochards were rarely recorded before the mid-1960s, but have appeared with increasing regularity since. At least two-thirds of the records have involved drakes.

This pattern suggests that most records have involved birds escaped from captivity and the monthly distribution of reports does little to alter this conclusion. There is

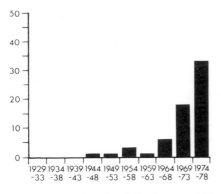

Five-yearly Totals of Red-crested Pochard, 1944-78

a spring peak in May and a marginal autumn peak in October, but the distribution over the seven months July to January is suspiciously even.

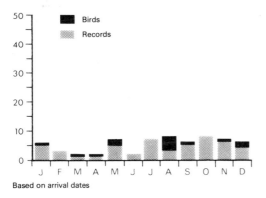

Based on arrival dates

Monthly Distribution of Red-crested Pochard, 1867 and 1944-78

Most individuals have remained only a day or two, but birds made prolonged stays in 1964/5 and in each year from 1972 to 1976, with up to four at Draycote between August 17th and October 26th 1973.

The most favoured localities have been Belvide (a total of about 19 birds in eight years since 1949), Draycote (15 birds in six years since 1972), Gailey (records in 1867 and in five years since 1964) and Blithfield (single birds in five years since 1955). A further 16 localities have produced between one and three records. Despite comments

by Smith it seems probable that the bird at Gailey in late October 1867 was a genuine vagrant, as this occurrence pre-dated the widespread establishment of the species in collections. A young bird ringed at London on July 18th 1952 was recovered at Trentham a month later.

Pochard
Aythya ferina

Fairly numerous and widespread winter visitor. Scarce breeding species.

The Pochard's breeding habitat requirements of large pools fringed with dense emergent vegetation are rather restricted and, unlike the Tufted Duck, the species has not taken much advantage of man-made habitats such as gravel pits and reservoirs. During this century, however, there has been a slow increase in the numbers breeding in England (Sharrock).

Norris referred to nesting in Edgbaston Park in 1866 and Smith said it bred in small quantities in Staffordshire, but the Pochard is still a very scarce breeding species in the West Midlands and nests regularly at only two localities. At Chesterton Mill Pool two or three pairs have bred in most years since 1946, while at Alvecote around five pairs have bred since 1950, with nine broods totalling 30 young in 1969. The species also bred at Westwood in at least four years between 1966 and 1971 and at Oakley in 1978 and 1979, while single pairs have nested sporadically at several other localities. During the *Atlas* projects (1966-72), breeding was confirmed or suspected in ten and six 10-km squares respectively, the breeding sites being widely scattered and involving all three shire counties. Even so, the regional breeding population is probably no more than about 10 pairs, which is in the middle of the range 2-20 pairs estimated by Lord and Munns.

196

Table 29: Five-yearly Means of Annual Maxima of Pochard at Selected Localities, 1954–78

	1954–58	1959–63	1964–68	1969–73	1974–78
Alvecote	222	365	242	262	244
Belvide	267	236	240	204	225
Bittell	18	97	45	50	79
Blithfield	326	312	179	449	690
Chasewater	176	—	104	99	102
Draycote	—	—	—	326	694
Gailey	200	266	184	122	85
Kingsbury	—	—	156	203	208
Shustoke	89	212	112	228	197

As a winter visitor the Pochard is much more widespread and numerous. At many localities wintering numbers have shown no significant trend during the past twenty-five years, though Blithfield has recently experienced an increase and the relatively new Draycote Water has attracted very large numbers in the last few years.

From June to September moulting flocks, consisting mainly of drakes at first, appear at several localities and at Belvide up to 150 birds have occurred at this time during recent years. The principal immigration of winter visitors, though, takes place during late October and November and large transient flocks are occasionally reported during these months at waters normally holding relatively few birds. The principal wintering waters, especially the larger reservoirs, frequently experience peak numbers at this time. The largest recorded counts were 1,200 at Blithfield in November 1976; 1,255 at Draycote in January 1978; and up to 1,700 in the Kingsbury – Lea Marston area during the winter of 1980/1.

As with Shoveler, two birds ringed at Middleton Hall during the first twenty years of this century provide the only recoveries. The first of these was ringed in February 1913 and reported from Mecklenburg (Germany) in the following August,

Table 30: Mean monthly concentrations of Pochard at Selected Localities, based on all available counts from 1953/4 to 1977/8

	Sept	Oct	Nov	Dec	Jan	Feb	Mar
Alvecote	11	57	98	129	111	176	117
Belvide	46	100	110	97	78	37	25
Bittell	3	24	28	19	35	21	14
Blithfield	47	149	251	225	176	127	60
Chasewater	6	42	45	27	23	25	10
Gailey	44	71	90	122	49	75	31
Shustoke	4	26	77	50	86	57	28
Westwood	12	32	64	63	49	39	24

whilst the other was ringed on November 1st 1919 and then recovered at Burton-on-Trent in January 1921.

Ring-necked Duck
Aythya collaris

Very rare vagrant. Five records.

During 1977 and 1978 Britain experienced a remarkable influx of Ring-necked Ducks (Rogers *et al*. 1979), a species which has been expanding eastwards across North America during the past thirty years (Cramp and Simmons 1977). Three or four birds were recorded in the Region, and there was a further occurrence in 1979.

A male was at Blithfield from November 20th to 22nd 1977 (*Brit. Birds* 72:515); an immature male visited Draycote from December 24th 1977 to February 22nd 1978 and then transferred to Brandon, where it remained until April 5th (*Brit. Birds* 72:515 and 73:500); an immature male was at Kingsbury from January 7th to 15th 1978 and a male at Alvecote on January 14th and 19th was probably the same bird (*Brit. Birds* 72:515 and 73:500); and an adult drake was at Kingsbury from May 3rd to 9th 1979, but this record is still subject to the *British Birds* Rarities Committee's acceptance.

Ferruginous Duck
Aythya nyroca

Rare visitor. Fifteen records.

Between 1949 and 1978 there were 13 records of the Ferruginous Duck, with one report involving two birds. It is doubtful whether all these records involved wild birds, though a significant proportion occurred during the autumn months. There were two additional records in 1979.

One was at Earlswood on February 13th 1949; a drake visited Alvecote from October 10th to 29th 1950; an immature male was at Edgbaston Reservoir on August 2nd 1957; an immature male was at Gailey from February 21st to 29th 1960 (*Brit. Birds* 54:182); an immature male was at Edgbaston Park from April 30th to May 13th 1960 (*Brit. Birds* 54:182); a duck visited Bittell on October 3rd 1961 (*Brit. Birds* 55:569); a drake was seen at Bittell on September 9th 1968 (*Brit. Birds* 62:465); an immature duck was shot at Sambourne on November 20th 1969; an immature male was at Draycote on December 9th 1972; another immature visited Shustoke on February 18th 1975; two immatures were at Chasewater on September 14th 1977; a drake was at Belvide from July 12th to 16th 1978; an immature male stayed at Draycote from November 12th to December 16th 1978; a male was at Wormleighton on December 16th 1979; and an immature male, perhaps the same bird, at Draycote on December 26th 1979.

Tufted Duck
Aythya fuligula

Widespread and fairly numerous breeding species and widespread and numerous winter visitor.

The Tufted Duck was unknown as a breeding species in Britain until the mid-nineteenth century. During this century it has expanded significantly and now nests on islands and amongst rank waterside vegetation in many places. A few pairs breed in urban parks, but in particular it has taken full advantage of the new artificial habitats provided by gravel pits and reservoirs (Sharrock).

Smith regarded breeding as widespread in Staffordshire, but, apart from an isolated case in 1893, nesting was not recorded in Warwickshire until 1934 (Norris) nor in Worcestershire until 1936

41. Once scarce in the Region, the Tufted Duck is now well-established as a breeding species and is the most numerous diving duck in winter. *M. C. Wilkes*

(Harthan). By 1968 breeding was occurring in at least twenty-two 10-km squares (29 per cent) within the Region (Lord and Munns), while during the BTO *Atlas* project breeding was confirmed in forty-three squares (56 per cent), covering all areas except the Carboniferous limestone district of north Staffordshire and the Lower Lias beds south of the Avon, where enclosed waters are scarce. Overall the combined *Atlas* projects revealed confirmed or suspected breeding in fifty-four 10-km squares (70 per cent), which, at a density of 3 or 4 pairs per square (Sharrock), points to a regional population of 150–220 pairs, compared to Lord and Munns' estimate of 20–200 pairs.

The more important breeding sites include Alvecote (22 broods totalling 128 young in 1972), Belvide (24 broods totalling 123 young in 1970) and the Kingsbury area (31 broods totalling 180 young in 1977). In recent years this species, along with other wildfowl, has tended to use Belvide as a post-breeding moulting ground. In 1980, 718 birds were present during August and this proved to be the highest concentration of the year at that locality.

During the winter, the Tufted Duck is the most widespread and numerous diving duck and can be found at most reservoirs, gravel pits and ornamental lakes. A few also visit rivers, canals and urban parks. At several localities numbers have shown signs of an increase during the past decade. More remarkably, the relatively new Draycote Water has attracted unprecedented numbers in the last few winters, reaching 2,000 in January 1978,

Table 31: Five-yearly Means of the Annual Maxima of Tufted Duck at Selected Localities, 1954–78

	1954–58	1959–63	1964–68	1969–73	1974–78
Alvecote	111	101	54	215	195
Belvide	262	173	138	355	545
Bittell	42	86	135	146	160
Blithfield	365	390	242	316	93
Chasewater	81	—	—	197	396
Draycote	—	—	—	315	953
Gailey	260	224	271	296	321
Kingsbury	—	—	—	218	319
Shustoke	—	—	—	131	117

which trebled the previous regional maxima of 600 at Blithfield in January 1962 and 680 at Belvide in December 1976. Contrary to this trend, the numbers frequenting Blithfield have declined (cf. Shoveler). Prior to 1974 the annual maximum at this reservoir was usually around 300, but subsequently numbers have rarely reached three figures. As disturbance at Draycote is at least as great as at Blithfield, these trends are almost certainly related to food availability. Smallshire (*in litt.*) suggests that the mollusc population declined with the low water-levels during the autumns of the 1970s.

Peak concentrations occur significantly earlier in the season at some localities than at others, perhaps reflecting the varying degree to which they are used as late summer and autumn moulting grounds.

The most recent known West Midlands recovery of a Tufted Duck involved one ringed in London in January 1952 and found in Burton-on-Trent in December 1953. Prior to that the only recovery was one ringed at Molesey (Surrey) in September 1933 and reported from Kings Norton in March 1936.

Table 32: Mean monthly concentrations of Tufted Duck at Selected Localities, based on all available counts, 1953/54 to 1977/78

	Sept	Oct	Nov	Dec	Jan	Feb	Mar
Alvecote	28	48	63	82	88	91	83
Belvide	103	114	145	145	99	67	86
Bittell	34	58	53	49	46	45	34
Blithfield	124	172	172	170	158	137	174
Chasewater	26	78	51	91	123	114	66
Gailey	202	106	101	143	90	141	97
Packington	139	117	91	105	126	115	117
Shustoke	31	43	64	67	63	58	32

Scaup
Aythya marila

Scarce, but regular, winter visitor and passage migrant.

The Scaup was described by Harthan as an uncommon winter visitor to Worcestershire and by Norris as a rare straggler to Warwickshire. By contrast, Smith regarded the species as a regular winter visitor to Staffordshire, most often to Chasewater and Gailey, and this assessment was echoed by Lord and Blake (1962), who also cited Belvide and Blithfield as regular haunts.

Between 1934 and 1978 at least 437 Scaup were recorded in the Region. Numbers increased rapidly in the mid-1950s, declined dramatically in the following decade, and then recovered strongly in the 1970s. Currently, the species is appearing annually in both Staffordshire and Warwickshire, but rather less regularly in Worcestershire and the West Midlands County. Inland occurrences are known to vary in frequency, but no close association between numbers inland and on the coast has been demonstrated (Atkinson–Willes 1963), nor does there appear to be a strong correlation with severe winters. It is noteworthy, however, that a similar pattern of decline and recovery during the past twenty years has been exhibited by the divers. In recent years an increasing number of hybrids has created considerable confusion.

The immigration of Scaup to British coastal waters reaches a peak in late October, while the spring exodus commences in mid-March (Cramp and Simmons 1977). These movements are reflected by subsidiary peaks in the monthly distribution of arrivals in the West Midlands. The principal peak, however, occurs in December and January and, although cold winters do not always produce increased arrivals, this mid-winter

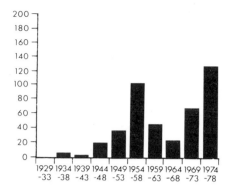

Five-yearly Totals of Scaup, 1929–78

peak is presumably related to periods of hard weather. Summer records are scarce, but increasing, and in 1978 a drake remained at Belvide from June 3rd to October 14th.

Most occurrences have involved between one and three birds, but there are nineteen records of between four and eight individuals. A party of ten appeared at Belvide on March 22nd 1963, while up to six at a time and perhaps as many as 12 different individuals were present at Draycote during the winter of 1978/9 and up to ten were at Blithfield in late November and December 1979.

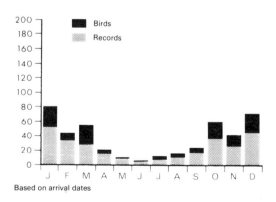

Based on arrival dates

Monthly Distribution of Scaup, 1929–78

Most birds remained in the area for only a day or two, but about 70 stayed for over a month. Less than half of the published records include data on sex, but, of those sexed, 43 per cent were recognisable drakes.

The species has occurred at Belvide in twenty-three years since 1939, Chasewater in twenty years since 1949 (and in every year since 1968), Blithfield in sixteen years since 1953, Bittell in sixteen years since 1934, Gailey in fifteen years since 1939, Bartley in thirteen years since 1938 and Draycote in seven years since 1969.

Eider
Somateria mollissima

Rare winter visitor.

This maritime duck was first recorded in both Worcestershire and Staffordshire during 1948. There have subsequently been 18 records totalling 34 birds, including the long-awaited first Warwickshire record in 1978. Thirteen of the occurrences were during the 1960s, with Chasewater, a favoured locality for maritime species, producing records in no less than seven years during the decade.

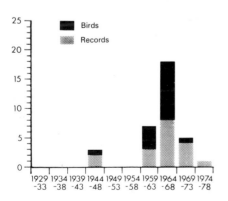

Five-yearly Totals of Eider, 1944–78

The incursion during the 1960s has been associated with a marked southerly extension of the wintering range (Dean 1971). The subsequent decline perhaps reflects

decreases suffered by breeding populations in the Netherlands, which provide most of the birds wintering on the south and east English coasts (Taverner 1959, 1967).

December and January have provided the majority of records, all of which have involved immatures or ducks, and this suggests birds seeking refuge from rough weather on the coast.

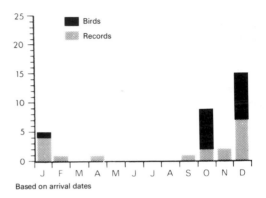

Based on arrival dates

Monthly Distribution of Eider, 1948–78

Most birds have remained only a few days, but seven have stayed for three weeks or more, with the longest stay involving an immature drake at Belvide from January 16th to April 19th 1964. This bird also visited Gailey on April 12th.

Chasewater has been by far the most favoured locality, with ten records totalling 26 birds between 1948 (the first Staffordshire occurrence, two birds from October 31st to November 27th), and 1972. All records of more than one bird have been from this reservoir, including four on December 27th 1961, seven on October 15th 1966 and five from December 16th to 31st 1968. Single birds have appeared three times at Belvide, once visiting Gailey; twice at Bittell (including the first regional and Worcestershire record on April 21st 1948); and once each at Adbaston, Branston and Draycote (the first Warwickshire record from November 14th to December 16th 1978).

Long-tailed Duck
Clangula hyemalis

Very scarce, but fairly regular, winter visitor. Rare in summer.

Previously a very rare visitor to the Region (Smith, Harthan and Norris), the Long-tailed Duck has occurred more frequently during the past thirty years and was described by Lord and Blake (1962) as a fairly regular winter visitor to Staffordshire. In the fifty years up to 1978 there were 58 records of single birds and seven records of two. Typically, only one or two individuals are recorded each year, and never more than seven, but the species appeared in all but six years between 1948 and 1978.

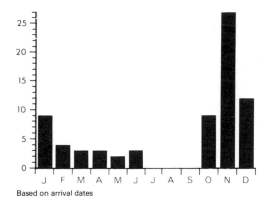

Based on arrival dates

Monthly Distribution of Long-tailed Duck, 1929–78

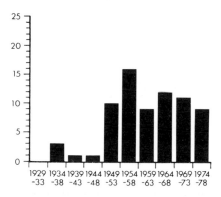

Five-yearly Totals of Long-tailed Duck, 1929–78

The main influx of Long-tailed Ducks into northern British waters occurs during October. Subsequent southerly movements are reflected by a distinct November peak in the monthly distribution of arrivals in the West Midlands.

In recent years there have also been several late spring and summer records, the most notable involving an adult drake at Alvecote from June 28th to July 13th 1969; an immature male, entangled with fishing line, in the Tame Valley between June 10th and August 9th 1973; and a bird at Draycote from June 13th to 16th of the same year.

Over a third of the birds remained only a day, but a significant proportion (25 per cent) stayed for a month or longer.

Nearly all Long-tailed Ducks occurring inland are immatures or ducks (Smout 1969) and only five West Midlands records have been published as specifically involving adult drakes.

The most frequently visited localities have been Belvide (16 birds in ten years since 1937), Blithfield (ten birds in seven years since 1954), and Chasewater (eight birds in six years since 1949). Bartley, Bittell, Draycote, Gailey, Shustoke and Upton Warren have each produced between three and five birds during the period of analysis.

Table 33: Duration of Visits of Long-tailed Duck, 1929–78

Duration of stay in days	1	2–5	6–9	10–19	20–29	30 and over
No. of birds	27	4	9	8	6	18

Common Scoter
Melanitta nigra

Scarce, but regular passage migrant.

Smith said the Common Scoter visited south Staffordshire in autumn and winter, sometimes in fairly large numbers, but Harthan called it irregular and Norris occasional. During the past fifty years, however, Common Scoters have appeared in the Region with steadily increasing frequency. This probably reflects the increase in both observers and suitable waters, though, by contrast, records in the London area have declined since the mid-1960s (Chandler and Osborne 1977).

recognised, with three distinct peaks; in April–May, July–August and October–November respectively (Spencer 1969). The spring movement, and more particularly the July–August passage, contain a high proportion of drakes, while in the late autumn immatures and ducks predominate. Occurrences in the West Midlands conform closely with this pattern, both in timing and in sex-age composition (Lord 1962, Hawker 1970, Dean 1971), though data on the latter have been included for less than half the published records.

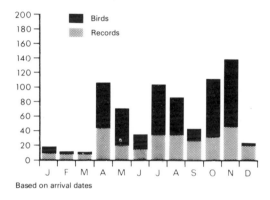

Monthly Distribution of Common Scoter, 1929–78

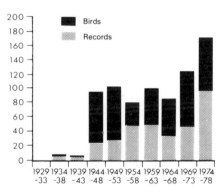

Five-yearly Totals of Common Scoter, 1929–78

Although primarily a maritime species, the Common Scoter is well known for its overland moult and passage migrations. In several inland regions of England purposeful east to west movements have been

Although records of single birds are the most frequent over the year as a whole, parties of two to five occur fairly regularly during the passage months. In fact, it is this tendency towards small parties, rather than a marked increase in the number of occurrences, which produces the prominence of the three seasonal peaks. Flocks of

Table 34: Monthly Variation in Sex and Age Composition (where recorded) of Common Scoter, 1929–78

	Jan	Feb	Mar	Apr	May	Jun	Jul	Aug	Sep	Oct	Nov	Dec
Recognisable Drakes	4	2	3	32	15	4	34	23	11	8	6	3
Immatures and Ducks	7	2	2	14	8	6	10	9	7	53	89	12

Table 35: Size of Common Scoter Flocks, 1929–78

Flock Size	1	2–5	6–9	10–19	20 and over
No. of Records	187	90	14	11	4
Total No. of Birds	187	250	103	130	105

more than half-a-dozen are scarce, but during the period 1929–78 there have been four involving 20 or more birds, namely 30 at Bartley on November 9th 1946; 25 at Belvide on October 30th 1949; 20 at Chasewater on May 5th 1969 and 30 at Belvide on August 20th 1970.

The localities most frequently visited by Common Scoters are Alvecote (records in ten years since 1957), Bartley (records in 23 years since 1934), Belvide (22 years since 1940), Bittell (17 years since 1938), Blithfield (19 years since 1954) and, particularly, Chasewater (26 years since 1951 and annually since 1961). There have been occurrences at Draycote in six years since 1972 and this locality, together with the south Staffordshire reservoirs, frequently produces several records in the course of a year.

Surf Scoter
Melanitta perspicillata

One record.

Tomes (1904) records an adult male shot some years before 1904 a few miles below Stratford-on-Avon. Norris noted that this bird was in Tomes' collection, though it does not receive mention in the *Handbook*. (Witherby *et al.* 1938.)

Velvet Scoter
Melanitta fusca

Rare and irregular winter visitor and passage migrant.

The Velvet Scoter is a decidedly rare visitor to the Region. A male at Bittell in May 1920 seems to be the first documented record, while the fifty years 1929 to 1978 produced only a further 30 occurrences involving 47 birds.

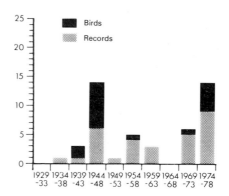

Five-yearly Totals of Velvet Scoter, 1929–78

The most productive years were 1947, when there were four records totalling nine birds, and 1976, which produced five records totalling eight birds.

Occurrences have been fairly evenly

Table 36: Duration of Visits of Velvet Scoter, 1929–78

No. of Days	1	2–5	6–9	10–19	20 and over
No. of Birds	30	10	1	3	3

spread between the months October to January, with rather fewer in spring. This suggests that visits of Velvet Scoters result from birds penetrating inland at the time of the autumn influx into British coastal waters as much as from hard weather movements.

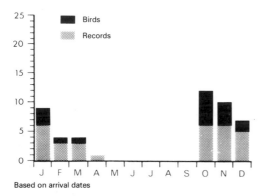

Monthly Distribution of Velvet Scoter, 1929–78

Nineteen of the 30 occurrences involved single birds, while there were six records of two, four records of three, and one record of four birds – at Bartley on October 19th 1947. Most individuals were immatures or ducks and only 11 drakes have been identified. Very few birds remained more than a week, but of three which arrived at Belvide on October 13th 1974, two remained until November 23rd and one until November 27th. These constitute the longest visits.

The most favoured localities have been Bartley (six records totalling 11 birds since 1947), Belvide (five records totalling ten birds since 1947), Bittell (one in 1920 and three records totalling six birds since 1941), and Chasewater (seven records totalling nine birds since 1956). Draycote, Gailey and Shustoke have each produced two records and single birds have appeared at Kingsbury, Tamworth and Upton Warren. Outside the period of this analysis, birds were reported from Sandwell Valley in January 1979 and Chase-

water in April 1979, this being only the second April record.

Goldeneye
Bucephala clangula

Regular winter visitor, not scarce.

The Goldeneye has always been a regular winter visitor to the Region, with Bartley, Bittell and the Staffordshire reservoirs being traditional haunts (Smith, Harthan, Norris, Lord and Blake 1962). During the last fifteen years, however, there has been a significant increase in the numbers frequenting certain localities. Before the mid-1960s, Chasewater was considered a poor water for Goldeneye, but numbers have subsequently increased steadily and, despite wide fluctuations due to disturbance by boating activities, over 60 are now recorded during most winters. At Belvide, always a good site for this species, unprecedented numbers have occurred in recent winters, including the regional maximum of 139 on January 18th 1976. Numbers in the Tame Valley have increased substantially during the past decade too, while Draycote has also proved attractive to the species.

Associated with this increase in numbers has been a general broadening in geographical spread. Nearly all waters of any size are now visited annually, though at many localities only one or two birds are involved. At Bartley, a drinking water reservoir from which birds used to be dis-

206

Table 37: Five-yearly Means of the Annual Maxima of Goldeneye at Selected Localities, 1954–78

	1954–58	1959–63	1964–68	1969–73	1974–78
Belvide	56	55	52	90	121
Blithfield	49	41	34	55	55
Chasewater	—	—	13	44	63
Draycote	—	—	—	13	46
Kingsbury	—	—	7	29	46
Shustoke	5	3	2	15	22

couraged, one or two Goldeneye are frequently the only resident wildfowl during the winter months. In contrast to this general trend, the numbers visiting Worcestershire waters remain small and very rarely reach double figures, though there were 21 at Bittell on January 7th 1951.

Over thirty-eight years the average date for first arrivals has been September 29th, whilst the average date over thirty-six years for the last birds to depart has been May 5th. The extreme dates were August 1st (1956 and 1973) and June 5th (1974 and 1975) respectively, though some individuals have summered in recent years. In the Region as a whole, numbers increase rapidly during late October and November and then stabilise towards the end of the year. At several larger waters, however, small but steady increases occur throughout the winter and numbers reach a peak in March. These late arrivals are presumably either passage migrants, or local birds deserting smaller waters and joining larger assemblies in preparation for their spring exodus. Large numbers remain well into April at some localities, such as Belvide.

Away from Belvide, noteworthy locality maxima have included 20 at Bartley on December 19th 1950, 70 at Blithfield in April 1973 and January 1974, 70 at Chasewater in December 1977 and 100 at Draycote during February 1979.

Isolated individuals are not infrequently recorded in the summer. There has been a total of 28 June/July records, involving ten localities, in eighteen years since 1951 and every year since 1968. Several birds have made prolonged stays and a drake has been seen at Kingsbury in each summer but one since 1971, though two different birds have probably been involved during this period.

Table 38: Mean monthly concentrations of Goldeneye at Selected Localities, based on all available counts between A 1953/54 and 1977/78 and B 1969/70 and 1977/78

	Oct		Nov		Dec		Jan		Feb		Mar	
	A	B	A	B	A	B	A	B	A	B	A	B
Belvide	4	4	28	33	38	56	45	77	47	70	52	76
Blithfield	3	—	18	—	23	—	24	—	31	—	34	—
Chasewater	2	1	6	12	19	31	23	34	21	34	16	22
Shustoke	—	—	9	—	8	—	8	—	5	—	3	—

Smew

Mergus albellus

Scarce, but regular winter visitor.

The Smew is a regular, but scarce, winter visitor to the Region, with the fifty years to 1978 having produced a total of 391 birds. With the exception of 1974, the species has appeared annually since 1935. The distribution of records, however, displays a distinct correlation with cold winters, with the 1950s and early 1960s being particularly productive while the subsequent succession of mild winters produced relatively few birds. The species has always been considered a regular visitor to south Staffordshire (Smith), though Harthan and Norris reported it as infrequent elsewhere in the Region.

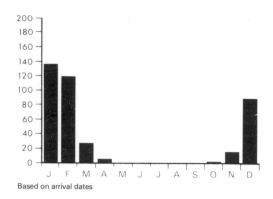

Based on arrival dates

Monthly Distribution of Smew, 1929–78

Belvide involving ten or more individuals, with the largest flock being of 12 birds on January 4th 1941. Prior to this period, 16 were present at the same locality on January 1st 1927.

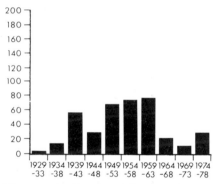

Five-yearly Totals of Smew, 1929–78

Reflecting the association with cold weather, the monthly distribution of records displays a peak in January and February, though birds have appeared in all months from October to April.

Most occurrences have involved from one to five birds, but, during the last fifty years, there have been three records at

Table 39: Flock-size of Smew, 1929–78

Size of Flock	1	2–5	6–9	10 and over
No. of Records	143	67	5	3
No. of Birds	143	181	35	32

Immatures and ducks far outnumber mature drakes, with only 67 of the latter receiving specific mention among the published records. Although transient birds are in the majority, a significant proportion of birds has remained in the area for long periods, with birds present at Belvide from January to April of each year from 1959 to 1962.

There have been records from Belvide in thirty-one years since 1933 (and every year

Table 40: Duration of Visits of Smew, 1929–78

No. of Days	1	2–5	6–9	10–19	20 and over
No. of Birds	253	20	14	23	81

from 1957 to 1968), Blithfield in fifteen years since 1955 (and every year from 1955 to 1964), Bittell in fifteen years since 1938, Gailey in fourteen years since 1935 and Alvecote in ten years since 1951. Bartley, Chasewater, Draycote and Shustoke have been visited in six, seven, four and seven years respectively.

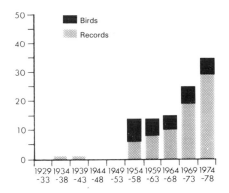

Five-yearly Totals of Red-breasted Merganser, 1929–78

Red-breasted Merganser
Mergus serrator

Scarce, but regular winter visitor and passage migrant.

Before the middle of this century the Red-breasted Merganser was a very rare visitor to the Region. Smith, Harthan and Norris included a total of only about 14 records, of which all but four were from Staffordshire. Since then, however, the species has become an increasingly regular visitor – a change probably associated with the marked southward extension of the breeding range (Dean 1971). In 1950 Mergansers began nesting in north-west England and by 1953 they had spread to North Wales (Parslow 1967). More recently, nesting has been proved or suspected in the Peak District and South Wales (Sharrock). In 1954 the species visited Warwickshire for the first time for fifty years (a pair at Alvecote on April 25th), while a bird at Belvide from December 28th to 30th was only the second Staffordshire occurrence since the early years of the century. Mergansers have subsequently appeared in the Region in every year except 1958, 1961 and 1964.

Birds have arrived in all months from October to May, but 57 per cent appeared in the period December to February. There have been 54 records of single birds, 15 records of two and three records of three birds, whilst five were observed at Chasewater from February 17th to 19th 1956 and seven at Alvecote on February 24th 1963. Both these records were probably the result of cold-weather movements and they contribute significantly to the February peak. Outside the main period of analysis, seven also appeared during the cold weather in mid-February 1979, whilst in 1980 an eclipse drake appeared at Belvide on September 20th, which is the first September record for the Region.

Most individuals have remained in the area for only a day or so, but, during the period 1929–78, 11 birds stayed for twenty days or more. At Blithfield, a drake was present with the Goosander flock in each winter from 1976/7, when it was in first-winter plumage, to 1978/9. It generally arrived in mid-December and departed during March.

Table 41: Duration of Visits of Red-breasted Merganser, 1929–78

No. of Days	1	2–5	6–9	10–19	20 and over
No. of Birds	65	20	1	8	11

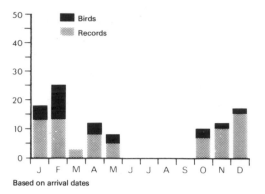

Based on arrival dates

Monthly Distribution of Red-breasted Merganser, 1929–78

Staffordshire remains the most frequently visited county, accounting for 45 of the 74 records; only eight are from Worcestershire. Blithfield has produced 18 records in eleven years since 1965, Belvide 16 birds in ten years since 1954, Chasewater 16 birds in six years since 1956, Draycote 13 birds in seven years since 1970 and Alvecote 11 birds in four years since 1954.

Goosander

Mergus merganser

Regular winter visitor, local but not scarce.

The Goosander has long been known as a winter visitor, with Tomes (1901) stating it was not rare on rivers, Smith quoting many Staffordshire records, Harthan describing it as a regular visitor and Norris

as fairly regular in small numbers. Today most of the larger lakes and reservoirs in the Region are visited annually by Goosanders, but there is a distinct bias towards waters in the Trent Valley sub-region and only at Blithfield is there a significant wintering population. Numbers at Belvide, once a favoured haunt, have declined during the last twenty-five years and the appearances of the species are now erratic. This has been attributed to either a decline in fish stocks or the transference of birds to Blithfield, (Smallshire and Richards 1976). The largest flocks at these two sites involved 65 at Belvide on January 14th 1951 and 148 at Blithfield during late February 1976.

Several other Staffordshire waters, especially Copmere and Aqualate, occasionally attract noteworthy concentrations. At the latter water 110 birds were reported on January 28th 1969, though in general the species is less regular than formerly.

Outside Staffordshire, Alvecote, Bittell, Kingsbury and Shustoke are among the more regularly visited sites, but the numbers involved are very small and the duration of visits usually brief. Even the large expanse of Draycote Water has as yet attracted few birds, though there were up to 50 during the cold spell in early 1979.

Over thirty-four years the average date for first arrivals has been October 29th and over thirty-one years the average date for last departures has been April 14th. Extreme dates have been September 12th (1962) and May 5th (also 1962). At Blith-

Table 42: Five-yearly Mean of Annual Maxima of Goosander at Belvide and Blithfield, 1954–78

	1954–58	1959–63	1964–68	1969–73	1974–78
Belvide	32	35	22	26	9
Blithfield	64	94	59	68	112

field numbers increase steadily until the end of the year. Smaller increases also occur during January and February, and numbers reach a peak in March – a pattern recalling that of Goldeneye.

Table 43: Mean monthly concentrations of Goosander at Blithfield, 1953/4 to 1977/8

	Nov	Dec	Jan	Feb	Mar
No. of Birds	5	24	43	50	55

Most birds wintering in southern Britain are considered to be of northern European origin. One of two birds at Chasewater on October 18th 1978, however, had been wing-tagged during the summer in Northumberland. It proved to be both the first sighting of the season of a marked bird, and the furthest south ever of the study group's birds. Most movements of individuals nesting in northern England are concentrated between North and WSW (Meek and Little 1977).

Smith quoted a summer record from Gailey in August 1928 and at Blithfield a bird summered in each year from 1970 to 1973, being joined by a second bird in 1974. Unfortunately they did not appear subsequently.

Ruddy Duck
Oxyura jamaicensis

Well-established and not scarce as a feral breeding species. Fairly numerous, but more local, in winter.

The Ruddy Duck is a North American species which has been breeding ferally in Britain since 1960. Three pairs were imported by the Wildfowl Trust in 1948 and began breeding the following year. Not all the young of succeeding years were pinioned and between 1952 and 1973 at least 70 juveniles escaped. During the cold winter of 1962/3 all free-flying individuals left Slimbridge. A feral breeding population was soon established and by 1975 over 50 pairs were nesting ferally in Britain (Hudson 1976).

The first definite sighting of the species in the West Midlands was at Belvide in September 1959, though there was an intriguing report of five "stiff-tails" at Aqualate in August 1954, at which time only two Ruddy Ducks were known to have escaped from Slimbridge. Breeding occurred for the first time locally in 1961, at both Belvide and Gailey, and the species has since nested at Belvide in most years. During the next decade small numbers of birds were reported at several sites in Staffordshire, Warwickshire and Worcestershire, though breeding was confined to Belvide and Copmere (1968). The first breeding records for Worcestershire occurred in 1971, at Upton Warren and Westwood Park, but it was not until 1976 that nesting was confirmed in Warwickshire, at Middleton Hall and Packington Park. In 1979, two broods were reared at Edgbaston Park, in the new West Midlands Metropolitan County and only 3 km from the centre of Birmingham.

To date, the species has bred at at least 20 sites involving a minimum of eighteen 10-km squares. At most sites only one or two pairs nest, though there were five pairs at Chillington in 1977; five broods totalling 31 young were seen at Upton Warren in 1978; probably eight pairs attempted to nest at Belvide in 1979, though several were unsuccessful; and five broods were noted at the same locality in 1980. In all, the breeding strength is probably over 100 pairs, compared to Lord and Munns' estimate in 1970 of three pairs.

The Ruddy Duck has a long breeding season and in the West Midlands young have appeared in all months from May to

42. Over half the British population of Ruddy Ducks winter in the West Midlands, mostly at Belvide and Blithfield reservoirs. *M. C. Wilkes*

early October. The species nests among emergent vegetation and tends to require a cupola over the nest (Cramp and Simmons 1977), so it seems possible that the timing of breeding may be at least partly related to vegetation type. At waters containing *Phragmites* beds, which provide adequate cover in the spring, nesting is often relatively early. At Belvide most nests are in *Glyceria*, which provides optimum cover later in the season, and at this locality young generally appear in late August and early September. (D. Smallshire *in litt.*)

After the breeding season, birds leave the smaller pools and congregate in large moulting and wintering flocks. The numbers frequenting Belvide and Blithfield indicate that birds from outside the Region use these reservoirs as wintering grounds. At the time, a gathering of 350 at

Belvide on October 22nd 1976 was the largest concentration of the species yet recorded in Britain, but 630 were counted at Blithfield on January 18th 1981, when 180 were also at Belvide. Thus the West Midlands held at least 800 birds at this time, or over half of the known British population.

An interesting pattern of movement has developed between Belvide and Blithfield. Moulting flocks gather at Belvide from July onwards and reach a peak between October and December. Towards the end of the year many birds transfer to Blithfield, where peak numbers occur during January and February. Since 1976, a second influx has occurred at Belvide during February to April. This too involves moulting birds – the Ruddy Duck being unusual in having two complete primary

Table 44: Maximum concentrations of Ruddy Duck at Belvide, Blithfield and Upton Warren, 1966–78

	1966	1967	1968	1969	1970	1971	1972	1973	1974	1975	1976	1977	1978
Belvide	10	16	25	23	6	35	62	59	110	181	350	188	313
Blithfield	8	10	7	22	15	12	35	67	74	100	119	200	215
Upton Warren	—	—	—	—	2	8	4	7	6	28	19	16	25

Table 45: Monthly Maxima of Ruddy Duck at Belvide and Blithfield in 1977 and 1978

		Jan	Feb	Mar	Apr	May	Jun	Jul	Aug	Sep	Oct	Nov	Dec
Belvide	1977	100	143	155	60	17	6	6	26	98	188	135	52
	1978	140	54	87	70	38	18	53	97	223	313	200	58
Blithfield	1977	200	133	61	8	10	5	1	2	5	2	175	200
	1978	99	200	150	73	8	3	3	7	15	20	55	215

moults each year. The rapidly increasing numbers and developing patterns of movement make mean monthly and yearly figures misleading. The annual maxima at Belvide, Blithfield and Upton Warren from 1966 to 1978, and the monthly maxima at Belvide and Blithfield during 1977 and 1978, however, illustrate well the colonisation of the area, the current seasonal trends and the correlation of numbers between the two Staffordshire sites.

In view of its special association with the West Midlands, the Ruddy Duck was adopted as the symbol of the WMBC in 1975.

Honey Buzzard
Pernis apivorus

Now a very rare vagrant. Formerly more common.

Smith, Harthan and Norris listed a total of 19 Honey Buzzards in the Region between 1831 and 1894, including a breeding pair at Waverley Wood in 1841 and a

bird taken near the nest at Coventry in 1867. There have been only four substantiated twentieth-century records, all involving single birds: near Wolverhampton on June 19th 1903; at Tamworth on September 30th 1908; at Pershore on September 2nd 1951; and at Warwick Park on June 26th 1966.

Red Kite
Milvus milvus

Rare visitor.

Formerly widespread, the Red Kite became extinct as a breeding species in England by the 1880s (Sharrock). In the West Midlands, the species bred near Allesley around 1800 (Tomes 1904) and at the Roaches until the early nineteenth century (Smith). Harthan (1946, 1961) recorded that a pair nested successfully at Alfrick as late as 1840 and that birds were occasionally noted over the Malverns until about 1850. There were seven dated sightings of single birds between 1848 and 1884, while during the present century there have been eight or nine occurrences of single birds as follows: Croome in December 1910; Malvern in January 1933; Chasewater on August 4th 1953; Bittell Reservoir on December 10th 1961; Ladbroke on March 26th 1972; Whitnash, possibly the same bird, on April 22nd 1972; Stourport on August 25th 1973; Shustoke on August 22nd 1975 and Cannock Chase on March 23rd 1978.

White-tailed Eagle
Haliaeetus albicilla

Formerly a rare vagrant, but only two records this century.

Between 1792 and 1945 there were six records of the White-tailed Eagle, involving seven or eight birds. All except the birds of 1879 and 1905 were originally reported as Golden Eagles but later adjudged to refer to this species (Harthan, Norris and Smith).

Two were on Cannock Chase in the spring of 1792; an immature was trapped and a second bird reported present at Knavenhill on November 22nd 1879; one was between Birdingbury and Leamington Spa on December 2nd 1885; one was near Stratford-on-Avon on January 24th 1891; an immature male remained in the Cannock–Sandon area from November 30th to December 4th 1905; and an immature was at Wannerton on December 25th 1945.

Marsh Harrier
Circus aeruginosus

Very scarce passage migrant.

Harthan and Norris provided details of six early reports of the Marsh Harrier, while Smith recorded one at Dovedale on November 7th 1916. There were no further occurrences until 1943, but the species has appeared in thirteen subsequent years and in all but one year from 1972 to 1978. Of

the 31 birds recorded during the thirty-six year period, 15 appeared between 1974 and 1978, presumably reflecting the recent improvement in the species national breeding status (Batten *et al.* 1979).

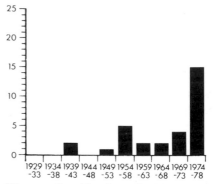

Five-yearly Totals of Marsh Harriers, 1929–78.

The species appears far more frequently during the spring than the autumn, with half the records falling in May. A bird remained in the Tame Valley from May 21st to June 26th 1975, but all other individuals have stayed only a day or so, although since the period of main analysis a bird remained at Belvide for ten days in August 1979.

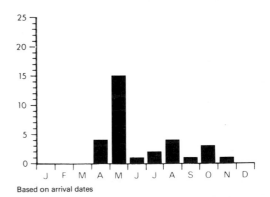

Based on arrival dates

Monthly Distribution of Marsh Harriers, 1929–78.

Most occurrences have involved immatures or females, with only three or four adult or near-adult males reported. Records tend to be concentrated at those

214

localities affording the species preferred marshland habitat, with Brandon providing nine birds, Kingsbury four birds, and Alvecote and Blithfield each three.

Hen Harrier
Circus cyaneus

Scarce winter visitor and passage migrant. Has recently bred in Staffordshire.

Today in the West Midlands the Hen Harrier is a bird of moors and heaths. Up to 1800 it bred throughout Britain, but, following agricultural changes and persecution, the population declined dramatically and by the beginning of this century it was confined to Orkney, the Outer Hebrides and Ireland. In the post-war period, reduced persecution and the increase in suitable habitat provided by afforestation has resulted in a marked recovery and, during the 1950s and 1960s, many areas of Scotland and northern England were recolonised (Sharrock). The history of the Hen Harrier in the West Midlands reflects this national pattern.

Tomes (1904) and Smith recorded that the species bred in the area until the early nineteenth century, but Harthan, Norris and Smith could cite only ten subsequent sightings of the species up to 1921. Appearances remained very few during the next three decades, increased briefly during the early 1950s, and then increased strongly during the late 1960s and 1970s. As a breeding species the Hen Harrier is erratic. In August 1974 a family party was observed on the North Staffordshire moors, breeding was attempted in the same area during 1975 and a pair also bred successfully in 1979. That threats are still posed to the species, however, is illustrated by birds found poisoned at Winnington Grange and at Gnosall during April 1976.

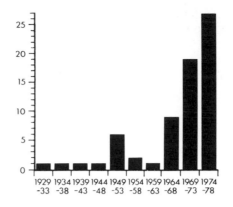

Five-yearly Totals of Hen Harriers, 1929–78

The passage and early winter periods have accounted for most records, though, excluding the breeding pair, birds have appeared in all months except July. Outside the breeding season, moorland or open heath are the most likely places for a bird to appear. The North Staffordshire moors have produced records in every year but one since 1968, while there have been nine individuals on Cannock Chase since 1953 and four in Sutton Park since 1949. Other areas visited in more than one year include Blithfield, Sheriffs Lench and Upton Warren.

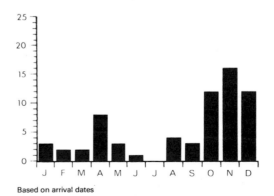

Based on arrival dates

Monthly Distribution of Hen Harriers, 1929–78
(A pair which attempted to breed during 1975 is excluded from the figure.)

Montagu's Harrier
Circus pygargus

Rare visitor. One record of breeding.

Apart from a strange record of an adult male shot at Sutton Coldfield during the winter of 1839/40 (Norris, Tomes 1901), all published records of the Montagu's Harrier have been during the last seventy years, including a pair which nested at Hewell Grange in 1926, raising three young (*Brit. Birds* 20:226).

The twelve twentieth-century records were as follows: a male in Sutton Park on October 11th 1911; a pair, which bred, at Hewell Grange in 1926 (the male first appeared on June 5th and the pair, together with their three young, were observed until October); one at Tettenhall on August 11th 1947; a pair at Enville during the spring of 1951; a male in Sutton Park on May 13th 1951; a male at Stratford-on-Avon on April 28th 1953; one at Dovedale on August 22nd 1955; a female at Churchhill on June 24th 1956; a male at Swallow Moss on May 23rd 1965; one at Droitwich on May 2nd 1969; one on the North Staffordshire Moors on May 9th 1971; and a male at Goosehill Wood on May 5th 1973.

Goshawk
Accipiter gentilis

Scarce, but recently reported from several localities and breeding in very small numbers.

Smith listed three records of the Goshawk, at Swythamley in 1853, Uttoxeter in 1857 and Rolleston in 1877. For a century these remained the only published sightings, though birds observed in Worcestershire in 1954 and 1958 were considered to be of this species by experienced observers.

During the 1960s it became apparent that Goshawks were re-establishing themselves as a British breeding species, probably derived largely from introduced and escaped birds (Sharrock). The recent history of the species has not been fully documented, and in the West Midlands there is little specific data. The WM *Bird Report* for 1978 however, mentioned sightings from at least eight localities involving three counties, two pairs were known to have bred in 1979 and the total number of pairs could be as high as five. Although the Goshawk frequents both broad-leaved and coniferous woodland, it does require extensive tracts of wooded country, so in a Region with one of the smallest percentages of woodland, suitable habitat is clearly limited.

Sparrowhawk
Accipiter nisus

Now again a fairly numerous and widespread resident, following a dramatic decline in the 1950s.

The Sparrowhawk is essentially a bird of woodland, preferring coniferous or mixed woods, especially those with mature larches. However, it also hunts over open country and farmland, where it frequently surprises prey by swooping over hedgerows. Occasionally it appears in the more-heavily-treed urban districts as well.

Until the 1950s it was a widespread and fairly common breeding species in Britain, and Norris (1951) regarded breeding as regular throughout the Region. During the late 1950s, however, the English population declined dramatically, probably as a result of poisoning by toxic chemicals (Prestt 1965). In the West Midlands the species became very rare and there were only one or two breeding records in each year from 1958 to the mid-1960s. The WMBC *Atlas* project hinted at the beginnings of a recovery by 1968, however, even though the nine 10-km squares (12 per cent)

43. The Sparrowhawk, seen here at its plucking post, has made a welcome resurgence in recent years.   *R. J. C. Blewitt*

in which breeding was confirmed or probable were all in Staffordshire. By the time the BTO survey had been completed in 1972, breeding had been confirmed in nineteen 10-km squares (25 per cent) and considered probable in a further ten (13 per cent). Again the majority of squares (66 per cent) were in Staffordshire, reflecting the greater extent of woodland, particularly around the Potteries, in the Churnet Valley, Needwood Forest and Cannock Chase. The remaining squares were mostly in Worcestershire, notably in the well-wooded countryside west of the Severn, and the species remained very scarce in Warwickshire. Recent breeding season reports suggest a continuing improvement and the species is now widespread and even quite numerous in several areas of Staffordshire. In 1976 up to six Sparrowhawks were observed together during spring display flights over Cannock Chase.

In Worcestershire and, more particularly, Warwickshire the species has not yet recovered to the same extent, though during 1977–8 there were breeding season reports from twenty-five areas of Worcestershire, mostly in the Severn and Teme Valleys, and eleven areas of Warwickshire, mostly in the Avon Valley. Encouragingly, during the same period there have been reports in various seasons from a dozen sites within the West Midlands County and in 1978 nesting was confirmed in Sutton Park. As an indication of the recovery, the current regional population is estimated at 200–300 pairs, compared to less than 20 pairs estimated by Lord and Munns as recently as 1970.

The earliest recoveries of Sparrowhawks involved six ringed as nestlings at Rubery on July 3rd 1911, four of which were recovered within a year at distances up to 10 km. This information was originally published in *British Birds* and helps to show just how exciting any recovery must have been in the early days of ringing. As in those early days, most Sparrowhawks are still ringed as nestlings and four have been recovered at distances up to 35 km. However, a first-year bird ringed at Leigh (Greater Manchester) travelled 55 km to Stoke-on-Trent.

Buzzard
Buteo buteo

Scarce and very local as a breeding species, passage migrant or winter visitor.

Today the Buzzard's preferred habitat seems to be a mosaic of open hillside or heath, pastures and wooded valleys. In this Region such countryside is largely confined to the Churnet Valley, south-west Staffordshire and more especially the Teme Valley. Until the nineteenth century, however, the Buzzard bred over much of Britain and at that time the species certainly nested in several areas of the West Midlands (Hastings 1834, Tomes 1904 and Smith). During the subsequent decades a substantial decline occurred and by 1915 Buzzards were confined to south-west England, the Lake District, Wales and Western Scotland. The onset of war and the consequent decline in the number of gamekeepers permitted a recovery, which was maintained until the mid-1950s and resulted in the highest density of Buzzards in Britain for more than a century (Moore 1957, Sharrock).

In 1944 a pair returned to breed in Worcestershire, for the first time since 1840 (Harthan), while nesting was again proved in Warwickshire during 1949 and in Staffordshire during 1954. The BTO survey of 1954 revealed 13 breeding pairs in Worcestershire and single pairs in Warwickshire and Staffordshire, a dis-

tribution reflecting both the high British population at that time and the position of the Region on the eastern fringe of a stronghold of the species in Wales (Blake 1955). The outbreak of myxomatosis in 1955/56 decimated the rabbit population upon which the high density of Buzzards depended and produced a second decline. This was further compounded towards the end of the decade by organochlorine poisoning, so that by 1958 no more than four pairs were breeding in Worcestershire (Harthan 1961) and nesting had ceased elsewhere in the Region.

Following the ban on organochlorine sheep-dips in 1966, the British population has again shown signs of recovery (Sharrock). During the WMBC *Atlas* project (1966–68) breeding was confirmed in only one 10-km square in the Region, but by 1972 field-work for both *Atlas* projects had confirmed breeding in three squares in Staffordshire and two in Worcestershire, while the total number of squares in which nesting was considered probable had increased from two to four. Even so, some earlier sites were not occupied in subsequent years and the regional population is probably no higher than five pairs. Lord and Munns placed it no higher than three pairs.

In the autumn and winter months isolated birds occur fairly regularly in widely scattered areas of Worcestershire and Staffordshire, but the species remains a relatively scarce visitor to Warwickshire, particularly to the less-sylvan terrain on the Lias clay south of the Avon. During 1976 there were sightings away from breeding sites and at various seasons from four northern and twelve southern localities in Staffordshire, eleven well-distributed localities in Worcestershire, four sites in central Warwickshire and four areas of the West Midlands County.

Rough-legged Buzzard
Buteo lagopus

Rare winter visitor.

Smith and Norris mentioned a total of ten dated records of the Rough-legged Buzzard between 1840 and 1897, and referred to four or five additional, but undated, occurrences. The next published record was not until 1949 and there then followed a further gap until 1966. The thirteen years from 1966 to 1978, however, produced ten records, including six during 1974, a year which witnessed exceptionally large arrivals on the east coast of Britain (Christie 1975).

These records, all of single birds, were as follows: Preston Pastures on March 3rd 1949; Whittington Sewage Farm on October 23rd and November 6th 1966; Alvechurch on October 29th 1966 (the first Worcestershire record); Cannock Chase on April 9th 1967; Brandon on February 24th 1974; Draycote on October 29th 1974; Mancetter, found shot during October 1974 and released on December 12th; Chasewater on November 2nd 1974; Cannock Chase from November 2nd to 5th 1974; Bromsgrove on November 7th 1974 and Coombes Valley in late February 1978.

1979 brought two further records, with one near High Green on March 15th and an immature, conceivably the same bird, found freshly dead at Belvide on March 31st. Subsequent analysis revealed death to be due to poisoning by alphachloralose in treated bait.

Golden Eagle
Aquila chrysaetos

No recent records.

Smith included one or two seventeenth-century records of the Golden Eagle from Staffordshire, though he did not consider the reports from Needwood Forest prior to

1844 and from Brakenhurst Covert about 1873 to be well authenticated. All subsequent reports of the species, which involved all three shire counties, were later attributed to White-tailed Eagles.

Osprey
Pandion haliaetus

Scarce, but regular, passage migrant.

Smith, Harthan and Norris documented about 17 nineteenth-century records of the Osprey, plus one at Stourport on October 6th 1906. There were no further records until 1953, when two were seen, but then followed occurrences in 1954, 1960, 1963 and every year from 1965 to 1978. The period 1953–78 produced a total of 64 birds, including nine in 1976, seven in 1973 and six during both 1970 and 1978. This encouraging pattern of increase is clearly associated with the recolonisation of Scotland since the mid-1950s, though the recent stocking of the larger reservoirs with trout may also be encouraging more birds to pause on migration.

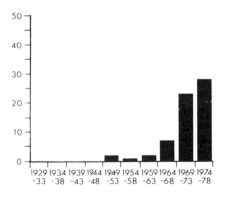

Five-yearly Totals of Ospreys, 1949–78

During this recent period Ospreys have appeared in all months from April to October, while a bird which arrived at Tittesworth in October 1976 remained until November 7th. In spring, April is the peak month, though May and June have been almost equally productive. During the autumn there is a more clearly defined peak in September.

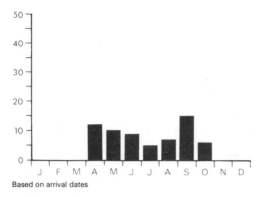

Based on arrival dates

Monthly Distribution of Ospreys, 1949–78

Although most occurrences have involved only single dates and birds, 18 individuals remained for a week or more and there have been five records of two birds.

Since 1953 six birds have appeared at Brandon, four at Bittell (where there was also one in 1883), four at Belvide and four at Draycote, but by far the most productive locality has been Blithfield, where the excellent fishing has attracted 13 birds and where the species has occurred in every year but one from 1970 to 1979. A further 23 localities have been visited once or twice.

Lesser Kestrel

Falco naumanni

One record.

The only record involved a male on the West Midlands County side of Chasewater on November 4th 1973 (*Brit. Birds* 67:319). Individuals were also recorded in Glamorgan and Sussex around the same date (Smith *et al*. 1974), but this was the first ever inland.

Kestrel

Falco tinnunculus

Widespread and numerous resident.

The Kestrel is by far the most widespread and numerous of the raptors breeding in the Region and might be seen anywhere, even in the centre of cities. Norris (1951) showed it nesting throughout the Region, while Smith, Harthan, and Lord and Blake (1962) all described the species as widespread, including developed areas.

As with other raptors, there was a decline in the late-1950s and early-1960s, probably associated with organochlorine poisoning. Harthan (1961) considered that the species had declined in Worcestershire since the 1930s and Lord and Munns also reported a decrease around 1960. Owing to the wide range of habitats frequented by Kestrels, however, this species was less seriously affected than most other raptors and quickly recovered. The raptor enquiry sponsored by the then Nature Conservancy resulted in 72 reports of confirmed or probable breeding in the Region during 1964–5 (Lord and Richards 1965, Lord and Munns). Records were widespread in Staffordshire and Warwickshire, but the species was absent from several areas of Worcestershire. During the WMBC *Atlas* project (1966–8), breeding was confirmed in forty-eight 10-km squares (62 per cent) and considered probable in a further eighteen (23 per cent), though again there were considerable areas of the Severn and Avon valleys, especially in Worcestershire, from which nesting was unreported. In the next few years these gaps were filled and breeding was confirmed or probable in all but two of the seventy-seven 10-km squares surveyed during the BTO *Atlas* project (1968–72). Assuming a density of 35–45 pairs per square (see Sharrock), the regional population would be 2,500–3,500 pairs, which is above the range 200–2,000 estimated by Lord and Munns.

A major asset of the Kestrel has been its

44. The Kestrel is the most widespread raptor, having adapted even to urban situations. Along motorways it can often be seen perched on fences or hovering overhead in search of prey. *R. J. C. Blewitt*

ability to colonise developed and industrial areas, while its adaptability is further illustrated by the increasing frequency with which it has been observed feeding along motorway verges. Indeed, hovering Kestrels have become synonymous with the new motorway network. There are, however, several recent references in the WM *Bird Reports* to birds being killed by traffic.

During 1978 a special study was carried out of Kestrels nesting in the West Midlands Conurbation (Pike 1979). Twenty-one definite and seven more probable nest sites were located within the Metropolitan County, plus a further site just outside the county boundary. The most common nest site was on commercial or industrial buildings (14 locations), but eight were in

church spires. A total of 59 young was hatched from 18 nests and 33 of these were ringed. Two birds were subsequently recovered, both in Yorkshire and in the same season, indicating unusual late-summer movements of 125 km and 220 km NNE respectively.

A large number of recoveries have also resulted from the ringing of nestlings elsewhere. Many of these recoveries occurred in the birds' first year of life, with movement in all directions of the compass and just over half of the 40 recoveries at distances over 100 km. The longest movement of a West Midland bird within this country was to Cornwall, but four have been found in various parts of France and another was on the west coast of Eire just three months after fledging. A Scottish bird ringed at Forres (Moray) was in Warwickshire in April 1977.

Red-footed Falcon
Falco vespertinus

Very rare vagrant. Four records.

There is one historical record and three recent ones of this rare vagrant from eastern Europe and Siberia. An adult male was at Welford-on-Avon in June 1870; an immature male was at Middleton Hall on May 14th and 21st 1967 (*Brit. Birds* 67:341); another immature male was observed on both the West Midlands County and Staffordshire sides of Chasewater between May 28th and June 6th 1973 (*Brit. Birds* 67:319) and an adult male was at Brewood on August 23rd 1977 (*Brit. Birds* 71:497). May and June are the typical months for this species to appear in Britain, but the August record was somewhat unusual (Sharrock and Sharrock 1976).

Merlin
Falco columbarius

Regular, but scarce, passage migrant and winter visitor. Very scarce breeding species.

As a breeding species, the Merlin frequents high moorland and blanket bog. The North Staffordshire Moorlands lie on the fringe of its breeding range and Smith reported that it attempted to breed every year. There was subsequently a dearth of records during a period when the species declined nationally (Sharrock), but in 1972 a pair again reared three young on the moors, apparently the first successful nesting for about fourteen years. In 1974 a pair again bred and the presence of three pairs in the area during 1978 suggests that nesting is currently annual, though it is doubtful if more than two pairs breed in any one year.

Elsewhere in the Region the species is a scarce, but fairly regular, passage migrant and winter visitor, mainly to heaths, reservoirs or open ground with wintering flocks of passerines. Harthan and Norris described its occurrences as rare and occasional respectively, but Lord and Blake (1962) regarded it as a regular visitor to Chasewater and, to a lesser extent, to Belvide and Cannock Chase. Following a decline around 1960, the species has been reported with increasing frequency during the last fifteen years. To some extent, this no doubt reflects increased observer activity, but hopefully it also indicates an improvement in the species breeding success.

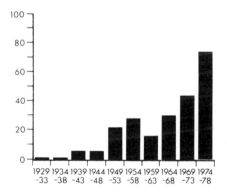

Five-yearly Totals of Merlins, 1929–78
(excluding breeding birds)

Away from the moors, the Merlin is most frequently observed during the period September to March, with a peak in November and December.

Excluding breeding pairs, the vast majority of birds have been recorded on only a single date, but there are several series of reports from around Chasewater that suggest a bird wintering in the area, while during the winters of 1977/8 and 1978/9 a bird was observed fairly regularly in the Tame Valley.

The most frequently visited localities during the winter and passage periods have been Chasewater, with records in

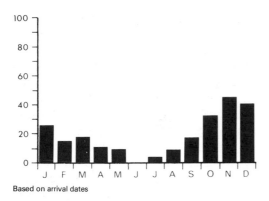

Monthly Distribution of Merlins, 1929–78
(excluding breeding birds)

eighteen years since 1948; Belvide, with records in fourteen years since 1950; Cannock Chase, with reports in twelve years since 1950; and Blithfield, with records in nine years since 1962. Otherwise Draycote produced occurrences in every year from 1974 to 1978, while other localities visited on a number of occasions include Brandon, Kingsbury and Sutton Park.

The only ringing recovery is that of a bird ringed as a nestling on July 3rd 1927 in east Cheshire and recovered in July 1929 at Leek.

Hobby

Falco subbuteo

Scarce, but regular, breeding species and passage migrant.

In the West Midlands the Hobby is a scarce summer visitor, principally to areas of open country with tall hedgerow trees, small woods, scattered copses and shelterbelts. Passage birds occasionally appear on heaths or even in less typical habitats, whilst in autumn birds frequently visit waters with emergent vegetation to prey on feeding or roosting hirundines. Smith, Harthan and Norris all described the Hobby as a not uncommon visitor to the Region during the nineteenth century and quoted seven breeding records between 1850 and 1883. During the first three decades of this century the species was recorded only rarely, however, but Hobbies have subsequently occurred with increasing regularity and, apart from 1961, have appeared annually since the mid-1940s.

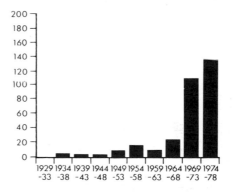

Five-yearly Totals of Observations of Hobbies away from Breeding Sites, 1929–78

Breeding was again established in Warwickshire and Worcestershire during 1964 and 1965 respectively. The WMBC *Atlas* revealed breeding in two 10-km squares in Warwickshire and one in Worcestershire between 1966–8, while by the end of the BTO *Atlas* survey (1968–72) breeding had been confirmed in five additional squares and was considered probable in a further four. With two exceptions, of confirmed breeding in Staffordshire and Worcestershire, all of these squares were within the Avon Valley, where coincidentally the *Atlas* surveys showed the Sparrowhawk to be virtually absent and the Kestrel to be comparatively scarce. The breeding history in Staffordshire is obscure. The first fully documented occurrence was in 1976, when juveniles were seen in a nest, but there is evidence that a pair bred elsewhere on at least two occasions in the early 1970s, with at least one juvenile being taken by a falconer (R. J. C. Blewitt *pers comm.*).

The West Midlands is on the northern edge of the Hobby's British breeding range, but in recent years around four pairs have nested annually, whilst even allowing for some duplication, the large number of casual sightings during the past decade suggests that the true number of pairs may be as high as ten. Lord and Munns gave the population in 1970 as less than three pairs.

Over the last eleven years the average date for first arrivals has been May 1st, while the average date of the last birds to leave over the past fourteen years has been October 1st. During the same period the earliest and latest dates were April 13th (1974) and October 12th (1976) respectively. Observations away from the breeding sites peak during the passage months of May and August–September. The not inconsiderable number of mid-summer sightings presumably involves birds breeding in the area and observed on feeding excursions, or roving, non-breeding immatures. Individual birds are frequently observed over considerable periods at good feeding sites, such as Swallow roosts, and in such cases only the first date of occurrence has been entered in the monthly distribution.

Among the more regularly visited feeding sites may be mentioned Belvide (records in fourteen years since 1936), Brandon (fifteen years since 1954), Draycote (nine years since 1969) and Upton Warren (at least eleven years since 1941).

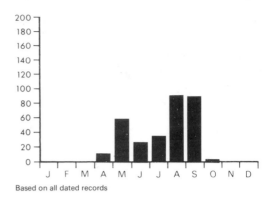

Based on all dated records

Monthly Distribution of Observations of Hobbies away from Breeding Sites, 1929–78

At all these sites the species has occurred annually in recent years.

As with other raptors, Hobbies are nearly always ringed as nestlings and one ringed as such in Warwickshire in August 1976 was found unable to fly at Havant (Hants.) in August 1977.

Gyrfalcon
Falco rusticolus

No recent records.

Two records of this Arctic vagrant gained acceptance in earlier works. One was shot on an unspecified date in Beaudesert Old Park (Garner 1844, Smith) and the other was shot at Quinton, Warwickshire, during the winter of 1852 (Tomes 1901, Norris).

Peregrine
Falco peregrinus

Very scarce winter visitor and passage migrant.

During the nineteenth century and earlier years the Peregrine was a regular visitor to the Region. A pair was stated to have bred at Warwick Castle in 1892 (Norris) while

Smith considered that the species probably nested formerly in north Staffordshire. There were few occurrences during the early years of this century, however and both Harthan and Norris noted that the visits of Peregrines were less regular than formerly. Owing to its depredations of homing pigeons, the Peregrine population of England was reduced by nearly half during the war years (Sharrock), but, surprisingly, its appearances in the West Midlands during this period were increasing, though this could reflect better documentation rather than a real increase.

Nationally, the Peregrine increased substantially in the post-war decade, but then decreased alarmingly in the late 1950s and early 1960s. This decline was generally attributed to poisoning by organochlorine pesticides (Ratcliffe 1963). The occurrences of the species in the West Midlands in recent decades have mirrored these national trends. Peregrines were recorded annually from 1945 to 1963, with peak numbers during the mid 1950s, but there were no appearances at all from 1964 to 1966 inclusive. Subsequently there has been an improvement, but, allowing for the great increase in observer activity, visits remain well below the level of the 1950s.

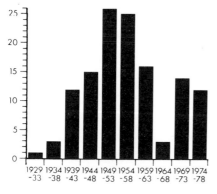

Five-yearly Totals of Peregrines, 1929–78

The species has been recorded in all months, but most occurrences have fallen between August and April, though there is a surprising lack of September arrivals.

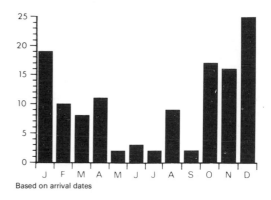

Based on arrival dates

Monthly Distribution of Peregrines, 1929–78
(Three records for which specific arrival dates are unavailable are excluded from the figure.)

Most individuals have been recorded on only a single date, but there are five records of birds remaining a week or more. Additionally, series of reports at Belvide in 1947 and Branston in both 1953 and 1956 probably involved birds which lingered in the general area of these respective localities. A Peregrine was also recorded regularly during February and March 1960 at Coven Heath, while a bird spent a month at Blithfield during August–September of both 1975 and 1976.

There have been occurrences at Belvide in twenty years since 1940, at Bittell in seven years since 1934 and at Blithfield in six years since 1954. Interestingly, birds have appeared over Birmingham and its suburbs in nine years since 1948, possibly attracted by feral pigeons.

Red Grouse
Lagopus lagopus

Fairly numerous resident, but restricted to the heather moors of north Staffordshire.

Plot (1686) believed that the Red Grouse

was less common than the Black Grouse in Staffordshire, but by the nineteenth century the converse was true and this situation has persisted to the present day. Birds were said to have occurred at Sutton Coldfield in the early nineteenth century, but not after 1868. Red Grouse were introduced to Cannock Chase, probably in the middle of the nineteenth century, but the only WMBC records are for the years 1950–62, with a maximum of eight on December 31st 1951. The severe weather of early 1963 finally exterminated this population. Red Grouse were extinct at Chartley by 1897, but Smith reported the continued presence of birds at Swineholes ridge and on the lower moors near Oakamoor and Whiston. Smith also quoted a few old lowland occurrences in severe weather, but there have been no such records this century.

Since 1963 the only records for the Region have come from the high moors of north Staffordshire, where Yalden (1979a) calculated a spring population of 982 pairs for the years 1969–72, which was thought to be about 10 per cent of the Peak District population. However, the Peakland population is known to have declined considerably in recent decades. For example, the annual bags for eight grouse moors fell from 25,019 in 1935/6 to only 3,226 in 1957/8 (Yalden 1972). This decline is primarily attributed to increased competition for heather by the larger upland sheep flocks, but a reduction in the number of keepers and the poorer management of grouse moors are thought to have exacerbated the decline. Yalden (1979a) found the species in forty-two 1-km squares in Staffordshire during 1969–72, mostly coinciding with the presence of ling, bilberry and crowberry growing on acidic, peaty soils often over underlying Millstone Grit.

Heather moor such as this will only support a typical flora and fauna, including good populations of Red Grouse, provided that correct management techniques are maintained. Old, rank heather provides poor nutrition and therefore sustains only a low breeding density, whilst systematic burning will maximise density. Other changes can be even worse. Afforestation or overstocking with sheep, for example, will both result in the disappearance of heather within a few years. In north Staffordshire, formerly potential threats almost became a reality recently, when in 1977 more than 480 ha of the Swythamley Estate were sold for intensive sheep-rearing and the future of its Red Grouse population (37–42 pairs in 1978; bags of 174–366 during 1971–5: Yalden 1979) and many other typical moorland birds came under serious threat. Fortunately, though, the Peak Park Planning Board has now successfully interceded to save this area, but the debate over the future use of Britain's uplands continues unabated. The future of our typical moorland avifauna rests in the hands of those concerned with upland forestry, subsidised marginal sheep-rearing, recreation and nature conservation.

Black Grouse
Tetrao tetrix

Resident, not scarce but confined to north Staffordshire.

In the nineteenth century, this species was considerably more widespread and abundant in the Region than at present. However, shooting pressure and habitat change caused a large decline and contraction of range around the turn of the century and outlying populations in the Wyre Forest and Sutton Park had been exterminated by 1895 and 1897 respectively. Other nineteenth-century records came from the Teme woods near Eastham, and from Aston (Staffs), Caverswall, Cheadle and Stoke, whilst birds were also present in Needwood Forest prior to its enclosure in 1801. There were a few records of lowland wanderers at Orslow, the Burton area, near Ripple in 1910, and at Chaddesley Woods and Hartlebury Common in 1915.

In addition to the northern moors, which are still frequented today, Smith gave records for the early 1900s for Bosley Cloud, Burnt Wood–Bishop's Wood, Chartley and seven localities in the Stoke area, ranging from Maer in the west to Ramshorn in the east. In the nineteenth century Cannock Chase was another major haunt and the best British daily bag of 252 was obtained here in about 1860. By 1897, however, the best day's bag was only 41 and the last two birds were seen on a tennis lawn near Rugeley in 1924.

In 1925 Masefield and Smith said the species was still quite numerous on the moors north of Leek, with daily bags of 15–20 still being achieved. There was a further contraction in range, however, until about 1945, since when suitable habitat and consequently numbers have remained fairly steady. Black Grouse prefer a "patchwork quilt" of heather moor, birch scrub, pasture and small conifer plantations, and loss of such habitat is thought to be the cause of a decline in the species elsewhere in the Peak District. In 1973–5, 56 out of a total Peakland population of about 66 pairs were thought to be present in north Staffordshire (Lovenbury *et al* 1978).

Since 1950 birds have been reported from 14 localities in a horseshoe-shaped area between three and eleven kilometres to the north and east of Leek. Most of these lie within the boundary of the Peak District National Park, where it is hoped the essential mosaic of habitats will be conserved, thus maintaining a healthy population of Black Grouse.

Red-legged Partridge
Alectoris rufa

Numerous and widespread resident.

Although the first introductions of Red-legged Partridge in England took place during the seventeenth and eighteenth centuries, it was not until the nineteenth century that any introductions were made in this Region. Then Lord Foley released at least a hundred brace at Witley in about 1820, of which only two were seen again, and Garner (1844) also mentioned introductions at Teddesley. Less than a dozen subsequent records were published for the remainder of the nineteenth century, however, though two of these did refer to nesting – at Evesham in 1878 and Kings Bromley in 1886.

Numbers appear to have increased steadily in the first half of this century and most areas have now been colonised. The Red-legged Partridge prefers agricultural land and especially the drier, sandy arable areas, and it is still absent from the wet moorlands of Staffordshire and the heavily built-up parts of the Conurbation. It is also relatively scarce in some dairy districts, notably around Stoke-on-Trent and south of Birmingham, and has only recently spread into the Lias clay lands south of the Avon as these have increasingly been cultivated for arable crops. This distribution pattern was first revealed by Norris (1951) and later confirmed by the WMBC and BTO *Atlases*. Lord and Munns also noted little evidence of any recent change in status, despite Harthan's (1961) observation that the species was declining in

Worcestershire due to agricultural change. Overall the *Atlas* projects (1966–72) showed confirmed breeding in forty-seven 10-km squares (61 per cent) and suspected breeding in a further twelve squares (16 per cent), whilst the national density from the CBC farmland plots in 1978 was 1·90 pairs per km^2. Using these figures and extrapolating for the regional area of crops and grass (about 4,900 km^2), a population estimate of 8,000–10,000 pairs is obtained. This accords well with the range of 2,000–20,000 pairs postulated by Lord and Munns.

Although the Red-legged Partridge is found primarily on farmland, it also frequents rough grassland, such as that around power stations, reservoirs and gravel pits, and also woodland rides. Coveys of up to 20 are frequently noted between late summer and February, although most records and the largest coveys occur during October–December. The largest reported was about 60 birds at Belvide on December 15th 1977.

Grey Partridge
Perdix perdix

Widespread resident; numerous or abundant.

The Grey Partridge is a common farmland bird throughout the Region. Unlike the Red-legged Partridge, it is equally at home on dairy farms or in areas of high rainfall and can even be found nesting in marginal areas such as the heather moors of Cannock Chase and the Peak District. The spread of urban development has left relict populations isolated in green "islands" of derelict land and birds have probably existed for decades in such circumstances in the Black Country and the Potteries.

The overall distribution pattern has probably remained stable for much of this century and perhaps longer, though Smith thought the species used to be more numerous on the northern hills, where it has never been as common as in the lowlands. The decline reported nationally in recent decades (Sharrock) has not been apparent from WM *Bird Reports*. However, in this Region at least, it is not known to what extent artifical rearing and introductions have counteracted the damaging effects of poor spring weather, less stubble and rough grasslands, fewer hedgerows; and increases in the use of pesticides, spring barley areas and autumn ploughing. Breeding success is known to be poor after cold, late springs, but, although some years such as 1972 have indicated declines, others such as 1967 and 1970 have suggested the converse and no consistent trend is apparent.

The *Atlas* surveys (1966–72) revealed confirmed or probable breeding in seventy-five 10-km squares (97 per cent) and the CBC for 1978 showed 2·78 pairs per km^2 nationally and 4·88 pairs per km^2 regionally. Related to a regional area of 4,900 km^2 of crops and grass, such densities imply a population of some 14,000–24,000 pairs, which is at or slightly above the upper end of Lord and Munns' range of 2,000–20,000 pairs.

Over most of the Region the Grey Partridge is more widespread and numerous than the Red-legged Partridge and coveys of up to 40 birds may be seen at any time outside the breeding season, with most between September and January. Exceptionally, up to 80 birds have been found within a small area.

Quail
Coturnix coturnix

Scarce and erratic summer visitor and passage migrant. Very rare in winter.

In the West Midlands, the Quail is an erratic summer visitor to open farmland, especially cereal crops in river valleys. It used to occur more frequently than today in long grass as well, but early mowing and cutting for silage often cause nest desertion or destruction (as with the Corncrake). However, these deleterious effects have probably been more than offset by an increase in cereal crops, especially spring barley which the Quail often favours.

Norris believed that the species was far more numerous in the nineteenth century, although Smith and Harthan thought that it had always been uncommon and irregular. A decline in the late-nineteenth century, however, would be entirely consistent with a reduction in the tillage area consequent upon the agricultural depression. There were virtually no records for the first three decades of this century, but subsequently Quail have occurred in all but eight of the years since 1933 and have been annual visitors since 1964. Usually there have been less than half-a-dozen records each year, but the well-known "Quail years" of 1940, 1953, 1964 and 1970 were exceptionally good. In 1970 there were 34 records in the Region, involving at least 54 birds. Fewer records, but more than average, also occurred in 1947, 1948, 1960, 1961, 1971, 1972 and 1977. Such influxes appear to be associated with warm, dry springs and southeasterly winds. In all, the fifty years 1929–78 produced a total of 137 records involving about 300 birds, although two-thirds of these were in the last two decades. During the *Atlas* surveys (1966–72) breeding was confirmed in seven 10-km squares (9 per cent) and suspected in a further twelve (16 per cent), but even so there were only 13 confirmed and 34 probable instances of breeding during 1929–78 and the Quail remains a very scarce breeding species.

In good years small concentrations may be encountered, such as the eight calling from a 1·6 ha field of barley near Pershore on August 12th 1970. Most calling birds are heard in June and July, however,

though bevies, presumably containing juveniles, are sometimes encountered in early autumn. Apart from two very early records at Sutton Park on April 5th 1980 and Coven Heath on April 7th 1960, the earliest arrival has been on May 9th (1948), while the latest bird occurred on October 4th (1970). In addition, Smith records a bird shot near Tutbury on December 15th 1856, while in 1950 there were February and March reports from the Harvington area and a female was later shot there on December 25th. Two unusual records referred to birds heard on nocturnal migration: at Bromsgrove on May 27th 1943 and at Sheriff's Lench on the same date in 1946.

Pheasant
Phasianus colchicus

Numerous and widespread resident.

Pheasants are found predominantly in woodland, parkland or wooded farmland. Rank vegetation is often frequented and the relatively bare, upland areas hold very few. Birds are sometimes encountered in the more sylvan suburbs, however.

The species was probably first introduced into England by the Normans in the late-eleventh century and was fairly well distributed by the sixteenth century (Sharrock). These early introductions were of the nominate subspecies from the Caucasus, but since about 1785 most introductions have involved the Ring-necked Pheasant *P. c. torquatus* from China. Despite much interbreeding, the white neck-ring of the latter is often present on feral males.

Pheasants are now widely, though unevenly, distributed throughout the Region. The undoubtedly self-sustaining feral population is greatly augmented by the liberation of reared birds on sporting estates, however, and the extent to which the population is dependent on these introductions was illustrated by the scarcity of birds during wartime, when the rearing effort was much reduced. During the *Atlas* surveys from 1966–72, breeding was confirmed or suspected in all but one 10-km square (99 per cent). The exception was the centre of Birmingham, where the isolated population of Edgbaston Park appears to have been exterminated by the winter of 1963, although birds still remain in small numbers in Sutton Park. The national CBC for 1978 showed a farmland density of $2 \cdot 28$ pairs per km^2 and this, together with Sharrock's assumed density, would suggest a regional population of some 11,000–13,000 pairs compared to the 2,000–20,000 estimated by Lord and Munns.

Pheasant rearing and protection have had quite a profound effect on the rural landscape of England and the West Midlands is no exception. Estate management has often revolved around game production and many of the by-products, especially in habitat maintenance, have benefited wildlife. Others, like raptor persecution, though, have been less desirable, but with sensible management it has been shown that losses due to predators can be minimised. For example, in 1975 Barn Owls nested within 50 m of a rearing pen at Packington Park without causing disturbance.

Smith believed that Pheasants could detect low-frequency sound and/or air disturbance and quoted examples in Staffordshire of birds crowing immediately prior to bomb explosions and Zeppelin arrivals as well as during the North Sea battle of January 24th 1915 and the earthquake of January 16th 1916. Perhaps not surprisingly, birds often crow at – or even before – gunshot.

Golden Pheasant
Chrysolophus pictus

Very rare as a feral species.

This species was admitted to the British and Irish List by the British Ornithologists' Union in 1971, since when the following records have been published; one in Edgbaston Park between October 1971 and March 1972, one at Droitwich on March 19th and 22nd 1972, one at Weston-under-Lizard in November 1976, two females at Packington Park on March 17th 1978, a male at Wheaton Aston on April 8th 1980 and another at Sugnall in autumn 1980. All of these records presumably refer to escapes or introductions rather than vagrancy from feral populations elsewhere in England.

Water Rail
Rallus aquaticus

Not scarce as a winter visitor or passage migrant, but scarce as a breeding species.

The Water Rail has apparently always been a fairly scarce visitor to the Region's wetlands, although records have increased considerably in recent decades, due in part, at least, to increased habitat in the form of flooded river-valley gravel pits. Its favoured habitat is the muddy edges of reed and reedmace beds around pools or along slow-flowing watercourses, although bulrush and other types of emergent vegetation are also frequented. In winter, sewage farms, marshy fields and overgrown ditches are visited.

Smith quoted only two instances of nesting prior to 1929; Harthan and Norris none. Of course Water Rails are secretive birds, usually heard and not seen, and it is very difficult to prove nesting in their aquatic habitats. However, during 1929–78 there were at least twenty instances of confirmed breeding, at Barford, Blakedown, Brandon (up to three or four

45. Though nowhere common, the secretive Water Rail can sometimes be glimpsed in wet areas, especially in winter.   *R. J. C. Blewitt*

pairs), Coombe, Edstone (prior to 1939), Hillmorton, Kings Bromley, Kingsbury, Ripple (at least two pairs), Wilden and Wormleighton. There were also at least ten instances of probable breeding, at sites including Alvecote, Chesterton, Lady-walk, Leamington Spa, Packington and Stubbers Green, together with well over 50 summering records. The heavy preponder-ance of Warwickshire records is sur-prising, though this was redressed to some extent by the more even distribution revealed during the *Atlas* surveys (1966–72) when breeding was confirmed in seven 10-km squares (9 per cent) and sus-pected in a further eleven (14 per cent). Although it is hard to estimate numbers with such a secretive species, the regional population is probably still the same as

that suggested by Lord and Munns, namely 3–20 pairs. Subsequent to the main period of analysis, breeding occurred at Branston in 1979 and was suspected at Belvide in 1980.

In winter the Water Rail is more numerous and widespread, with birds from the continent arriving mainly in October and leaving again in the following March or April. In the autumn of 1908 there was apparently a large migration over Birmingham, with several shot or picked up exhausted and some found in the streets (Smith), and in November of that year there were eight at Hamstall Rid-ware. Outside the breeding season, small parties like this are not unusual, with Baggeridge, Brandon, Copmere, Lady-walk, Northwick, Oakley, Tixall, Westminster Pool and Westwood Park having held maxima of four to six, whilst in recent winters at least eight and twelve have occurred at Upton Warren and Kingsbury respectively.

Spotted Crake
Porzana porzana

Very scarce passage migrant and winter visitor. Rare in summer, but has bred.

The Spotted Crake was apparently much commoner in the nineteenth century than in the first half of this century. It bred in Sutton Park in 1886 at least, and may have nested regularly in Warwickshire (Norris). Harthan, however, could give no evidence of breeding in Worcestershire, although he said it occurred occasionally in summer, whilst Smith gave at least a dozen nineteenth century Staffordshire records, but again with no evidence of breeding. For the early part of this century, Smith quoted seven further occurrences between 1904 and 1918 and Norris gave two additional records for the same period, but Harthan none. Despite a national resurgence during 1926–37 (Sharrock), there were no further records until 1947.

Between 1947–78 there were 33 records, mostly from marshy situations similar to those frequented by Water Rails. The records were as follows: Brandon (8), Belvide (4), Ladywalk (4), Alvecote (2), Earlswood (2), Westwood Park (2), Baginton, Bickenhill, Blithfield, Branston, Chillington, Draycote, Frankley, Kingsbury, Knightwick, Perton and Tixall.

Apart from a summering bird at Brandon in 1975, all records fell between July 8th and April 10th, with the monthly distribution as shown in Table 46.

Thus 69 per cent of the Region's records fell during August to November, during which period very few have been recorded in the London area (Chandler and Osborne 1977). All records referred to single birds, except in September 1966, when two occurred at Belvide and up to three were at Westwood Park.

A recent record of one at Chillington on July 11th 1979 is of interest in the light of a dead juvenile found there on August 26th 1978.

Little Crake
Porzana parva

One record.

An adult male remained at Ladywalk from November 7th to December 12th 1974. This rare visitor from central and east Europe occasionally winters in Britain (*Brit. Birds* 68:315).

Corncrake
Crex crex

Formerly bred, but now a rare summer visitor and passage migrant. Very rare in winter.

The national decline of the Corncrake is a classic tale of avian misfortune due to changing farming techniques. With the advent of mechanical mowing, many nests were disturbed or destroyed and in conse-

Table 46: Monthly Distribution of Arrivals of Spotted Crakes, 1947–78

	Jan	Feb	Mar	Apr	May	Jun	Jul	Aug	Sep	Oct	Nov	Dec
No. of records	1	3	1	2	—	—	1	4	9	5	5	1

quence the population was depleted and the range contracted. Even as early as the mid-nineteenth century the species had declined considerably in the intensively farmed areas of the English lowlands and by the turn of the century it was becoming scarce in the West Midlands. The decline was particularly marked in the early years of this century (Smith, Harthan and Norris) and has continued to the present day. Even the great increase in observer activity has not increased the numbers being found and the mean number of birds recorded per annum is as shown in Table 47.

Table 47: Mean Number of Corncrakes per Annum, 1933-78

1933–48	1949–58	1959–68	1969–78
4·4	4·0	4·2	1·9

Traditionally Corncrakes were found in mowing grass, notably the undrained, damp Lammas meadows of the Avon, Severn and Trent valleys, where, following the cutting of hay, common grazing rights were exercised until the winter floods came. In the 1930s and 1940s the meadows of the Severn and lower Avon valleys were particularly well-known as the regular haunt of Corncrakes, but birds disappeared rapidly from the Avon Valley during the early 1950s. Another regular haunt was Whittington Sewage Farm, where ten birds were calling in 1937, but there have been records from all parts of the Region and from alternative habitats such as clover, cereals and nettlebeds.

Since 1935 there have been only five confirmed instances of nesting, at Eckington in 1942, Atch Lench in 1946, Wick in 1967 and Chesterton in 1968 and 1969. The *Atlas* surveys showed suspected breeding in a further five 10-km squares between 1966–72, but, notwithstanding this, the Corncrake can no longer be regarded as a regular breeding species in the West Midlands.

The first arrivals are often silent and passage birds, which migrate typically at night, may be rarely detected. Certainly the majority of the 173 or more birds recorded since 1933 were heard "craking" and, although the earliest of these was heard at Rickerscote in the early hours of April 5th 1980, most have been noted from May to August, with a few in September and a late bird which was found stunned at Lea Marston on October 17th 1965. There is an exceptional winter record of a fresh corpse found at Great Chatwell on January 5th 1908.

Moorhen
Gallinula chloropus

Widespread and abundant resident. Migrant of unknown status.

The Moorhen is one of the two most abundant and widespread aquatic non-passerines in the Region, the other being the Mallard. Providing there is sufficient marginal cover, it can be found along all waterways from drainage ditches to main rivers and canals, and around all enclosed waters from farm ponds to large lakes. Sewage farms are also frequented. Indeed there are few areas of freshwater without Moorhens, although the species is scarce on the polluted waterways of the Black Country and the Potteries and tends to avoid the fast-flowing streams of the northern moors. In agricultural areas it may be regarded locally as a pest, due to its grazing habits, although Coot can create worse problems. It also predates eggs and chicks in game-rearing areas.

Previous authors all described the Moorhen as abundant and generally distributed. During the *Atlas* surveys (1966–72) breeding was confirmed in every 10-km square and in 1978 the CBC farmland plots showed a national density of 3·62 pairs per km^2 and a regional density of 2·85 pairs per km^2. While some birds remain shy and secretive, others have become tame and nest in urban parks quite oblivious of human presence. Well-vegetated pools and waterways hold the greatest breeding densities, however. At Brandon, for example, there were at least 30 pairs in 162 ha of marsh and gravel workings in 1971 (a density of 18·5 pairs per km^2), whilst 46 territories were found along 57 km of the fast-flowing River Teme in 1979 (J. J. Day *in litt.*). Such figures suggest a regional population in excess of 20,000 pairs, which would be above the range of 2,000–20,000 pairs postulated by Lord and Munns, but it is difficult to be more precise.

Concentrations outside the breeding season may be of considerable size, with 550 at Blackbrook Sewage Farm on December 5th 1956 the largest on record. At Belvide up to 200 are regularly present in autumn. Severe winters can substantially reduce Moorhen populations, although numbers had probably returned to normal within three years of the hard winter in 1963. During very cold weather birds often feed away from water, such as in the branches of trees at heights of 15 m or more above ground level.

British breeding Moorhens tend to be sedentary (Cramp and Simmons 1977) and a bird ringed at Middleton in June 1915 had certainly not moved far when it was recovered in Erdington five months later. However, immigrants from the Baltic and the Low Countries are known to winter in Britain (Cramp and Simmons 1977) and some of these reach the West Midlands. One ringed at Flushing (Holland) in March 1928 was found on the River Teme eleven months later and in January 1978 a West German ringed bird was recovered near Rugby.

Coot
Fulica atra

Widespread and numerous winter visitor and passage migrant. Fairly numerous or numerous breeding species.

The Coot is widespread, common, and likely to be found on any water larger than 0·5 ha. Sometimes it even occurs on smaller pools and sluggish rivers. There is no evidence of any noticeable change in status from that given by previous authors, but its numbers must have increased with the colonisation of newly-created sites. In particular, well-vegetated, shallow gravel pits with islands have provided ideal breeding sites for a territorial species like the Coot. Winter populations have certainly increased.

During the *Atlas* surveys (1966–72) breeding was confirmed in seventy-three 10-km squares (95 per cent), with birds absent mainly from the Carboniferous limestone district in the extreme north and the Jurassic limestone district in the extreme south. Neither of these districts, of course, contains suitable habitat. Information on density is scant, but there were 15–30 pairs on 162 ha at Brandon from 1972–6 and a similar area at Alvecote held 15 pairs in 1980, whilst at Kingsbury there are usually about 30 pairs on 250 ha. Belvide possibly has the highest density in the Region and in 1979 there were 110 breeding pairs along the 4 km of natural shoreline. Such densities suggest a regional population of 1,000–2,200 pairs, which is at the upper end or slightly above the range of 200–2,000 pairs estimated by Lord and Munns.

When feeding and nesting conditions are good, the spring and summer populations mainly comprise breeding birds. By late June, though, birds are beginning to congregate at favoured feeding and moulting sites and peak numbers are reached between late August and November, when immigrants arrive from the con-

46. The Coot is widespread and numerous, though its numbers vary with fluctuations in water level. *M. C. Wilkes*

tinent. Sites such as Blithfield and Draycote, with good grazing meadows nearby, may also hold large flocks throughout the winter, but by the end of March much smaller numbers remain and these again represent the potential breeding population.

The supply of food plants and their accessibility can be greatly influenced by water levels, so reservoirs in particular hold very variable numbers of Coot. For example, the counts at Belvide in 1979 illustrated extreme conditions. During the severe weather of January numbers were depleted to a mere 100, which were grazing in nearby meadows, but from late February to early June, when water levels were very high, 215–250 were present. These represented an optimum breeding population of 110 pairs, though the

normal population is less than 40 pairs. By September 1st, numbers had increased to 1,850, but the water level then fell by three metres and numbers plummeted until only 60–100 remained during November and December.

Constant high water-levels in late summer and autumn guarantee a good food supply for moulting flocks, which consist mainly of passage migrants or winter visitors from north-east Europe and the USSR (Cramp and Simmons 1977), although there is also much local movement of British breeding birds. During the late 1960s and early 1970s, grazing flocks of Coot were cannon-netted at Blithfield and this has resulted in ten recoveries of birds from all directions and up to 170 km away. One, which was ringed in December 1973, was found at Loiret (France) less

Table 48: Five-yearly Means of Annual Maxima of Coot at Principal Localities, 1944–1978

Brackets denote mean of two years data or less.

	1944–48	1949–53	1954–58	1959–63	1964–68	1969–73	1974–78
Alvecote		(80)	416	252	396	890	424
Belvide	391		517	333	350	794	1450
Bittell	(185)		(56)	215	115	118	196
Blithfield		(450)	823	878	822	732	419
Chasewater	(250)	397	424	172	198	247	462
Draycote						770	1401
Gailey			263	(150)	300	275	343
Kingsbury					(155)	(775)	639

than four months later.

Monthly counts at 26 waters in the Region revealed maxima of 2,809 on November 13th 1955 and 2,442 on November 4th 1956. That numbers have since increased is shown by the fact that peak counts in 1977 and 1978 at only eight waters, two of which did not exist in 1955–6, were 2,926 and 3,731 respectively, both occurring in September.

The most favoured sites, with their respective maxima, have been Belvide (2,050 in September 1980), Draycote (1,865), Blithfield (1,480), Copmere (1,000), Alvecote (975), Chasewater (900), Kingsbury (850) and Gailey (610).

Crane
Grus grus

Very rare vagrant. Four records.

The only early record was of a pair shot at Knowle on December 1st 1903 (Norris). In recent times, one circled Blithfield on October 1st 1971 before departing to the south-west (*Brit. Birds* 65:331); three, including two adults, remained in meadows alongside the Avon in the Welford–Weston–Milcote area from January

3rd to February 13th 1977 (*Brit. Birds* 71:498); and one paused at Belvide for a short period on May 5th 1979, before leaving to the west (*Brit. Birds* 73:504).

Little Bustard
Tetrax tetrax

Very rare vagrant. Two or three old records.

Chase (1886) recorded that one had occurred at Thickbroom, near Tamworth, but Norris placed this record in square brackets. Smith quoted two records of single birds, one shot at Birchfield many years prior to 1893 and the other at Warslow around 1899.

Great Bustard
Otis tarda

One old record.

One was shot near Worcester a few years prior to 1828 (Harthan). The species last bred in England in 1832, being unable to adapt to the enclosure of its former open, treeless breeding habitat.

Oystercatcher
Haematopus ostralegus

Scarce passage migrant and very scarce breeding species.

Smith gave several passage records prior to 1929, Harthan none and Norris four, including one of five birds at Sutton Park on November 25th 1894. However, this species has appeared more regularly in recent decades as Table 49 shows.

Eighty-two per cent of these 431 records, which involved 702 birds, were divided equally between spring (late-February to late-May) and autumn (late-July to late-September).

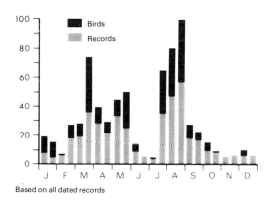

Based on all dated records

Half-monthly Distribution of Oyster-catchers, 1929–78

Temporal segregation between migrants of different breeding populations, ages, and sometimes sexes is a common feature of wader movements. In the Oystercatcher the spring passage is abnormally pro--

longed for a wader and suggests movement by two different populations. Late March and early April records probably refer to birds returning to their breeding grounds in northern Britain, while the birds recorded a month or so later are probably *en route* for more distant areas. Interestingly the May influx was not apparent in Mason's (1969) analysis of Leicestershire records. In autumn, adults return from late July onwards and juveniles follow about a month later. A bird ringed at Heacham (Norfolk) in March 1971 was recovered six months later at Minworth, but otherwise nothing is known of the origins of birds that visit the West Midlands.

Most records refer to ones and twos seen only in flight, but very occasionally small parties are seen or birds remain in an area for a week or more. The largest parties to have been noted were 17 at Belvide on March 22nd 1965 and 11 flying north over Upton Warren on January 23rd 1977. Summer records have increased in recent years, with 12 of the 15 June records occurring during 1975–8, and, in common with some other Midland counties, a few pairs have bred in recent years, mainly at flooded gravel pits. Circumstantial evidence of nesting was obtained during the filling of Draycote Water in 1969 and the following year breeding was confirmed at Branston, where a pair also nested unsuccessfully in 1979. Circumstantial evidence of nesting was obtained at another east Staffordshire site in 1978 and 1979.

Table 49: Ten-yearly Distribution of Oystercatchers, 1929–78

	1929–38	1939–48	1949–58	1959–68	1969–78
No. of records	3	15	78	98	237
No. of birds	3	17	120	183	379

Black-winged Stilt
Himantopus himantopus

One record.

There is just one record of this vagrant from southern Europe. A bird which remained at Belvide from June 11th to 16th 1968 was seen by many observers (*Brit. Birds* 62:470).

Avocet
Recurvirostra avosetta

Very scarce passage migrant, mainly in spring.

Plot (1686) referred to a record of eight birds at Aqualate, Hastings (1834) to one shot near Worcester Bridge a few years prior to 1832 and Garner (1844) to one shot on the Dove shortly before 1844. There were no further records until a bird appeared at Belvide in 1958, whilst the first Warwickshire record came shortly afterwards, in 1960, when a party of eight visited Polesworth Sewage Works on April 3rd and one was at Alvecote on the same day.

After two records in 1963, there have been 21 further records involving 28–33 birds in nine years during 1967–80. This increase has, of course, mirrored the consolidation of East Anglian breeding colonies. All records that were dated occurred between March 31st and November 1st, but there was a heavy spring bias, with 84 per cent before June 8th and most in early April and late May.

About half of the recent records have come from Warwickshire, including seven from the Tame Valley, five from Brandon and three from Draycote. Otherwise, Upton Warren has provided three records, Belvide two and Blithfield, Cannock Chase, Pirton Pool, Westwood Park and Yoxall one each.

Stone-curlew
Burhinus oedicnemus

Very rare visitor, last recorded in 1956. Formerly bred in Worcestershire.

Hastings (1834) wrote that "a few pairs bred among the stony barren parts of the Broadway and Bredon Hills", and Tomes (1901) reported two specimens from Eardiston in the Worcester Museum. It is thought that Stone-curlews ceased to breed in Worcestershire soon after 1840 and there have been no subsequent records from that county.

Norris knew of three records, namely one at Wilmcote on October 19th 1847, one near Weston-on-Avon on January 1st 1853 and one with Lapwings in a wheat field near Sutton Coldfield in early May 1861. There have been only three or four twentieth-century records for the Region, with one at Harborne on August 9th 1927, a bird probably of this species heard calling near Alveston on April 22nd 1951, one near Handsworth on May 3rd 1952 and finally one at Blackbrook Sewage Farm on December 28th 1956, which remains the only Staffordshire record.

Little Ringed Plover
Charadrius dubius

Regular summer visitor, breeding species and passage migrant. Currently local, but not scarce.

The spread of the Little Ringed Plover into Britain through its exploitation of transient habitats such as sand and gravel-pits, quarries, slag heaps and industrial wastes is one of the avian success stories of the twentieth century. The species first bred in England in 1938 and has done so annually since 1944, whilst the first record for the West Midlands was of a bird at a gravel-pit near Coventry on May 22nd 1945. The next was of unsuccessful breeding in 1952 at

Branston, where nesting has been regular since 1960. Since 1959 breeding has occurred in every year and during the *Atlas* surveys (1966–72) it was confirmed or suspected in twenty-one 10-km squares (27 per cent) embracing parts of the Severn, Avon and Trent Valleys. A peak of 32 pairs at 16 localities was reached in 1977 and currently between 25 and 30 pairs nest each year.

Breeding was initially concentrated in the Tame, Anker and Trent Valleys, where birds rapidly colonised gravel-pits as they were excavated. Indeed, at least two-thirds of the recorded nest sites have been at gravel-pits, but smaller numbers have been reported from sites such as open-cast coal workings, power station fly-ash lagoons, industrial wasteland, sandpits, rubbish tips and even building sites – often with very little water nearby. That Little Ringed

Plovers will quickly use newly-exposed sites was shown by their colonisation of Draycote Water whilst it was under construction and of Belvide, Bittell, Blithfield, Chasewater and Tittesworth while their water levels have been temporarily low for one reason or another. Most sites hold less than three pairs, but Kingsbury held seven pairs in 1979, Brandon has held up to six pairs and Packington gravel-pits up to five pairs.

In all, breeding has been confirmed at 48 localities and suspected at a further nine, but several of these are no longer attractive to breeding birds. Most quarried or opencast areas are now filled and returned to agriculture or put to amenity use, urban sites may be built upon, and, if nothing else happens first, then steadily encroaching vegetation will finally deter birds from nesting.

Little Ringed Plovers are early returning

47. The Little Ringed Plover is an opportunist which, during the past thirty years, has been quick to exploit the increasing number of gravel pits. *S. C. Brown*

migrants, with regular March occurrences. Its extreme arrival and departure dates have been March 2nd (1975) and October 27th (1972), though the averages for first and last birds over fourteen years have been March 25th and October 4th respectively. Birds are widespread on passage, especially in late July and early August, when parties have totalled up to 25 at Chasewater, 27 at Blithfield and a maximum of 34 at Kingsbury on August 2nd 1979. There is sometimes a small influx of juveniles, perhaps of continental origin, in early September.

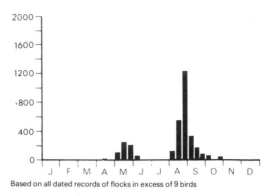

Based on all dated records of flocks in excess of 9 birds

Distribution of Ringed Plovers by Thirds of a Month, 1929–78

Ringed Plover

Charadrius hiaticula

Regular and fairly numerous passage migrant, but very scarce as a breeding species and rare in winter.

Previous authors considered the Ringed Plover a regular migrant in small numbers. During this century, with the provision of further aquatic feeding sites in the form of drinking-water reservoirs and gravel-pits, it appears to have become steadily more numerous. Indeed, most wader species have benefited greatly from the appearance of such sites. The Ringed Plover is one of the commonest passage waders inland, but during its migratory pauses it is usually restricted to areas of exposed mud. For this reason it occurs less widely than the other two common small waders, Dunlin and Common Sandpiper.

Most records fall between March and November and, during 1958–78, there were only four records in December, three in January and six in February, excluding an exceptional record of a bird which remained at Shustoke from December 9th 1974 until March 3rd 1975. The peak spring passage occurs in May and early June, at a time when large numbers are moving through western Britain *en route* for Ice-

land and Greenland. These birds are often accompanied by Sanderlings, especially in the last two-thirds of May which is the time when the largest spring flocks – up to 35 at Chasewater, 36 at Belvide and 59 at Blithfield – have occurred.

Although the first returning birds may begin to trickle through at the end of June, sizeable movements do not occur before August. Indeed most of the larger autumn flocks have been noted in the last third of August, when the sometimes vast expanse of mud at Blithfield has attracted totals of 165 in 1962, 140 in 1965 and a record 166 on August 23rd 1970. At this time too, Chasewater has held up to 80 birds and Belvide up to 50, while 44 arrived from the north-east at Bittell on September 2nd 1956. These influxes may include birds from north-east Europe as well as the Greenland birds which pause to feed *en route* to their wintering grounds in North Africa. Some proof of the origin and destination of autumn migrants is given by one of only five adults to have been dye-marked in Greenland in the summer of 1972 and which was then seen at Draycote on August 20th; and by a bird ringed at Blithfield in August 1963 that was recovered in Portugal in October 1967. Occasional influxes occur in late September, when Mason (1969) noticed a distinct peak at Eyebrook Reservoir (Leics.), and early October, but November records are unusual.

Ringed Plovers are reputed to have nested at Bittell in 1902 (Harthan). A pair was seen with a nest scrape at Chasewater on May 1st 1976, but apparently nesting did not take place. However, on June 24th 1979, a juvenile that was unable to fly was seen with two adults at Branston, at least three pairs nested in the Tame Valley in 1980 and further colonisation of the Region seems likely. A pullus ringed at a Nottinghamshire breeding locality on June 23rd 1980 was found dead or dying at Rocester thirty-four days later.

Kentish Plover
Charadrius alexandrinus

Rare passage migrant. Nine records.

There have been only nine records of the Kentish Plover, all since 1940. Five of these occurred in spring, between April 22nd and May 14th, and four in autumn, between July 15th and September 16th. All records have been of single birds, namely a juvenile at Bittell on July 15th 1940 which remains the only Worcestershire record (*Brit. Birds* 34:90); a male at Belvide on May 2nd 1948; one at Belvide on September 13th 1950; an immature at Blithfield from September 11th to 16th 1970; one at Draycote on August 21st 1974; a female at Draycote on April 22nd 1976; a female, possibly the same bird, at Chasewater on April 28th 1976; a male at Kingsbury on May 14th 1977; and a male which flew from Drakelow (Derbys.) across the Trent into Staffordshire on May 7th 1979.

Dotterel
Charadrius morinellus

Rare passage migrant. Five recent records.

Smith, Harthan and Norris listed several early records of birds being shot or killed,

namely an immature at Welland in 1861, reports of birds being shot on the Dove near Tutbury prior to 1863, one at Great Barr on September 4th 1867, ten on Cannock Chase on May 15th 1875, one or two at Perry Barr in 1882, a male some years before 1901 in a part of the Avon Valley that used to be in Gloucestershire and one at Clifton Campville one spring around 1908 which was curiously reported as an immature.

Smith also quoted J. E. Smith who had seen "Dotterel duns" made from the feathers of birds obtained at Cauldon Low and the Weaver Hills, and the evidence suggests that these uplands may once have been a regular migration resting place. On May 13th 1922 a male hit wires near Draycott-le-Moors and another was found dead at Sandon.

Subsequently there have been only five records, namely an adult at Chasewater on August 24th 1950; one at Walton Hill, Clent, on September 24th 1964; an adult at Chasewater from August 16th to 26th 1971; a pair on Bredon Hill on April 26th 1976; and a trip of nine, comprising four males, three females and two indeterminates, in a field at Grove End on May 13th 1978.

Golden Plover
Pluvialis apricaria

Numerous, but local, winter visitor and passage migrant. Scarce or very scarce breeding species on the higher moors.

As a breeding bird the Golden Plover is restricted to the highest parts of the North Staffordshire Moors, where breeding was first proved in 1925 and up to five pairs have been reported in most subsequent years. Birds are normally present in and around their breeding areas of cotton-grass or heather moor from March until July or August.

241

48. Dotterel are often tame and afford good views on the rare occasions when they visit the Region's uplands or reservoirs. *G. W. Ward*

The species is much better known, however, as a winter visitor and passage migrant to the Region, though even then it has a decidedly local distribution. It is invariably much scarcer than the Lapwing, although it is usually found alongside that species and the Black-headed Gull. Even so, Golden Plovers are apparently more numerous now than at the beginning of the century, although previous authors were able to provide only scant information on status and distribution. The accompanying map summarises the distribution and maximum size of the thirteen flocks that have been regularly recorded since 1929 and also shows the location of other isolated flocks. How this distribution pattern differs from that of previous years is largely unknown, but changing agricul-

tural practices must have had some effects – perhaps even beneficial ones – on the density of earthworms, which form a major winter food item of the Golden Plover.

Most winter flocks inhabit flat or gently undulating farmland where the soils are well-drained, neutral or basic, medium-textured loams. Such situations are capable of holding large populations of earthworms and other macro-invertebrates. Crops vary from pastures, including quite often high-yielding leys which are in arable rotation, to cereals and even potatoes, especially in south Staffordshire and north Warwickshire. In all these situations the soil receives much organic matter, though this is least under cereals, and it is therefore ideal for several earthworm species. While

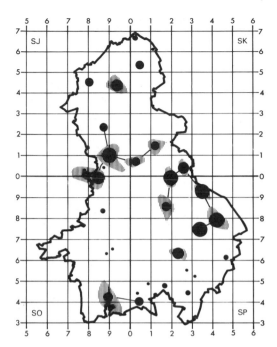

Flock size:

- •   0–99   ● 400–799
- • 100–199   ● 800–1,200
- ● 200–399

▓▓▓– Flock range, with probable directions of inter-mixture

Distribution and Maximum Size of Golden Plover Flocks, 1929–78

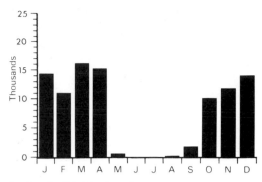

Based on all published dated records

Monthly Distribution of Golden Plover, 1929–78

this offers an explanation for the presence of Golden Plovers in certain areas, however, it does not explain the scarcity of birds in several other, apparently suitable areas.

Winter visitors are usually present from August to early May, though the average dates of first and last birds over thirty-one and thirty-two years respectively have been August 12th and April 24th. There have been only three June and two July records from the lowlands. All of these involved single birds, except for a flock of 27, of which 26 were recognisable as northern form birds *P. a. apricaria* (after Hale, 1980), which roosted on June 17th 1978 at New Invention. This was a regular roost

site until 1978 and was interestingly on grassland whereas most birds roost on cultivated land. As the histogram indicates, the largest numbers are present from October to April. In common with other species that prey on soil invertebrates, Golden Plovers are very susceptible to snow cover or freezing conditions and will soon move to warmer climes. Such hard-weather movements may be widespread and involve large numbers, although individual flocks are usually quite small. For example, a total of 1,000 in parties of 50–100 flew south-west at Branston in one hour on November 14th 1954. The low overall total for February suggests that many birds desert regular localities at this time. However, the average size of reported flocks for February is the highest for any month (278 birds) and this suggests that birds may concentrate at fewer feeding localities, as well as moving out of the Region. Several flocks have reached 1,000 birds, though none has exceeded the 1,200 at Essington on April 11th 1971. Overall it is likely that the Region's total of wintering Golden Plover is in the order of 5,000–7,500 birds.

In spring and autumn, small numbers of passage birds are widely reported and the histogram suggests an influx in March and April, by which time British breeding birds are settling into their territories. This spring passage, therefore, presumably comprises birds breeding in northern Europe and many, indeed, are clearly recognisable as northern form birds. Although not separable in winter plumage, it is generally thought that the vast majority of wintering birds are continental immigrants of the northern or intermediate forms.

Grey Plover
Pluvialis squatarola

Scarce passage migrant.

Garner (1860) and McAldowie (1893) included the Grey Plover on their lists of Staffordshire birds, but gave no details, and the only dated nineteenth century record was of one near Birmingham on October 3rd 1899. There were several records from the Chasewater area around the turn of the century, whilst one was shot near Ashbourne (Derbys.) in 1916 and two occurred at Belvide on September 30th 1923.

During 1929–78 there were 200 records, which involved about 395 birds, in thirty-five of the fifty years. Most of these occurred after the war and there were records in every year after 1953, except in 1968. On average, less than two records per year were reported until the mid-1950s, then up to five per year until 1969. However, the last decade produced even more records and no less than 45, involving at least 95 birds, were received in the drought years of 1975 and 1976, when many water levels remained low.

Nearly half of the Grey Plover records fall in September and October, when juveniles arriving in England from Siberia may encounter bad weather and pause inland. Although most records refer to less than three birds, and usually to only one, there have been some larger parties in

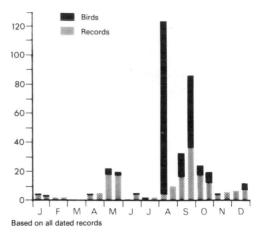

Based on all dated records

Half-monthly Distribution of Grey Plovers, 1929–78

autumn, notably 10 at Bittell on September 2nd 1956 and August 10th 1975; 13 at Blithfield on September 21st 1975; and an exceptional flock of about 110, including many adults, which flew over Belvide from the east on August 12th 1969. 17·5 per cent of the records occurred in May, representing a small, but fairly regular, spring passage. There have been very few June and July records and these have mostly involved non-breeding birds in first-summer plumage. Interestingly, most Derbyshire records occur in September, October, March and May (Frost 1978), but there have been no March records at all in the West Midlands, whilst in Leicestershire Mason (1969) noted a May peak, but very few in autumn.

Birds occasionally stay for a week or two and exceptionally as long as eight weeks when water levels permit. There have also been a few records of birds on agricultural land, where they associate with Lapwings or Golden Plover.

White-tailed Plover
Chettusia leucura

One record.

An adult remained at Packington Gravel

Pits from July 12th to 18th 1975 and this constituted the first record for Great Britain and Ireland of this vagrant from the Middle East, southern USSR and Asia. This record followed fourteen previous records for Europe, of which seven occurred between March and July 1975 (*Brit. Birds* 69:334–5; 70:465–71).

Lapwing
Vanellus vanellus

Numerous breeding species and abundant winter visitor and passage migrant.

The Lapwing has long been a common breeding bird and winter visitor to the Region's farmland. Smith considered that there had been local declines in breeding numbers, such as on moorland near the Churnet Valley between the wars, and Harthan described a decline with agricultural intensification during and after the 1939–45 war. Norris, however, made no reference to any change in status or distribution, but the ploughing campaign of the war certainly resulted in the loss of much permanent pasture and this, together with the subsequent increase in the use of machinery and artificial fertilisers, was probably the cause of numerous other reports of declines in breeding density until

about 1955. Comparative information on the current status is sadly lacking.

Little is known of the densities in different cropping situations, but estimates of breeding density on various farms in the Region between 1956 and 1963 averaged 2·67 pairs per km^2, compared with the 1978 CBC average of 3·66 pairs per km^2 nationally. Applying these densities to the regional area of crops and grass suggests a breeding population in the order of 14,000–18,000 pairs, compared to the range of 2,000–20,000 estimated by Lord and Munns. Lapwings were confirmed or suspected of breeding in every 10-km square during the *Atlas* surveys (1966–72). In fact, they will breed in a variety of agricultural situations providing the vegetation is short, though damp meadows are perhaps their traditionally favoured haunts. Many pairs now nest in such situations around reservoirs, gravel pits and on derelict land, and Alvecote has held up to 10 pairs, Belvide 18, Blithfield 15, Brandon 12 and Whittington Sewage Farm 38; whilst four unspecified fields at Griff held a scattered colony of about 30 pairs in 1971. In 1975 and 1976 a pair even nested on a traffic island on the A435 in Warwickshire.

Lapwing chicks ringed in the Region have been recovered in winter months in Devon, Ireland, France, Spain, Portugal and Morocco, indicating a winter range to the south and south-west of Britain. Equally, from mid-summer until early spring the Region hosts many immigrant Lapwings from the Baltic area and the USSR. Ringing has shown the West Midlands to be a winter haunt for birds ringed as chicks in Cheshire, Sweden, Denmark and the Netherlands. Birds ringed in late summer and autumn in recent years have been recovered in Sweden, the Netherlands, France, Hungary, Portugal and the USSR, including one ringed on July 24th 1977 at Coventry and recovered unusually far east at 67°E on July 21st 1978. Birds ringed in winter have been reported sub-sequently from the USSR (two birds at 35°E and 39°E), Finland, the Netherlands, France and Suffolk.

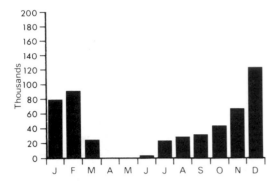

Based on all published dated records

Monthly Distribution of Lapwing, 1929–78

Small parties may be noted as early as the end of May and during June flocks may be up to several hundred strong. From late June onwards continental immigrants stream into the Region, with moulting adults especially noticeable amongst the first arrivals, and the histogram indicates that this immigration continues until December at least. Feeding flocks of up to 3,000 are noted regularly, with occasionally as many as 5,000 and a record 8,000 at Kingsbury on March 3rd 1980. Daytime migration, mainly to the west and south-west, is often noted in autumn and movements have involved up to 4,000 birds. Radar studies have shown that the migration of many species of bird, including waders, also takes place at night.

Like Golden Plover, Lapwings are very susceptible to ice and snow cover and spectacular hard-weather movements can frequently be seen between December and February. The largest of these involved 9,000 passing to the south and south-west over Leamington Spa on January 30th 1972 and 10,000 moving south over Belvide on February 9th 1974. With milder weather in February and March, the

return can be rapid and the movements almost as spectacular, with, for example, 5,000 moving north-east over Belvide on March 11th 1970.

Knot
Calidris canutus

Scarce passage migrant.

Smith quoted Garner (1844) and Mosley (1863), who regarded the Knot as occasional, and reported two other pre-1929 records, of one shot at Tittensor in December 1892 and two at Belvide on October 12th 1913. However there were no Worcestershire records until 1929 and none from Warwickshire until 1951.

Between 1929 and 1978 there were 246 records of Knot involving some 419 birds. About three-quarters of these birds were seen in Staffordshire, where the muddy margins of the larger reservoirs seem particularly attractive. The majority of records fell between late July and September, during which period two distinct influxes occur. The first influx in late July is small and comprises adults, which are often still in breeding plumage, whereas the second influx in late August and September, which accounts for 44 per cent of all records, coincides with the arrival of juveniles in Britain, although some adults may still be detected. Interestingly, the late July influx was less obvious in analyses carried out for Leicestershire (Mason 1969) and the London area (Chandler and Osborne 1977). The vast majority of Knot occurring in Britain originate from breeding grounds in Greenland and north-east Canada, but influxes of Siberian birds may occur on passage (Prater 1974).

Further small influxes occur in early November and February. The former is probably of adults arriving from their moulting grounds on the Dutch Wadden-

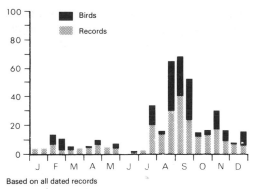

Half-monthly Distribution of Knot, 1929–78

see, while the latter may be birds returning there. Many ringing recoveries show that wintering East Coast birds go to the West Coast, particularly Morecambe Bay, before migrating north in spring. It is presumed they fly across country at high altitude, so they remain undetected. After this cross-country movement, however, a very weak spring passage is evident in April and May, at which time many Knot are concentrated at Morecambe Bay building up fat reserves for their long flight back to their breeding grounds (Wilson 1973). There is only one June record, of two birds at Belvide on the 17th 1961.

Although most records refer to single birds, small parties of up to ten occasionally occur, and 13 at Chasewater on September 29th 1957 is the largest party on record.

Sanderling
Calidris alba

Scarce passage migrant. Very rare and exceptional in winter.

Harthan quoted a record of a Sanderling on the Teme in 1826, while Smith knew of birds shot near the county boundary at Walton-on-Trent (Derbys.) and in the Burton area and reported Coburn's findings that the species was "regular and

fairly numerous" at Chasewater in 1908 and 1909. Norris, however, knew of no Warwickshire record before 1947, when several birds appeared.

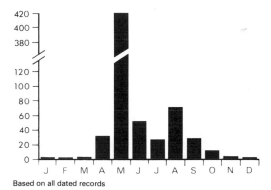

Monthly Distribution of Sanderling, 1929–78

Between 1929 and 1978 there were at least 280 Sanderling records involving 666 birds. About 60 per cent of these records (76 per cent of birds) occurred between late April and early June, with the majority between May 13th and 25th (at least 80 records and 318 birds). Thus the Sanderling can be seen to be principally a spring migrant to the West Midlands and at this season small parties, the largest of which was 22 at Blithfield on May 21st 1956, may pause briefly at sites with suitable exposed mud. Most records have come from Belvide, Blithfield, Draycote and especially Chasewater. This passage is part of a concentrated movement through western Britain of birds *en route* from south and west Africa to their breeding grounds in Greenland. At times birds will also settle on artificial banks, such as the causeway at Blithfield, but they are primarily lovers of sandy beaches and many more Sanderlings must pass over without stopping due to the scarcity of suitable feeding areas of exposed mud at West Midland waters in spring.

By contrast, the autumn passage is much smaller, more leisurely and dispersed, with only 101 records (34 per cent) having occurred in the period July to September. Of these, 85 per cent referred to single birds, with adults passing through between mid-July and early August and juveniles about a month later. Records between mid-June and early July, or in October, are unusual, whilst those between November and March are rare. Low water levels early in 1976, however, led to three unusual winter records of Sanderlings accompanying Dunlin flocks at Blithfield, Draycote and Shustoke.

Little Stint
Calidris minuta

Scarce passage migrant; occasionally not scarce.

Norris knew of two nineteenth-century records and Garner (1844) vaguely included the species on his list, whilst Coburn found Little Stints at Chasewater and Earlswood in the early years of this century. Four birds then occurred at Belvide in September 1911 and another in September 1923, whilst Bittell produced the first Worcestershire record, of five birds, on September 27th 1921 and another in August 1928.

Only single birds were noted between 1935–49, but there has undoubtedly been a great increase in records in recent decades, with small parties turning up frequently, especially in autumn. In common with other species nesting in sometimes inhospitable tundra regions, juvenile Little Stints may be quite numerous in some autumns, while in others very few appear. During the last decade, the autumns of 1973 and 1976 were particularly good for this species, with at least 86 and 66 birds being noted respectively.

The autumn passage often commences with a few adult birds in late July and early August, followed by immatures between late August and early October. Peak numbers, especially in good years, usually appear in the second half of September, though occasionally, as in 1975, they may occur in late August or early September. Flocks have reached double figures on ten occasions, with a maximum of 26 at Blithfield on September 24th 1973. The favoured localities have been Belvide, Bittell, Brandon, Blithfield, Chasewater, Draycote and Ladywalk. When water levels have remained low, such as in 1943, 1962 and six years in the past decade, up to five birds have lingered into November and December, but there have been no records between December 16th and March 22nd.

Single birds occurred at Alvecote and Blithfield on the early dates of March 22nd 1975 and April 8th 1963, respectively, but all the remaining 29 spring records fell between April 25th and June 2nd, with most in mid-May. All spring records, which have been annual in recent years, have referred to ones and twos, except for eight at Chasewater on May 25th 1958. There have been seven records between June 9th and 27th, all since 1960, but none between June 28th and the first "autumn" records on July 13th.

Temminck's Stint
Calidris temminckii

Very scarce passage migrant.

Harthan recorded a bird shot at Bittell in 1888 and Smith one that was shot near (Great) Wyrley on an unknown date. Subsequently Temminck's Stints appeared in 1935, 1946, 1950, 1955, 1959, when two late birds at Middleton Hall on November 2nd provided the first Warwickshire record, and in fourteen subsequent years.

Between 1929–78 there were 34 records involving 44 birds. Half of these records

(24 birds) fell in spring, between April 29th and May 31st, and half in autumn, between July 17th and November 2nd. In spring birds stayed from one to four days, with an average of two days – a duration typical of most inland waders in spring. In contrast to the West Midland's pattern, Chandler and Osborne (1977) found birds to be much scarcer on spring passage in the London area.

All but three records referred to single birds, but there have been two "pairs" and an exceptional party of six, which remained at Upton Warren from May 25th to 27th 1975. In all there have been six records each from Belvide, Blithfield and Upton Warren, four each from Bittell and Chasewater, and singles from Alvecote, Blackbrook Sewage Farm, Brandon, Draycote, Kingsbury, Ladywalk, Middleton, Packington, Wilden and near Wyrley.

Least Sandpiper
Calidris minutilla

One record.

A bird which stayed at Chasewater from August 9th to 11th 1971 was the eighteenth British record, the earliest in autumn and the first inland of this North American vagrant (*Brit. Birds* 65:333).

White-rumped Sandpiper
Calidris fuscicollis

One record.

An immature at Blithfield from November 10th to 12th 1979, and what was presumed to be the same bird again from December 8th to 11th, is the only occurrence of this North American vagrant.

49.   Autumn is the most likely time for transatlantic vagrants like the Pectoral Sandpiper to make their rare appearances.   *G. W. Ward*

Pectoral Sandpiper
Calidris melanotos

Rare vagrant. Eighteen records.

One at Alvecote from June 4th–6th 1957 was the first record for the Region. It was also an unusual date for this, the most regular of the transatlantic vagrants. One occurred at Blithfield on September 23rd 1962 and this was followed by a further 16 records during 1967–79. All of these fell between July 24th and October 4th, with half of the arrivals concentrated during the period August 29th–September 8th. Details of these sixteen records are: one at Blithfield from August 12th–17th 1967; one at Whitacre on September 2nd 1967; one at Belvide from August 31st to September 20th 1970; one at Blithfield from September 5th–20th 1970; one at Upton Warren from September 8th–22nd 1970; one at Ladywalk on October 3rd 1970; one at Upton Warren from July 24th–26th 1971; one at Wilden between August 29th and September 2nd 1971; one in breeding plumage at Upton Warren on July 24th 1972; one at Blithfield on September 22nd 1972; two at Blithfield from August 29th to September 5th 1973, with one remaining until September 15th; one at Draycote from August 30th to September 18th 1973; one at Wilden from September 4th to 17th 1973; another at Draycote on October 4th 1973; one at Bittell from September 18th to 21st 1978; and one at Kingsbury on October 1st 1979. It is interesting that 12 of these 16 records occurred in just four years from 1970–3 inclusive, during which time there were at least two records in every year.

Curlew Sandpiper
Calidris ferruginea

Scarce passage migrant, occasionally not scarce.

After two nineteenth-century records of birds shot at Stratford in 1869 (Norris) and Cofton Reservoir in 1885 (Harthan), both of which were in September, Curlew Sandpipers have been detected with increasing regularity. There were records in thirty-five of the fifty years 1929–78 and in all but two years since 1948.

Like the Little Stint, the Curlew Sandpiper has good and bad autumns. 1969, 1975 and 1978 produced 35, 53 and 54 birds respectively and were particularly good autumns when the combination of good breeding seasons and easterly winds were responsible for record numbers of juveniles reaching Britain (*cf.* Stanley and Minton 1972). The first autumn records, in July and early August, refer to adults which are often in breeding plumage and there have been 29 such records, involving 45 birds, which is slightly more than for Little Stint. The following influx of juveniles is usually larger and occurs roughly between August 20th and September 20th, which is somewhat earlier than that of the Little Stint. As would be expected of a species wintering in southern Africa, few occur after early October, although the latest record was on November 2nd.

Blithfield is consistently the most regular haunt in the Region and holds the record with a regional maximum of 37 on September 10th 1978. Otherwise Belvide has held up to 19, Chasewater 14 and Kingsbury 16, whilst numbers less than ten have appeared at many localities.

Curlew Sandpipers are very scarce in spring and there have been only 18 records of 24 birds in sixteen of the fifty years 1929–78. Although these have spanned the period April 2nd to May 29th, 15 of the records were concentrated into the period April 25th to May 20th.

Purple Sandpiper
Calidris maritima

Rare passage migrant.

The first definite record for the Region was of a bird at Chartley in March 1909, although others were said to have been obtained on Burton sewage farm (Smith). Since 1930 there have been a further 21 records in sixteen years, of which 57 per cent have occurred since 1968. One appeared in early February, three in mid-May, one in July, two in August, seven in September including five between the 13th and 17th, three in October, three in November and one in December. In contrast, there have been few autumn records before October in the London area (Chandler and Osborne 1977). The most unusual record was that of a bird at Kingsbury on July 7th 1978.

All records referred to single birds, apart from two at Baginton on September 13th 1951 and at least four at Walsall Sewage Farm on October 2nd 1930. In addition to the localities already mentioned, birds have been seen at Alvecote, Bartley, Belvide (three), Bittell (two), Blithfield (four), Brandon, Chasewater (four), Draycote and Edgbaston Reservoir.

Dunlin

Calidris alpina

Fairly numerous passage migrant; usually scarce in winter.

Dunlin have never been found breeding in the Region, although birds have bred on the moors at Axe Edge, just 200 m across the border into Derbyshire. It appears that their preferred habitat in the Peak District is tracts of wet cotton-grass moor above 400 m and this is not to be found in Staffordshire (D. W. Yalden *in litt*).

The species has long been known as a passage visitor to the lowlands, however, and is often the most numerous wader around our waters. Although alluvial mudflats are preferred, birds occur in a wide range of habitats and will sometimes feed in short vegetation around lake margins and on temporarily flooded grassland. Parties up to 70 were noted several decades ago at canal-feeder reservoirs, such as Belvide, Bittell and Chasewater. However, the advent of large drinking-water reservoirs, with their greatly fluctuating water-levels, saw the appearance of some much larger flocks, notably at Blithfield where the regional maximum of 260 occurred on November 4th 1973. In addition, the chains of gravel pits along the main river valleys have attracted many more small parties of Dunlin on passage.

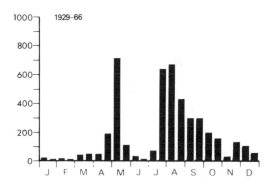

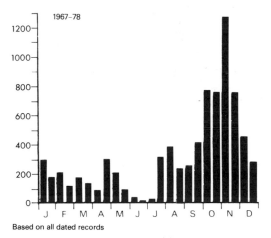

Based on all dated records

Half-monthly Distribution of Dunlin, 1929–78

The histograms illustrate several influxes or peaks and also reveal a shift in emphasis of these peaks in recent decades. Dunlin movements in the West Midlands are complex, as indeed they are in the rest of Britain, and interpretation is complicated by the different movements of both the adults and, in autumn, juveniles of the three main sub-species, although it is not yet known to what extent, if any, *C. a. arctica* from Greenland occurs in the Region.

Spring passage has always been very evident, but in the last decade or so it has peaked in late April rather than in early May, as previously. Spring flocks have totalled up to 70 at Belvide, Bittell, Chasewater and Draycote.

The first autumn influx usually occurs in late July or early August and is believed to comprise many birds of the southern race *C. a. schinzii*, which breeds in Britain, the Baltic and Iceland (Hardy and Minton 1980). In some years this influx may be large and widespread, although the high water-levels of early autumn often deter migrants. The largest example was 175 at Blithfield on August 2nd 1974. Since 1967, the three further influxes in late September–early October, early November, and late November–early December have been much larger than in previous years. Indeed Blithfield has often held over 100 Dunlin for several weeks in late autumn. These influxes involve the northern form *C. a. alpina*, which breeds in northern Europe and Siberia, and are presumed to comprise both birds arriving from their continental moulting grounds, like the Waddensee, to winter in Britain and birds moving south and west to their wintering grounds after having moulted on British estuaries (Hardy and Minton *op cit*). This presumption is supported by three birds from a wintering flock at Draycote in February 1976, of which two had been ringed on the Wash the previous August, when they were presumably moulting, and the other was subsequently retrapped on autumn migration at Ottenby (Sweden) in July 1978.

If Dunlin desert the reliable saline feeding conditions of estuaries for inland wintering sites, they may have to cope with inhospitable, frozen reservoir margins and rising water-levels, neither of which is conducive to successful over-wintering by a mud-loving wader. Nevertheless, there has been a dramatic increase in successful over-wintering in the Region over the last decade, which has coincided with a succession of mild winters and periods of low water-levels at reservoirs, notably in the drought years of 1975–6. For example, in early 1976 there were flocks of up to 30 at Bittell, 69 at Blithfield and 110 each at Chasewater and Draycote. Some fluctuations occur in winter flock numbers and by no means all of the birds are remnants from autumn influxes. Flocks disperse in March.

In common with other coastal waders, inland occurrences of Dunlin can often be correlated with weather factors. Without doubt, larger numbers of waders, and other maritime species such as Common Scoters and Arctic Terns, overfly the Midlands undetected and only when flights are upset by adverse weather conditions will they break their journeys inland.

Buff-breasted Sandpiper
Tryngites subruficollis

Very rare vagrant. Two records.

One at Chasewater on September 14th 1978 was the first record for the Region of this, the second commonest transatlantic vagrant to Britain (*Brit. Birds* 72:523). A second bird remained at Blithfield from September 7th to 22nd 1980. September is the classic month for this species to appear in Britain (Sharrock and Sharrock 1976).

Ruff

Philomachus pugnax

Regular passage migrant, not scarce. Very scarce in winter.

The Ruff is predominantly a wader of freshwater mudflats, but it is occasionally seen on grassland or even cultivated land. After only three dated records in the late nineteenth century (Smith and Norris), birds appear to have become steadily more numerous through the decades of this century and in excess of 2,000 were reported between 1929–78. About three-quarters of these occurred in autumn, when large numbers pass along the western seaboard of Europe *en route* from their north-eastern breeding grounds to their winter quarters in West Africa. Nearly half of the West Midland records have fallen in late August-early September, when the largest flocks usually occur. Brandon, Upton Warren and especially Blithfield are the most favoured localities, with the latter regularly holding double figures, including the regional maximum of 33 on August 14th 1977.

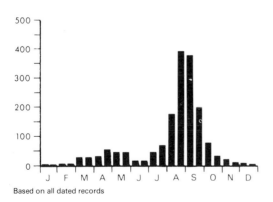

Half-monthly Distribution of Ruff, 1929–78

Based on all dated records

In spring Ruff apparently take a direct great-circle route to their northern breeding grounds, which takes them across central Mediterranean regions rather than Britain. Spring passage in the West Midlands is thus much smaller than in autumn and accounts

for only 15 per cent of the birds recorded. The largest spring flock was 12 at Belvide on May 25th 1953. While March records have increased in the last decade, perhaps as a result of the recolonisation of England, it would appear that May records have become scarcer. Since 1959 there have been a few winter records, especially during 1973–5 when there were five records in January and four in February and during 1980/1 when at least five wintered in the Tame Valley. These records accord with Ruff having wintered in increasing numbers in a few select areas of England during the last decade or two (Prater 1973).

Jack Snipe

Lymnocryptes minimus

Not scarce as a passage migrant and winter visitor.

The Jack Snipe has long been known as a regular and fairly common winter visitor in small numbers (Smith, Harthan and Norris), but it is more local than the Snipe owing to its more specialised habitat requirements. It prefers wet grazing meadows or marshes of grass, rush and sedge, with plenty of dead or dying vegetation for camouflage. Birds also invariably sit tight and are therefore much overlooked.

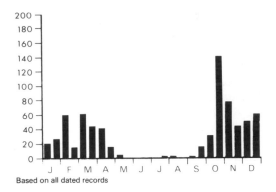

Based on all dated records

Half-monthly Distribution of Jack Snipe, 1937–78

The first autumn arrivals from Scandinavia and Siberia appear in late September and over thirty-seven years the average date of the first arrival has been September 30th, with the earliest recorded on August 26th (1973). Immigration culminates in a considerable influx in late October and maximum numbers are usually recorded then or in early November. These have included up to 10 each at Sutton Park and Upton Warren, 12 at Ford Green, 14 at Belvide, 27 at Chasewater and 30 at Tillington Marsh on October 24th 1971. Most birds probably move on to milder climes, such as Ireland, as soon as the first severe frosts make probing for food impossible, but small concentrations may still be found at ice-free sites during hard weather, such as 20 along a stream at Draycote on February 1st 1976.

In spring there is evidence of a small return passage in March and early April, when 12–14 have occurred at Chell Heath, Ford Green, Upton Warren and Whitmore Heath. Over thirty-three years the average departure date for last birds has been April 17th, but a few have sometimes lingered into early May, with the latest being on the 13th (at Chasewater 1958). There have been two exceptional summer records, of single birds at Chasewater on July 27th 1952 and Belvide on August 14th 1962.

Snipe
Gallinago gallinago

Widespread and numerous as a passage migrant and winter visitor. Fairly numerous as a breeding species.

Extensive drainage operations in the early nineteenth century considerably reduced the numbers of breeding Snipe. The population then appeared to recover somewhat in the first three or four decades of this century, but with the post-war intensification of drainage operations, notably in the last decade, it has since suffered a further marked decline.

Smith said the species was well distributed in large numbers on the moors and in the Trent Valley, but both Harthan and Norris said only a few pairs bred, although they regarded it as a common or abundant migrant and winter visitor. Snipe still breed widely only in the wet areas of the northern moorlands and their adjacent rough pastures, and even much of the latter is threatened by agricultural improvement. In the lowlands, birds breed locally in marshy or boggy areas along the main river valleys, although the vast majority are concentrated into the Trent–Sow–Tame–Anker river system. An exceptionally good breeding density has been reported in recent years from Wilden, however, where there were at least 20 pairs in 1972.

During the *Atlas* surveys (1966–72) breeding was confirmed or suspected in fifty-two 10-km squares (68 per cent), but many held just a few pairs. A survey of Staffordshire in 1978 revealed probable breeding in eighty-three 1-km squares of the uplands to the north and east of the Churnet Valley, but in only forty squares of the lowlands to the south and west. The county population was estimated at about 500 pairs, of which less than 100 were in the lowlands (F. C. Gribble *pers comm*). It is unlikely that the breeding populations of Warwickshire and Worcestershire exceed 40 and 100 pairs respectively, giving a regional population of between 600–700 pairs, which is well above the range of 20–200 pairs estimated by Lord and Munns.

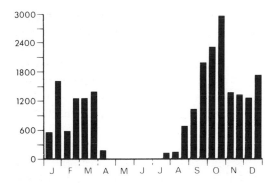

Summation of all Dated Snipe flocks, 1929–78

After the breeding season few remain on the moors, but in the lowlands the Snipe is widely known as a passage migrant and winter visitor to wet pastures, marshes, reed-beds and sewage farms. Peak autumn migration from the continent occurs in October, but, as with Jack Snipe, numbers fall with the first frosts of the winter, when birds move on to milder areas. Sites which do not readily freeze, however, especially old-fashioned sewage works, may hold large numbers in mid-winter. The largest concentrations have been 600 at Tillington Marsh in October 1971 and January 1977, 500 at Whittington Sewage Farm and 400 at both Blackbrook Sewage Farm and Rickerscote. In addition, Alvecote, Baginton, Belvide, Bittell, Blithfield, Branston, Curdworth, Eckington, Holt, Kingsbury, Latherford, Leam Meadows, Middleton, Radford, Upton Warren, Whitmore Heath and Wilden have all held peaks of 150–300 birds.

A number of foreign recoveries have resulted from ringing Snipe in the Region, especially at Tillington in the autumn. A bird ringed at this locality in November 1976 was killed by a cat in Finland the following May. Other ringed birds have been reported between October and February in Eire, France, Italy, Morocco, Portugal, USSR and West Germany. In addition, Snipe ringed in Denmark, Germany and Sweden have been recovered in the West Midlands. Within Britain, birds have been recovered in Dyfed and Dorset during the year of ringing, but the only nestling to be recovered was ringed at Cheadle in 1913 and found at Stoke-on-Trent in November 1916.

Great Snipe
Gallinago media

Very rare vagrant. Two records since 1916.

Norris recorded five or six birds between 1875 and 1894 and Smith suggested that at least a dozen birds had occurred in Staffordshire, with the latest being shot in 1916. Most of these early records, and certainly the credible ones, concerned birds that were shot. Localities included Burton, Chillington Estate, Clifton Campville, the Leek moors, Loynton, near Lutterworth (Leics.), Sandon, near Stafford, near Stratford, Sutton Park, near Tamworth and Wolf Edge. The six dated records occurred in January, August, September, October and November (two).

Due to the difficulty of correctly identifying Great Snipe, several records claimed in recent years have proved inconclusive and only two have been published as acceptable, namely one in Sutton Park on May 6th 1950 and one at Whittington Sewage Farm on April 25th 1954.

50. The Woodcock is well camouflaged on the ground and can best be seen during its roding flights in late evening.   *S. C. Brown*

Woodcock
Scolopax rusticola

Widespread and fairly numerous both as a breeding species and winter visitor.

Woodcock frequent deciduous, coniferous or mixed woodland and young forestry plantations. They prefer open canopies, rides or clearings and a combination of wet feeding sites and dry nesting sites with suitable ground cover for camouflage. Cannock Chase and Wyre Forest are fine examples of Woodcock habitat, with the latter perhaps holding one pair per 80 ha (Harthan), but all areas of extensive woodland appear to be occupied (Fincher 1955) and even copses as small as 2 ha may support roding birds. The estimation of breeding populations is now known to

be extremely difficult and counts of roding birds, although unreliable population indicators, are usually the only records available. Roding is usually noted from March to July, exceptionally a month earlier or later.

J. D. Wood estimated about 30 pairs in Warwickshire following an enquiry in 1934-5 (Norris). This included at least 12 pairs near Atherstone, presumably at Merevale, where the keeper estimated at least 50 pairs in 1966. However, during 1946-50 regular breeding was recorded at only six or seven sites in Staffordshire and Warwickshire and two in Worcestershire. The position had apparently improved by the time of the *Atlas* surveys (1966-72), although it must be remembered that crepuscular species such as the Woodcock are much overlooked and their status generally underestimated. Woodcock were confirmed or suspected of breeding in fifty-seven 10-km squares (74 per cent) during the *Atlas* years, with the main gaps being in the West Midlands Conurbation, the Severn Valley and the Lias clay lands south of the Avon. From this a very tentative estimate of 500–1,500 pairs may be made for the regional population, which accords well with Lord and Munns' figure of 200–2,000 pairs.

The apparent increase in breeding birds has probably been due in part to afforestation, although greater observer coverage has no doubt resulted in a clearer picture of the bird's status.

Woodcock are much more widespread in winter, with continental immigrants arriving mainly in October and November. Like Snipe and Jack Snipe, their arrival often coincides with a full moon, although fog may result in disorientation and birds sometimes turn up in unusual localities, such as suburban gardens. One such was ringed at Handsworth Wood in October 1969 and recovered at Atherstone three months later, while a Swedish ringed bird was recovered in Birmingham in December 1927. A nestling ringed at Swynnerton in 1928 was recovered in Shifnal (Shropshire) during the winter of the following year.

Estimates of the wintering numbers are difficult to obtain, but shooting bags and keepers' reports give some indications. Norris quoted an average seasonal bag of 50–59 shot on the 6,677 ha of five large Warwickshire estates; up to 60 have been flushed during shoots at Bentley, where 67 were shot in the 1961–2 season; and up to 200 have been seen in a day at Castle Hill (Worcs.) (Harthan).

Black-tailed Godwit
Limosa limosa

Scarce, but regular, passage migrant.

A bird shot near Tamworth around 1860 was the only record prior to 1929, but there were 22 records involving 30 birds in the eleven years between 1929 and 1948. Records then increased dramatically during the next three decades, coincident with a national increase in wintering Icelandic Black-tailed Godwits *L. l. islandica* (Prater 1975). However, it is likely that a few recent records have involved the nominate sub-species, which has recolonised a few suitable sites in England.

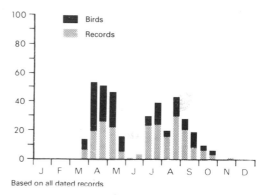

Half-monthly Distribution of Black-tailed Godwits, 1929–78

There were 219 records involving at least 385 birds during 1929–78 – all occurring between mid-March and late November. Of these, 37 per cent (47·5 per cent of birds) occurred during March–early June, with most in April and early May. The remaining records conformed to a bimodal pattern, with peaks in late July and late August that represent influxes of adults and immatures respectively. The former influx was much less apparent in Mason's (1969) analysis of Leicestershire records. Parties up to five have been recorded in autumn (average 1·46 birds), but some larger parties have occurred in spring (average 2·26 birds), with 15 at both Blithfield on April 13th 1967 and Belvide on April 15th 1972, and 10 at Draycote on May 21st 1977. On average between 1969–78 there were about a dozen records per year, with a range of 6 to 21. Alvecote, Blithfield, Brandon and especially Belvide have been the most favoured localities.

Bar-tailed Godwit
Limosa lapponica

Scarce passage migrant.

Smith reported that Bar-tailed Godwits had occurred several times in the Burton area, but these were the only records for the Region prior to 1929. Following single records in 1930, 1935 and 1936, however, birds subsequently appeared in most years from 1948. On average, there were about 10 records per year during 1969–78, ranging from a minimum of 4 to a maximum of 19.

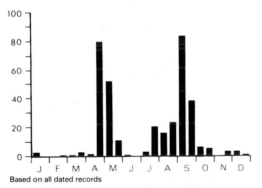

Based on all dated records

Half-monthly Distribution of Bar-tailed Godwits, 1929–78

Between 1929–78 there were 150 records involving 355 birds, most of which appeared in adverse weather conditions. There is a good spring passage and 29 per cent of records (37 per cent of birds) occurred in late April and early May, when large numbers move north-eastwards through the English Channel. (*Brit. Birds* 65:313; 66:319; 67:400). A few small parties have occurred in the Region at this time and an exceptional flock of 54 appeared at Draycote on April 27th 1976. Most autumn records (47 per cent of all records, 51 per cent of birds) have fallen between late July and September, with the majority during the latter month. Immatures arriving in Britain in early and mid-September are sometimes disorientated by bad weather conditions and three parties, the largest of which contained 16 birds, have been noted moving north-west at this time. The largest autumn party was 22 at Blithfield on September 5th 1975.

Whimbrel
Numenius phaeopus

Scarce, but regular, passage migrant.

Earlier opinions on the Whimbrel's status varied. Smith regarded it as a regular migrant in spring and autumn, but Harthan called it occasional and Norris rare. Nineteenth-century records were certainly scarce, with very few shot and thus correctly identified. This is not surprising, as very few migrant Whimbrel alight in the Region. Indeed, of the 413 records (840 birds) reported during 1929–78 the vast majority referred to birds calling in flight, usually over wetland areas. The distinctive call has been heard at night over various parts of the Region and undoubtedly many more overfly the Midlands than is generally realised.

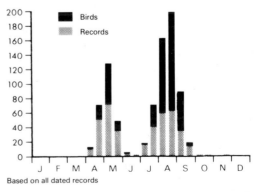

Based on all dated records

Half-monthly Distribution of Whimbrel, 1929–78

All occurrences have fallen between April and November, with extreme dates of April 2nd (1939) and November 17th (1963). However, records in early April, June, October and November are rare. From 1969-78 there were on average over 21 records per year, with extremes of 16 and 28. About one third of all records (30 per cent of birds) came in spring, when most birds are heading between north-west and north-east *en route* for Iceland and north-east Europe. No spring party has ever reached double figures, although 11 parties have done so in autumn, the largest being 46 at Belvide on August 23rd 1953 and 37 there on August 10th 1970. Nearly half of all records (62·5 per cent of birds) have fallen between late July and early September, when most birds are heading between south and west *en route* to West Africa.

Curlew
Numenius arquata

Fairly numerous breeding species and passage migrant; not scarce, but very local, in winter.

Prior to about 1920 the Curlew was virtually restricted as a breeding bird to the moorlands of north Staffordshire. Between 1917 and 1960 however, the species underwent a sensational expansion in breeding range, first colonising lowland meadows, especially along river valleys, and later arable land. At least 20 pairs colonised the lower Severn, Teme and Avon valleys and the Tenbury-Bockleton area during 1917-40. By 1947 birds had spread further into central Worcestershire and up the Avon as far as Stratford. During the 1950s many agricultural areas of Worcestershire and Staffordshire were colonised (Harthan 1961, Lord and Blake 1962), along with the Tame Valley in north Warwickshire. A further spread appears to

have occurred in the early 1960s, although agricultural intensification and urban growth caused some local declines. These declines have accelerated during the past decade, though, because of the earlier cutting and grazing of traditional hay meadows and the drainage and improvement of damp pastures. For example, the riverside meadows of the Avon and Severn in south Worcestershire now hold very few Curlews, or Redshanks, whereas ten years ago there may have been a hundred pairs (G. H. Green *pers comm*).

During the *Atlas* surveys (1966-72), breeding was confirmed or suspected in fifty-seven 10-km squares (74 per cent) covering most areas except the Birmingham Plateau and the heavy clay lands to the south of the Avon. On the moors in 1978 there were four pairs on 461 ha of the Swythamley Estate, plus a further four pairs on adjacent land (Yalden 1979), but this density of about 0·9 pairs per km² is unlikely to be matched in the agricultural areas of the Region. Currently there may be in the order of 250 pairs in Staffordshire, about half of which are on the moors, and perhaps 50 pairs in Warwickshire and Worcestershire. Thus the regional population is perhaps 250-400 pairs, which is at the lower end of Lord and Munns' estimated range of 200-2,000 pairs.

Lowland breeding birds often return in February, though moorland birds may arrive up to a month later. Flocks are often noted *en route* for the moorlands. For example, up to 75 have been seen at Tittesworth and 100 flew north-east over Eccleshall on March 25th 1956. Breeding grounds are deserted again during late June-early August, when concentrations of up to 72 have been noted at Middle Hills. An indication of the movement of local birds is provided by two nestlings ringed at Chartley in 1951. One was subsequently recovered on the River Parrett (Somerset) in December 1952, while the other was recovered in

Vendée (France) in December 1954. Such dispersal, often as far as Iberia, is frequent in British-bred birds (British Ornithologists' Union 1971).

During autumn and winter Curlews regularly roost at one or two sites in the Region. The early spring roosts of up to 70 on the flashes at Upton Warren and the early autumn roosts of up to 54 at Ladywalk may comprise either British or continental breeding birds, but the two largest concentrations of Curlew to be recorded in the Region, at Blithfield and Whittington, were probably entirely of continental origin. Since its flooding, Blithfield has held both an autumn and winter roost. The autumn roost totalled 130 on August 20th 1957 and the winter roost 110 on February 16th 1958, while on the evening of April 2nd 1956 a flock of 210 gathered. Something of the origin of these birds is perhaps revealed by an adult which was ringed there in August 1963 and recovered in Sweden in June 1968. Since about 1965, Curlews have been more erratic at Blithfield and numbers have not exceeded 50. Whittington Sewage Farm has also long been known as a regular autumn and winter haunt of Curlew. Flocks in August and September usually total 100–165, although 200 occurred in 1975. Some 60–120 birds usually winter, remaining until about March, though there was an exceptional roost of 340 in December 1975. Away from these localities, very small numbers are noted occasionally at scattered places in the lowlands during winter, but these rarely number as many as 50.

Upland Sandpiper
Bartramia longicauda

One old record.

The first record for the British Isles of this North American vagrant was of a bird shot by Lord Willoughby de Broke at Compton

Verney on October 31st 1851 (Witherby *et al* 1938).

Spotted Redshank
Tringa erythropus

Passage migrant, usually scarce, but sometimes not scarce. Rare in winter.

Spotted Redshanks have increased dramatically as migrants to the Region in recent decades. Harthan reported one shot on the Avon on August 15th 1848 and Norris two shot at Stratford in November 1875 and one observed on September 15th 1902 at an unknown locality. Singles then occurred at Bittell in September 1923 and May 1938, but it was not until the 1940s, when nine single birds occurred in autumn at Belvide, that any records were received for Staffordshire. During the last three decades the totals as shown in Table 50 have been reported.

Table 50: Ten-yearly Distribution of Spotted Redshanks, 1949–78

	1949–58	1959–68	1969–78
No. of records	51	101	258
No. of birds	71	171	453

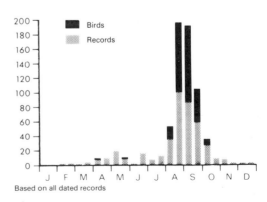

Based on all dated records

Half-monthly Distribution of Spotted Redshanks, 1929–78

The Spotted Redshank is mainly an autumn visitor to the Region and during 1929–78 only 50 of the 419 records (56 of the 704 birds) occurred in spring, mostly between April 24th and May 26th. Since 1959 there have been 25 records between June 13th and July 12th, probably of females arriving to moult in Britain, as males then are left behind on the breeding grounds to tend their young. However, by far the largest numbers have occurred between mid-August and early October, with 44 per cent of records and 55 per cent of birds concentrated into a peak period in late August–early September, when many juveniles are recognisable. A bird ringed at Blithfield on August 15th 1955 was caught twenty days later at Perpignan, Pyrenees Orientale (France). Small parties are sometimes noted and there have been up to eight at Kingsbury and Middleton, nine at

Packington, 10 at Draycote and a record 12 at Alvecote from August 28th to September 9th 1972.

Although there were four December records and two February ones between 1959 and 1978, the first prolonged winter stay occurred in the Kingsbury area in 1980/1.

Redshank
Tringa totanus

Declining, but not scarce, as a breeding species. Widespread and fairly numerous passage migrant, but scarce and local in winter.

Redshanks are usually found in rushy fields, wet meadows and the similarly vegetated margins of reservoirs and gravel

51. Following progressive drainage, the Redshank has declined markedly as a breeding species, though it still visits most wetlands on passage. *R. J. C. Blewitt*

pits. They occupy their breeding territories from late February to early July.

After a huge national decline in the breeding population following drainage operations in the early nineteenth century, it began a comeback in this Region at the end of that century. Dickenson (1798) and Pitt (1817) quoted Aqualate Mere as the only Staffordshire resort, while Lees (1828) recorded the Redshank as a local resident in Worcestershire. For much of the nineteenth century the species was encountered only occasionally in the Region, but by 1893 birds were breeding at Burton and during 1895–8 at Rugby. The Dove, from Sudbury upstream, was colonised from 1896 and a further 12 localities in south and central Staffordshire had been colonised by 1910, along with Sutton Park in 1905 and Hampton-in-Arden in 1907. This spread continued in Staffordshire and Warwickshire, but it was 1926 before breeding recommenced in Worcestershire, in the Stour Valley. The *Atlas* surveys (1966–72) showed breeding confirmed in twenty-three 10-km squares (30 per cent) and suspected in a further seven (9 per cent). These encompassed the northern moorlands, where up to three pairs usually breed, and the valleys of the Trent, Dove, Sow, Tame, Anker and lower Avon and Severn. Favoured breeding localities include Alvecote, with up to six pairs, and Belvide, with up to ten pairs, while about 12 pairs bred in the Great Wyrley area in 1952.

The last decade has seen further drainage and agricultural improvement of many traditionally wet meadows and the current regional breeding population is probably no higher than 100–150 pairs, which is in the middle of Lord and Munns' range of 20–200 pairs. The flood meadows of the Avon and Severn valleys in south Worcestershire probably now hold less than a quarter of their population ten years ago, which may then have been as high as 200 pairs (G. H. Green *pers comm*). Although a survey of Staffordshire in 1978 revealed breeding Redshanks in twenty-seven 1-km squares, all but one of these were south and west of the Churnet Valley and most held only one or two pairs. The county population was probably around 40 pairs, while Warwickshire probably holds around 20 pairs.

Three birds ringed in the Region as chicks have been recovered subsequently. One ringed at Bickenhill in 1912 was recovered in Norfolk in September of that year; one ringed at Hampton-in-Arden in 1914 was found in Cornwall in January 1915; and one ringed at Charlecote in 1975 was found dead in Essex on November 21st of that year.

As a passage migrant the Redshank is widespread and visits most wetlands. Early spring influxes frequently total double figures, and the largest number recorded was 30 at Belvide on March 30th 1968. In autumn, plumage and size variations indicate that some migrants may be of the Icelandic race, *T.t.robusta* and parties of 17 at Chasewater on August 23rd 1961 and 25 at Blithfield on August 5th 1965 may well have been of this sub-species. Wintering birds, again perhaps involving *T.t.robusta*, were very scarce until the 1970s, when perhaps up to 20 per year have at least attempted to winter. Low water levels at reservoirs have assisted wintering, but severe frosts mean that the most frequented localities are those which remain ice-free longest. The Tame Valley area has proved to be popular and up to 10 have occurred at Whitacre, 13 at Shustoke and 15 at Kingsbury in winter.

Marsh Sandpiper
Tringa stagnatilis

One record.

One at Belvide on June 22nd 1974 was the nineteenth British record and the first in

June of this east European wader (*Brit. Birds* 68:317).

Greenshank
Tringa nebularia

Regular and not scarce as a passage migrant. Very rare in winter.

The Greenshank was very rarely recorded in the nineteenth century and Smith, whilst describing it as not rare on migration in south Staffordshire, admitted that it used to be reckoned scarce everywhere. Harthan knew of only one or two occurrences in most years and Norris classed it as very scarce, though possibly regular. With more reservoirs and flooded gravel pits to provide suitable feeding areas, observations have increased considerably this century.

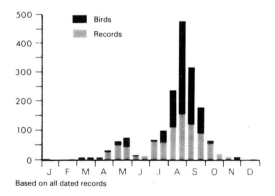

Based on all dated records

Half-monthly Distribution of Greenshank, 1929–76

During 1929–78 there were over 904 records involving 1,839 birds. Only 17 per cent of these (12 per cent of birds) occurred in spring, mostly in May, and no more than six have ever appeared together at this season. In autumn, though, about 30 parties have reached double figures. Blithfield has probably been the most consistently attractive locality, although the usual high water levels there in spring

result in relatively few wader records, including Greenshank, at that season. Belvide, Bittell, Ladywalk, Upton Warren, Wilden and especially Alvecote, Brandon, Draycote and Kingsbury have also been well-frequented. Twenty-six at Blithfield in mid-July 1976 is the regional maximum.

About three-quarters of all records (87 per cent of birds) have appeared in autumn, mostly in late August–early September, when peak numbers of juveniles are on passage. A smaller, but distinct, influx of adults takes place in late July. There are few records between late October and mid-April, with mid-winter occurrences being particularly rare.

Lesser Yellowlegs
Tringa flavipes

One record.

An adult in worn breeding plumage was seen by many observers at Blithfield between September 15th and 30th 1979. The date is typical for British appearances of this North American vagrant (Sharrock and Sharrock 1976).

Green Sandpiper
Tringa ochropus

Widespread and not scarce as an autumn passage migrant. Scarce in winter and spring.

Previous authors gave a somewhat confusing picture of the Green Sandpiper's status in the Region. Smith said it was not uncommon in winter, but Harthan called it an irregular autumn migrant and Norris an uncommon passage migrant. As with other migrant waders, records have increased considerably in recent decades with the appearance of many more suitable wetlands and a greater number of bird-watchers. Passage birds are likely to turn up at suitable sites anywhere in the Region, even at the muddy edges of farm ponds and small streams. They are extremely fond of running water and even at reservoirs and gravel pits will often choose to feed along the feeder streams.

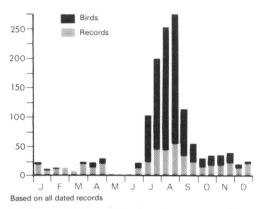

Half-monthly Distribution of Green Sandpipers, 1929–78

Most records have always come during the autumn passage period, but overwintering has increased. Records between May and mid-June remain very scarce. Green Sandpipers are one of the earliest migrants to return after the breeding season, with small numbers typically appearing in the last week of June. Like the Spotted Redshanks which appear at this time, these birds may be females which have left their mates to tend the young on their breeding grounds (Harrison 1975). Without doubt it was summer appearances such as these that gave rise to suspicions of nesting at Hamstall Ridware and Stretton around the turn of the century (Smith). Nearly three-quarters of all the birds recorded during 1929–78 occurred between July and early September, with the largest numbers in August. Parties of 10–15 have occurred at Baginton, Belvide, Blackbrook Sewage Farm, Kingsbury and Shustoke, whilst the largest have been 23 at Blithfield, 20 at Brandon, 22 at Upton Warren and a regional maximum of 26 at Grimley on

August 21st 1977.

A bird ringed at Blithfield in July 1972 was killed in September 1975 in Spain, which, together with France, is the normal area of recovery for British ringed Green Sandpipers.

Sewage farms, saline flashes, alluvial river banks and other sites which resist freezing longest provide Green Sandpipers with suitable overwintering sites. In particular, the Tame Valley has proved very attractive to wintering birds and in recent years it has become the haunt of perhaps 10–15 birds. Even so, it is unlikely that the Region holds more than 25 wintering birds in total.

Green Sandpipers nest early and the small peak in records in late March and April presumably represents an early and very meagre spring passage.

Wood Sandpiper
Tringa glareola

Scarce passage migrant. Formerly very rare.

Prior to 1929 there were only three records, all of single birds. One was shot at Barr on August 26th 1858 (Smith), another was shot at Castle Bromwich Sewage Farm on November 12th 1898 (Norris) and one was seen at Bittell in August 1926 (Harthan).

In 1946 there were two records at Baginton, while there were six records at Belvide during 1942–8. The subsequent increase in records has been dramatic as Table 51 shows.

Table 51: Ten-yearly Distribution of Wood Sandpipers, 1949–78

	1949–58	1959–68	1969–78
No. of records	71	109	184
No. of birds	120+	153	235

This increase has paralleled an apparent increase in Wood Sandpipers in the northern Palaearctic forest belt, where they are the most abundant and characteristic wader species (Ferguson-Lees 1971).

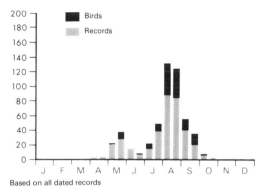

Half-monthly Distribution of Wood Sandpipers, 1929–78

During 1929–78, all birds appeared between April 9th and October 18th. Spring passage is mainly in May and early June, when 17 per cent of all records and 15 per cent of birds occurred. The most recorded in spring was seven at Brandon on May 26th 1957. Autumn passage is much stronger and at least 80 per cent of all birds appeared between July and September, mostly during August. The largest party at this season was 10 at Kingsbury on August 27th 1972, but up to nine have occurred at Blackbrook Sewage Farm, whilst Belvide and Blithfield have each held eight and Alvecote seven.

Common Sandpiper
Actitis hypoleucos

Widespread and fairly numerous as a passage migrant, but scarce and local as a breeding species and very scarce in winter.

The breeding range of Common Sandpipers in the Region appears to have con-

tracted considerably during the last fifty years. Smith described "many" birds in the Staffordshire uplands on the Dane, Hamps, Manifold, Churnet and upper Dove valleys up to Axe Edge and Morridge and also at Tittesworth and Knypersley. Two miles (3·2 km) of the upper Manifold held up to six breeding pairs, but even in those days human disturbance at Dovedale was sufficient to force birds to nest further downstream. Indeed, birds then bred in the lower Dove and Trent valleys in some numbers and were apparently common in the Burton area. Isolated nesting was also confirmed at Madeley in 1890 and at Meaford in 1919. Nest sites were occasionally far from water and included woods, railway cuttings, gardens, strawberry beds and a quarry.

In Warwickshire nesting was suspected in Sutton Park in 1917 (Norris), but Harthan reported more definite instances for Worcestershire. Two or three pairs nested regularly until 1910 at Bittell, where the species also bred in 1924. Nesting was probably regular along the Teme as well, and a brood was seen at Shelsley Kings in 1943. Finally a pair which bred on the Avon near Evesham in 1944 comprised the most recent breeding record for Worcestershire.

The species declined in Staffordshire,

such that by 1962 Lord and Blake merely referred to its breeding "sparingly in the northern moorlands area". During the *Atlas* surveys (1966–72) breeding was confirmed or probable in nine 10-km squares (12 per cent) in the northern moorlands and lower Trent and Dove valleys and was probable in a further four squares in the Trent-Sow-Penk basin and two in the Avon and Leam valleys between Rugby and Leamington. Most of the Staffordshire birds were in the north-east of the county, where about six pairs were located on the moors in 1970. Confirmed breeding in the lowlands was restricted to perhaps only three pairs on either the Trent or the Dove below Uttoxeter. Reports suggest that numbers have subsequently declined still further and the Common Sandpiper, which is essentially a bird of fast-flowing upland streams and rivers, now has a very tenuous hold as a breeding species in the Region. There are now less than five pairs left in the moorlands, mainly in the less disturbed areas. Human pressures will probably decide the future of the Common Sandpiper as a breeding species in the Region. Certainly the current population is at or below the lower end of the range 3–20 pairs suggested by Lord and Munns.

As a passage migrant, the Common Sandpiper is one of the most frequently encountered waders. However, numbers are typically small, although birds may gather to form small flocks at dusk. In spring and in autumn birds may turn up at almost any stretch of water, including not only lakes, reservoirs and gravel pits, but also rivers, canals and ornamental park pools. As well as using natural shorelines, they will also feed along artificial shorelines which other waders generally shun. For example, Common Sandpipers feed avidly along the stonework of reservoir embankments.

The Common Sandpiper is a fairly early migrant in both spring and autumn. Over forty-two years the average date for first

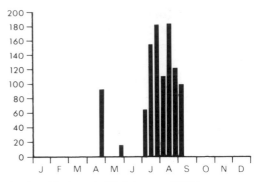

Based on all dated records of flocks in excess of 17 birds

Distribution of Common Sandpipers by Thirds of a Month, 1929–78

arrivals has been April 7th, with the earliest being on March 9th (1944), although in recent years the presence of wintering birds has made early and late migrants difficult to determine. Most spring passage takes place between mid-April and mid-May, when small parties can often be seen along the main rivers. The largest spring flocks were 27 along the River Severn at Bewdley on April 24th 1966 and 29 at Belvide on April 22nd 1978.

Autumn passage typically begins in the last days of June and there were as many as 14 at Gailey on June 28th 1964. It then reaches its climax between mid-July and early September, with peaks especially in mid-to-late July and mid-August. At this time favoured sites, with their maxima, have included Bartley (17), Belvide (35), Bittell (20), Chasewater (28), Draycote (25), Kingsbury (30), Upton Warren (18), Westwood (33) and Wilden (23), whilst Blithfield frequently holds 20–40 and hosted 50 on August 5th 1970. A migrant ringed there on August 18th 1964 was recovered at Otley (West Yorkshire) on September 11th 1965 and there have been two recent re-traps from Draycote, one of which was ringed at Moorhouse NNR (Cumbria) on July 9th 1978 and recovered on July 20th 1979, whilst the other was ringed as a pullus at Middleton-in-Teesdale

(Durham) on July 4th 1979 and recovered on August 9th of that year.

Few birds are noted after September, though the average date for the latest apparent migrants over thirty-eight years has been October 19th. Late passage birds may include some continental drift-migrants and potential winterers, however, as the latter often first appear in late October. There have been 41 records between November and February inclusive, of which 66 per cent have occurred in the last decade, when conditions at some reservoirs have been particularly attractive. However, there have been only half a dozen or so instances of successful over-wintering, many birds having disappeared at the onset of severe weather. Complete overwintering has taken place at least once at both Blithfield and Branston, twice at Chasewater and several times in the Tame Valley, notably at Ladywalk and Shustoke.

Spotted Sandpiper
Actitis macularia

Very rare vagrant. Two records.

A bird in summer plumage remained at Draycote from May 8th–10th 1977, constituting the 35th record for Britain, but only the seventh in spring (*Brit. Birds* 71:504). In common with other Nearctic waders, most British records of this species have occurred in autumn. It is interesting to note that this bird was observed display-ing to its Palaearctic counterpart, the Com-mon Sandpiper. A second bird, also at Draycote, remained from September 10th to 25th 1980.

Turnstone
Arenaria interpres

Regular, but scarce passage migrant. Formerly very rare.

Chase (1886) described the Turnstone as very rare in his list of Birmingham birds, but gave no precise details, whilst Smith noted that a specimen had been obtained in the Chillington area and that the species was said to have occurred near Burton prior to 1893. Further Turnstones were re-corded at Chasewater in August 1904, 1907 and 1908, at least, and the first Worcester-shire birds appeared at Bittell in May 1929. The subsequent fifty years provided some 305 records of 546 birds in the Region. Over half of these came during 1969–78, when there were annual totals of up to 20 records and 47 birds and means of 16 and 29 respectively.

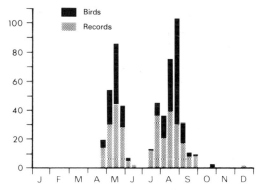

Based on all dated records

Distribution of Turnstone by Thirds of a Month, 1929–78

The Turnstone is a passage visitor in small numbers to reservoirs and gravel pits. Virtually all records have been between late April and late September, with a third fall-ing in May and about a half between mid-July and early September. The spring peak occurs in mid-May, while on return passage there is a peak of adults in late July followed by a more pronounced one of both adults and juveniles in mid-to-late August. Five parties of double figures have occurred in the second half of August, the largest of which was 20 flying over Leacroft on August 31st 1975. Birds are rare after September and there have been only eight records for the period

October–December and just one exceptional winter record of a bird at Draycote during February 8–14th 1976.

Birds from two separate populations, namely Greenland and north-east Canada on the one hand and Scandinavia and west Russia on the other, are known to occur in Britain, though the latter appears mainly on passage and winters in West Africa (Branson *et al* 1978). The origins of birds encountered in the West Midlands are unknown, but the spring migration pattern is very similar to that of Ringed Plover and Sanderling and, like those species, our Turnstones are probably heading northwest. As both populations reach Britain at about the same time in autumn, only ringing will reveal the origins and destinations of visitors to the Region.

female in summer plumage at Upton Warren on July 8th 1972; one at Blithfield on September 12th 1973; one at Chasewater on May 27th 1974; one at Bittell from September 8–10th 1974; an immature at Belvide on August 31st 1975; and one at Bittell on August 20th 1978.

Red-necked Phalaropes appear on both spring and autumn passage. Of 17 dated records, there were four during late May–early June, singles in late June, early July and late July, nine between mid-August and late September, and one in early November. The majority stayed just one day, but there were two records of birds staying three days.

Single unspecified Red-necked or Grey Phalaropes occurred at Lifford Lake on October 3rd 1968 and at Belvide on October 13th–14th 1977.

Red-necked Phalarope
Phalaropus lobatus

Rare passage migrant. Eighteen records.

There are four old records of Red-necked Phalaropes, namely one at Tamworth prior to 1886, one shot at Handsworth on August 24th 1887, two shot at Moseley on September 9th 1893 and an adult male shot at Morton Bagot on September 11th 1910 (Smith and Norris).

A further 14 birds then occurred in eleven of the years between 1952 and 1978 as follows: one in juvenile plumage at Belvide on September 28th 1952; another at Belvide on November 1st 1952; one in almost full summer plumage at Belvide on July 22nd 1956; one in summer plumage at Brandon on June 10th 1959; one at Blithfield on September 1st 1963; one in full summer plumage at Blithfield on June 28th 1964; one in full breeding plumage at Belvide from May 30th to June 1st 1965; one at Chesterton on June 3rd 1972; a

Grey Phalarope
Phalaropus fulicarius

Rare, storm-blown autumn migrant. One summer record.

There were at least 14 records of single birds and several other vague references prior to about 1910 (Smith, Harthan and Norris). A further 28 records of single birds followed in twenty-two of the years 1933–80, with six at Blithfield, five at Belvide and Bittell, two at Draycote and one each at Alvecote, Blackbrook Sewage Farm, Blakedown, Chasewater, Hagley, Kingsbury, Leamington Spa, Lickey and Napton. The Grey Phalarope is almost exclusively an autumn visitor. Of 34 dated records, all but one occurred between late August and December 7th with the majority between mid-September and early November. The exception was an adult female in full breeding plumage at Blithfield on June 15th 1976.

Pomarine Skua
Stercorarius pomarinus

Rare storm-driven visitor. Eight records.

One of the less frequent skuas, even on the coast where it occurs on passage to and from its circumpolar breeding grounds, this species is noted inland only after severe coastal gales.

The earliest record for the Region concerns a bird shot at Shipston-on-Stour in September 1869. A dark-phase bird was obtained at Oldbury in 1879 and an adult shot at Chasewater sometime prior to 1910 was mounted for use as a lady's hat! A further bird was shot at Shugborough on October 30th 1912, after it had fed on a duck. Of three skuas at Bittell on October 21st 1936, one was identified by its twisted tail feathers as a Pomarine.

After strong north-westerly gales an exhausted, pale-phase adult was found at Belvide on October 21st 1970; unfortunately it died shortly afterwards. A bird flew over the border from Drakelow (Derbys.) to Branston on September 20th 1977. The latest record concerns an immature at Draycote on December 31st 1978 and January 2nd and 3rd 1979. This is the only dated record to fall outside the more usual passage months of September and October.

Arctic Skua
Stercorarius parasiticus

Very scarce passage migrant.

The commonest of the skuas on passage around our coasts, the Arctic Skua has become a tolerably frequent visitor to the Region in recent years, usually, but not invariably, coinciding with gales.

There are four or five nineteenth-century records of this species: one near Worcester in 1846; two immatures shot in

the Burton area prior to 1863; one picked up at Ladywood in October 1879; and one shot near Hamstall Ridware about 1898. An injured bird found at Aldridge on September 2nd or 3rd 1909 was apparently one of two specimens shot earlier at Chasewater.

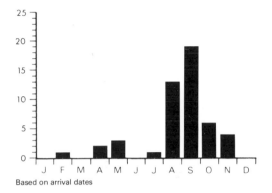

Based on arrival dates

Monthly Distribution of Arctic Skuas, 1929–78

No further birds occurred in the Region until one in 1957, but then two records followed in 1960 and during 1966–78 there were 30 occurrences involving 41 birds, almost invariably at the larger, well-watched reservoirs. Of all dated records, 74 per cent (78 per cent of birds) occurred between August 16th and October 25th, with over half of these in late August and early September. For those records where details have been published, 15 referred to dark-phase birds, but only six to pale-phase. However, there have been 13 adults and 14 immatures, excluding seven adults which flew south-west over Belvide on September 14th 1980. A similar number, including six adults, flew east at Chasewater on August 21st 1973. These records apart, all reports have concerned ones and twos. No more than three records have fallen in a year except in 1973, when there were nine records totalling 17 birds.

There are four records of single skuas not positively identified, namely an Arctic or Pomarine at Alvecote on November

16th 1961 and probable Arctics at Bartley on March 10th 1963, and at Blithfield on September 28th and December 6th 1970.

Long-tailed Skua
Stercorarius longicaudus

One record.

An immature was shot on Lichfield race-course on October 7th 1874.

Great Skua
Stercorarius skua

Rare storm-driven visitor. Sixteen records to the end of 1980.

There are two early instances of Great Skuas being shot, at Chasewater in September or October 1896 and at Fillong-ley in September 1909.

Nearly fifty years elapsed before the next occurrence, at Belvide on November 9th 1957. One then appeared at Blithfield on November 5th 1967, whilst birds were seen at no less than four sites (Bartley, Bittell, Blithfield and Upton Warren) on September 26th 1971. There were subsequent records at Blithfield on August 27th and September 21st 1972; Chasewater on October 12th 1973; Belvide on September 7th 1974; Draycote on September 4th 1976; Bartley on September 11th 1976; Blithfield on September 22nd 1977; and Draycote on September 13th 1980.

Thus, up to 1980 all 16 records had fallen in autumn, between August 27th and November 9th. [However, following unusually severe weather for late April, there were three exceptional spring records in 1981; at Draycote on April 25th–26th, Blithfield on May 3rd and Belvide on May 10th. *Ed.*]

Mediterranean Gull
Larus melanocephalus

Rare visitor. Ten records.

With increased attention being paid to gull roosts and the clarification of identification points, the detection of occasional Mediterranean Gulls in the Region has not been altogether unexpected. However, birds could still be missed in the hordes of Black-headed Gulls roosting in the Region.

There were nine records of single birds during 1971–6: at Draycote on July 6th 1971 (adult); Blithfield on September 20th 1972 (second winter); Chasewater on May 5th 1973 (first summer), December 4th 1974 (second winter) and February 25th–28th 1975 (first winter); Blithfield on February 9th 1975 (second winter) and November 22nd 1975 (adult); Belvide on December 27th 1975 (adult); and Chasewater on October 23rd 1976 (adult).

There has been one subsequent record of an adult at Blithfield on February 17th and 24th 1980.

Little Gull
Larus minutus

Previously rare, now scarce passage migrant.

There are three or four old records of Little Gulls being shot in the Region; at Upton-on-Severn prior to 1828, along the Worcestershire Avon prior to 1901, at Bidford prior to 1904 (possibly the same bird) and near Chasetown on December 2nd 1911.

Twenty-five years then elapsed before one appeared at Curdworth on May 9th–10th 1936, to be followed two years later by one at Bittell on May 3rd. There were no further occurrences until 1952, but between then and 1968 came 37 records involving 50 birds. Of these, 66 per cent (75 per cent of birds) fell in August and early September and only three were spring records. By contrast the next decade saw 180 records involving 297 birds, of which only 31 per cent (26 per cent of birds) came in August and early September, but 34 per cent (47 per cent of birds) fell in April and May.

This recent surge in records has been paralleled elsewhere in Britain, and is probably due to an increase in the breeding population east of the Baltic (Hutchinson and Neath 1978). The spring records are indicative of potential colonists and indeed there were three records of attempted breeding in England during 1975-8.

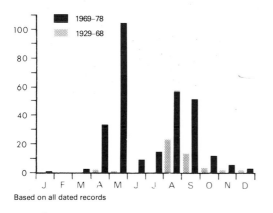

Monthly Distribution of Little Gulls, 1929-78

Of the 28 birds during 1936-68 for which age details are available, only six (21 per cent) were adults. In the following decade, however, 54 (42 per cent) of the 130 that were aged were adults and 27 (50 per cent) of these occurred in spring when only 23 (30 per cent) of the immatures were recorded. The histogram below shows a tendency for adults to appear slightly earlier in spring (and perhaps autumn) than immatures – a feature which has also been noted in the Netherlands.

There are few records of second year birds.

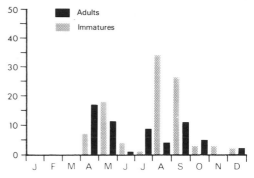

Based on all dated records

Monthly Distribution of Little Gulls by Age, 1929-78

A total of 34 Little Gulls occurred in the Region on May 4th-5th 1974, including 15 at Blithfield on the 4th, and about 20 appeared at Draycote on May 17th 1975. The largest autumn party was 14 immatures at Blithfield on August 18th 1979.

Sabine's Gull
Larus sabini

Very rare vagrant. Three records.

This Arctic breeding gull was once considered to be quite a rarity to British waters, but in recent years the great increase in the number of competent birdwatchers has shown it to be a much more frequent visitor than formerly believed. Despite the greater number of coastal sightings, however, it remains very much a vagrant to inland Britain and there are only three records of this species to the Region, namely an immature bird picked up at Coleshill on October 2nd 1883, an adult at Bittell on August 15th 1948 and an adult at Blithfield on September 11th 1976.

Black-headed Gull
Larus ridibundus

Usually scarce breeding species. Abundant winter visitor and passage migrant.

Although the number of Black-headed Gulls breeding in England and Wales almost tripled between 1938 and 1973 (Gribble 1976), it is the phenomenal increase in wintering numbers of this species which has really brought "seagulls" to the notice of even the disinterested urban dweller. Ringing has shown that large numbers of Black-headed Gulls from expanding colonies in the Baltic area now winter in the Region, attracted by plentiful food supplies at open refuse tips and on farmland, and safe roost sites at large lakes and reservoirs.

For breeding Black-headed Gulls, however, this Region has seen mixed fortunes. For over 132 years several hundred pairs nested at a succession of sites in the Norbury area of west Staffordshire and up to 2,500 juveniles were rounded up annually, raised on a diet of offal and finally sold for the table. The traditional site near Norbury continued to be used for a few years from 1662, birds then moving to Offley Moss, near Woodseaves, for about three years, thence to Aqualate for two years and finally to Shebdon, where the colony persisted until about 1800. By 1816 the sites at Shebdon and Norbury had been drained and enclosed and no further nesting in the area is known of until 50–80 pairs were discovered breeding at Doley Common, Gnosall, during

52. As a breeding species, the Black-headed Gull has declined in the Region, but its numbers have increased phenomenally in winter, when it frequents reservoirs, rubbish tips and farmland.
R. J. C. Blewitt

1969–72. In 1973 150–180 pairs bred there, but drainage in 1974 caused the site to be deserted. It is perhaps no coincidence that in 1974 150 birds took up residence at Norbury, where three pairs had bred in 1972, but sadly human persecution soon led to desertion of this site as well and it too was finally drained in 1976.

Persecution was also the reason for desertion at a colony of up to 12 pairs at Curdworth Sewage Farm, which had been in existence for "some years" prior to 1913 and was finally dispersed in 1942. Breeding has also taken place at Alvecote (one pair in 1947), Coleshill (an unsuccessful attempt in 1955), Branston (two–five pairs during 1955–61 and 12 nests in 1966), Belvide (unsuccessfully in 1962), Stretton Gravel Pits (about eight pairs in 1966), Brandon (unsuccessfully in 1970), Wilden (one pair in 1970, 1975 and possibly 1974, and four unsuccessful pairs in 1976) and at Elford (at least eight nests in 1979 rising to over 40 nests in 1981). Pulli ringed at the last site in July 1980 were recovered during the following winter at Wrexham (Clwyd) and Bournemouth (Dorset).

The breeding record for SP 06 in the BTO *Atlas* was apparently in addition to the 1970 Wilden (SO 87) record, which was not shown.

As winter visitors, Black-headed Gulls were scarce in the early twentieth century over much of the Region, especially in upland areas, where parties reaching double figures were unusual. In 1926 even a concentration of 150 at Belvide in December was considered notable. However, in the 1940s larger numbers suddenly began roosting and further surges in numbers were apparent in the late 1950s –

Table 52: Maximum Roosting Numbers of Black-headed Gulls at Principal Localities

Locality	Pre-1955	1955–68	1969–80
Aqualate	2,500	4,700	5,000
Alvecote	—	—	8,000
Bartley	3,000	12,000	10,000
Belvide	5,000	7,000	23,000
Bittell	2,000	10,000	8,000
Blithfield	—	6,000	19,000
Brandon	—	—	3,000
Chasewater	1,100	4,000	10,000
Chillington	—	—	5,000
Copmere	—	—	2,000
Draycote	—	—	80–100,000
Edgbaston Reservoir	1,500	—	—
Kingsbury	—	—	17,000
Ladywalk	—	4,000	8,000
Shustoke	1,000	3,500	25,000
Westwood	—	—	5,000
Estimate of Regional wintering population	10–15,000	30–50,000	150–250,000

early 1960s and in the early 1970s. Table 52 indicates the maximum roosting numbers at main localities.

Thus, the wintering numbers have increased more than ten-fold during the last forty years. However, there has been an indication of a recent decline in Black-headed, Herring and Lesser Black-backed Gulls at some roosts, apparently correlated with the practice of covering refuse promptly after tipping, in a direct attempt to discourage gulls. The closure of some tips and the advent of a few incinerators has also caused local reductions in food supply.

Since the 1976/7 winter an extensive study has been made of gulls at rubbish tips in Warwickshire and Worcestershire (Green 1978). Overall, this has shown 54 per cent of all the birds caught, and 80 per cent of all the gulls, to be Black-headed, although counts indicate that the percentage of Black-headed to other gulls is significantly less in the late autumn. The number feeding at tips, however, is dependent on the weather and it seems this species prefers to take invertebrate food from arable or pastureland and mainly resorts to refuse tips when this is unavailable because of frost or snow. Green

(1977) found 2 per cent of the trapped birds contained salmonellas. Some individuals wander over a wide area and have been caught at different tips in the same winter, but others have been re-trapped at the same site in subsequent years, indicating that some at least will winter in the same district each year if the food supply is adequate.

Ringing has shown that our wintering birds have mixed origins, with some breeding in south-east England and others in northern Europe. Recoveries mostly relate to birds ringed during the 1970s and foreign recoveries are summarised in Table 53.

Many of the birds from abroad were ringed as nestlings and this proves their origin. Interpretation of recoveries within Britain needs to be treated with care due to there being relatively few localities where gulls are caught. However, West Midlands wintering Black-headed Gulls have now been recovered in Britain in all months of the year. British-ringed nestlings caught in the Region during winter had been ringed at colonies in Shropshire (1 bird), Lincolnshire (1), Yorkshire (1) and Kent (2). One particularly interesting recovery is that of a

Table 53: Summary of Black-headed Gull Movements to and from the Region by Country and Month of Ringing or Recovery

	Jan	Feb	Mar	Apr	May	Jun	Jul	Aug	Sep	Oct	Nov	Dec	Total
Finland	—	—	—	2	4	6	2	—	—	—	—	—	14
Norway	—	—	—	1	1	3	—	—	—	—	—	—	5
Sweden	—	—	—	—	—	3	3	—	—	—	—	—	6
Baltic States	—	—	—	1	1	7	—	—	—	—	—	—	9
Poland	—	—	—	—	3	2	1	—	—	—	—	—	6
Denmark	—	—	—	3	1	16	—	1	—	—	—	—	21
Germany	—	—	1	2	1	3	3	—	—	—	1	—	11
Holland	—	—	1	1	5	6	2	1	—	2	—	1	19
Belgium	1	—	1	—	1	—	1	—	—	—	—	—	4
France	—	—	—	—	—	—	—	—	—	—	—	1	1

nestling ringed at Gnosall in June 1970 and caught in Neubrandenburg (East Germany) in March 1978.

The first substantial autumn arrivals occur in August (e.g. 25,000 at Draycote on August 16th 1977) and continue until numbers peak in December and January. Currently, Draycote holds by far the largest roosting total, with estimates of up to 100,000, although the potential problems involved in assessing such concentrations are considerable. There is a rapid dispersal with fine weather in March, leaving much reduced numbers of mainly immature birds during the summer months.

Common Gull
Larus canus

Not scarce as a passage migrant and winter visitor, locally numerous.

The Common Gull has never lived up to its name in the Region, despite being quite numerous in winter on the well-drained pasture lands of the Cheshire Plain, the East Midlands and Gloucestershire.

It appears that Common Gulls were particularly scarce in the nineteenth and early twentieth centuries. In 1947 an exceptional 150 roosted at Bartley in early February and 140 paused there briefly on April 9th. Since then roosting numbers have been small, with 150 at Blithfield on November 3rd 1963 being the most away from Draycote, where some very large estimates have been made in the last decade, including about 2,000 on January 7th, 1978 and about 10,000 on March 24th 1979. The problems of finding and counting Common Gulls spread liberally through a carpet of Black-headed Gulls are enormous. It is likely that the majority of the Common Gulls roosting at Draycote hail from feeding grounds to the east, where they are known to be quite numerous. On

rubbish tips small parties up to ten are occasionally seen and during the cold weather of February 1978 nine were captured, but nothing is known of their origins or movements (Green 1977, 1978).

Summer records have always been scarce, but birds occur more widely on passage in spring and early autumn. From late March to early May, in particular, small parties quickly pass through the Region, generally in an easterly or north-easterly direction, presumably bound for their Scandinavian and Baltic breeding grounds. This movement, probably from the Severn to the Wash, is most obvious in Warwickshire, where up to 200 have appeared at Brandon and flocks are frequently seen following the plough.

Lesser Black-backed Gull
Larus fuscus

Numerous passage migrant and winter visitor, fairly numerous in summer.

The British race of the Lesser Black-backed Gull, L. f. graellsii, was known primarily as a scarce passage migrant through the Region until numbers increased and roosting and wintering began in the 1950s. Totals rarely reached double figures until 1955 and birds were particularly scarce outside the passage periods of April–May and August–November. However, small numbers of mainly immature birds summered along the major rivers and appearances after coastal gales were also noted.

In 1955 there were roosts of 61 at Belvide and 40 at Chasewater and within fifteen years winter roosts had increased dramatically to a Regional total of around 10,000 birds by the late 1960s. An even more dramatic increase, though, occurred by the mid 1970s, with around 25,000–35,000 present in winter roosts. Up to 10,000 roosted at Kingsbury in 1975, and on December 31st 1974 about 12,000 were estimated in the Draycote roost. In the late 1970s there was evidence of a decline in total numbers, to perhaps less than 20,000, this being most apparent at the major roosts. Lesser Black-backed Gulls appear to rely heavily on refuse tips for food and the now prompt covering of rubbish at tips is probably the major cause of this decline. Indeed, around rubbish tips it is noticeable that whilst birds will spend many hours roosting and preening in the fields, they will seldom search for food on farmland. Nevertheless, the study at rubbish tips in Warwickshire and Worcestershire (Green 1977, 1978) has shown 14 per cent of all gulls ringed to be Lesser Black-backed, so this is still the second-commonest gull. Only 0·7 per cent contained salmonellas. Arrival of adults begins in early autumn and numbers reach their peak in November. During December and January there is a decline, followed by a second peak in February, and by the end of March most birds have left. Roosts show a similar pattern. Immatures

increase in late spring and peak in early summer.

Since roosting began, peak numbers have always been at the end of the year. During 1955–68, November and December totals were similar, but in the following decade November totals were almost double those of December. Roost numbers decline over the winter until a resurgence with the spring passage, when considerable movement to the north and west is evident. Midsummer records of mainly immature birds have increased in recent years, exemplified by concentrations of 2,000 at Little Packington rubbish tip and 3,000 at Ladywalk in June 1974. Return movement begins in late June and by August quite large numbers are present.

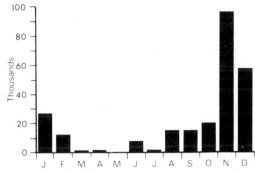

Based on all dated records

Monthly Summation of all Lesser Black-backed Gull Records, 1955–78

Ringing has shown that Lesser Black-backed Gulls wintering in the West Midlands originate mainly from Northern England and Scotland, with recoveries from Preston, Tarnbrook and Walney Island (Lancs.), the Farne Islands (Northumberland) and Mallaig (Inverness). There is also some association with the colonies at Skokholm (Dyfed) and Copeland Island (County Down), whilst one was found dead at Cobham (Surrey) and another, ringed as a nestling on Anglesey (Gwynedd) in 1967, was found dead at Hallow in September 1978.

Table 54: Peak numbers of Lesser Black-backed Gulls at Selected Localities

Locality	1955–68	1969–78
Aqualate	—	500
Bartley	400	3,000
Belvide	340	1,150
Bittell	150	100
Blithfield	3,500	6,000
Brandon	—	4,000
Branston	1,000	—
Chasewater	2,000	4,150
Chillington	—	2,000
Draycote	—	12,000
Kingsbury	—	10,000
Ladywalk	1,800	2,000
Packington	1,200	3,000
Shustoke	400	4,000
Westwood	200	1,200

Since 1942, very small numbers of adults recognisable as the Scandinavian race, *L. f. fuscus*, have been identified. There were over 60 records up to 1978, totalling more than 75 birds, not including "many" at Chasewater on November 17th 1965 and 50 at Shard End gravel pit on December 23rd 1959. All records have fallen between August 28th and May 8th, about two-thirds having occurred between October and January.

Herring Gull
Larus argentatus

Numerous winter visitor, scarce in summer. One breeding record.

Prior to the mid 1940s, Herring Gulls were known only as very scarce migrants or storm-blown visitors, although they were apparently more frequent in the Trent and lower Severn valleys. About 150 gulls, chiefly Herring, were seen on meadows near Okeover in November 1905, but this was quite exceptional.

Twenty roosted at Belvide in 1936, increasing to 250 by 1942, and 40 roosted at Bartley for the first time in 1944. During 1949-51 there were around 600 roosting in the Region, this figure rising dramatically in the late 1950s to 3,000-4,000; 5,000 by the mid-1960s; and over 15,000 by the early 1970s. A decline to about 10,000-12,000 was apparent in the late 1970s, mirroring that of Black-headed and Lesser Black-backed Gulls. 5,000 at Draycote on November 11th 1973, is the largest roost to date.

After returning to their breeding grounds in March and April, Herring Gulls become scarce in the Region, but there is a surprising record of a nest with two eggs which was found on a large rubbish tip at Bromsgrove in 1969. Apart from records at Gloucester, this was the only inland breeding record for the Midlands shown in the BTO *Atlas*. Most summer records refer to immatures, although increased numbers of adults were reported at Packington in the summer of 1977. By far the largest numbers are present between November and February, with the mid-winter peak

Table 55: Peak numbers of Herring Gulls at Selected Localities

Locality	Pre-1955	1955-68	1969-78
Bartley	200	200	2,500
Belvide	350	2,800	1,800
Bittell	12	200	1,000
Blithfield	—	2,000	4,000
Brandon	—	—	3,000
Chasewater	50	2,000	3,500
Chillington	—	—	150
Draycote	—	—	5,000
Edgbaston Reservoir	30	—	—
Ladywalk	—	2,000	—
Kingsbury	—	—	1,000
Packington	—	—	650
Shustoke	50	—	600
Westwood	—	—	350

being a month or two later than for Lesser Black-backed Gull.

Hume (1978a) described the relative abundance, colour and structure of four types of Herring Gulls wintering at Chasewater. About 90 per cent of the roost comprised small, pale-mantled birds, probably *L. a. argenteus,* of British and perhaps south Scandinavian origin; there was more or less a gradation to large, dark-mantled birds (*L. a. argentatus* from north Scandinavia), which arrived from early November and peaked at 100 or more by January; less than five per winter were variant, or hybrid, birds with dark mantles and extensive white wing-tips with small, black sub-terminal areas; and finally white headed, very dark-mantled, yellow legged birds of Siberian, Mediterranean or perhaps hybrid origin (*Brit. Birds* 74:349–53) were noted in very small numbers between late October and early March, after the first in November 1973. A few of this and the previous type have been noted at Blithfield and there are also a number of records of presumed variant,

53. Fifty years ago the large gulls, like this immature Herring Gull, were rare inland, but now they are commonplace at rubbish tips and reservoirs. *M. C. Wilkes*

albinistic, leucistic or hybrid birds from several other localities.

Although this was the third commonest gull on rubbish tips in Warwickshire and Worcestershire, it accounted for only 6 per cent of the those ringed (Green 1977, 1978). Unlike the Lesser Black-backed, though, its numbers remained fairly constant from late October through to February. There is some evidence to suggest that the incidence of salmonellas may be higher in Herring Gulls than in other species, but the number so far sampled is too small to draw a definite conclusion. It also seems that the measurements of birds trapped are larger than those of British Herring Gulls, being similar to those of Scandinavian and Russian birds, and one bird ringed as a nestling on Great Ainov Island (SSR) in 1961 was recovered in Warwickshire in February 1965. Other Herring Gulls found in Warwickshire had been ringed as nestlings at Walney Island (Lancs) and Flat Holm (Glamorgan), whilst those ringed on rubbish tips have been found in Cheshire (3 birds in August), Shropshire (4 in December), Gwynedd (2 in July) and Angus (1 in May).

Iceland Gull
Larus glaucoides

Scarce winter visitor.

This species first wandered to the Region from its northern Nearctic breeding grounds as recently as 1949, when one appeared at Belvide on January 11th (Lord and Blake 1962). Singles at Belvide in 1957 and 1959 were followed by birds at Blithfield and Sutton Park in 1962, Bartley in 1963 and Woodseaves in 1967.

Iceland Gulls then appeared annually from 1969, with about 70 birds in the subsequent decade, mostly at roost sites. As the histogram shows, the majority (80 per cent) of the records have fallen between

January and March and all records between November 4th (1967) and May 1st (1970).

About 61 per cent of birds have been adults, compared with only 35 per cent of Glaucous Gulls. It is difficult to be certain

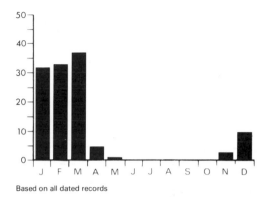

Based on all dated records

Monthly Distribution of Iceland Gulls, 1949–78

of the total numbers involved, but up to 1978 there had been at least 21 birds at Chasewater, 17 at Blithfield, 16 at Belvide, five at Bittell, four at Brandon, two each at Bartley, Kingsbury and Wednesfield, and singles at Bridgtown, Great Wyrley, Hatherton, Leacroft, Upton Warren and Westwood. There have been a dozen instances of two Iceland Gulls together at a roost and there were three at Blithfield on February 11th 1973. Like other gulls, it has declined recently, with only 16 records during 1977–80 compared with 36 in the previous four years.

Glaucous Gull
Larus hyperboreus

Scarce winter visitor.

Like the preceding species, the Glaucous Gull did not appear in the Region until roosts had become established at reservoirs. Single, different, birds appeared at

Belvide on February 20th and 27th 1949. Then followed single records in 1958 and 1959 and three in 1962. Since 1966 Glaucous Gulls have been annual visitors, with about 120 birds in the decade 1969–78.

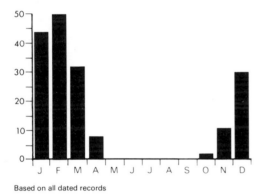

Based on all dated records

Monthly Distribution of Glaucous Gulls, 1949–78

All records have fallen between October 11th (1958 and 1970) and May 1st (1979), with about three-quarters of the occurrences between late December and early March. Thus Glaucous Gulls tend to arrive and depart slightly earlier than Iceland Gulls. Adults and immatures are of roughly equal abundance at the end of the year, but in the early months immatures are consistently about twice as numerous.

There have been ten instances of two birds together and three occurred at the Blithfield roost on February 10th 1974 and again on March 6th 1977. In fact Blithfield has attracted the most Glaucous Gulls (a total of about 36), followed by Chasewater (21), Belvide (17), Bittell (8), Brandon (7), Draycote (6), Bartley and Branston (5 each), Westwood (4), Kingsbury and Packington (3 each), two each at Great Wyrley, Hammerwich, Kenilworth, New Invention and Shustoke, and singles at Gallows Green, Ladywalk, Redditch, Romsley, Shire Oak, Sutton Coldfield, Upton Warren and Worcester. Subsequent to the main period under review, an exceptional five together were noted in the Blithfield roost on January 20th 1980.

Great Black-backed Gull
Larus marinus

Not scarce as a winter visitor, but scarce in summer.

Being a more strictly maritime bird than the other large gulls, Great Black-backed Gulls have always been relatively scarce visitors. There are very few dated records prior to 1940, after which very small numbers appeared more frequently, but parties into double figures were virtually unknown until 1970.

In the 1970s roosts increased considerably and there were records of up to 50 at

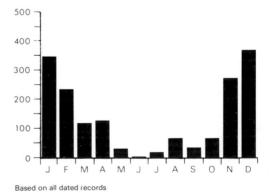

Based on all dated records

Monthly Summation of all Great Black-backed Gull Records, 1946–78

Bartley, 73 at Chasewater, and 100 at Blithfield (February 1st 1976), Brandon (November 2nd 1972), and Draycote (January 7th 1978). The wintering population during that decade may have been in the region of 400 birds. As with many other species, generally only maxima are published in annual reports, but the histogram probably reflects with some accuracy the monthly distribution of records.

Kittiwake
Rissa tridactyla

Scarce passage migrant.

Kittiwakes are the most maritime of Britain's breeding gulls, nesting on the coast, feeding offshore and performing transoceanic flights. Nevertheless, small numbers now occur frequently inland and by no means all occurrences coincide with stormy conditions.

Breeding numbers were severely depleted by human predation in the nineteenth century, but the subsequent recovery has been quite dramatic, with a rate of increase estimated at 50 per cent per decade (Coulson 1974). It is therefore puzzling that Kittiwakes were once believed to be "numerous" in Worcestershire (Hastings 1834) and "the commonest gull around Burton and Tutbury" (Mosley 1863). Likewise Whitlock (1893) and Jourdain (1905) indicated that the species was frequent in the Trent Valley, at least in March and April. Conversely, Norris knew of only about a dozen Warwickshire records prior to 1947. In view of what is now known about the Kittiwake's status in the Region, considerable doubt must be cast on the early statements of abundance, and confusion with Black-headed and Common Gulls is assumed to be the explanation.

There are only about 22 dated records prior to 1929, followed by over 200, involving about 375 birds, in the following fifty

years, over half of which occurred in the last decade. During 1929-78, 48 per cent of the records fell between February and May with other influxes apparent in August and November-December. The evidence for a spring passage of mainly adults through the English Midlands was given by Hume (1976) and precisely the above pattern was illustrated. As if to emphasise the fact, 17 of the 18 records for the Region in 1979 fell between February 11th and June 3rd and included a record flock of 42 adults at Blithfield on March 11th. Oddly, the latter flock departed south, whereas 37 of the 39 adults at Chasewater on March 5th 1975 flew off north-east. While spring birds may occur in fine, calm weather, gales often accompany records later in the year, including flocks of 25-30 at Chasewater on November 5th 1967 and 11 (ten adults) moving north through Blithfield on November 13th 1977.

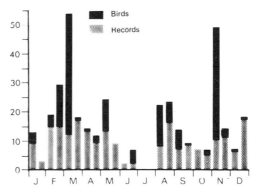

Based on all dated records

Half-monthly Distribution of Kittiwakes, 1929-78

Of all aged birds, 77 per cent were adults, with the proportion rising to 88 per cent during February-May and 82 per cent in November, but falling to 35 per cent during August-September. The latter influx of immatures presumably involves post-breeding dispersal, after which many birds spend three years in the north-west

Atlantic before returning to their natal colonies (Coulson 1974).

Gull-billed Tern
Gelochelidon nilotica

Very rare vagrant. Four old records.

Breeding in Denmark and Southern Europe, a few passage birds are noted in Britain each year, mainly at coastal sites. However, inland records are exceptional (BOU 1971), so the four birds that were shot in the Region in the nineteenth century are of particular interest. They were at Wormleighton Reservoir on April 24th 1876, an adult female in breeding plumage at Shirley on August 8th 1896, Coleshill in 1899, and an undated specimen from Cofton Reservoir.

Strangely, none of these records is given in the *Handbook* (Witherby *et al* 1938–41).

Caspian Tern
Sterna caspia

Rare vagrant. Seven records.

This spectacular tern is a recent addition to the Region's avifauna and the seven records since 1968 are due in no small way to greater observer coverage at our waters. The Caspian Tern is a nearly cosmopolitan species, but most birds straying to Britain are believed to originate from the expanding Baltic colonies. However, there are British ringing recoveries from both Sweden and Lake Michigan; indeed, a North Amercian origin is perhaps more likely for the atypical October bird below. All records have referred to single adults and visits have been typically brief, apart from the abnormally long stay of the Kingsbury bird. The seven records were as follows: Belvide on July 20th 1968 (*Brit.*

Birds 62:473); Upton Warren on July 29th 1971 (*Brit. Birds* 65:337); Blithfield on July 16th 1972 (*Brit. Birds* 66:344); Chasewater on October 14th 1973 (*Brit. Birds* 67: 327); Kingsbury from July 6th to 11th 1975 (*Brit. Birds* 69:342); Draycote on June 23rd 1976 (*Brit. Birds* 70:427); and Sandwell Valley on July 25th 1979.

Between 1958 and 1972, July was the peak month of occurrence in Britain (Sharrock and Sharrock 1976), most probably reflecting the wanderings of failed breeders or non-breeders from the Baltic area.

Sandwich Tern
Sterna sandvicensis

Scarce passage migrant.

After early records at Alveston and Hampton-in-Arden in April 1876, at Edgbaston Reservoir in early November 1877 (still the latest on record), and one at Castle Bromwich prior to 1913, there were singles at Bittell on April 12th 1923 and April 23rd 1925. In addition, there is rather dubious hearsay evidence of three at Chasewater in September 1909 and several records for Earlswood prior to 1910.

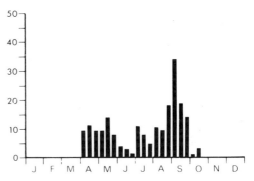

Based on arrival dates

Distribution of Sandwich Terns by Thirds of a Month, 1929–78

Increasing numbers of Sandwich Terns then appeared during 1929–78, with the five decades producing totals of one, five, 25, 16 and 58 records respectively. The anomolous 1959–68 total is puzzling when viewed against the background of a national increase in that period. Of the 105 records during 1929–78, Bittell provided 19 and Belvide, Blithfield, Chasewater, Draycote and Kingsbury each 10–14. There have been records of up to six together and an exceptional 24 flew over Coventry on September 9th 1953. Nearly a third of the records have fallen in April–May and almost a half in August–September. The species arrives earlier in spring than other terns, with the earliest in the Region to date being two at Belvide on April 3rd 1971. Edgbaston Reservoir has provided the two latest records, with one on October 15th 1948 in addition to the November record above.

Roseate Tern
Sterna dougallii

Rare passage migrant. Fourteen records.

With less than 1,000 pairs breeding around the coasts of Britain and Ireland in 1979 (Dunn 1981), this most elegant of the "sea terns" is probably our rarest breeding seabird. Its overall scarcity and apparent aversion to inland migration are reflected in the mere handful of records in the Region since the first on June 4th 1941, when an exceptional party of nine appeared at Belvide.

Subsequently single birds have appeared at Blithfield on May 7th 1955; Belvide on May 28th 1955; Blithfield on August 8th 1957; Alvecote on May 21st 1959; Budbrooke on September 28th 1965; Draycote on May 4th 1969; two at Kingsbury on May 31st 1969; singles at Belvide on June 10th 1971 and June 24th 1972; at Westwood

from May 28th to June 1st 1972; Coombe on August 31st 1973; Blithfield on June 2nd 1974; and Belvide on May 7–8th 1980. The Budbrooke bird, which was found dying, had been ringed as a nestling on Anglesey (Gwynedd) on July 7th 1963.

The ten records in May and early June reflect the rather late return of this species to its British breeding sites.

Common Tern
Sterna hirundo

Regular and not scarce as a passage migrant. Scarce or very scarce breeding species.

During the last century, when the true identity of many birds was established only when shot, both this and the next species were regarded as rare or scarce passage migrants to the Region. Records indicated that the Common Tern was scarcer than the Arctic Tern, judging by birds shot and, therefore, presumably identified correctly. Migrants were occasionally recorded along major rivers and at lakes and canal reservoirs. However, in recent decades both species have been observed with greater regularity, notably at the larger expanses of open water created by river-valley gravel extraction and drinking-water reservoirs. During 1929–78 the annual totals of Common Terns were exceeded by Arctic Terns in only eight years. WM *Bird Reports* for these years reveal minimum totals of 2,400 Common and 2,173 Arctic Terns, although in the last decade better identification standards have revealed that Common Tern is only 5 per cent more numerous than Arctic Tern (see Table 56).

The histograms show that Common Terns have, in fact, been only about two-thirds as numerous as Arctic Terns in spring in the last decade. However, in autumn, when both species are generally

Table 56: Mean Annual Totals (with ranges) of Common and Arctic Terns.

	1929–38	1939–48	1949–58	1959–68	1969–78
Common Tern: Mean	1·8	4·9	21·0	40·8	171·5
: Range	(0-8)	(0-29)	(2-48)	(18-99)	(66-305)
Arctic Tern : Mean	0·9	24·9	11·6	17·0	162·9
: Range	(0-4)	(1-218)	(4-26)	(5-57)	(46-565)

less numerous, Common Terns are twice as numerous.

The earliest record of a Common Tern is for April 4th, but the peak spring movement occurs during May. As most British breeding birds have returned to their breeding grounds by early June, the relatively high numbers occurring in the Region in

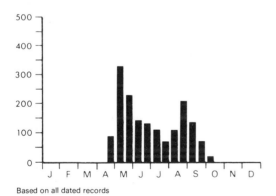

Based on all dated records

Half-monthly Distribution of Common Terns, 1969–78

June must involve mainly non-breeding adults and second-year birds. A very small number of first-summer Common and Arctic Terns, which retain the basic characteristics of first-winter plumage, has occurred in recent years, although most birds apparently remain in the southern hemisphere during their first year (Sharrock 1980). Peak autumn passage occurs in the latter half of August, and few birds are seen after September (51 records during

1929–78), with the latest being on November 3rd. The earliest and latest records of "Commic" Terns (Common or Arctic Terns not specifically identified) both come from Blithfield, on March 12th 1967 and November 21st 1954 respectively. Common Terns tend to travel in smaller parties than Arctic Terns and only five flocks have exceeded 20 birds, compared with 18 flocks of Arctic. The largest party was 45 at Draycote on May 4th 1974.

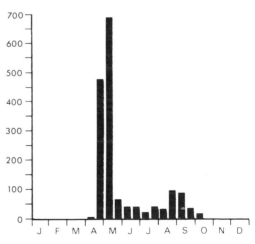

Based on all dated records

Half-monthly Distribution of Arctic Terns, 1969–78

The advent of flooded gravel-pits in the valleys of the Trent and Tame has enabled Common Terns to attempt to colonise areas previously unsuitable for nesting. In

1952 the first breeding record for the Region was confirmed at Branston Gravel Pit, where up to three pairs have bred with varying success in at least seventeen years up to 1979. At nearby Stretton Gravel Pit, two pairs bred successfully in 1956 and one pair in 1966, while at Kingsbury Water Park one or two pairs attempted to breed annually between 1969 and 1974, with success in three of those years, three or four pairs bred again in 1980 and in 1981 about a dozen pairs raised 37 flying young. In 1979 birds bred successfully at another Tame Valley site. Both human and avian predation, disturbance and flooding appear to be major factors limiting success and expansion of these incipient colonies. A reliable source of small fish is also necessary for rearing juveniles, although it is interesting to note that passage birds feed almost exclusively on invertebrates, especially emerging *Chironomid* midges. The provision of islands or artificial nesting rafts at undisturbed, unpolluted river-valley gravel-pits should result in a more stable breeding population.

Arctic Tern
Sterna paradisaea

Regular and not scarce passage migrant, sometimes fairly numerous.

In the nineteenth century Arctic Terns were considered to be more common than is generally believed nowadays. In more recent decades numbers were much under-estimated because of the problems of field identification. Undoubtedly the species is sometimes numerous during bad weather in spring, but many birds have been merely labelled "Commic" Terns or, worse still, logged as Common Terns. Practical field identification techniques (e.g. Hume and Grant 1974) have been determined only very recently and it is now possible to

identify these terns with some accuracy and confidence.

In southern England during late April and early May there is a considerable, rapid northerly and north-easterly movement of Arctic Terns heading for northern Britain, Iceland and Scandinavia. When this movement coincides with a passing depression, the associated rain and strong winds may result in the brief appearance of surprising numbers of terns inland. In the West Midlands this passage is most marked between April 20th and May 9th, a week or two earlier than the spring peak of Common Tern. Birds stay only a matter of hours and feed almost continuously before moving on, usually to the north and east. Sometimes flocks accumulate to give quite large concentrations and simultaneous appearances at many localities suggest movement on a broad front across the Region. It is likely that many terns annually overfly the Midlands and only appear at ground level if they encounter inclement weather.

In early May 1842 large numbers of Arctic Terns appeared at waterways throughout the Region. Birds came up from the Severn Estuary and hundreds were to be seen along the Avon. Many were shot and 40 were taken to one Evesham taxidermist. Another large influx occurred in 1947, when at least 218 were identified, of which six were found dead; during April 25–29th up to 33 appeared at Belvide, 50 at Gailey and 55 at Bittell. In 1971 108 birds arrived at Belvide on April 24th, and good numbers were also recorded in the Region in 1974, 1976 and 1978. In 1974 up to 30 appeared at Draycote and 37 at Belvide, while larger unspecified flocks of 50 at Blithfield and Draycote and 70 at Chase-water in early May were almost certain to have been Arctic Terns. In 1978 about 535 birds occurred between April 18th and May 12th, with peaks of 44 at both Draycote and Kingsbury and 58 at Belvide during May 2nd–6th.

It is believed that each autumn Canadian

54. Arctic Terns sometimes appear in good numbers on spring passage, but they seldom stay more than a few hours.   *T. C. Leach*

Arctic Terns (and Leach's Petrels) cross the Atlantic to European waters before turning southwards, and such birds probably occur in the Region at least occasionally. For example, after strong westerly winds in early September 1967 many terns were concentrated in Liverpool Bay and at least 52 Arctic Terns appeared in the West Midlands at that time. However, not all autumn migrants have such distant origins. A juvenile found dead at Westwood Park on August 5th 1951 had been ringed as a chick in Sweden two months earlier.

Arctic Terns tend to arrive earlier and stay later than Common Terns and there have been seven records in early April and six in November, the extreme dates being April 8th and November 15th.

Sooty Tern
Sterna fuscata

One record.

The single record of this extreme rarity is one of the Region's most unusual occurrences, since it involved a bird on the Dove near Tutbury in 1852 that was killed by a stone thrown by a deaf and dumb boy named Ault. This comprised the first European record of this tropical species. The date is unrecorded, but Mosley (1863) referred to "summer", Whitlock (1893) to "about October" and a note in the *Zoologist* dated December 20th 1852 records "killed about four months ago." For some reason the *Handbook* (Witherby *et al*

1938–41) decided upon October as the month of capture.

Little Tern
Sterna albifrons

Scarce passage migrant.

Little Terns were noted as rare stragglers to the Region's waterways before 1929, with at least half-a-dozen birds having been shot or found dead. The species was considered irregular at Chasewater, although four appeared there on June 6th 1909. At Bittell it was more regular and birds occurred in May 1923, 1934, 1938 and 1939. After this its appearances in the Region gradually increased and there have been up to eleven records (1974) annually since 1950.

Between 1929 and 1978 there were 143 records, with a steady increase in frequency interrupted only by a paucity during 1962–69. However, the national breeding total declined from a peak around 1930 to no more than 1,600 pairs by 1967 (Norman and Saunders 1969) and there has been no subsequent increase to match the increase in regional records from a mere 19 during 1962–9 to 53 during 1970–8.

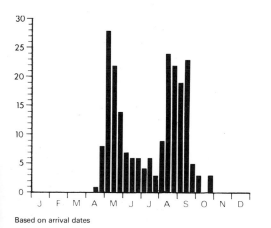

Based on arrival dates

Distribution of Little Terns by Thirds of a Month, 1929–78

As the histogram shows, most records refer to spring and autumn passage, with 34 per cent in May and 21 per cent each in August and September. The earliest and latest birds were at Alvecote on April 19th 1961 and Belvide (three) on October 30th 1954 respectively. Up to six have occurred together, with an exceptional 12 at Brandon on August 30th 1963. Little Terns have appeared on 30 occasions at Belvide, 21 at Blithfield, 19 at Bittell, 16 at Alvecote and ten or less at many other waters.

Whiskered Tern
Chlidonias hybridus

Very rare vagrant. Three or four records.

In most springs a few of these marsh terns overshoot their south European nesting grounds and appear in Britain, but few occur in autumn.

The first record for the Region came as recently as 1969, when an early bird paused briefly at Belvide on April 27th (*Brit. Birds* 63:280). The second occurred in the following year, when an unprecedented national influx brought a bird to Blithfield on May 10th (*Brit. Birds* 64:355). The two subsequent records probably refer to the same bird, which was seen on May 24th 1976 at Draycote and the following day at Kingsbury, 35 km to the north-west (*Brit. Birds* 70:426).

Black Tern
Chlidonias niger

Passage migrant, not scarce.

Discounting a few recent nesting attempts, the Black Tern ceased to be a regular breeding species in east and south-east England around the middle of the nineteenth century, with no certain records of nesting between 1858 and 1966. However, it continued to occur regularly on passage in spring and autumn.

Allowing for the paucity of observers during the nineteenth century, birds appear to have been fairly frequent visitors to the Region's waters and the records include a flock of 40 at Gailey on August 15th 1887. Numbers apparently declined in the early part of this century, at least in Warwickshire, for which Norris could find only six records for the period 1909–46.

However, Black Terns have occurred annually in the Region since 1934, with periodic large influxes. About 5,000 birds were noted during 1929–78, 2,000 of these being in the last decade. Over the fifty years as a whole 47 per cent appeared during April–June, but whereas from 1929–68 more birds were noted in Spring (56 per cent during April–June) than autumn (44 per cent during July–November), the colder springs of the following decade saw fewer influxes and the corresponding figures were 34 per cent and 66 per cent.

Flocks of ten or more birds have been reported on 115 occasions. Of these 42 occurred in May, 28 in August and 33 in September. There have been few April records, the earliest being on the 11th at Westwood in 1979. Records are also scarce in mid-summer and after mid-October,

with the latest being two at Blithfield on November 5th 1967.

Warm, south-easterly or easterly winds are the usual ingredient for a Black Tern influx and the following notable incursions have occurred in the Region: 246 around May 18th 1948; about 90 in mid-May 1950; about 100 on May 9th 1954; about 200 on August 5th 1954, including 100 at Bittell; at least 275 around May 23rd 1959; at least 330 in early June 1963, including 180 flying east in small parties at Belvide on the 1st and 120 at Blithfield on the 3rd; around 160 during August 30th–September 1st 1963; at least 300 during September 12–15th 1974; and about 100 on May 26–27th 1977. In the typical fashion of terns inland, parties generally pause at individual sites for a matter of hours rather than days.

White-winged Black Tern
Chlidonias leucopterus

Rare vagrant. Fifteen records.

Very small numbers of this east European species attach themselves to wandering parties of Black Terns and subsequently appear in Britain. The identification of

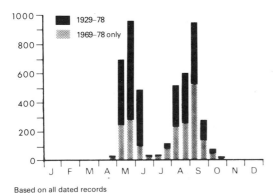

Based on all dated records

Half-monthly Distribution of Black Terns, 1929–78

immature and winter-plumage birds has been fully resolved only in the last twenty years and now about 70 per cent of British birds are recorded in autumn. That the last ten records in this Region refer to autumn birds is in no small way due to an increase in observer coverage and expertise and this contrasts markedly with earlier records.

The first four birds were all in spring – at Welford-on-Avon on May 8th 1884, Packington on May 8th 1909, Edgbaston Park on May 27th 1927, and Westwood on May 14th 1969 (*Brit. Birds* 64:368) – while a fifth was at Alvecote on the unusual date of July 12th 1969 (*Brit. Birds* 64:368).

The last ten records refer to immatures unless stated otherwise: at Draycote on August 24–25th 1970 (*Brit. Birds* 64:355); Belvide on September 8th and Blithfield on September 27th 1970 (*Brit. Birds* 64:354), with an adult and an immature at Belvide the following day (*Brit. Birds* 65:351); Upton Warren on September 26th 1971 (*Brit. Birds* 65:336); an adult at Blithfield on September 20–21st 1973 (*Brit. Birds* 67:327); Belvide from September 15–17th (not 18th as stated in *Brit. Birds* 68:321) 1974; Chasewater from September 27th to October 4th 1976 (*Brit. Birds* 70:426); Edgbaston Reservoir on August 17th 1977 (*Brit. Birds* 71:509); and Blithfield on September 7th 1978 (*Brit. Birds* 72:528).

In addition, Alexander (1929) mentioned an undocumented record of one at Bittell in April 1883.

Guillemot
Uria aalge

Very rare vagrant. Six or seven records.

Inland occurrences of all auks generally relate to strange and exceptional circumstances and perhaps none more so than the bird which Tomes (1904) reported as being shot from the roof of a thatched cottage in Warwickshire. Unfortunately, neither the locality nor date is recorded and Norris doubted the specific identity and placed the record in square brackets.

Better substantiated records concerned one near Stoke in 1841; singles shot at Gailey in June 1889 and June 1901; one found injured after flying into telegraph wires near Stone on May 19th 1920; a corpse found at Bartley on December 16th 1976; and one found at Wordsley on December 30th 1980, which subsequently died in Dudley Zoo.

Razorbill
Alca torda

Very rare vagrant. At least five records.

Tomes (1904) mentioned locally obtained Razorbills being taken to Warwick and Stratford for preservation, but gave no further details. There are three other records: one found dead at Harborne on July 25th 1890; an injured bird at Sandwell golf course on November 11th 1912; and one found dead in a rabbit trap near Overbury on May 10th 1953.

Little Auk
Alle alle

Rare vagrant: 25–30 records.

This tiny, pelagic species is a scarce winter visitor to northern Britain. In England it is most frequently seen off the east coast during late autumn and winter gales, though during particularly severe weather birds may be blown well inland and "wrecks" occur periodically. One such wreck in 1912 brought birds to Freeford on January 20th, Stone on February 2nd, Burton, Handsworth and Small Heath the following day, and near Evesham sometime during February.

The earliest record was of two birds near Worcester in November 1841. This was followed by several shot on the Trent after a storm around 1844 and singles at Malvern in November 1850; Water Orton on November 4th 1863; near Birmingham in February 1873; Handsworth in November 1875; Wheaton Aston in February 1901; Rugby on November 5th 1912; and near Cheadle on December 15th 1917. There were also undated specimens from Great Alne many years prior to 1904, somewhere in Staffordshire prior to 1907 and two from the Chillington area prior to 1911.

More recent occurrences involved two birds at Alveston in early February 1947 and singles at Evesham on October 6th 1947; Sheriffs Lench on February 11th 1950 and at Tredington and near Tenbury the following day; Chasewater on November 8th 1971; and Penkridge on November 1st 1974.

Thus, all of the twenty dated records fell between October 6th and the end of February, with no less than ten in the latter month and seven in November.

Puffin
Fratercula arctica

Rare vagrant. At least eight records.

Though they are very numerous at many of their coastal colonies, Puffins spend from August to March at sea, usually well out of sight of land, and are seldom blown inland by storms.

Tomes (1901, 1904), vaguely referred to immatures appearing in Warwickshire and Worcestershire after gales and Smith quoted an undated specimen from the Aqualate area. One was picked up in Broad Street, Birmingham, around 1880 and other birds appeared near Hagley in July 1936 and at Earlswood on February 1st 1953. Three records then followed in 1963,

with an adult at Rugeley in June and immatures at Redditch on August 30th and Bartley on October 18th. Finally, the latest record was also of an immature, at Longdon (Staffs) on November 4th 1967.

Pallas's Sandgrouse
Syrrhaptes paradoxus

Very rare vagrant. About a dozen records prior to 1909.

In response to snow covering its winter food supply in the Ural-Caspian steppes, this mainly sedentary, southern Asiatic species has made periodic invasions into Western Europe. Between 1859 and 1909 birds reached Britain in at least twelve years, with the largest influxes in 1863 and 1888, but there has been a virtual absence of records since, probably due to a marked desiccation of the western part of the species' range (Brit. Birds 60:419).

In the 1863 invasion, three birds were shot out of a flock of about 20 at Eccleshall on May 22nd. A further bird was shot at Swinfen in 1866, but the majority of records occurred during the 1888 influx as follows. A male was seen at Littleton on May 18th and a female shot at Rough Hill, Wolverhampton, five days later. Another was shot at Radway on June 22nd, one out of nine near Kineton was shot in July and similar fates met two out of four at Ipstones in September and four out of five at Cofton Hall on December 28th. Others were also seen, shot and preserved, or eaten, in the Edge Hill area sometime in 1888. The last record in the Region concerned a bird at Hamstall Ridware on December 18th 1908.

Despite documentation, none of the above records is given in the Handbook of British Birds (Witherby et al 1938).

Feral Pigeon
Columba livia

Numerous or abundant resident.

Feral Pigeons are now a familiar sight in many centres of population in Britain, not least of these being Birmingham and the other large towns and cities in the Region. Their original ancestor was the Rock Dove and all domesticated forms have been selectively bred from this species. It appears to be only a short step from the cliff-nesting sites of wild Rock Doves to the artificial ledges and cliffs of the urban environment now exploited so successfully by Feral Pigeons. However, pigeons have been domesticated by man at least since the ancient Egyptian era and in Britain they were kept in dovecotes as a source of meat from Roman times.

It has been calculated that in 1650 there were probably 26,000 dovecotes in England, perhaps containing 10 million birds. The keeping of dovecote pigeons declined during the eighteenth century as agricultural techniques improved and other sources of meat became available. Today this use of dovecotes has virtually disappeared. However, domestic pigeons are still with us in the form of ornamental varieties and the many thousands of racing birds kept by pigeon fanciers. Current Feral Pigeon flocks stem from the descendants of the dovecote pigeons joined by escapes and strays from pigeon fanciers.

Each flock has its own territory with roosting and feeding areas separate from neighbouring flocks and birds are most unlikely to leave their own territory and enter that of another flock, even if they can see food there. Some birds are capable of breeding all the year round, while others seem to maintain a continuous non-breeding condition. Diet varies with location, but bread is the staple diet in towns, now often supplemented by more exotic foods, and weed seeds are obtained from derelict land. Although some Feral Pigeons spend their lives within urban areas, others fly out to surrounding agricultural land to feed on newly-sown cereals and to glean stubble, and flocks of several hundred may occasionally be encountered. A curious twist of fate concerned the recent records of Feral Pigeon squabs being taken from inside old warehouses in the Potteries by Kestrels to feed to their nestlings.

The population in towns is largely dependent upon food provided accidentally or deliberately by people. Thus, urban Feral Pigeon populations are high and considerable economic problems occur from their accumulated droppings, feathers and nests. The cost of cleaning buildings is expensive and permanent damage can occur, particularly to the softer sandstone buildings and statues that characterise many Midland towns, due to the corrosive effect of the pigeons' droppings. Droppings falling onto wet pavements produce a slippery and dangerous surface.

Previous regional literature has included no reference to the occurrence of this species. Indeed, the publication of the BTO *Atlas* saw the first recognition of this fact. Not surprisingly, the *Atlas* showed a distribution centred on the West Midlands Conurbation and the Potteries area, with breeding confirmed in twenty-one 10-km squares (27 per cent), probable in ten (12 per cent) and possible in ten (13 per cent). The only other data available concerns an estimate of 5,000 birds in Birmingham in 1968, when attempts were being made by the Corporation Health Committee to reduce the number. Nevertheless, casual observations suggest that the number frequenting St. Philip's Churchyard has remained fairly stable in recent years.

Stock Dove
Columba oenas

Numerous resident.

Over the last two centuries, at least, the Stock Dove has probably always been less common in the Region than the Wood-pigeon. This is due primarily to the scarcity of the nest-sites preferred by Stock Doves, namely holes or cavities in old trees or buildings. The species now breeds sparsely in the Region's rural areas, its density increasing with that of mature timber. The landscaped parklands of large estates are particularly well-favoured.

There was evidence of a decline after the severe 1962/3 winter, but subsequently the size of flocks appears to have increased considerably. The *Atlas* surveys (1966–72) revealed probable or confirmed breeding in seventy-three (95 per cent) of the 10-km squares in the Region. CBC plots held a national average of 1·77 pairs per km^2 in 1978, from which density a regional population of 5,000–10,000 pairs is suggested. This is at the lower end of the range 2,000–20,000 pairs suggested by Lord and Munns. Not surprisingly, birds are virtually absent from the urban and suburban centres, and breeding densities are also low in the most intensive arable areas and the uplands of the north which

55. Stock Doves occur wherever old trees or buildings afford nest sites and they can be seen in suitable habitat well inside the urban area. *R. J. C. Blewitt*

are similarly deprived of old timber. None the less, there are semi-rural "islands" within the Conurbation, such as Edgbaston Park and Castle Hill, Dudley, that suit the Stock Dove's requirements and Warley Park holds a high concentration of breeding pairs (Teagle 1978).

Like Woodpigeons, Stock Doves congregate at good feeding sites, especially outside the breeding season, but are almost invariably far less numerous. Indeed, flocks of over a hundred birds have been reported on only a dozen or so occasions. A flock of up to 500 in the Bredon Hill area in the 1974/5 winter is by far the largest on record. The largest flocks gather during November to February, with fewer in March. Gatherings of up to 50 birds, however, have been found even in mid-summer on occasion.

A Stock Dove ringed as a nestling in June 1933 at Evesham was recovered at Landes (France) in November 1933, but no significance ought to be attached to this as by the end of 1977 it remained one of only two foreign recoveries of this species known to the BTO. The only other recoveries of note concerned two birds ringed at Binton on May 9th 1970, one of which went to Mickleton (Glos.) and the other just 12 km to Wellesbourne.

Woodpigeon
Columba palumbus

Abundant resident and winter visitor.

The Woodpigeon, as its name implies, was originally a bird of deciduous woodland, feeding on acorns, beechmast, various seeds and green vegetation. With such conditions, breeding densities were relatively low and stable. When woodland was cleared for agriculture, however, the Woodpigeon adapted to the new environment, nesting in spinneys, tall hedgerows and single trees and feeding partly on agricultural crops such as cereals and brassicas. In this new environment the population increased and studies have shown that numbers are controlled by the winter food supply. In the nineteenth century the cultivation of turnips and the technique of undersowing cereals with clover provided a greatly increased winter food supply, which assisted survival and led to a much higher population. In the last decade a further change in agricultural practice has led to a decline in the amount of clover grown and the Woodpigeon population is apparently falling. However, the simultaneous introduction of oilseed rape as a major crop to the Region has once again provided a plentiful winter food supply and the population may return to its previous high level. During the *Atlas* surveys (1966–72) breeding was confirmed in every 10-km square and with the suggested densities of 1,000–5,000 pairs per 10-km square suggested by Sharrock, the Region's population may be in the range 77,000–385,000 pairs. Lord and Munns merely placed the population in excess of 20,000 pairs.

The extent to which winter immigrants from elsewhere in Britain and abroad swell the numbers of resident birds is not fully understood. However, large immigrations of continental birds occur in Britain at least occasionally and such incursions may be reflected in this Region. For example, flocks totalling over 1,000 flew south-west over Earlswood on November 25th 1959, when immense numbers were arriving on the Norfolk coast (Seago 1977). Certainly large feeding and roosting flocks congregate outside the breeding season and many of these birds must be local residents. Flocks of up to a thousand or more are frequent from October to April, with most in January, November and especially December. The largest flocks on record are of 4,000 at Coven Heath on December 25th 1962 and at Westlands on November 25th

1977, and 4–5,000 roosting at Amington Woods in late February 1961. A flock of perhaps 5,000 on stubble near Stratford in December 1903 may well have comprised winter immigrants.

Serious economic damage to brassicas, peas, cereals and some fruit crops are the main problems caused by Woodpigeons and in periods of frost and snow whole fields may be devastated. 150 per day were shot from flocks in cabbages at Bonehill in late January 1963 and an incredible 110,000 were shot on eight large vegetable farms in the Vale of Evesham during the severe 1962/3 winter. Shooting Woodpigeons is a popular sport and helps to scare birds away from vulnerable crops, but it is considered unlikely to affect the overall population. Shooting pressure has increased during the last decade due to a demand for Woodpigeons from the continent and birds are exported from a canning factory in Evesham.

About 150 years ago Woodpigeons started to colonise urban areas in Europe and now they can be seen exploiting garden crops in many of Britain's conurbations. Within the city limits of Birmingham, Woodpigeons may join Feral Pigeons to feed on scraps of food deliberately provided. Birds are generally only absent from new suburbs and industrial areas, but otherwise are noted widely and commonly throughout the Region.

The only foreign recovery is of a bird ringed as a nestling in Warwickshire in September 1968 and found in southern France in October 1969 (nearly all British Woodpigeons recovered abroad have been found in France). Distant recoveries within this country concern a bird coming to this area from Penrith (Cumbria) in 1929 and others going to Ashford (Kent) in 1956 and Falmouth (Cornwall) in 1969.

Collared Dove
Streptopelia decaocto

Numerous resident. First records in 1961.

The colonisation of Europe by the Collared Dove during the last fifty years has been one of the most dramatic success stories on record. The species originated in Asia, where it is widespread from the Red Sea eastwards. At the turn of the century it maintained only a slender foothold in south-east Europe, but a north-westward spread at an explosive rate became apparent in the 1930s, perhaps the result of a mutation in the western peripheral population (Mayr, 1951). France, Belgium and Norway were reached during 1950–52 and the first British nesting occurred in Norfolk in 1955.

The first records for the Region concerned breeding near Kings Bromley and Worcester in 1961. The pattern of spread followed colonisation of widespread nuclei at farms, villages and mature suburbs, with subsequent infilling which was still apparent in the 1970s after more than a decade of exponential growth. There were reports of up to 50 near Worcester within three years of the first records and birds were reported from over 50 localities in the late 1960s.

Breeding has been reported in all months of the year and such fecundity has undoubtedly been a major factor in the species' spread. The *Atlas* projects revealed confirmed or suspected breeding in sixty-one 10-km squares (79 per cent), but birds were present in a further ten squares and the only obvious gap in the distribution was in the uplands of north Staffordshire. Allowing for further infill and increase since then, and bearing in mind that the CBC showed 2·39 pairs km^2 in 1978, the current regional population must be at, or perhaps even above, the upper end of the "numerous" category, probably between 15,000–25,000 pairs.

From 1969 onwards there were reports of

flocks in excess of 100 birds, typically around sites with a plentiful supply of grain or at overnight roosts. The largest flock so far reported was 350 at Barton-under-Needwood on September 29th 1975, but 200 roosted at Lanchester Polytechnic, Coventry, in November 1972 and 300 at Bunkers Hill in December 1975, whilst 85 frequented a Pershore grainstore in 1971, 100 were counted at Kidderminster cattle market the following year and 100 were shot in 1974 at a site in Warwickshire to prevent grain loss. In view of the economic losses being suffered, particularly at grain-stores, the Collared Dove lost its special legal protection in 1967 and only a decade later was admitted to the list of "pest" species.

Observations at several well-watched localities, such as Belvide, have revealed a small, but distinct passage of birds in spring, especially in May. This is known to be the time of peak dispersal in Britain (Hudson 1965) and influxes are then particularly obvious at coastal sites such as Bardsey Island (Gwynedd).

In the late 1970s there were suggestions of local declines for unknown reasons, but the extent, or indeed reality, of any decline is impossible to judge on the scant evidence. After the initial surge of interest attached to the appearance of any such rare bird, the Collared Dove has come to be regarded with indifference by most bird-watchers and much less endearance by those suffering grain loss, or indeed those kept awake by incessant cooing.

Turtle Dove
Streptopelia turtur

Numerous summer visitor.

The Turtle Dove is a fairly common bird of mainly arable farmland wherever tall hedges, spinneys and coppices are available for nest-sites. The disturbed soil of arable land hosts various annual flowers such as fumitory, the seeds of which may account for up to half the Turtle Dove's diet. The species appears to have increased considerably in the Region in the latter half of the nineteenth century, before which it was very scarce. Around 1800 the amount of arable land was increasing to keep pace with the expanding urban society, which in turn provided "modern" farm machinery to increase production. This period also saw the maturing of many hedgerows planted as a result of the Enclosure Acts. So, by about 1850, good, stock-proof hedgerows and an abundance of arable weeds must have made farmland a much more attractive prospect for Turtle Doves.

There is no evidence to suggest that the agricultural depressions prior to the two world wars led to any clear changes in numbers and the species apparently main-tained its then fairly common status. The ploughing campaign of 1939–45 and more recently a reduction in grassland in the Region released more land for arable crops and if anything conditions must have favoured further increase. Certainly, the national CBC index has risen steadily since 1962, so the surge in herbicide usage for arable weed control cannot have been too great an obstacle for the species. Neverthe-less, there have been a few recent reports of scarcity in Staffordshire, though in parts of Warwickshire an increase has been noted as Collared Doves have declined (D. Marland pers comm.).

The Atlas surveys (1966–72) revealed probable or confirmed breeding in all but three 10-km squares (96 per cent). The main gaps in distribution were predictably the West Midlands Conurbation and the high pasture and moorland of north Staf-fordshire. Sharrock suggested a density of possibly 100 pairs per occupied 10-km square, which along with a national CBC farmland density of 1·65 pairs per km^2 in 1978 indicates a Regional population of 5,000–7,500 pairs. This is considerably

56.   The Turtle Dove is a summer visitor, with a liking for tall hedgerows and woods, and its distinctive song is often the first indication of its presence.   *S. C. Brown*

higher than the 200–2,000 pairs estimated by Lord and Munns.

The Turtle Dove arrives rather late in spring, mostly in late April and May, with the mean date of first arrival over forty-three years being April 26th. There have been only a handful of records for mid April, a week or more after the earliest records at Belvide on April 6th 1974 and Bromsgrove on April 7th 1969. Birds have taken readily to breeding in conifer plantations in otherwise suitable arable areas. For example, 50 pairs were estimated in the Million Plantation of nearly 3 km^2 at Enville in 1966. Small flocks may form in summer and autumn and of the fifteen to be reported that consisted of more than 20 birds, two were noted in June, nine in July, three in August and one in September. The largest flock was of 89 birds on wires over potatoes at Wellesbourne on July 25th 1953.

The mean latest date over forty-three seasons has been September 24th, but there have been a dozen or so October records up to the 20th and the latest of all was at Lady-walk on November 11th 1972. There is, however, an exceptional record of over-wintering, involving three birds which remained with Collared Doves at Sutton Coldfield during the 1968/9 winter, when a similar event occurred in at least three other English counties (Hudson 1973). The recovery of three ringed birds in France and Iberia on autumn passage indicates the route taken by migrants to the presumed wintering grounds in west Africa.

Cuckoo
Cuculus canorus

Numerous summer visitor.

The familiar notes of the Cuckoo's song are to be heard widely in the Region from the spring arrival until song ceases in June. The logging of the first Cuckoo is something of a national obsession and records have always been dogged by poorly documented reports of song, particularly since the potentially confusing notes of the Collared Dove began to assail the ears of the inexperienced. The earliest acceptable records for the Region concern possibly the same bird, at Exhall, near Coventry, on March 12th 1957 and at Whitnash two days later. These constitute the fifth and sixth earliest British records up to 1972, at least (Hudson 1973). There are only two other March records and a handful in early April, and the mean first date over forty-two years has been April 11th.

Most adults have departed by the end of July and subsequent records almost invariably refer to juveniles. The average date for last birds over forty-one seasons has been September 5th, but September records are not common. The latest bird recorded to date was at Belvide on October 11th 1975.

Perhaps because of its distinctive voice, the Cuckoo has been known from time immemorial and its behaviour and activities are inextricably woven into myths and legends, but it was not until quite recent times that some of the fact has been separated from fiction. It was actually in part of the Wyre Forest that Chance (1922) first began to unravel the mysteries of the species' breeding habits.

Past literature and recent records tend to be vague about status and distribution, but no great changes are intimated for the past century despite many reports of local fluctuations from year to year. Certainly some local declines have occurred periodically when habitats for particular hosts have disappeared, such as that associated with the loss of Meadow Pipit habitat at Upper Welland in 1950. In the *Atlas* surveys (1966–72) breeding was confirmed in sixty-five 10-km squares (84 per cent) and thought probable in the remaining twelve. National CBC results gave a density of 1.34 pairs per km^2 for farmland and 2.81 for woodland plots in 1978, when a figure of 4.8 pairs per km^2 was calculated for woodland plots in this Region. Thus, the national density of 5–10 pairs per 10-km square suggested by Sharrock may be too low and a Regional population in the order of 4,000–6,000 pairs is probable. Lord and Munns placed the population at between 200 and 2,000 pairs. Small concentrations of Cuckoos are sometimes noted in spring and early summer, but 30 present at Packington on May 21st 1972 must be regarded as exceptional.

Birds are most obvious in large expanses of monotonous habitat where large numbers of a single host species are available for parasitism. Thus, the commonest host in upland areas and lowland heath is the Meadow Pipit, while the Dunnock is the favourite elsewhere in the lowlands. Pied Wagtail, Tree Pipit and Reed Warbler are locally important and even the nationally rare Marsh Warbler does not escape parasitism. Less frequent hosts have included Skylark, Robin, Wren, Spotted Flycatcher, Sedge Warbler, Willow Warbler, Grey Wagtail, Yellowhammer, Reed Bunting, House Sparrow, Blackbird, Mistle Thrush, Starling and even Pheasant. Cuckoos are relatively uncommon in suburban areas, but if they do occur, then the Dunnock is usually the host.

The rare rufous, or "hepatic", phase female has been encountered on only eight occasions, six of these occurring after 1966.

There is just one ringing recovery of a bird ringed in a nest at Stafford on July 26th 1948 which was at Loos, Pas de Calais (France) on October 10th of the same year.

Barn Owl
Tyto alba

Fairly numerous resident.

The Barn Owl became the typical owl of open areas when England's ancient forests were cleared for agriculture and then enclosed. In this Region it was generally regarded as a common bird until the nine-teenth century, hunting for small mammals over farmland and rough, open sites. Even the bird's name, of course, indicates its strong ties with agriculture, although with a steady loss of suitable out-buildings it is possible that the majority now nest in the traditional hollow tree. A few pairs in the Region are known still to nest in that other classic site, the church tower.

An apparently steady and widespread decline in the Barn Owl population occurred around the turn of the century. Harthan considered this to be in the order of 75 per cent and several authors predicted future rarity if not total extinction. All "hook-bills" were then regarded almost

57. Like all nocturnal species, the Barn Owl is easily overlooked, but nowhere is it as common as formerly because of the reduction in its rodent-rich hunting grounds.

S. C. Brown

universally as vermin and human persecution contributed considerably towards that decline. However, it is likely that the intensification of agriculture around that time effected a reduction in rodent-rich hunting grounds. More recent changes have virtually banished Barn Owls to hunting along hedgerows and over rough grassland, open scrub and small pockets of derelict land. Deaths from collision with cars, trains and overhead wires have become frequent and even drowning in water and sewage tanks has been reported. The advent of persistent insecticides led to further declines as toxic compounds accumulated in prey items. However, in this Region the effects on status were less apparent than those on the Sparrowhawk. Nevertheless, national concern was sufficient to merit the entry of both species to Schedule I of the Protection of Birds Act in 1967.

The exceptionally severe winters of 1946/7 and 1962/3 caused a sharp reduction in Barn Owl numbers, but subsequent mild winters, special legal protection and a more enlightened public attitude have given rise to a reasonably healthy situation at present. The detection rate of this basically nocturnal species must be fairly low, even though birds may hunt in daylight at times and some individuals habitually do so. Population estimates must therefore be viewed with some caution. Blaker (1943) indicated that in 1932 roughly 60 per cent of the Region held densities of 11–25 pairs per 100 square miles (0·4–1·0 pairs per km²) and the remainder held in excess of that range. The WMBC 1950 Breeding Bird Distribution Survey, which was based mainly on opinion, suggested that the density of the species was less than 10 pairs per 50 square miles (130 km²) in 23 of the 38 Rural Districts in the Region and 10–100 pairs per 50 square miles in eleven others. One pair per four square miles (10·4 km²) was suggested in 1947 for north-east Warwick-

shire and other estimates, possibly biased towards areas with high densities, included three pairs each in about 21 km² at Preston-on-Stour in 1949, in 4 km² at Sheriffs Lench in 1950, and in 2·6 km² at Eccleshall in 1959. Thus, at a time when Barn Owls were considered to be at a low ebb, a regional density of over one pair per 10 km² was probable, exceptionally reaching one pair per km², with a total population of around 800 pairs.

The *Atlas* surveys (1966–72) revealed confirmed or suspected breeding in fifty-nine of the Region's 10-km squares (77 per cent), with the most obvious gap in distribution being the West Midlands Conurbation, although a few birds are known to persist even there in "green wedges", derelict land and rubbish tips. Barn Owls were reported at more than 240 localities during 1967–76. Allowing for under-recording in many rural areas, it is likely that the current regional population is at least that of the 1950s and certainly higher than Lord and Munns' estimate of 20–200 pairs.

Although Barn Owls have been ringed as nestlings in reasonable numbers, recoveries show no consistent subsequent dispersal. Five have been recovered at distances greater than 100 km, with birds going to Cambridgeshire, Hertfordshire, Suffolk and Wiltshire and one coming from Yorkshire.

Little Owl
Athene noctua

Fairly numerous resident.

Little Owls from the European mainland were introduced into Yorkshire in 1842 and later into other counties, though it was not until extensive introductions of Dutch birds were made into Northamptonshire during 1888–90 that the species finally became

58.   Although it was virtually unknown in the Region until this century, the Little Owl is
now firmly established in most rural areas.   *M. C. Wilkes*

established as a breeding bird and began to spread to other parts of Britain. In this Region it was virtually unknown before the turn of the century, the earliest records being in Worcestershire, near Knightwick in spring 1897, at Eardiston "on some previous date" and at Blakeshall in December 1901. The first Staffordshire record was at (Great) Chatwell in October 1906 and the first for Warwickshire only a month later near Stratford. A rapid expansion in range and population was apparent in the decade or two following the first breeding record in 1912.

By 1929 the species had become so numerous, at least in some areas, that a keeper near Welford was able to kill over one hundred in a single year. Such unjustified persecution persisted for several decades at least, for in 1961 five were reported on a keeper's gibbet at Arbury. Although set back for a few years by the hard winters of 1946/7 and 1962/3, Little Owls had reached most areas by the *Atlas* survey years and their presence was confirmed in all but two of the Region's 10-km squares, with breeding confirmed or suspected in seventy-three squares (95 per cent). The only obvious gap in the distribution was not surprisingly the Conurbation of the West Midlands, where only a very few penetrate to perhaps occasionally appear in gardens.

Little Owls are very much birds of agricultural land that is well-endowed with hedgerow trees, although mature parkland, old orchards and even farm buildings are also favoured. Their partly diurnal behaviour and characteristic habit of perching conspicuously, on roadside poles for example, makes recording of this owl much easier than other species. Indeed, during 1967–76 Little Owls were reported at some 365 different sites, compared with only 240 for the more strictly nocturnal Barn Owl. The 1955 WM *Bird Report* commented that the former was the commonest owl in most areas and this still appears to

be the case. However, because of different detection rates the situation is not as it first appears. Indeed, national CBC figures reveal that Tawny Owls outnumber Little Owls by about 2:1. In 1950 four pairs bred on 142 ha at Sheriffs Lench; in 1967 a census of owls over 100 km^2 in the Coventry–Rugby area revealed about 0·4 birds per km^2; and in 1978 at least 0·4 territories per km^2 were located over 30 km^2 centred on Wheaton Aston. The latter two densities are probably representative of much of the Region's suitable Little Owl habitat, although they are in excess of the 0·28 pairs per km^2 nationally, as shown by the CBC in 1978. Thus, an overall density of at least 10 pairs per occupied 10-km square, compared with Sharrock's conservative estimate of 5–10 pairs, would indicate a regional population of at least 750 pairs. The figure might be in excess of 1,000 pairs, but is still undoubtedly below that for Tawny Owl. This accords well with Lord and Munns' estimate of 200–2,000 pairs.

The only record of a Little Owl moving out of the Region concerned a bird ringed in 1926 at Malvern, which was subsequently found at Moreton-in-Marsh (Glos.) in March 1927.

Tawny Owl
Strix aluco

Numerous resident.

The Tawny Owl is generally the commonest and certainly the most widespread owl of the Region. Earlier works indicate that there has been little notable change in its status in the Region over the last century or more, although Smith believed there had been declines in the lower hill-country of Staffordshire following timber felling. Indeed the species is rarely found far from trees and so sizeable fellings are bound to result in territory shift, if not population decline. All owls have suffered from persecution, especially in the nineteenth century, but the Tawny Owl seems to have prospered despite this.

Although Tawny Owls are common in broad-leaved and mixed woodland, they also permeate farmland, where they frequent spinneys and shelter-belts, and conifer plantations, mature suburbia and even the centres of cities such as Birmingham. Smith noted that the species competed successfully with the Long-eared Owl, except perhaps in the highest, matted and stunted conifer stands in the north, and it is known that the latter may even fall prey to the more powerful Tawny Owl (Mikkola 1976). In addition to the typical sites in old nests, squirrel dreys and hollow trees, nests are occasionally noted in tree roots and rabbit burrows. Smith recorded nests among rocks and in caves in the northern hills and described Tawny Owl and Kestrel nesting close together on a rock ledge in the Dovedale district.

Wood mice and bank voles are the staple diet in woodland, while House Sparrows and brown rats predominate in towns, but earthworms undoubtedly form an important constituent at times of high rainfall. A study of pellets from a roost in mixed farmland near Birmingham from December 1968 to May 1969 showed short-tailed voles, followed by birds and wood or yellow-necked field mice, to be the main quarry. The remains of an adult Magpie and a warty newt in a nest at Sheriffs Lench in 1948 exemplify the range of food items taken by this species. In 1950 a bird took Blackbird eggs from a nest at Studley in addition to removing large flower-bed labels!

Despite being quite a noisy bird, the Tawny Owl is none the less strictly nocturnal and consequently suffers from the under-recording typical of all nocturnal species. Thus, data on breeding density are scarce and population estimates are subject to large errors. Norris believed the species

to be less abundant than the Little Owl in Warwickshire, but appearances were probably deceptive and the *Atlas* surveys (1966–72) showed the Tawny Owl to be more widespread over the whole of the Region, with breeding confirmed in sixty-nine 10-km squares (90 per cent) and suspected in a further seven (9 per cent). National CBC data, which showed $0 \cdot 5$ pairs per km^2 in 1978, presumably underestimates the density of Tawny Owls, but even this suggests that the species is about twice as numerous as the Little Owl. Regional estimates of density have included the following: about one pair per two square miles ($0 \cdot 2$ pairs per km^2) in several parts of north Warwickshire in 1947; at least three pairs in half a square mile ($2 \cdot 3$ pairs per km^2) at Selly Oak in 1948; and four pairs in 200 ha at Sheriffs Lench in 1950, whilst a census of owls in the Coventry–Rugby area in 1967 revealed at least $1 \cdot 0$ pairs per km^2. These are all well in excess of the 10–20 pairs per 10-km square suggested by Sharrock and indicate a Regional population of 1,500–15,000 pairs. The true figure probably lies around the middle of this range, which accords with the 2,000–20,000 pairs postulated by Lord and Munns. Although Tawny Owls must have been ringed in reasonable numbers since ringing started in 1909, no recovery of more than 20 km is known.

Long-eared Owl
Asio otus

Scarce resident and winter visitor.

This species has always been very scarce and local in Warwickshire and Worcestershire, although in Staffordshire it was more numerous and widely distributed around the turn of the century than thirty years later. The bird's apparent preference for breeding in coniferous woodland in this Region may be forced upon it by competition with the Tawny Owl, which increased in England during 1900–30 while Long-eared Owls declined. There is no evidence to suggest that Tawny Owls increased regionally at this time, but such an increase could well have gone unnoticed and this seems to be the most likely cause of the Long-eared Owl's demise. Smith was aware that the latter suffered from interspecific competition, although then, as now, both species could occasionally be found inhabiting the same wood.

Prior to 1929 the few confirmed breeding records away from Staffordshire included the Wyre Forest in the 1890s, near Barnt Green in 1907, Rugby in 1927 and the Lickey Woods prior to 1928. In Staffordshire it was considered to be present wherever suitable conifer woods existed and known localities included the plantations north of the Churnet Valley and near Cheadle, pine-woods on the moors and in the upper Dove Valley, Huntley Woods, Dilhorne, Draycott (-le-Moors), the Stone area and Weston-under-Lizard.

Since 1929 breeding has been proven as follows: at Rugby in 1930, Sutton Park in 1937, Exhall near Coventry in 1954, Enville in 1966, 1967 and 1977, different parts of Cannock Chase in 1967, 1968, 1973, 1974 and 1979, Coombes Valley in 1973 and at various localities on the North Staffordshire Moors in at least seven years, including no less than five sites in 1973. It is noteworthy that several recent records from the latter area have referred to the use of old corvid nests in hillside hawthorn scrub. Long-eared Owls are undoubtedly scarce birds, but much under-recorded and the scatter of isolated summer records suggests that they may be less scarce than generally supposed. Indeed, summer reports have emerged from over 25 localities in the last thirty years. The *Atlas* surveys (1966–72) showed confirmed breeding in four 10-km squares (5 per cent) and suspected breeding in a further four. The regional population is still probably in the range of 3–20 pairs

as suggested by Lord and Munns. Birds usually form a greater part of the diet than in other owl species and prey items identified at nests have included Woodlark, Whinchat and Meadow Pipit.

Away from breeding areas, the species has been recorded increasingly in the last decade or two. The location of birds at a daytime roost is largely a matter of luck and the increased observer coverage could more than account for this apparent increase. Indeed, some roosts have been discovered at well-watched sites such as Alvecote, Brandon, Chasewater and Kingsbury. Roosts are often located in willow scrub or evergreens in areas with suitable rough grassland or heathland nearby for hunting purposes. Even in winter competition with territorial Tawny Owls is likely. Winter visitors apparently originate from the Continent and a bird ringed at Hopwas on February 17th 1973 was probably on its breeding ground when it was recovered in Czechoslovakia on May 24th 1976. This is the only British-ringed Long-eared Owl to be recovered in that country to date. An unprecedented influx occurred in Britain during the winter of 1975/6, with at least 30 occurring in this Region, including no less than 14 roosting at Ufton Hill on February 23rd 1976.

Short-eared Owl
Asio flammeus

Scarce breeding species and winter visitor.

Short-eared Owls have apparently been scarce winter visitors for many decades, although there were only four Warwickshire records during 1936–45 and only eight in Worcestershire during 1939–61. Most winter visitors originate from Scandinavia and are present in the Region from October to April. Their autumn arrival coincides with that of the Woodcock and for this reason the species has often been known in the past as the "Woodcock Owl". The numbers reaching Britain are variable and correlate with the breeding success of the preceding summer. Breeding success is limited by prey numbers and peaks in the rodent population result in the successful rearing of many young owls. In the autumns of such years "invasions" of Short-eared Owls reach Britain.

Fifteen birds disturbed by hounds near Stretton (Hall) on January 16th 1895 comprised not only the largest roost on record for the Region, but also the only sizeable concentration prior to 1970. Less than six birds per year were reported until records increased in the late 1960s and then around 20 or more per year appeared in the 1970s. Over 25 birds were noted in the Region during the 1970/1 winter, including five at Belvide and six at Priors Hardwick in December 1970. Then a very large influx occurred in the 1978/9 winter, when over 80 birds were reported, including five at Brandon, Stratford and Tamworth, six at Nuneaton and seven at New Invention. Birds are typically found in open habitats where small mammals can be obtained in quantity, such as heaths, commons, rough pasture, marshes, reservoir margins, gravel-pits and disused airfields.

Large tracts of undisturbed country, such as moorland, are required for nesting and the North Staffordshire Moors now form the southern extremity of this species' Pennine breeding range. Smith reported that Short-eared Owls were known to have bred on the Morridge moors, but gave no date. No further confirmation of breeding on the moors came until 1960, since when pairs have bred or summered in several areas, reaching peaks of possibly five pairs in 1971 and eight pairs in 1977, of which two pairs were confirmed as breeding. The recent increase in the British breeding population has been due largely to the proliferation of new conifer plantations in upland areas, which provide good nest-sites and short-tailed voles in plenty. Such a

plantation at Gib Torr has been used regularly by one or two pairs in recent years, but several other pairs still frequent moorland and marginal pasture. At least some birds remain on or near the moors in winter, but the extent to which further dispersal occurs is unknown.

There are periodic summer records from the lowlands, which perhaps involve lingering winter immigrants. From April to June 1966 a pair frequented Defford Aerodrome and in 1974 a pair bred successfully at a disused airfield at Perton. Very few, if any, other breeding records are known for the lowlands of the English Midlands. Up to six birds remained at Perton until the following spring, when housing development forced them to disperse.

Tengmalm's Owl
Aegolius funereus

One old record.

According to J. W. Lloyd writing in "The Field", a bird of this species was caught in a pole trap at Wolverley on November 17th 1901. The bird was kept alive until February 25th 1902, when it was identified and subsequently cared for by the writer. The record was not included in the *Handbook of British Birds*, although four other of these Scandinavian vagrants appeared in eastern Britain during the autumn of the above occurrence. The record coincided with the first records in the district of the Little Owl, when a "similar" bird to the above was shot little more than one kilometre away at Blakeshall on December 16th 1901. That this Little Owl was seen and identified as such by Lloyd adds credence to the correct identification of the Tengmalm's Owl.

Nightjar

Caprimulgus europaeus

Summer visitor, not scarce but very local.

At the turn of the century the Nightjar was a familiar breeding bird throughout most of Britain, occurring in a wide variety of habitats such as dry, sandy heaths, commons, moorland, open woodland and even dense coppice. Since around that time, however, there has been a widespread drastic decline, the cause of which remains obscure. Loss of habitat, increased human disturbance, temperature fluctuations and even the spread of the grey squirrel have been postulated, but none fully explains the demise.

The decline in this Region began around fifty years ago, before which the bird had been locally common and fairly widespread. Staffordshire appears to have had most, Warwickshire least. Many bred on the northern moorlands, where there have been no recent reports, and birds were fairly common in open woodlands and heaths in the lowlands. Birds had been plentiful in Needwood Forest sometime before 1863 and the Lickey, Monk, Trench and Shrawley Woods held many before 1900. By the 1930s Nightjars were known to frequent Willoughbridge, Maer, Madeley, Trentham, Down's Banks, Oakamoor, Hopwas, the Lenches, Randan Wood, Kinver and near Rugby, with isolated records from various other localities. Also at this time the extensive conifer plantations on Cannock Chase were becoming ideal for Nightjars, with trees around 1–3 m high. Foresters considered that the population may then have been in the order of 50 pairs (B. Craddock *pers comm.*). Perhaps the woodlands of Wyre Forest were similarly attractive, for an estimate of about one pair per hundred acres ($2 \cdot 5$ pairs per km^2) was made in 1930.

Subsequent conifer plantings in other areas quickly attracted Nightjars, if only for restricted periods. For example, after replanting with conifers at Enville around 1950, the population increased from around five to 12–16 pairs by 1957–67, but only one or two pairs persisted in the area up to 1975. Likewise, up to five churring birds appeared at Hanchurch Hills around 1960, but only one or two remained a decade later. Similar plantations at Bagot's Park held a peak of about 14 churring birds in 1959.

As these plantations have matured there have been few further sizeable plantings to replace the lost Nightjar habitat and so by the 1970s birds were virtually restricted to their stronghold on the heathland of Cannock Chase. A census there in 1976 revealed 21–25 pairs, a population which seems to be stable at present, but subject to very considerable human pressures and associated fire hazards. Some shift in territories is currently taking place as conifer stands are clear-felled and replanted, but it is hoped that this action will help to maintain the Nightjar population of the Chase, which is now perhaps the healthiest remaining in the English Midlands. Two or three pairs at Kinver Edge and perhaps the odd pairs still lingering at Highgate Common and in the Kidderminster area are the only remaining regular birds in the Region away from Cannock Chase. The current population of less than 30 pairs compares most unfavourably with counts made during 1957–9, when over 60 churring birds were located at 15 sites apart from Cannock Chase, where a further 15 birds were found with only partial coverage.

The erratic and isolated appearance of churring birds and occasional nesting attempts at new sites in the Region indicate that birds may be desperately seeking suitable areas for colonisation. Since 1965 such occurrences have been reported from Weeford in 1965, Consallforge and Oakamoor in 1967, Brandon Wood (breeding attempted) and Packington in 1968,

59. Away from their regional stronghold on Cannock Chase, Nightjars have declined almost everywhere. *R. J. C. Blewitt*

Idlicote, Scotch Hill and Wellesbourne Wood in 1970, Coombes Valley (bred) and Fradley Wood in 1971, a new site in Worcestershire in 1978, and Dimmingsdale, Loggerheads and Meriden in 1979, when a churring bird in Sutton Park was the first there since 1959.

Smith noted that birds arrived in late April and May and departed by September, with one at Burton on September 24th being the latest on record. The mean date of first arrival over thirty-eight years was May 16th, with the earliest at Burton Hastings on April 20th 1952. The mean latest record over twenty-nine years was August 22nd, though singing birds have been noted in early September and a well-feathered juvenile was still at Highgate Common on September 13th 1977. Birds at that site six days later and at Sheriffs Lench on the

same date in 1955 are the latest in recent decades. A few passage birds have turned up at atypical sites such as on main roads, in gardens and on the reservoir dams at Belvide and Chasewater.

Swift
Apus apus

Numerous and widespread summer visitor.

This most aerial of birds is a common sight in summer over cities, towns, villages and open country throughout the Region. On warm evenings parties of Swifts can be seen spiralling ever upwards, their piercing screams quite audible even after they are lost to sight, whilst on dull, cold days or in wet, windy weather large numbers whirl and wheel over meadows or open stretches of water, often within a few feet of the ground. Nests are usually beneath the eaves or in a small crevice of an old building and the Swift's colonial breeding habit brings it back to the same traditional sites year-after-year.

Tomes (1901) observed that, whilst the hirundines had so seriously diminished in numbers, the Swift was quite as plentiful as ever it was and Harthan reiterated this opinion. Norris described it as a common and generally distributed summer migrant, whilst Smith said that it delighted in hawking insects from over the highest hills and many bred in the holes and fissures of the limestone rock-faces. Frost (1978) also mentioned breeding in crevices at Raven's Tor in the Dove Valley in 1932, but there have been no subsequent reports of such sites being used and it would be interesting to know if they still are.

The same widespread distribution was confirmed by the *Atlas* surveys (1966-72), when breeding was confirmed in seventy-three 10-km squares (95 per cent) and suspected in three more. Only on the bleak Millstone grit moorland was breeding not even suspected. Virtually nothing is known about density or long-term population trends, but, as there are few suitable nest sites amongst the new buildings that have replaced the old Victorian and Edwardian suburbs since the last war, many colonies may have been eliminated. Furthermore, the Swift often feeds at altitudes which are above atmospheric pollution and will also travel great distances to feed, so it has not benefited from the Clean Air Act to the same extent as the House Martin. Nevertheless there is no evidence of a decline and at Sharrock's assumed density of 40 pairs per 10-km square, the regional population would be around 3,000 pairs. This is at the lower end of Lord and Munns' range of 2,000-20,000 pairs, but is probably an under-estimate.

The Swift is one of the most punctual of spring migrants and during the years 1952-78 first arrival dates ranged only from April 15th to 29th, except for 1961 when one appeared on April 10th – the earliest date on record. Over forty-four years the average date for first arrivals has been April 23rd. Numbers then build up quickly as a strong passage develops in early May. Several movements of a hundred an hour have been reported, but the heaviest passages involved 250 moving north-westwards through Chasewater in an hour on May 5th 1957 and 2,000 passing through Gailey in six hours on April 29th 1961. Unlike the hirundines, whose gatherings are invariably largest in autumn, the biggest concentrations of Swifts are often seen in spring, particularly over water in cold, wet weather. Up to two thousand have gathered at several localities and even larger counts have included 3,000 at Belvide at dawn on June 3rd 1973, 5,000 at Blithfield on June 10th 1978 and 10,000 at Draycote on May 17th 1975.

The latter occurred in heavy rain, when low-flying insects were being hawked from across the reservoir. Such behaviour is typical of Swifts in cold, wet weather, but on warm, dry days they soar to great heights to feed on aerial plankton. Indeed weather has a strong influence on the behaviour of Swifts and radar studies have shown they may travel up to 150 km to feed on swarms of insects that gather at the junctions of cold and warm fronts (Flegg 1981). Birds travelling such great distances to an abundant food source could easily be mistaken for migrants and this might explain the occasional reports of passage on unexpected dates, such as 400 per hour at Sheriffs Lench on June 27th 1954. On other occasions birds have been seen moving ahead of thunderstorms to feed on swarming insects, or congregating at infestations. For example, 4,500 fed on greenfly at Blithfield in May 1967 and 2,000 on midges at Belvide in July 1969; several flocks up to 200 fed on swarms of ladybirds rising above oilseed rape in 1976; and 1,000 fed on blackfly over beans in 1977.

Breeding colonies are occupied from early May until late July or early August, when return passage begins. The main exodus occurs in the first half of August, when gatherings up to 2,000 strong can again be seen at reservoirs. The heaviest passages to be recorded were 2,000 moving WSW along the Tame Valley in fifteen minutes on August 15th 1951 and 4,000 moving in the same direction within two hours at Wilnecote on August 13th 1952. Migration continues until mid-September and over forty-four years the average date for last birds has been September 24th. Last dates are extremely variable, however, with October records in thirteen years and even November ones in 1956, 1965 and 1976–8. The latest dates were November 24th (1978) and November 28th (1976).

Ringing has shown the Swift to be a very long-lived species, with one West Midland bird still alive after thirteen years. This can mean a long wait for recoveries, but this must be worthwhile when, as in the case of a bird ringed at Castle Bromwich in 1963, the recovery comes twelve years later from Mozambique. Other African recoveries have come from Malawi and Tunisia, whilst European ones have come from Spain and Denmark. Several have been recovered in this country too, mostly at or near their place of ringing. The most distant British recovery was at Cupar (Fife) just one year after ringing at Wolverhampton.

Several aberrant birds have been reported with white or pale patches in their plumage and one at Westlands in September 1978 resembled an Alpine Swift. For a species that spends so much of its life on the wing, it is interesting to note that at Smethwick in 1954 two were observed sheltering in a tree during a severe hailstorm.

Alpine Swift
Apus melba

One record.

A single bird, which paid a brief visit to Upton Warren on May 6th 1973, (*Brit. Birds* 67:329), remains the only record for the Region of this rare visitor to Britain from southern Europe, Asia and Africa. Its appearance was part of a distinctly unusual influx into Britain during that month (Smith 1974).

Kingfisher
Alcedo atthis

Fairly numerous and widespread resident.

In the West Midlands Kingfishers might occur at some time of the year on any sluggish stretch of river, small stream, canal, gravel pit, reservoir or ornamental lake that is not too polluted to support fish life. If suitable nest sites are scarce, they will even nest in a bank that overhangs polluted water, so long as there is somewhere else nearby for them to feed. Small, still waters with overhanging perches are preferred for feeding, but these are often the first to freeze in cold weather, so in winter the Kingfisher is sometimes forced onto the larger reservoirs and faster-flowing streams and rivers.

Some early authors inferred that the species had been more common in the nineteenth century. Garner (1844), for example, said it was numerous in the Trent meadows and Tomes (1901) ascribed its decrease in Worcestershire to late spring floods destroying nests and to its being shot for its plumage, which was used to decorate ladies' hats. According to Smith, the Kingfisher was increasing in certain districts, though he too admitted that it had been more common formerly and he pointed out that since the Trent had become a dirty

60. Except where heavily polluted, most of the rivers, streams, canals and pools might be graced by a Kingfisher, but they are much scarcer after a severe winter. *M. C. Wilkes*

river upstream of Great Haywood, so Kingfishers had become fairly scarce. On the northern streams and rivers, though, Smith said it bred in fair quantity, being sometimes abundant along the upper Dove and Dane, except in Dovedale where it was not common. This, even in 1938, was thought perhaps to be owing to tourist traffic! Harthan also said the Kingfisher was not so common as formerly, though it was reasonably so where the situation suited its needs. Along the Avon, where the low banks generally restricted the choice of nest sites, he said that it could still be found in embankments near to watermills. In Warwickshire, Norris considered the status of the species to have remained stationary through the previous fifty years and said that it was resident in small numbers, with perhaps a pair or two in summer along every 10 km of river bank, although this was a far from unfailing rule.

The Kingfisher's fortunes over the years have been greatly influenced by man. On the debit side, its plumage was much sought after to decorate ladies' hats, fishing interests persecuted it and an expanding industrial society polluted many of its watercourses. Against this, the canal system, water-supply reservoirs and gravel pits considerably increased its available habitat and in recent times it has prospered from more sympathetic public attitudes and protective legislation.

Sometimes these factors have counteracted one another. For example, as rivers have become more polluted so canal traffic has declined and where the two run parallel it is not unusual to find Kingfishers nesting in the natural river banks and feeding along the nearby canal. In 1967 a pair nested on the polluted River Churnet and their flights to and from their feeding pool almost 1 km away regularly took them through an intervening wood.

Today polluted waters and hard winters are the Kingfisher's worst enemies. It is very susceptible to prolonged freezing and

the winters of 1946/7 and 1962/3 decimated the population. Afterwards any sighting was a rare event and many usual breeding areas remained unoccupied for some years. Indeed, during the WMBC *Atlas* survey (1966–8) breeding was confirmed in only eleven 10-km squares and suspected in just fifteen more. Heavy flooding in May 1968 caused further losses, but the Kingfisher fortunately has a high reproductive rate and single pairs may rear three broods in a good season, as at Brandon in 1969. Consequently numbers are soon replenished.

Thus, by 1972, the combined *Atlas* surveys showed breeding confirmed in fifty-two 10-km squares (68 per cent) and suspected in twelve more (16 per cent), but the general absence of birds from the Conurbation was a sure sign that polluted watercourses were still restricting its distribution. Nevertheless birds have been seen along the very urbanised River Cole at Hall Green and, as recently as 1971, at Shirley, whilst breeding occurs sporadically in Edgbaston Park. Aside from the Conurbation, the species was absent only from Warwickshire's Cotswold fringe, where suitable habitat is scarce. Even the Tame Valley, with its grossly polluted parent river, supports a thriving population, with pairs nesting amongst the numerous gravel pits as well as along the cleaner tributaries.

In 1946 the Kingfisher was found to be regularly distributed along the Severn, with one pair to every 3 km, but it was said to be less numerous along the Teme. The most comprehensive survey, however, was that carried out on the Teme in 1979 (J. J. Day *in litt.*). Although this recorded Kingfishers in 15 of the 51 km, some consecutive sightings were clearly of birds in a single territory and Day's impression was of 5–10 territories, but probably seven, at a density of 1·2 per 10 km. Whilst this is very similar to the general assessment for Warwickshire made thirty years earlier by Norris, it did follow immediately after the cold winter of 1978/9 and Day considers that the carrying capacity of the Teme might be three times this level. Most of the territories were on the lower, more sluggish reaches and no correlation was apparent between Kingfisher sightings and the distribution of shallows or earthbanks, though the latter are of course extensive anyway. Away from the rivers, the gravel pits at Brandon and Kingsbury each held five pairs in 1970. On the basis of Sharrock's suggested 3–5 pairs per 10-km square, the regional population is probably some 200–350 pairs, which is slightly above Lord and Munns' estimate of 20–200 pairs.

During the breeding season Kingfishers are strongly territorial and even the young are driven away soon after fledging. This causes some dispersal, but on the whole the species is not very mobile. A nestling ringed at Alderley Edge (Cheshire) in May 1937 was at Stoke-on-Trent by October 1938, whilst the longest recovery was of a juvenile which had moved from Minworth in September 1961 to Beeston (Notts) by January 1962.

Bee-eater
Merops apiaster

Very rare vagrant. Four records.

This brilliantly coloured bird from southern Europe, southern Asia and northwest Africa has reached the Region on just four occasions. The first of these was in May 1886, when a pair appeared at Red Hill. The female was shot on May 29th and on dissection was found to contain five or six eggs, whilst the male suffered a similar fate on June 2nd. Then, in 1955, one was seen near Minworth on May 14th and two, probably of this species, were watched making flights from an aerial at Kidderminster on September 2nd. The fourth and most recent record relates to a bird near

Redditch from September 22nd to 27th 1970 (*Brit. Birds* 64:357). All these records are generally in accordance with the pattern of vagrancy shown by Sharrock and Sharrock (1976).

Roller
Coracias garrulus

Very rare vagrant. Two records.

Smith quoted two records for this vagrant from south and east Europe, western Asia and north-west Africa. One was seen at Berkeley between 1856 and 1863 and another was shot by a keeper at Patshull in June 1908. A third record, of a bird near Droitwich in 1934, was placed in square brackets, but since the only evidence appears to have been a brief mention in the *Birmingham Post* for September 17th, it is hardly admissible.

Hoopoe
Upupa epops

Very scarce, but regular, passage migrant.

The Hoopoe has a sufficiently distinctive and exotic appearance to be noticeable even to the inexperienced and its habit of feeding on garden lawns means that it is often reported by non-birdwatchers. Following the first record, of a pair at Whitmore in 1830, there were sixteen subsequent records from the last century, many of which were obituaries of birds that had been shot. In this century there were another half-a-dozen records prior to the Second World War, but since the war the Hoopoe has emerged as a regular passage migrant, with appearances in twenty-six of the thirty-two years 1947–78.

During this time there were 58 records, all of single birds except for two together at Kidderminster from May 9th until late May or June 1950. Indeed persistent rumours suggested that one, or possibly both, of these birds stayed until the end of July, whilst another individual remained at Burton Green from late May until mid-August 1962. These were exceptionally lengthy stays for a species which seldom remains more than a week and is usually seen on just one day.

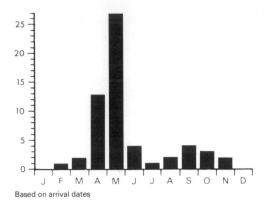

Based on arrival dates

Monthly Distribution of Hoopoes, 1929–78

This suggests that the small autumn passage involves birds that have been displaced westwards whilst moving south (Sharrock 1974).

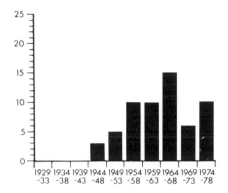

Five-yearly Totals of Hoopoes, 1929–78

The temporal distribution of records is remarkably even. In fact the most to appear in any one year was five, in 1968, whilst 1954, 1964 and 1974 all produced four records. Conversely there were six blank years.

As might be expected of a species that breeds further south in Europe, the majority of records appear to involve birds over-shooting their range on spring migration. This is why most are seen in the Severn and Avon sub-regions (61 per cent) and fewest in Staffordshire (20 per cent).

Although the records span from February 16th (1974) to November 18th (1967), the vast majority (61 per cent) occurred in late April and May. This pattern is very similar to the national one (Sharrock 1974) and that for the London area (Chandler and Osborne 1977).

Only 19 per cent of all records occurred in autumn, but it is worth noting that almost half of these came from Staffordshire and fewest from Worcestershire.

Wryneck
Jynx torquilla

Regular, but very scarce, passage migrant. Formerly a common summer visitor.

Although the Wryneck is a very scarce visitor to the West Midlands today, in the early nineteenth century it was still considered to be a common summer visitor (Sharrock) and its familiarity was well demonstrated by its many local names. Several of these, such as "Cuckoo's messenger" and "Cuckoo's marrow", arose through its arrival coinciding with that of the Cuckoo. Nowadays, though, it is mostly seen in autumn, when it sometimes visits gardens to search for ants on paths and lawns.

In 1844 Garner considered it not rare in north Staffordshire, yet by 1938 Smith said he knew of only one northern record in recent years. Elsewhere he said it used to frequent Forest Banks and the orchards around Tutbury, but again in recent years it had become uncommon in this, the Burton area. The Wryneck appears never to have been numerous in Staffordshire, however, though Smith thought it common in the Severn Valley just across the border in Worcestershire. In fact Worcestershire, with its wealth of old cider orchards, was undoubtedly the regional stronghold. Even here, though, Tomes (1901) noted that it had become much less frequent in the Evesham district about 1890. Subsequently it declined steadily and Tomes blamed this on the destruction of old orchards, but Harthan, who observed that its disappearance began in south-eastern districts and spread northwards, doubted this was the cause. Around Malvern the Wryneck was still common until 1900 and it persisted in this district until the early 1940s. Indeed, Malvern provided the last breeding records for the Region, in 1938 and 1941, and the only ringing recovery, of a nestling ringed in 1914 which returned to the same place the following year. In Warwickshire Norris said it had once been fairly common, coming regularly to Edgbaston Park until 1875, but by the beginning of this century it had decreased and was local.

Since the Wryneck disappeared as a breeding species, it has become very scarce even as a passage migrant. Excluding the breeding records of 1938 and 1941, there were only 53 records of 56 birds during the fifty years 1929–78. Roughly half of these were from the Malverns or the Severn and Avon sub-regions.

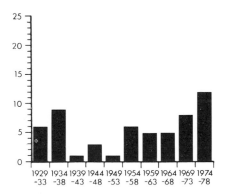

Five-yearly Totals of Wrynecks, 1929–78
(excluding breeding season records)

Despite being very scarce, the Wryneck has none the less remained a regular visitor, with appearances in thirty of the fifty years 1929–78. During this time its occurrences were fairly evenly distributed, with no more than three birds in any single year, except for 1976 when there were eight. More recently it has virtually become an annual visitor, having appeared in every year from 1968–78, except for 1975.

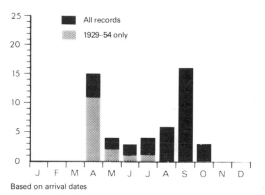

Based on arrival dates

Monthly Distribution of Wrynecks, 1929–78

All records have fallen between April 12th (1962) and October 9th (1973), but there has been a noticeable shift in seasonal distribution over the past half century. During the period 1929–53, when for part of the time there was still a lingering breeding population, every record was in spring or summer, with a pronounced peak in late April that tailed off into early May. In contrast, during the period 1954–78 the spring passage has been small and the majority of records (69 per cent) has been in autumn, with an even spread from late August to early October. Although conclusive evidence is lacking, these autumn birds are almost certainly drift migrants from Scandinavia. A similar, sharply defined autumn passage has been noted in the London area (Chandler and Osborne 1977), though the reasons for it are unclear. The unprecedented influx of eight birds in 1976 coincided with "falls" of Wrynecks along the East Coast.

Green Woodpecker
Picus viridis

Widespread and fairly numerous resident.

The Green Woodpecker regularly feeds on the ground, especially on ants. Because of this it favours more open country than the other resident woodpeckers and it is equally at home in farmland, orchards, parkland and other situations with scattered, old timber, as it is in mature deciduous woodland. On dry heaths, commons and open hillsides with short swards and scattered trees it is the commonest woodpecker, but in many areas, especially the more wooded districts, it is outnumbered by the Great Spotted Woodpecker.

Smith said the Green Woodpecker occurred throughout Staffordshire, even high amongst the hills, but that it had decreased in many former strongholds owing to timber felling. Harthan called it common and added that it was especially numerous around Bredon Hill and other areas of rough pasture covered in anthills, whilst Norris described it as a widely distributed resident throughout the agricultural parts of Warwickshire, particularly the south and east, but scarce in the Birmingham area.

The same widespread distribution prevails today and during the *Atlas* surveys (1966–72) breeding was confirmed in fifty-four 10-km squares (70 per cent) and suspected in seventeen more (22 per cent). Birds were absent only from the densely built-up parts of the Conurbation and the area between there and Cannock Chase. Of the woodpeckers, the Green suffers most in harsh weather because it feeds mainly from the ground and its numbers were severely reduced by the 1946/7 winter and less so by that in 1962/3. Recovery from such setbacks appears to take four or five years at least and reports of birds penetrating into the Staffordshire moorlands in 1972 and 1973 may have simply indicated a return to this bleak area after an absence of ten years. Sharrock thought the rate of recovery might be associated with the fortunes of the ants on which the Green Woodpecker feeds.

Examples of density range from one pair sporadically in 40 ha at Fradley Wood and five pairs in 200 ha of mixed countryside at Sheriffs Lench in 1950 to one pair on 460 ha of moorland in 1978. The regional CBC farmland plots showed $0 \cdot 4$ pairs per km^2 in 1978, but Sharrock pointed out that Green Woodpeckers have large territories, so that CBC results do not give an accurate indication of breeding densities and he suggested 10–20 pairs per occupied 10-km square. This would point to a regional population of 700–1,500 pairs compared to the much lower estimate of 20–200 pairs made by Lord and Munns shortly after the 1962/3 winter.

As the Green is the most conspicuous of our woodpeckers, it is sometimes thought to be the commonest, but it is doubtful whether this is true in the Region as a whole. It is certainly so in areas like the Malvern Hills and the open tracts of Cannock Chase, however, where the species is frequently encountered, and it is generally the case on the Lias clays south of the Avon, especially in permanent pastures where grazing sheep help to maintain a short sward much to the liking of ants. Although it is rare within the Conurbation, the Green Woodpecker is by no means unknown, having been recorded in Edgbaston Park, Hall Green and Saltwells Wood, where two birds in August 1964 were unusual. As with most woodpeckers, records of more than two or three together are unusual, but five were noted in the Sherbrook Valley on November 1st 1968 and ten were at Ufton Hill on August 1st 1975. An adult ringed at Fernilee (Derbys.) in July 1961 and recovered dead 69 km to the south, at Stretton, in January 1963 is the only notable movement.

61. The Great Spotted Woodpecker prefers deciduous woodland or other areas with mature trees. *S. C. Brown*

Great Spotted Woodpecker
Dendrocopos major

Widespread and numerous resident.

Decidedly more arboreal than the Green Woodpecker, this species prefers deciduous woodland, but can also be found in conifers, except perhaps the densest plantations. It avoids open agricultural country with few trees, but is regularly encountered in such situations if there are plenty of mature hedgerow trees, small copses or spinneys. In recent years it has also become a much more frequent visitor to suburban parks and gardens, often coming to bird tables during the winter.

Smith said that Great Spotted Woodpeckers occurred in most well-timbered districts and that locally their numbers were increasing, Harthan described it as more numerous than formerly and suggested that its increase was due to the cider orchards providing plenty of nesting sites, and Norris said it was a widely distributed, but rather local, resident, which had increased in numbers during the first half of this century. He added that it was frequently noted in the Birmingham area – a statement which is still true today.

The distribution of the Great Spotted Woodpecker has changed very little since and during the *Atlas* surveys (1966–72) breeding was confirmed in fifty-seven 10-km squares (74 per cent) and suspected in a further sixteen (21 per cent). Fincher (1955) thought this was the most common of the three resident woodpeckers and this is probably still true, although there are local variations arising from habitat composition. For example, in 1950 there were just three pairs on 200 ha at Sheriffs Lench compared to five pairs of Green Woodpecker. Of the CBC woodland plots, Fradley Wood held three or four pairs in 40 ha prior to the felling of its mature oaks in 1968, since when there have been only one or two pairs, whilst one or two pairs are usually present in Edgbaston Park's 16 ha. In 1978 there were five pairs in 3 km^2 of the Chillington estate.

Sharrock suggested 15–20 pairs per 10-km square in 1972, but since then the national CBC index has increased by 119 per cent. This has been attributed by the same author to Dutch elm disease making available a greater supply of larval and adult invertebrates. If so, a considerable population increase would have been expected in the West Midlands, where in parts of Warwickshire and Worcestershire as many as eighty per cent of the trees were elms and all have died. Strangely, there is no documented evidence of such an increase, although there is for the Lesser Spotted Woodpecker. Nevertheless it seems that 30–40 pairs per 10-km square might now be expected, giving a regional population of 2,000–3,000 pairs, which is above the range of 200–2,000 pairs postulated by Lord and Munns soon after the 1962/3 winter.

A number of factors affect the population. Past declines were often blamed on the increase in Starlings and grey squirrels and the Great Spotted Woodpecker is known to be susceptible to cold weather, though less so than the Green. In some areas it apparently took ten years to recover from the arctic winter of 1946/7, though that of 1962/3 had a less severe impact. Conversely the succession of mild winters and the outbreak of elm disease during the 1970s have probably brought about a population peak both for this species and the Lesser Spotted Woodpecker. Of the three resident woodpeckers, the Great Spotted is the more mobile and one ringed in Sheffield (South Yorks) in

1947 was found at Cheadle in December of the same year.

Lesser Spotted Woodpecker
Dendrocopos minor

Widespread and fairly numerous resident.

The Lesser Spotted Woodpecker favours deciduous woodland and parkland similar to the Great Spotted, but it also frequents old orchards and mature streamside alders. Hedgerow trees are visited as well, especially those affected by Dutch elm disease. Where the two species occur together, the Great Spotted is more frequently noted on the trunk and lower branches whereas the Lesser Spotted tends to feed and nest amongst the topmost branches. This, together with its small size and undemonstrative nature, makes it relatively inconspicuous and easily overlooked. Even so, it is undoubtedly the scarcest of the three resident woodpeckers.

Smith referred to several earlier writers who had described the Lesser Spotted Woodpecker as scarce, though he himself thought it frequently overlooked and not as rare as it was supposed to be. In Worcestershire it was said to be commoner than the Great Spotted at the turn of the century (Tomes 1901), yet Harthan thought it was the least common and Norris echoed this opinion for Warwickshire, commenting that it frequented gardens, orchards and parks even near the centre of Birmingham, but was definitely scarce in country districts, with perhaps only one to five or six Great Spotted Woodpeckers.

The WMBC *Atlas* survey (1966–8) showed the Lesser Spotted Woodpecker to be even scarcer than was thought, with breeding confirmed in only seven 10-km squares (9 per cent), though it was suspected in a further thirteen. It is possible, but not certain, that the 1962/3 winter was partly responsible for this very restricted distribution, as by the time the BTO *Atlas* work had been completed in 1972 breeding had been confirmed in twenty-five 10-km squares (32 per cent) and was suspected in nineteen more (25 per cent). Few of the regional CBC plots have held more than a single pair and none has consistently held the species, so little is known about density or long-term population trends. Sharrock assumed 5–10 pairs per occupied 10-km square in 1972, but since then the species has undoubtedly increased in this Region and twice this density would seem more appropriate. Thus the regional population is probably some 500–1,000 pairs, which is considerably more than Lord and Munns' estimate of 20–200 pairs.

In the West Midlands the Lesser Spotted Woodpecker appears to prefer lowland situations, since the vast majority of breeding records have come from the Trent, Severn and Avon sub-regions. Even those from the northern uplands have usually referred to birds breeding in deep, wooded valleys. Although present in parts of the Conurbation, including some mature gardens, it was not found breeding during the *Atlas* surveys. The species has definitely benefited from the outbreak of Dutch elm disease during the 1970s and it spread widely once the beetle infestation reached its peak around 1975, as Table 57 shows.

This spread was particularly evident in the Avon sub-region, where the elm was once so abundant.

Amongst the less usual occurrences, Lesser Spotted Woodpeckers have been reported feeding in reed-beds on more than one occasion and in 1959 a pair bred in a hop pole. As a rule birds are seen only singly or in pairs, but six were noted at Kinver in April 1947 and five pairs were estimated to be breeding at Arbury in 1954. There have been no ringing recoveries, but the species is believed to be largely sedentary.

Woodlark
Lullula arborea

Rare passage migrant. Formerly a local, but not scarce, resident.

The Woodlark favours dry heaths, bare hillsides or open land with a thin, short sward, scattered trees and patches of rank grass for nesting. Such sites are scarce in the West Midlands and over half the records since 1929 have come from a relatively few suitable areas, notably the Clent, Lickey and Malvern Hills, the heaths of Enville and the Kidderminster district, Kinver and the Lenches. The Enville–Kinver area was for many years the stronghold of the species.

Judging from early authors, the Woodlark's distribution was much wider during the last century. For Staffordshire, Dickenson (1798) referred to it without comment, Pitt (1817) noted casually that it sang at night and early morning, and Garner (1844) considered it not uncommon. In the

Table 57: Breeding Season Records of Lesser Spotted Woodpecker, 1969–78

	1969	1970	1971	1972	1973	1974	1975	1976	1977	1978
No. of breeding season localities	20	16	27	32	17	20	39	59	58	57

other counties Tomes (1901 and 1904) said its chief Worcestershire haunt was the Teme Valley, whilst in Warwickshire he noted it as uncommon and local, or even rare in the north.

Its fortunes in Britain have fluctuated markedly this century. Parslow (1973) described a gradual increase after the 1920s, with some expansion in range, followed by a marked increase in the 1940s, which reached its peak in many areas about 1951. From about 1954 onwards, however, there was a decrease which became very marked after 1960 and an almost total collapse in many Midland counties after the cold 1962/3 winter. A very similar pattern was apparent in the West Midlands, though there was no documentary evidence of any increase before or during the Second World War, perhaps because of a lack of observers. Smith described the Woodlark as very local and scarce and Harthan, too, said it was very local, though as well as frequenting the Teme Valley it could be found on the hills south to Malvern and eastwards via Clent to the Lickeys. He also added that a few pairs nested on Bredon Hill and in the Lenches. Norris, though, could quote only one definite Warwickshire record, of a bird at Edge Hill in April 1898.

Although it was always very local, the sudden increase and dramatic decline of the Woodlark appear to have been unexpected and not all records were fully documented. Consequently it is impossible to produce histograms. Immediately after the war, however, there was a tremendous upsurge in records with a noticeable peak in 1950–2, when there were 16 pairs in the Enville–Kinver area alone. By 1954 numbers had fallen substantially and there followed a depressing series of last records, from Malvern in 1957, the Kidderminster district in 1960, Clent in 1962, the Lenches in 1964 and the Lickeys in 1965. In the Enville–Kinver area birds were seen annually until 1961 at least and the last record was in 1967.

Sharrock discussed the various reasons advanced for this decline and concluded that whilst severe winters were clearly the cause of short-term fluctuations, they were not the sole cause of long-term change. Two other factors that seem to have had some bearing are climatic change, with cooler and wetter springs and summers, and habitat change, because myxomatosis had decimated the rabbits and so swards were less closely cropped. In the West Midlands most Woodlarks have certainly occurred in or near to the Severn sub-region, where they were most likely to find the optimum combination of a favourable climate and suitable habitat. It may not be without significance that the steady disappearance of both the Woodlark and the Wryneck can be traced progressively from south to north through Worcestershire, whereas nationally the species retracted in the opposite direction.

Although the Woodlark was recorded annually until 1965, so severe has been its decline that the only Worcestershire record subsequently was of one at Leigh Sinton in late July and early August 1972. Outside its restricted former breeding range, the species has been almost entirely absent this century. Single birds at Earlswood in 1950 and 1958 and three at Packington in 1976 remain the sole Warwickshire records since that of 1898, though birds were seen in Sutton Park (now the West Midlands County) in 1976 and 1977. In Staffordshire two were seen at Hopwas in 1950, one at Abbots Bromley and two at Chasewater in 1954 and two at Weeford in 1969. Apart from the Chasewater birds, which were seen on July 19th, these were all spring records, falling between the extremes of February 18th and May 30th. This implies a pre-breeding rather than post-breeding dispersal and the presence of a singing bird on Cannock Chase for about a fortnight from July 4th 1979 raises hopes that the species might one day breed again.

Skylark

Alauda arvensis

Abundant resident, passage migrant and winter visitor.

The Skylark is one of our commonest and most widespread birds, whose song is a familiar sound to everyone. It prefers open, tree-less country and is ideally suited to agricultural land, being commoner amongst the large fields and low hedges of arable districts than amongst the smaller fields and tall hedgerow timber of dairy areas. None the less, for feeding it prefers short grass or ground that is almost bare and even in arable areas it often favours long-term grass leys to cereals (Robson and Williamson 1972). Outside farming districts, the Skylark is common on heaths and open hillsides, whilst in urban situations it can be found on allotments, playing fields and waste ground. The breeding population is largely sedentary, but is augmented each autumn by large numbers of immigrants from the Continent and during winter months sizeable flocks can sometimes be seen feeding on stubbles or around reservoirs and sewage works.

Smith, Harthan and Norris all agreed that the Skylark was abundant in agricultural districts and especially on arable land. Smith expanded this by saying that in summer it was as common on the northern hills as in the lowlands, but that it gradually slipped away after the breeding season and though wandering droves might appear on stubbles in autumn, few remained anywhere among the hills in winter. He added that vast numbers appeared in October and November, possibly from far-off regions, to pass the winter on lowland flats and sewage farms, unless the weather was severe enough to drive them elsewhere. Harthan too said that large flocks appeared in winter and that these must include a number of migrant visitors from elsewhere. Norris, though, portrayed a slightly more accurate picture, commenting that

numbers increased in early autumn, but that in late October and November large numbers left and comparatively few were seen from December to February, particularly in a hard winter.

The distribution is just as widespread today, with breeding confirmed in seventy-four 10-km squares (96 per cent) during the *Atlas* surveys (1966–72) and suspected in the remaining three. Of these three, only the square covering central Birmingham could conceivably have lacked breeding birds and the failure to prove breeding in the other two squares must have been due to inadequate coverage. Densities vary considerably, even in typical farmland habitats. Random counts showed six pairs on 162 ha at Sheriffs Lench in 1952, but only two on 400 ha in 1957; 28 pairs on 105 ha of farmland around Brandon in 1966, which had increased to 35 pairs by 1968; 14 or 15 pairs on 81 ha at Cofton Richards from 1965–7 and normally 12 pairs on 81 ha at Wilnecote during the late 1960s and early 1970s. On the heathland of Cannock Chase there were 49 pairs in 89 ha of the Sherbrook Valley in 1967, whilst on the moors there were 10 pairs in 456 ha of the former Swythamley estate in 1978. In this latter area though, the species is more numerous on the Carboniferous limestone plateau than it is on the Millstone grit.

The CBC provides more detailed data. Numbers on the 184 ha plot at Moreton Morrell declined from 10 pairs in 1972 to only five in 1975, but then recovered to 11 pairs in 1977 and 1979. Such fluctuations may partly be due to annual variations in the cropping regime. On another CBC plot of 95 ha at nearby Wellesbourne, the population varied from 40–80 pairs per km^2 during the period 1962–70 (Hardman 1974). Whatever the population, though, Hardman always found the highest densities where hedgerow trees were absent (cf. Corn Bunting). Overall in 1978 the Skylark was the second commonest bird on

62. The Skylark prefers open country with large, arable fields and few trees. Resident birds are joined each autumn by continental immigrants and large flocks sometimes occur in winter.   *M. C. Wilkes*

the CBC farmland plots, along with the Dunnock and Wren, and it accounted for 8 per cent of all territories. Only the Blackbird was more numerous. These figures may overstate the general strength of the Skylark in the Region, however, since the Wellesbourne plot held 65 pairs, or almost two-thirds of the regional CBC total, at one of the highest densities known in Britain. Nevertheless the regional CBC farmland density in 1978, at $20 \cdot 8$ pairs per km^2, compared favourably with the national figure of 20 pairs per km^2. Both are comfortably within Sharrock's range of 500–1,000 pairs per 10-km square, so the regional population is probably near the centre of the range 40,000–75,000 pairs. Lord and Munns merely placed the popula-

tion in excess of 20,000 pairs. Several birds have been heard singing over derelict sites in the Conurbation, such as Aston, Kings Heath and the Wolverhampton–Walsall–Bilston area, and one or two were still doggedly breeding in the open-plan front gardens of an estate at Leamington five or six years after the houses were first occupied.

After the breeding season large numbers of Skylarks arrive from the Continent. Passage usually begins in early September, exceptionally late August, and then continues intermittently until November, but in most years it peaks in October. Movements up to a hundred an hour are often reported, usually to the south or south-west, but occasionally westwards or even north-westwards. During a study of diurnal migration near Walsall in September and October 1952, 17 per cent of the birds recorded were Skylarks, making this the second commonest migrant behind Starling. The heaviest recorded movements in the Region were 300 passing SSW in fifteen minutes at Upton Warren on October 22nd 1967, 1,500 passing S or SW along the Tame Valley in two hours on November 6th 1951, 150 an hour passing east over Nuneaton on October 23rd 1950 and 140 per hour passing SW through Bartley on October 18th 1959.

As Norris observed, many of these autumn migrants pass further south or west, perhaps to Iberia, so despite the heavy passage relatively few feeding flocks can be seen in October and even fewer in November. Even those that are found tend to be small, seldom exceeding 200. During the colder days from December to February, however, flocks are both more numerous and larger. Indeed a third of all reported flocks are in December, when concentrations up to 200 are by no means unusual, 500 have twice been seen and a record 600 were at Bartley on December 27th 1962, when an enormous passage was also noted at Sheriffs Lench. Although much smaller, a flock of 200 in inner Birmingham, at Aston in February 1977, is also noteworthy.

The Skylark's habit of congregating into a relatively few select areas often leads to the mistaken belief that it deserts the Region in winter, but only in the most severe weather conditions is this likely to happen. Hard-weather movements are a regular feature, however, and these can be even more impressive than the vigorous autumn passage. For example, 250 passed over Edgbaston Reservoir in an hour during a snow-storm on December 15th 1952, 350 passed SW through Sutton Park in one hour on January 13th 1960 and 1,000 moved NW over Wilnecote in half-an-hour on January 16th 1950. All are

eclipsed, however, by a staggering 5,000 that passed westwards over Sheriffs Lench in just one hour on December 27th 1962.

Spring passage is less marked, with 300 moving east through Bartley on March 3rd 1963 and 800 doing likewise through Belvide after heavy snow on March 4th 1970 the most reported. In most years birds have returned to their breeding territories by early March, though in a very cold spring some of the moorland sites may not be re-occupied until April. Very few Skylarks have been ringed, but one that was at Malvern in 1934 was recovered at Chepstow (Gwent) in March 1935. Otherwise the only recovery concerns one of several birds that were trapped at Wellesbourne early in 1967 and deported to various parts of north and west England. This surprisingly returned to Wellesbourne from St Helens (Merseyside) (Hardman 1974).

Shore Lark
Eremophila alpestris

Very rare vagrant. Three definite records.

Although a few Shore Larks regularly winter along the eastern coasts of Britain, they are rare inland. Smith referred to a female which had been shot near Enville prior to December 9th 1879, Harthan mentioned one seen on the Malvern Hills on January 14th 1920 (see also *Brit. Birds* 1921) and one was present at Blithfield on November 11th and 12th 1972.

In addition to these three definite records, one of a bird near Crabbs Cross on March 1st 1947 was placed in square brackets as it was only seen in flight and the description, though good, was felt to fall short of the complete proof required for so unusual an occurrence.

Sand Martin
Riparia riparia

Numerous and widespread, but local, summer visitor and passage migrant.

The Sand Martin is the least widely distributed of the three hirundines. It usually nests colonially in the vertical faces of sand and gravel quarries or those river banks that are dry and soft enough for it to excavate holes, but it is quite adaptable and there are several references to its having nested in the drainage pipes of retaining walls and banks and some to less usual situations. Feeding birds can also be seen hawking insects over reservoirs and other stretches of open water, especially during migration. In autumn it also roosts communally in reed-beds, often in association with Swallows.

Smith said that in north Staffordshire the Sand Martin must have either changed its nesting grounds or greatly decreased in numbers, since many of its former strongholds were deserted and it was much commoner in the river valleys, marl holes and gravel pits of the lowlands than in the quarries and stream-banks of the hill country. In 1948 it was said to be the commonest hirundine at Belvide. Harthan noted that there were colonies along the Severn and Teme and in the sandpits around Bromsgrove and Kidderminster, but he added that elsewhere it was very scarce. Norris described it as fairly common, but somewhat local, and concluded that it had decreased considerably in the first half of this century, citing as an example Sutton Park, where the species was common in 1896 but had almost gone by 1937.

Quarries and even river banks can change dramatically from year-to-year, however, so the Sand Martin has to be an opportunist, ready to desert former haunts that have deteriorated in favour of new ones in optimum condition. Thus its distribution is liable to vary markedly, though in the West Midlands the proliferation of sand and gravel workings since the last war has vastly increased the amount of suitable habitat available. During the *Atlas* surveys (1966–72) breeding was confirmed in fifty-eight 10-km squares (75 per cent) and suspected in two more (3 per cent). The only complete gaps in distribution were the heavily built-up areas of the Conurbation and the Cotswold fringe in the south, where in both cases there are few suitable nest sites. Birds are generally scarce right across the Lias deposits, though, as there are few quarries and the low banks of the

Table 58: Censuses of Sand Martin Colonies along the Rivers Severn and Teme

| | Length km | No. of nest holes | | | | | |
		1942	1943	1944	1953	1976	1979
Severn							
Upton – Worcester	16	31	11	31	50		
Worcester – Stourport	16		31	50			
Teme							
Mouth – Ham Bridge	24	164	273	222			
Mouth – Knightsford	19					67	195
Ham Bridge – Stanford Bridge	6		94	110			
Knightsford – Tenbury Wells	32					262	

Avon and its tributaries are prone to flooding and unsuitable for excavation.

The most suitable river banks for nesting occur where the Severn and its tributaries traverse the Keuper deposits and Old Red Sandstone. In the early 1940s the colonies along the Severn and Teme were censused by H. J. Tooby and more recently the Teme, which is a most important river to Sand Martins, has been resurveyed (J. J. Day *in litt*). The results of these surveys make interesting comparison.

Day has also provided some valuable data on the breeding ecology along the Teme, where the Sand Martin is the most numerous of the riparian species. In 1976 there were 329 nest holes in 28 colonies along the 51 km stretch from Tenbury Wells to the confluence with the Severn. The majority of suitable nesting habitat (74 per cent) was found in the 19 km downstream of Knightsford, where 31 per cent of the course had earthbanks. By comparison only 11 per cent of the 32 km upstream had earthbanks suitable for nesting. Yet the 1976 survey revealed only 67 nest holes (20 per cent) in the stretch below Knightsford compared to 262 (80 per cent) in the less favourable upper reaches. For the lower reaches there was no significant correlation between Sand Martin density and the length of earthbanks, but for the upper reaches the correlation was highly significant. Day concluded from this that density was most likely related to the availability of food rather than suitable habitat. Although birds do spend much of their time hawking over water, it has been shown elsewhere that aquatic invertebrates do not form a main component of the Sand Martin's diet (Waugh 1979). It seems they feed over a wide area, so the mixture of adjacent habitats is important and the upper reaches of the Teme are flanked by much more extensive tracts of woodland than the lower reaches.

In 1979 Day resurveyed the lower reaches and discovered 195 nest holes – an increase of 291 per cent in just three years. Some of these were in new colonies 20–30 pairs strong and many were in banks that had been in existence for several years. Of course this increase coincided with the recovery from the very low population levels of 1972–5, but the upper reaches were not resurveyed, so it is not known whether they experienced the same increase. It seems likely, however, that the stretch downstream of Knightsford is suboptimal habitat, which in consequence would be prone to wider fluctuations in population.

Some sand-pit colonies are equally large, such as 240 nest holes at Curdworth in 1942, 160 at Whittington in 1953 and 144 at Claverdon in 1954. The largest colony, however, appears to be that at Beckford, where 100 nests in 1954 had increased to 225 in 1976 and 265 in 1977. Nationally numbers have declined in the last two decades and this has been attributed to the drought in the Sahel zone of West Africa (Winstanley, Spencer and Williamson 1974). In the West Midlands, though, declines and desertions were first mentioned in 1964 – four years before the onset of the drought – and numbers reached their lowest level during the years 1973–5. Sharrock thought an appropriate density prior to the population crash to be 100 pairs per 10-km square, but considered that afterwards it might have only been 25 pairs per square. Recent years have brought a slight recovery and the current regional population is probably between 2,000–4,000 pairs, which is above the estimate of 200–2,000 pairs made by Lord and Munns. Amongst the less usual nest sites, birds have been reported at collieries, stone quarries, an iron foundry and power station fly-ash beds, whilst 30 pairs nested in a bank bordering a main road at Leek in 1964. Nesting holes are often taken over by Tree Sparrows and Starlings.

Sand Martins are amongst the first migrants to arrive each spring. The earliest ever recorded was on March 7th (1977) and over forty-three years the average date for first arrivals has been March 23rd. In favourable conditions spring passage can be heavy and at least 1,500 were noted at Belvide in 1947, 1961 and 1968. Return passage also begins early, often towards the end of June, and from late July onwards concentrations of a thousand or more can be seen and up to 3,000 have been recorded twice, at Blithfield on August 28th 1969 and Kingsbury on August 14th 1977. The heaviest passage was also observed at Blithfield, when 2,500 moved SW in an hour on August 28th 1962. Roosts are much smaller than those of the Swallow, but up to 2,000 have gathered on several occasions, 3,000 were at Bedworth on August 16th 1967 and 4,000 assembled at Two Gates on July 27th 1951 and at Brandon on August 31st 1971. Over forty-three years the average date for last birds has been October 9th, but individuals have been seen in early November in four years, with the latest on November 13th (1977).

West Midland ringers participated in the BTO's Sand Martin enquiry in the 1960s and this, together with subsequent ringing, has brought a substantial number of recoveries from adjacent counties and south-east England as the map shows.

There have also been nine Scottish recoveries, but a remarkable lack of recoveries from northern England, indicating that birds from that area do not visit the Region. The only foreign breeding season recovery is of a bird in southern Norway in June 1978 and this was only the second British-ringed Sand Martin to be found in Norway. Birds on spring passage have been recorded from Morocco (two in April), France (three in May) and Jersey (one in May). In autumn several returning birds have been reported in France during August and September and one was in Belgium in August. The only winter record concerns a bird recovered in Senegal in February 1971.

Each dot represents one recovery
Sand Martin Movements between the West Midlands and other British Counties.

Swallow
Hirundo rustica

Widespread and abundant summer visitor and passage migrant.

The Swallow is among the most familiar of our summer visitors and its return is eagerly awaited each spring. It is essentially a bird of farmland, nesting in barns and outbuildings and hawking insects from around cattle or across meadows. In arable areas it is scarcer. Unlike the martins, the Swallow is not colonial, but it does congregate in large numbers either to feed over water or to roost in reed-beds in autumn.

63. Swallows often perch on overhead wires, especially just prior to migration.

M. C. Wilkes

Lines of birds resting on overhead wires are also a characteristic feature of migration.

Previous authors all commented on a widespread decline which began before the turn of the century. Tomes (1901) said that within the last few years the Swallow had become a comparatively rare bird, to be counted in scores where formerly there were thousands, and Smith also said that its numbers were less than formerly, though it still visited every part of Staffordshire, with no elevation being too high for it. Harthan thought Tomes had been referring particularly to the Vale of Evesham, where corn, grassland and livestock had given way to the intensive cultivation of fruit and vegetables, and, whilst he did acknowledge that there had been a general decrease, he still described the species as common. Norris, however, only called the Swallow fairly common and said its decrease had continued, with numbers seldom counted in hundreds, or even in tens.

Such pessimistic statements are hard to reconcile with subsequent events, as within three years of Norris' remarks, for example, 50,000 birds were noted at a Warwickshire roost. Certainly the Swallow is now widely distributed and the *Atlas* surveys (1966–72) confirmed breeding in every 10-km square. Its density though is variable. Around Blithfield in 1961 there were 76 pairs in 15 km^2, whilst 34 km^2 of the Nuneaton area held 151 pairs in 1962 and 182 pairs in 1963, but only 138 pairs in 1964. Sharrock implied a mean density of about 150 pairs per 10-km square, but the national CBC index has shown a steady decline since 1968, with a marked drop in

1974, and this was also evident in the West Midlands. None the less, by 1978 the farmland CBC showed 2·4 pairs per km^2 regionally and 3·4 pairs nationally. Such densities point to a regional population of 7,000–12,000 pairs, which is in the middle of Lord and Munns' estimated range of 2,000–20,000 pairs. Breeding success can be gauged from the survey of the Nuneaton district mentioned above, which showed 5·9 nestlings per pair in 1962, 6·1 per pair in 1963 and 5·3 in 1964. Amongst the less usual nests were several 2 m below ground on the iron steps leading into a sewer at Baginton in 1949.

The first Swallows often appear in March, when they are especially vulnerable to late cold spells. Smith mentioned an exceptionally early arrival at Madeley on March 8th 1893, but with such an early date the possibility of confusion with a Sand Martin cannot be eliminated. Nevertheless there were March records in nineteen years during the period 1934–78, with the earliest on March 16th (1971). Over forty-four years though, April 1st has been the average date for first arrivals. Passage then continues through April and into May, but heavy movements and large concentrations are unusual.

Birds begin to leave again in August and, though passage usually peaks during September, it continues until late October or early November. Over forty-two years November 2nd has been the average date for last birds, but during the last twenty years there has been a noticeable tendency for Swallows to stay later that hitherto. In fact, since 1957 there have been November records in every year, except 1975, whilst in 1967 one remained at a motor works at Shenstone from late July until December 23rd and in 1968 one was seen at Sherbourne on December 7th. The latest bird ever, however, appeared at Selly Oak on December 24th 1935 following a snowstorm.

During autumn several hundred birds per hour are often reported moving to the south or south-west and the strongest passages on record were 5,000 in half-an-hour at Wilnecote on September 21st 1951, 1,200 in an hour at Bartley on September 17th 1957 and more recently 1,000 in an hour at Northwick Marsh on October 1st 1973. Even more impressive numbers of

Table 59: Summary of Swallow Movements to and from the Region by Country and Month of Ringing or Recovery

	Jan	Feb	Mar	Apr	May	Jun	Jul	Aug	Sep	Oct	Nov	Dec	Total
British Isles													
0–100 km	—	—	—	1	10	13	2	20	19	—	—	—	65
101–250 km	—	—	—	—	1	7	2	6	7	1	—	—	24
251–400 km	—	—	—	—	2	1	1	5	1	—	—	—	10
Norway	—	—	—	—	1	—	—	—	—	—	—	—	1
Holland	—	—	—	—	—	—	—	2	—	—	—	—	2
North Sea	—	—	—	—	—	1	—	—	—	—	—	—	1
Belgium	—	—	—	—	1	—	—	—	1	—	—	—	2
France	—	—	—	1	3	—	1	—	3	2	1	1	12
Spain	—	—	1	—	—	—	—	—	—	—	—	—	1
Italy	—	—	—	—	1	—	—	—	—	—	—	—	1
North & Central Africa	1	—	—	1	1	1	—	—	2	3	1	1	11
South Africa	9	8	2	1	—	—	—	—	—	—	5	5	30

10,000 or more roost together in autumn in reed-beds or other emergent vegetation. Various sites have been used, either regularly or sporadically, but Bedworth Slough has been the most consistent, though even this has not been used every year. There have been eleven roosts in excess of 10,000. No fewer than seven of these were at Bedworth, where 12,000 congregated in 1950 and between 20–30,000 assembled in 1958, 1959, 1966, 1968, 1973 and 1974. The others were a record 50,000 at Two Gates on September 25th 1950, 15,000 at Wychall in 1953, 20,000 at Baginton in 1960 and 25,000 at Alvecote in 1967. Since the population decline in 1974 roosts have been smaller, with the largest in 1975 when up to 7,000 gathered at Brandon and a similar number resorted to a maize crop at Marston.

Ringing recoveries are summarised in Table 59.

The North and Central African recoveries include single birds in Algeria, Zaire, Ghana, Nigeria and Uganda and six in Morocco. Otherwise the table shows that West Midland Swallows winter in South Africa and indicates the countries they fly through on migration. Several of the distant recoveries from within Britain have been from Scotland.

Red-rumped Swallow
Hirundo daurica

One record.

The only record is of a bird which was seen amidst a large gathering of Swifts and hirundines at Draycote on May 27th 1972 (*Brit. Birds* 66:346). The date is typical for vagrancy in Britain of this south and east Eurasian species (Sharrock and Sharrock 1976).

House Martin
Delichon urbica

Widespread and numerous or abundant summer visitor and passage migrant.

The House Martin is a familiar summer visitor to almost every town and village throughout the Region. It is gregarious, nesting usually in small colonies beneath the eaves of houses, though there are a few large colonies on single buildings or structures. Cliff nesting has seldom been reported in the West Midlands, though it occurs in Derbyshire (Frost 1978). Away from their nest sites, House Martins can frequently be seen feeding over water, sometimes in large numbers.

Smith (1938) said it had apparently decreased in numbers, but Harthan (1946) thought it had increased in recent years, whilst Norris described it as common and generally distributed, adding that there seemed to be no very noticeable alteration in its numbers through the years. Smith also commented that it was much persecuted by House Sparrows, which usurp its nests, and this is still the case today. During the *Atlas* surveys (1966–72) breeding was confirmed in every 10-km square, but there is no available data for the West Midlands on breeding density. Sharrock quoted densities of between 2·72 and 5·23 nests per km^2 for similar areas, however, and it seems likely there are at least 200–300 pairs per 10-km square. This suggests a regional population of 15,000–23,000 pairs, which is at the upper end or slightly above the range of 2,000–20,000 pairs suggested by Lord and Munns.

Generally speaking the House Martin is the last of the three hirundines to arrive each spring, though sometimes it appears before the Swallow. The earliest recorded arrival was on March 26th 1896 (Smith), but in more recent times birds were noted on March 28th in 1965 and again in 1967. Over forty-four years, however, the average first arrival date has been April 9th. Many of the earlier birds pass straight through and it is usually early May before nesting colonies are re-occupied. The number of nests in a colony often varies markedly from year-to-year, but the House Martin is an opportunist quite ready to colonise new sites, so such fluctuations do not necessarily mean an overall decline. Birds sometimes nest on houses still under construction so as to take advantage of the mud that can be found on most building sites. They are also faithful to traditional sites and one of these, the water tower at Hanchurch, provides the only insight into long-term population trends. Here there were 75 nests in 1960, 134 in 1963, 77 in 1976, 52 in 1977 and 60 in 1978. Such fluctuations are typical of many sites, but they reveal no consistent long-term trend. Some colonies have declined. That on the power house at Minworth sewage works, for example, fell from 119 nests in 1942 to just 23 in 1950, but the factories of Birmingham were spreading in the direction of Minworth at the time and this may simply have been a reflection of increasing atmospheric pollution. Since the passing of the Clean Air Act in 1956, birds have steadily penetrated deeper into urban areas, with colonies established in Handsworth and Harborne in 1961 and Edgbaston in 1970. The largest known colony was one of 200 occupied nests out of 277 on Ragley Hall in 1960. References to cliff nesting are few, but MacAldowie (1893) mentioned 100–999 nests on the Roaches and Seebohm stated that thousands used to nest in Dovedale, though Jourdain thought this an error. Nevertheless it seems likely that nesting does occur on the crags of the dales.

Amongst the more intensive autumn movements were 4,000 passing WSW over Wilnecote in two hours on September 18th 1951, 500 moving through Bartley in five minutes on September 3rd 1960 and the same number passing south over the Malvern Hills in one hour on August 5th 1970. Migrating birds also gather at reservoirs to feed and congregations of a thousand or more have been fairly regular at Blithfield and Draycote and 1,500 were at the former locality on August 31st 1975. Over forty-four years the average date of last birds has been October 31st, but during the period 1959–78 there were November records in eighteen out of twenty years and an exceptionally late bird was noted on December 5th 1962.

A bird ringed at Bredon in August 1925 was recovered in south-west France at the end of the following September and one ringed in Holland during May 1967 was re-caught at Temple Grafton in May 1970. Otherwise the relatively few recoveries for this species have involved autumn passage of birds within this country.

Richard's Pipit
Anthus novaeseelandiae

Very rare vagrant. Four definite records.

The history of this Asiatic vagrant is confused. Norris admitted it on the strength of a bird possibly seen at Welford at the end of last century, though he placed the record in square brackets with another reference that, according to the *British Association Handbook* of 1913, it had twice been recorded in Warwickshire, although in neither case were date or locality given. The next Warwickshire record – at Earlswood on March 6th 1949 – was also placed in square brackets. Certainly this is a time of year when the species is least likely to occur on the coast, let alone inland (Sharrock and Sharrock 1976). Smith mentioned one in a collection, which was believed to have been obtained in Staffordshire, though the collection had been dispersed and all trace of the bird lost.

More convincingly an adult male was taken near Hednesford on October 21st

1887 (Smith) and in recent times there have been three records. The first of these was again in spring, when one appeared at Blithfield on April 8th 1963 (*Brit. Birds* 57:275). Then a large influx into Britain during the autumn of 1967 brought the first Worcestershire record, with one at Upton Warren on October 7th (*Brit. Birds* 61: 353), and a remarkable series at Blithfield, with one on October 15th followed by two on October 29th and November 4th and three on November 5th (*Brit. Birds* 61:353).

Tawny Pipit
Anthus campestris

One record.

The only record for the Region of this Eurasian vagrant is a single bird seen at Tutbury on the most unusual date of December 29th 1953. Although the Tawny Pipit is an annual autumn visitor to Britain, December records are virtually unknown.

Tree Pipit
Anthus trivialis

Widespread and fairly numerous summer visitor and passage migrant.

Open glades or the edges of woods, parkland, or heaths and hillsides with scattered trees for song posts are the Tree Pipit's favoured habitat. Fincher (1955) said it could also frequently be found under the closed canopy of sessile oak, usually where the shrub layer was least developed. More recently it has taken to nesting in young

64.   Heaths or hillsides with scattered trees and the edges of woods are the usual haunt of Tree Pipits, particularly in the north and west, where sessile oak and birch abound.

R. J. C. Blewitt

conifer plantations as well, providing there are a few taller trees for song posts.

Smith said the Tree Pipit was universally numerous, except on the bare hills and moorlands, and Norris called it common, though rather local in distribution, adding that it avoided the industrial parts of Birmingham and Coventry and was least numerous in south and east Warwickshire. In Worcestershire, however, Harthan (1961) said it had decreased for the same reasons as the Woodlark. The present situation is much the same. The *Atlas* surveys (1966–72) revealed a widespread distribution, with breeding confirmed in fifty-one 10-km squares (66 per cent) and suspected in a further fifteen (19 per cent). The species was absent from Birmingham, the Black Country and parts of the Severn and Avon Valleys. It was also scarce or local across the heavy clay plains of the south and east. On the higher ground and light, sandy soils to the north and west of the Region, however, the Tree Pipit is common and quite numerous amongst its favoured heaths and hillsides, with their scattered birches and sessile oakwoods.

Tree Pipits arrive quite early in spring, but seldom as early as the one at Alvecote on March 14th 1967, which was the third-earliest ever in Britain (Hudson 1973). Late March or early April is more typical for the first arrivals, with April 8th the average date from forty-one years' records. The passage in 1967 appears to have been unusually strong and counts of 25 at Blithfield on April 6th and 50 at Belvide on April 25th are the highest on record. Migrants occasionally appear in urban situations and in 1961 one sang in allotments at Moseley.

Apart from its distinctive song flight, this is a relatively unobtrusive species which is probably under-reported. From the records available it is impossible to identify any long-term population trends, but there have been several references to declines since the mid-1950s. In common

with many summer migrants, numbers are liable to fluctuate quite markedly from year-to-year. For example, there were 20 pairs at Coombes Valley in 1977, but only eight pairs the following year. Most data is simply random counts of singing males, of which the largest were 40 in the Enville–Potters Cross area in 1951, 140 at Enville in 1967 (a particularly good year), 20 on the Malvern Hills in 1971 and 14 in the Wyre Forest in 1978. The only census data showed 26 pairs in 89 ha of the Sherbrook Valley–Brocton Field area of Cannock Chase in 1967 and usually one or two pairs (range none to four) in the 40 ha of Fradley Wood between 1966–79. Sharrock postulated 25–30 pairs per 10-km square, which would imply a regional population of 1,500–1,800 pairs. This is at the upper end of the range of 200–2,000 pairs suggested by Lord and Munns.

Tree Pipits leave again in August and September and autumn passage can be quite marked, with 20 per hour, for example, passing through Chasewater on August 26th 1951 and 200 in the Sherbrook Valley on September 7th 1979. Over thirty-nine years the average date for last birds has been September 20th, but October records have become more prevalent of late, with the latest on October 14th in 1972.

Meadow Pipit
Anthus pratensis

Numerous summer visitor, passage migrant and winter visitor.

As a breeding species, the Meadow Pipit is largely a bird of the uplands; open, hilly country; heaths and commons, being most abundant on the Staffordshire Moors. A few pairs do breed in the lowlands amongst rough grassland, colliery slag-heaps, derelict land, sewage works, reservoir margins, damp meadows and occasionally even cultivated land, but in such situations it is much better known as a passage migrant or winter visitor. It is also well known as a host to the Cuckoo in many areas.

Smith said the Meadow Pipit occurred in some parts of Staffordshire at all seasons, but that it was scarce on the uplands in winter except in very mild times. During the summer, though, he said the uplands were its chief breeding ground and nowhere in low-lying districts could it be found nesting so abundantly, though a few pairs did nest on the Potteries' slag-heaps. Tomes (1901) called it a common resident in Worcestershire, but Harthan described it as local, nesting freely on the commons at Castlemorton and Hartlebury, but being very scarce elsewhere. Norris said it was a local resident in small numbers and also a passage migrant.

The breeding distribution shown by the *Atlas* surveys (1966–72) confirmed these earlier assessments, but showed the species to be more widespread in Staffordshire and Worcestershire, with several pairs nesting regularly on the Clent and Malvern Hills. Overall, breeding was confirmed in forty-nine 10-km squares (64 per cent) and suspected in a further eight (10 per cent). The main gaps in the breeding distribution were along the Severn and Avon valleys, particularly on the heavy Lias clays. Both Norris and Sharrock had commented on this.

Long term population trends are obscure. There were 40 pairs on the Clent Hills in 1946, but only 20 in 1978, and fewer were said to be on the moors after the hard winter of 1962/3, but no firm conclusions can be drawn from such isolated reports. On the moorlands the Meadow Pipit is the most abundant species. Yalden (1979), for example, diffidently estimated 96 territories on 456 ha of the former Swythamley estate. This represented 26 per cent of all territories. On Cannock Chase 44 pairs were located on 89 ha of the Sherbrook Valley–Brocton Field area in 1967. Such figures accord favourably with the densities for prime habitats mentioned by Sharrock and from which he deduced an average of 1,000 pairs per 10-km square. In the West Midlands, though, such a density could only apply to the seven or eight squares of prime habitat. The species was absent from all the CBC plots and most of the remaining fifty or so occupied squares hold fewer than fifty pairs each. The total regional population is probably some 8,000–12,000 pairs, which is in the middle of Lord and Munns' estimated range of 2,000–20,000 pairs.

Meadow Pipits return to their territories in late March and April, though some may not reach the highest moors until early May. Small parties can be seen leaving the hills as early as late July (Smith), but many remain longer and 400 were at Morridge on August 28th 1968. In the lowlands there is a strong southerly passage during September and October, which has frequently been observed to reach 100 per hour. In the Tame Valley 275 were noted in forty-five minutes on September 24th 1950, whilst 500 were seen on September 20th 1951. During autumn parties up to 250 are regularly noted and an exceptional 600 were at Belvide on October 2nd 1959. Ringing provides a good indication of the destinations of autumn migrants. One of the earliest recoveries ever for this species was of a bird ringed in Warwickshire in September 1911 and found in Lisbon (Portugal) in December of the same year. Two birds ringed at Malvern in October 1934 were both recovered in south-west France, one in Landes in November 1934 and the other in Basses Pyrenees in October 1935, and another ringed at Malvern in 1937 was at Drogheda, Meath (Eire) in January 1942.

Passage fades in November, but during the winter months parties up to 100 can still be found, especially at sewage works. In this respect Hartshill and Whittington have been most favoured, with the former holding 200 in January 1963. Roosts of 50–100 birds can also be found in winter in a variety of habitats including long grass, conifers and marshes. Small numbers are also seen in winter on wasteland well within the urban area and one or two may even visit gardens in severe weather. Northward passage in spring is concentrated into late March and April. Movements are just as impressive as those in autumn, with 100 per hour passing over Sheriffs Lench on March 25th 1955 and 800 moving through Belvide on April 4th 1958. Parties up to 250 are also a feature, with 500 at Belvide on April 5th 1953 the maximum. Less is known about the destination of spring migrants, but one ringed in Worcestershire in March 1975 was found dead in Ayr just a month later.

Rock Pipit
Anthus spinoletta

Passage migrant, not scarce.

The normally coastal Rock Pipit, *A. s. petrosus*, is a passage migrant to watery places throughout the Region. Its status has apparently improved in recent years, though this must partially reflect the number of experienced observers that now regularly visit reservoirs, gravel pits and other favoured haunts.

Prior to 1948 there had been no definite records, although Tomes (1901 and 1904) said that it had appeared along the River Avon in both Warwickshire and Worcestershire, where it frequented weirs, and that several had been shot near Warwick some years earlier. Harthan, however, suggested that these birds were Water Pipits and Norris did not admit the records as no specimens could be traced. Smith made no reference to the species at all.

Following the first fully authenticated records, at Belvide on October 10th 1948 and Bartley seven days later, the Rock Pipit has appeared annually and about 300 birds

have now been recorded. Since 1970 there have been between 13 and 30 individuals each winter.

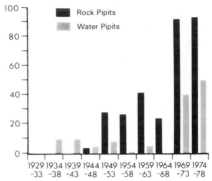

Five-yearly Totals of Rock and Water Pipits, 1929–78

The earliest autumn arrival was on September 19th (1970) and the latest spring departure on April 11th (also in 1970). Between these extreme dates birds have appeared in every month, though most have arrived in autumn with 58 per cent in October alone. There is, however, a small spring passage which peaks in late March.

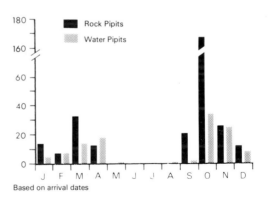

Based on arrival dates

Monthly Distribution of Rock and Water Pipits, 1929–78

The reservoirs of Staffordshire and Warwickshire have been the most favoured localities, especially Belvide (43 records and 79 birds), Blithfield (36 records and 58 birds), Chasewater (29 records and 43 birds) and Draycote (20 records and 33 birds). Although Worcestershire has had the fewest records, birds have been seen fairly regularly at either Bittell, Upton Warren or Wilden.

Most records have been of single birds, but up to three have not infrequently been seen together and during the autumn of 1959 up to six could be found at both Belvide and Blithfield. The same number also occurred at Belvide in October 1975, whilst eight were seen at Draycote on October 14th 1978. Many birds are present for only one day, especially in spring, but in autumn a few have lingered for a month or more. As yet, though, there have been no records of complete over-wintering.

The Scandinavian Rock Pipit, *A. s. littoralis*, has twice been positively identified, at Chasewater on April 9th 1975 and at Belvide on March 13th 1977. It is possible that the Chasewater bird had been present from at least April 3rd.

As a passage migrant, the Water Pipit, *A. s. spinoletta*, is also not scarce and in

recent years one or two have over-wintered. Although found in similar places, the Water Pipit is altogether scarcer than the Rock Pipit, though with more diligent searching by competent observers, it has been noted with greater frequency during the past decade. Indeed, of the 123 birds to occur in the period 1934–78, no fewer than 88 were reported in the eleven years 1968/9 to 1978/9, during which time between two and 17 appeared each winter.

The first record came from Bittell in

1919 and this was followed by five more from the same locality during the next decade (Harthan). These appear to have been the only records prior to 1934. One or two birds were then noted in most years during the 1930s and 1940s, mainly by H. G. Alexander, but these were followed by a dearth of reports from 1952 to 1967.

Autumn passage is the stronger and it peaks in October and early November, whilst the weaker spring passage peaks in late March and early April. Although most birds (37 per cent and 20 per cent respectively) occur at these times, the passage is less concentrated than that of the Rock Pipit and the records show a wider temporal spread. Despite its initial monopoly, there has been only one definite record from Bittell since 1940 and Belvide (33 records and 42 birds) and Chasewater (14 records and 18 birds) have been the most consistent localities. During the last year or two, though, three or four birds have remained throughout the winter at Wilden and this is undoubtedly today's most favoured haunt. Most records are just of single birds, though two or three have occurred together several times and up to five were at Chasewater during the 1971/2 winter.

Yellow Wagtail
Motacilla flava

Fairly numerous summer visitor.

The Yellow Wagtail is one of the earlier spring migrants to arrive and the bright males are a colourful sight as they daintily hawk insects from around reservoirs, gravel pits and sewage works, or from amidst cattle in tussocky water meadows and damp, marshy fields. However, not all birds breed in these situations, but frequently they will utilise growing crops well away from water, particularly in horticultural areas. They can sometimes be found in urban and industrial situations.

Towards the end of last century and in the early part of this, the Yellow Wagtail suffered a general and widespread decline. In Warwickshire, Norris said that numbers reached their lowest ebb between 1920 and 1940, after which there was a slight increase in migrants, though nesting birds were still recorded from only a few places. Harthan also indicated that it was not as common in Worcestershire as Tomes (1901) had earlier described, although he said that small numbers still bred in river meadows and on light land. This decline appears to have coincided with the decline in cereal growing that accompanied the agricultural depression between 1880 and the outbreak of war in 1939. Furthermore the Yellow Wagtail's subsequent revival has paralleled the resurgence in cereal growing despite the loss of habitat through drainage and improvement of its favoured wet meadows. In Staffordshire, where dairying has predominated throughout, there appears to have been less change. Smith, and later Lord and Blake (1962), described the species as a summer resident in all parts, including slagheaps in the Potteries and pit-mounds in the Black Country.

Reservoirs and gravel pits have helped to offset the loss of marshes and wet meadows and the horticultural intensification has been beneficial too. Of the pairs he found in crops, D. Smallshire (*in litt*) said that potatoes were preferred to beans, which in turn were preferred to oilseed rape. The *Atlas* surveys (1966–72) revealed a widespread distribution, with breeding confirmed in sixty-seven 10-km squares (87 per cent) and suspected in a further seven (9 per cent). Little is known about density, but there were five pairs on 95 ha of the National Vegetable Research Station at Wellesbourne in 1976 and two pairs on 64 ha of farmland at Willey in 1978. Sharrock suggested 25 pairs per 10-km square and on this basis the regional population is

probably between 1,500–2,000 pairs, which is at the upper end of Lord and Munns' suggested range of 200–2,000 pairs.

Birds have been noted in late March in nine years, with the earliest on the 20th (1966), but over forty-two years the average first arrival date has been April 1st. Passage usually peaks in mid-April and early May, when parties up to 50 are common and concentrations of 200 have twice been noted, at Belvide on April 17th 1955 and at Wilden on April 29th 1975.

Before and after breeding, Yellow Wagtails often roost together in reed-beds. Most spring roosts are small, usually no more than 50, but in August and September, when most birds are migrating, up to 100 are regular and even more impressive gatherings up to 500 have been reported, especially at Tillington–Doxey Marshes, where an estimated 1,000 roosted in both 1970 and 1972. At this time, too, parties up to 100 are often noted and 250 were at Upton Warren on September 6th 1968 and at Blithfield on August 27th 1972. The heaviest passage recorded, however, was of 1,000 birds which passed SW or SSE over Wilnecote in one hour on August 22nd 1950. Over forty-two years the average date for last birds has been October 10th. Occasionally birds have lingered much later, especially in recent years when 1974, 1977 and 1978 all produced November records. The latest recorded, however, was a bird at Blithfield on November 8th 1958.

All ringing recoveries of this species relate to its migration. On the south coast five birds have been recaptured at Radipole Lake (Dorset), two at Wick (Hants) and one at Thurlestone (Devon). Abroad there have been two recoveries which trace the migration route through Europe, with one report from Landes (France) and another from Beira-Litoral (Portugal). Both of these foreign recoveries occurred within two months of ringing.

In most years a few birds are claimed as

showing the characteristics of the main European race *M. f. flava*, the Blue-headed Wagtail. Many of these prove to be variants which cannot be definitely as-cribed to any specific race, but the Blue-headed Wagtail does occur as a scarce or very scarce migrant. Tomes (1904) said it had been met with at least three times near Welford-on-Avon, but these appear to be the only references prior to 1941. However, birds then appeared in twenty-five of the thirty-eight years 1941–78 and in every year since 1963, except for 1975. During this time 82 individuals were satisfactorily identified. The most in any one year was eight in 1978.

The vast majority occur on spring passage (87 per cent), though the paucity of autumn records is no doubt a reflection of the problem of satisfactory identification. Table 60 gives the monthly distribution of birds, based on their arrival date.

The earliest records were on April 9th (1947 and 1961) and the spring passage peaked in late April and early May, with a few stragglers into early June. There have been no records during mid-June, but birds have been seen in late June and July. The latest recorded was on September 7th 1952.

Practically all birds have been seen at wetland habitats. In the earlier years, old-fashioned sewage farms such as those at Baginton and Curdworth were particularly favoured, but in more recent times reservoirs and gravel pits have produced most records, with Belvide (21 birds), Blithfield (17 birds) and Upton Warren (8 birds) the most favoured localities. Most records referred to single birds, but two to-gether have been reported on seven occasions. One at Belvide on April 27th

1972 resembled the race *M. f. cinereoca-pilla*, or Ashy-headed Wagtail, from the central Mediterranean and another at Upton Warren on May 1st 1978 resembled *M. f. beema*, or Syke's Wagtail, from south-east Russia, but this latter bird was almost certainly a variant.

Grey Wagtail
Motacilla cinerea

Fairly numerous, but local resident. More widespread in winter.

The Grey Wagtail is primarily a bird of fast-flowing, rocky streams and rivers with gravelly shallows. Consequently most are to be found on the moorlands and uplands of Staffordshire or the hilly terrain west of the River Severn in Worcestershire. How-ever, a few pairs can be found along more sluggish watercourses, or even canals, where there are weirs, mill-races and locks. Outside the breeding season, birds are more widespread, visiting sewage farms, lakes and reservoirs and frequently occur-ring well within urban areas.

Smith said the Grey Wagtail was com-mon in summer throughout the northern hills, but that birds began to wander to the lowlands in early autumn, where they fre-quented the muddy confines of pools and rivers. He thought them more numerous along the upper reaches of the Dane, but in recent times there have been more records from the limestone than the gritstone country, as indeed there have in Derbyshire (Frost 1978). Harthan described it as a resident and winter visitor, which nested

Table 60: Monthly Distribution of Arrivals of Blue-headed Wagtails, 1941–78

	Jan	Feb	Mar	Apr	May	Jun	Jul	Aug	Sep	Oct	Nov	Dec
No. of birds	—	—	—	36	32	6	2	3	3	—	—	—

sparingly along all the tributaries of the Teme and Stour and the streams in the hilly northern parts of Worcestershire, and visited south-eastern districts in winter. Norris, however, knew the species only as a common passage migrant to Warwickshire, which occasionally remained through the winter. He added that it had nested, but could only quote five positive records.

Whilst this distribution still holds today, the Grey Wagtail is now known to be somewhat more widespread and it nests regularly in the few places in Warwickshire that are suitable, especially along the rivers Alne and Arrow. In recent years breeding has also been reported in urban localities, including some industrial ones, where birds were previously only winter visitors. At Edgbaston Park birds have bred sporadically since 1961. The species is known to suffer badly in severe winters (Sharrock) and during the WMBC *Atlas* (1966–8), which came soon after the arctic conditions of early 1963, breeding was confirmed in only seventeen 10-km squares and suspected in just a further four. By the completion of the BTO survey in 1972, though, these figures had increased to thirty-seven (48 per cent) and four (5 per cent) squares respectively and in many cases fieldwork for the BTO *Atlas* had revealed birds breeding in squares where only a year or two earlier they had not even been recorded as present. The main gaps in the present distribution are largely the Birmingham plateau, the Warwickshire Avon and the lower Severn Valley.

In April 1976, 15 were counted along 11 km of the River Dove, whilst during August and September 1979 25 were found along 57 km of the River Teme and from this a breeding density of 1·75–2·63 pairs per 10 km was estimated (J. J. Day *in litt*). Apart from this, little is known about density. Sharrock quoted the low densities on Midlands watercourses revealed by the BTO Waterways Survey, so presumably it would be realistic to assume a density at the lower end of his suggested range of 10–20 pairs per 10-km square. Doing so would imply a regional population of perhaps 400–600 pairs. Although this is higher than Lord and Munns' range of 20–200 pairs, it must be remembered that their estimate was made when the population was at a low level.

Grey Wagtails disperse widely in autumn (Sharrock 1969) and from September to November birds appear around the margins of lakes and reservoirs, along canals and even in industrial areas and shopping centres right in the heart of Birmingham and Coventry, where they sometimes feed in rain-water puddles on flat roofs. An indication of this dispersal is given by a juvenile which was ringed in May 1948 in Ayr and recovered the following November in Staffordshire. Some of these autumn migrants settle to stay for the winter. This species is the least gregarious of the wagtails and seldom are more than six noted together. However, in 1959 there was a strong autumn influx and 12 were at Earlswood on November 11th and 10 at Leamington Spa Sewage Farm on November 28th, whilst in 1971 up to 17 were reported from Whittington Sewage Farm on December 4th. Small roosts of a dozen birds have also been reported, from along the River Sowe in 1976 and Cannon Hill Park in 1977.

Pied Wagtail
Motacilla alba

Widespread and numerous resident, summer and winter visitor.

Though commonly found along streams, rivers, reservoirs and lakes, the Pied Wagtail is by no means confined to waterside habitats, but occurs widely in all types of country. Roadsides, farmyards, waste ground, suburban lawns and playing fields are all frequented and birds quite often nest

65. Pied Wagtails occur in all types of country, but are commonly found near water and can often be seen around sewage works or canal locks. *M. C. Wilkes*

on buildings in more urban settings. Outside the breeding season small groups often gather on sewage-works' filter-beds and birds also roost communally in reed-beds or factories.

Smith said Pied Wagtails bred commonly throughout Staffordshire, but that during winter comparatively few occurred in the lowlands and still fewer on the hills. Harthan found it common around farm buildings, large gardens and other inhabited places, whilst Norris described it as a fairly common and generally distributed resident, but observed that it seemed to be decreasing.

The species is just as widespread today and during the *Atlas* surveys (1966–72) breeding was confirmed in seventy-five 10-km squares (97 per cent) and suspected in one more. Only in the undulating terrain of the extreme south-west of Warwickshire was breeding not proved. Numbers decline after hard winters and very few were noted during a survey along the River Teme after the 1978/9 winter (J. J. Day *in litt*). Not many are present on the regional CBC plots, so there is little information on density, but Sharrock suggested 150 pairs per 10-km square and in 1978 the regional farmland plots as a whole held 1·42 pairs per km^2 compared to 1·94 pairs nationally. These figures imply a regional population of some 6,000–12,000 pairs, which fits comfortably in the middle of the range 2,000–20,000 pairs suggested by Lord and Munns.

Although Pied Wagtails are present throughout the year, there have been several references to diurnal migration, mainly in a southerly or south-westerly direction between August and October, and ringing has shown that the population changes. One locally bred bird has been recovered in the Region in winter and other recoveries also show that some birds are resident. However, another locally bred bird was recovered in Portugal and birds caught between July and September have been recovered in France (three), Spain and Portugal, which points to some being summer visitors. Equally, during the late 1960s several were caught in a Sparkhill factory and this showed that wintering birds can come from as far afield as north Scotland. Eleven birds ringed in the West Midlands have been recovered in Scotland, Cumbria and Northumberland.

Roosts of Pied Wagtails are widely reported every autumn and winter. Many of these, particularly in autumn, are in beds of reed or reedmace and usually number between 50 and 200 birds, though they may be considerably larger and 1,500 roosted in reedmace at Holt gravel-pit on October 15th 1978. In winter, though, the largest concentrations are usually found in industrial premises, where it is warm. The best known roost, which was in existence in the mid-1950s and was still in use in 1967, was that already mentioned in the roof of a Sparkhill factory, where up to 1,500 birds congregated each winter. A study of this roost proved that it attracted birds from a 16 km radius (Minton 1968a). Up to 1,500 were also estimated to be roosting at Brierley Hill steelworks in 1977, but the largest roost yet recorded was in the cooling towers at Courtaulds factory in Coventry, where 1,943 were counted on January 1st 1971. In addition to roosts, loose gatherings of up to 100 birds can be found in winter, usually at good feeding sites such as reservoirs and sewage works.

The continental White Wagtail *M. a. alba* is a scarce passage migrant to the Region's wetlands, especially its reservoirs. Smith quoted several records and thought it probably occurred every spring and autumn, but was perhaps frequently overlooked. Harthan knew of none prior to 1929 and said it was an irregular passage migrant, mainly in spring, and Norris traced about a dozen records, but doubted whether these gave a true idea of its status and called it a scarce passage migrant. The vast majority (98 per cent) are spring records, but in part this reflects the difficulty

Table 61: Monthly Distribution of Arrivals of White Wagtails, 1929–78.

	Jan	Feb	Mar	Apr	May	Jun	Jul	Aug	Sep	Oct	Nov	Dec
No. of birds	1	—	18	306	73	10	—	1	7	1	—	—

of autumn identification. Birds were recorded in all but six of the fifty years 1929–78 and in every year since the war. Unfortunately not all records have been fully documented, so it is impossible to compile complete histograms. However, at least 472 birds were recorded during these fifty years and in recent times there have been between 20 and 30 per annum. The most in any one year was 31 in 1958. Table 61 shows the monthly distribution of all dated arrivals.

The earliest record was on March 6th 1966, but although some birds occasionally arrive in late March, in most years none are seen until April. There then follows a concerted passage which peaks in mid-April and lasts until the end of May. Occasionally a few stragglers linger into June and in 1973 there were seven at Bartley on the 18th, whilst in 1975 a late bird was still at Brandon on 21st. Small parties up to ten regularly occur on spring passage and there were 20 at Chasewater on April 27th 1958 and 15 at Belvide on April 12th 1960. There have been only nine autumn records, all of which have fallen between August 4th (1949) and October 12th (1975). More surprisingly Smith referred to one at Belvide on January 7th 1935 and another was reported from Leamington Spa in the winter of 1975, whilst in June 1973 one was seen at a nest at Malvern.

Waxwing
Bombycilla garrulus

Erratic winter visitor, usually scarce but variable in number.

The Waxwing is a regular migrant to western Europe from its Scandinavian and Siberian breeding grounds and a few birds reach Britain each winter. These often wander inland in search of food and some may penetrate as far as the West Midlands. However, it is better known for its periodic irruptions, or invasions, when many thousands move westwards eating their way across the countryside. In such years it often occurs in considerable numbers and large flocks may descend on an area, feed for a few days and then move on. At such times birds will visit almost any habitat with berries. Haws, hips, mistletoe and rowan berries are the favoured indigenous fruits, but birds will regularly visit parks and gardens to devour firethorn, cotoneaster and other exotic berries.

Prior to 1929 there were 20 records as follows: 1827, 1830, 1835 (two), 1845, 1848, 1850 (two), 1859, 1860, 1868, 1882, 1892, 1893, 1896, 1901, 1913, 1919 and 1922 (two). In many cases actual dates and numbers are not known, but birds were said to be very common about Burton-on-Trent in 1827, 1835 and 1850 (Smith) and the same author referred to an invasion in 1913/4, although there was in fact only one record in this Region.

During the period 1929–78 there were 152 records involving some 1,272 birds. However, these include the major invasions of 1965/6, when there were 118 records (78 per cent) involving perhaps 761 birds (60 per cent), and 1970/1 when there were 35 records of some 264 birds. In addition, minor invasions occurred in 1946/7, when there were 13 records involving 38 birds, 1957/8 with 11 records and 27 birds and

1974/5 with five records and 28 birds.

Otherwise no year has produced more than 20 birds. Waxwings are erratic in appearance and a good year in one area is not necessarily good in another. For example, 1956/7 and 1959/60 brought minor invasions to the London area (Chandler and Osborne 1977), yet in the West Midlands there were none in the first of these two winters and only one bird in the second. Overall in this Region, birds were recorded in twenty-five of the fifty years 1929–78, but with more observers in recent times there have been fewer blank years. Since the war the longest period devoid of records was the five years 1952/3 to 1956/7, since when only the winters of 1960/1, 1964/5, 1968/9 and 1976/7 have been blank.

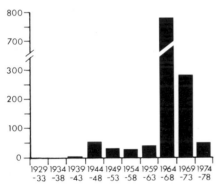

Five-yearly Distribution of Waxwings, 1929–78

The earliest autumn record came from Moseley on September 22nd 1971, but usually birds do not appear until mid-November and December is the main month of arrival. In most years the number of fresh arrivals then slackens in late December and early January, but builds up again to its main peak during February and early March. This suggests that birds might pass through the Region in December and return again in February, though the February arrivals could equally well be birds wandering further afield in search of food. Few birds are seen after mid-March,

but there have been occasional April records and two during May – from Maple Hayes, Lichfield, on May 5th 1976 and Selly Oak on May 14th 1979, when five were seen. During major invasions, such as those of 1965/6 and 1970/1, the pattern may be somewhat different. The initial build-up of birds is still typically in December, but it may continue well into the new year. In such years, though, it is difficult to know which are fresh immigrants to the Region and which are simply birds moving around.

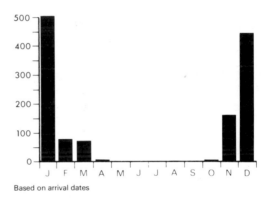

Based on arrival dates

Monthly Distribution of Waxwings, 1929–78

Waxwings are frequently seen in urban and suburban situations, especially gardens. Indeed about two-thirds of all records have come from localities within built-up areas. In good years birds regularly occur in small flocks up to 20, there have been one or two records of flocks of 30 and the most reported were 40 at Rugeley from December 19th 1970 until January 2nd 1971 and 50, at Rugeley again, in December 1965. Most birds are seen just for a day or two, but so long as their food supply subsists they may remain for several weeks. At Four Oaks birds were present from January 23rd to March 23rd 1971 and at Penkridge from January 21st to March 23rd 1975, although in the latter case numbers fell during this time from 23 to six.

66. The Dipper is a bird of the unpolluted, fast-flowing streams that occur particularly in northern and western districts. *A. Winspear Cundall*

Dipper
Cinclus cinclus

Resident, not scarce but local.

The Dipper is very much a bird of fast-flowing, boulder strewn streams and rivers with shallow, gravelly reaches in which it feeds. Such watercourses are largely confined to the moorlands of north Staffordshire and the undulating country west of the Severn and these are the two main breeding areas. It is highly territorial, even in winter (Hewson 1967), and can often be seen in the vicinity of weirs, waterfalls, bridges and mills. However, it does migrate up and down tributaries in response to the main river level.

Smith said the Dipper abounded in the nesting season on the Hamps, Manifold,

upper Dove, Dane and many of their tributaries and that large numbers bred annually. However, he added that they were scarcer than formerly in the Churnet Valley. Away from the hills he noted that birds were normally present along the middle reaches of the Dove, from whence they wandered irregularly as far downstream as Burton, that a few pairs bred near Stone and also in the Stour Valley, and that it was not uncommon at Weston-under-Lizard, but that it was only infrequently recorded from north-west Staffordshire. He also observed that many retired from the hills to lowland rivers and pools in winter, though curiously he went on to say that they were little affected by severe cold and even in the most arctic seasons would repair to the open water of rapid streams and rivers when those of a more sluggish nature carried too much ice. Possibly the movements that Smith observed were in response to low water-levels in autumn and that birds returned to their upland streams again with the onset of winter rains. Lord and Blake (1962) said the Dipper bred in the northern districts of Staffordshire, but only very sparingly in the Cheadle and north-west districts. Apart from the middle reaches of the Dove, the *Atlas* surveys (1966–72) reinforced these earlier distributions and breeding was confirmed in eleven 10-km squares and suspected in one more. These embraced all of the moorlands and most of the Central Staffordshire Plateau as far west as the Potteries, but excluding the urban area. Counts along the River Dove have consistently revealed about one pair per 1·5 km during the breeding season. Speaking of Derbyshire, Shooter (1970) thought the number breeding on gritstone streams varied with the severity of the preceding winter, whereas those on the limestone streams were limited more by territorial space and this is probably equally true of Staffordshire.

For Worcestershire, Hastings (1834) said the Dipper was of infrequent occurrence and Lees wrote in 1856 that Sapey Bridge was the only site in the county where he had encountered the "Water Ouzel", though Tomes (1901) said it was by no means scarce on the tributaries of the Severn and Teme. Harthan concluded that it had increased, but was a local bird which could be found on all the streams joining the River Stour above Kidderminster, along Dowles Brook where there were about ten pairs (*cf.* seven pairs in 1969), on the brooks between Bewdley and Holt Fleet, further south on the Teme and its tributary Leigh Brook, and in the extreme south-east where an isolated pair nested on the River Isbourne. In all he thought there were not more than thirty pairs in the county. The *Atlas* surveys found birds to be still present in all these situations, with breeding confirmed in nine 10-km squares. The Dipper's stronghold in Worcestershire is the Teme Valley and 57 km of the river were subject to a special study in September 1979 for the Worcestershire Nature Conservation Trust (J. J. Day *in litt*). Day found the overall Dipper density to be 0·5 birds per km, but this varied from 0·05 per km on the least suitable lower reaches to 0·7 in the reaches above Knightsford and 0·9 per km in the upper 16 km. Shallow water was found to be most important and the presence of tributary streams had some bearing on distribution. Of forty-two tributaries, sixteen were known to hold Dippers or have suitable habitat and a further fifteen could support the species. Only a quarter of all territories lacked a tributary and these all held extensive shallows. It seems that most birds hold breeding territories on the tributaries, where suitable nest sites are often more numerous, but leave them for the parent river in summer or autumn when the water level is low. Indeed Day concluded that the actual population of the Teme itself is probably less than ten pairs, mostly in the upper reaches, but that the river fills an extremely valuable role in the yearly social life of the Dipper.

Warwickshire has few streams and rivers suited to this species, apart from very limited stretches of the Stour and its tributaries where they flow from the Cotswold scarp and one or two reaches on rivers like the Alne where they leave the Birmingham plateau. Tomes (1904) knew of a few records from the Alne, Leam and Sherbourne Brook, whilst Norris quoted several isolated records, mostly in winter, from the Birmingham area, where birds were seen at Handsworth in 1882, Hay Mills in 1894/5, Sutton Park in 1914, Edgbaston Park in 1934 and Northfield in 1936, together with one in July 1936 near Willoughby. Overall, though, he described it as only an occasional visitor that had bred on the River Alne in 1937 at least. The Dipper may be better established than this. Breeding still occurs on the Alne in the vicinity of Wootton Wawen, whilst in the south of the county pairs have recently bred in the vicinity of Stourton (R. H. Smith *pers comm*) and at Tredington, whilst there is a regular nesting site at Burmington (J. A. Hardman *pers comm*). This latter site was not recorded in the *Atlas* surveys, but fieldwork did reveal the presence of birds on streams draining from the East Warwickshire plateau between Coleshill and Bedworth.

Severe winters and droughts have some impact on the population, but there appears to be little long-term change. Sharrock assumed 15 pairs per 10-km square, but some squares in this Region have insufficient habitat to support such a density. However, Lord and Munns' estimate of 80 pairs may have been somewhat conservative and a range of 100–150 pairs seems more likely. In winter birds sometimes penetrate high up the tributary streams and in recent times there have been records from outposts like Bittell and Halesowen. Since 1975 birds have also been seen in winter on Cannock Chase and a summer record in 1978 gives rise to the hope that the species might one day breed there.

Wren

Troglodytes troglodytes

Widespread and abundant resident.

The Wren is probably the most common of all our nesting birds and there is nowhere in the Region, from gardens to woods or riverside to moor, that it cannot be seen energetically exploring dense or rank vegetation, exposed tree roots and rock crevices in its search for insects. In particular it favours woods and the overhanging banks of rivers and streams.

Smith said the species was universally distributed and common, being found even on the moors at all seasons. Harthan called it common everywhere, though not often seen in town gardens, and Norris added that it was a fairly common and widespread resident. Today it is undoubtedly a regular garden visitor in most built-up areas, where it often nests in unlikely situations such as old jackets hanging in garden sheds, but otherwise these simple statements still hold good and during the *Atlas* surveys (1966–72) breeding was confirmed in every 10-km square bar one, where it was probable.

Smith also noted that even on the moors Wrens breed amongst heather, where it overhangs banks and quarries. Though still true, moorland is the least suitable habitat and in 1978 460 ha held only 21 pairs, most of which were in scrub woodland. Gorsey heath is more suitable and in 1967 89 ha in the Brocton Field–Sherbrook Valley area of Cannock Chase held 28 pairs. In 1978 the Wren was second only to the Blackbird in abundance on the farmland CBC plots, accounting for 8 per cent of all territories at a mean density of 21 pairs per km^2, whilst it was easily the commonest species in the woodland plots, holding 16 per cent of all territories at a mean density of 148 pairs per km^2. Nationally in 1978 the CBC showed 26 pairs per km^2 on farmland and 80 pairs in woodland, whilst Sharrock suggested a mean density of 3,000 pairs per 10-km square. To estimate the population of such a numerous species is difficult, but it seems likely there are between 130,000–230,000 pairs in the Region.

Hard weather severely depletes the population though, and the above estimate represents a peak level. For example, after the cold winter of 1978/9 the number of territories fell from 43 to 21 at Moreton Morrell, from 35 to 14 at Edgbaston Park and from 46 to just three at Fradley Wood. Fortunately such losses are quickly made up and its recovery after the notorious 1962/3 winter was unmatched by any other common species (Sharrock). To counteract cold weather, Wrens roost communally in many places such as old nests of their own or other species, cavities beneath the eaves of buildings and formerly in haystacks. In fact the largest roost in the West Midlands was of 100 in a haystack at Branston in January 1958.

After a severe winter, the order in which the Wren re-occupies available habitats tells us much about its preferences. Woodland and waterside are colonised first, followed by gardens and orchards, and finally hedgerows (Williamson 1969). Birds are also mobile and there have been recoveries from Lincolnshire and Glamorgan of individuals ringed in the West Midlands, whilst a Nottinghamshire bird was recorded in Warwickshire in 1977. Because of their nest sites, Wrens are rarely ringed as nestlings, but one that was at Malvern in 1928 was found in Hampshire the following October.

Dunnock
Prunella modularis

Widespread and abundant resident.

The Dunnock is quietly ubiquitous and can be seen almost anywhere, anytime, whether in urban gardens and parks, hedgerows, woodland edges and glades, or amongst the gorse, heather and bracken on heath and moor. It is perhaps most numerous in those woods that were once managed as coppice-with-standards.

Previous authors had little to say about the Hedge Sparrow, as it was then known. Smith said that at all seasons it occurred commonly everywhere, even in the more barren districts, and that it was numerous in the low-lying parts of Staffordshire, Harthan dismissed it simply as being as common as the Robin and Norris stated only that, though it was not abundant, it

could be found in most suitable places throughout Warwickshire. Harthan added that it was one of the three species most frequently victimised by the Cuckoo, but Smith considered it a rather uncommon foster parent. However, Glue and Morgan (1972) have since confirmed Harthan's observation.

The *Atlas* surveys (1966–72) confirmed breeding in every 10-km square, but there is considerable variation in density between habitats. Birds are least numerous on moor and heath and most abundant on farmland and in woods. Typical densities are five pairs in 460 ha of moorland in 1978, three pairs in 89 ha of heathland at Cannock Chase in 1967, 25 pairs on 64 ha of farmland at Willey in 1978, where it was the commonest species, and nine pairs in 4 ha at Tocil Wood, again in 1978.

Numbers also fluctuate markedly from year-to-year, but show no long-term evidence of any significant change. During 1972–9 at Moreton Morrell, for example, the number of territories ranged from 11 to 35, with a mean of 22 in 184 ha; at Edgbaston Park between 1966–79 the range was 2 to 21 pairs, with a mean of 10 in 16 ha; and at Fradley Wood during the same period the population varied from 10 to 33 pairs, with a mean for the 40 ha of 21 pairs.

In 1978 the combined CBC woodland census plots showed the Dunnock to be the fifth most numerous species, accounting for 8 per cent of all territories at a density of 74 pairs per km^2. On farmland it was the second most numerous species, behind Blackbird, accounting again for 8 per cent of all territories, this time at a density of 21 pairs per km^2. Interestingly, Williamson (1967) found a density of 28 pairs per km^2 from a sample of twenty-eight Midland farms. The comparative national densities in 1978 were 36 pairs per km^2 in woodland and 18 pairs per km^2 on farmland. Between them, such densities suggest a regional population of 70,000–120,000 pairs. Lord and Munns simply classed the species as abundant, with more than 20,000 pairs.

The Dunnock is a much ringed, but very immobile species and the only distant recovery is of a bird ringed at Alcester in December 1967 and recovered at Malton (North Yorks) in April 1968.

Alpine Accentor
Prunella collaris

One record.

Norris recorded one that was shot near Ettington a few years before 1904, though neither Tomes nor he could obtain definite data. The record did, however, gain acceptance in *The Handbook* (Witherby *et al* 1938).

Robin
Erithacus rubecula

Widespread and abundant resident.

This best known and most loved of British birds has endeared itself to man through its confiding ways. Although common in well-timbered hedgerows, gardens and parks, the Robin is primarily a bird of woodland. Both broad-leaved and coniferous woods are frequented, but it seems to prefer a coppiced understorey.

Smith declared the Robin to be universally common, but more frequent in the lowlands than on the bare hills, Harthan described it as abundant all the year round and Norris said it was a common and widespread resident. There has been no subsequent change in this widespread distribution and the *Atlas* surveys (1966–72) confirmed breeding in every 10-km square.

Like many sedentary small birds, Robins are susceptible to hard winters and these cause short-term fluctuations in their population, but there is no evidence of any long-term change. During 1966–79, for example, the annual number of pairs in 40 ha at Fradley Wood fluctuated from 19 to 44, with an average of 27, yet there were 43 pairs in 1967 and 40 in 1978. Likewise the population in 16 ha at Edgbaston Park ranged from 13 to 33 pairs, with an average of 20, but showed long-term stability with 22 pairs in both 1966 and 1978. Over the shorter 1972–9 period, the population on 184 ha of mixed farm and woodland at Moreton Morrell varied from 23 to 36 pairs, with an average of 28, though again there was little long-term change, with 24 pairs in 1972 and 23 in 1979. On average during these periods the Robin ranked second in abundance at Edgbaston Park, after Wren; third at Moreton Morrell, after Blackbird and Wren; and fourth at Fradley Wood, after Willow Warbler, Wren and Blackbird, which is a pattern typical of many woodlands in the Region. Indeed in 1978 the Robin was the commonest species in the 13 ha of New Close Wood and averaged over all CBC plots it was second in abundance to the Wren in woodland and sixth in abundance on farmland, with densities of 115 and 19 pairs per km^2 respectively. The comparable national densities in 1978 were 74 and 19 pairs per km^2. Such densities point to a regional population of 90,000–120,000 pairs. Lord and Munns merely placed the population in excess of 20,000 pairs.

The breeding population is not entirely static, since in autumn Robins apparently foresake upland woods (Yalden 1979) and Smith remarked that in autumn and winter birds arrived in the heart of large industrial centres. Certainly more can be found in urban parks and gardens in winter and there is some evidence that movements are not always local. A continental Robin *E. r. rubecula* was obtained near Malvern on December 3rd 1940 (Harthan) and another immigrant bird that had been ringed in Denmark was caught at Knowle in November 1976. Conversely some birds emigrate and a nestling ringed at Cheadle in 1913 was later recovered in Gers (France), a juvenile ringed at Sutton Coldfield in July 1964 was subsequently found in Spain and one at Packington in September 1975 had reached Weymouth (Dorset) two months later.

Nightingale
Luscinia megarhynchos

Summer visitor, not scarce but confined to the Severn and Avon valleys.

The West Midlands has long been astride the north-westerly limit of the Nightingale's range, but there is no doubt that the species was formerly more widespread and numerous than it is today. Better known through its reputation as a songster than by observation, the Nightingale is a bird of dense undergrowth in woods and commons, thorn thickets and overgrown hedges. It prefers pedunculate oakwoods with a rich ground flora, especially those with standard oaks above hazel coppice. Such woods are common in the Severn and Avon valleys. Stuttard and Williamson (1971) showed that coppicing on a twelve to twenty-five year rotation was most suitable, with the five to eight year old stools providing the optimum habitat. In 1924 nearly 27 per cent of the Region's woodland was coppiced, but by 1947 this figure had fallen to less than 10 per cent and by 1965 it was virtually nil. This fundamental change in woodland structure has probably

67. The West Midlands straddles the limit of the Nightingale's British range and this magnificent songster is now largely confined to the woodlands of the Severn and Avon valleys. *M. C. Wilkes*

been more significant in the Nightingale's decline than has the actual loss of woods, copses and hedges through felling and grubbing up. Simms (1978) also observed that the species was commoner in valleys below 100 m or so, possibly because of the higher summer temperatures, and this is also true in the West Midlands.

Historical records point to a steady southerly withdrawal. In Staffordshire, Smith recorded breeding in the Tutbury and Burton-on-Trent areas in the mid-nineteenth century and at Weston-under-Lizard in 1871. Even as late as 1904 it was still breeding in the Dove Valley north of Uttoxeter. By 1951 Norris was still able to record the species to the west of Stafford, but by 1962 Lord and Blake could report it as nesting only irregularly in the Enville district. For Warwickshire, Norris (1947, 1951) stated that it was irregular and rather local in the north, but common and fairly widespread in the south and east. Tomes (1901) said that it was plentiful in the valleys of the Severn, Avon and Teme and such parts of their tributaries as run through low, fertile places. Harthan, however, noted it was uncommon west of the Severn and most abundant in the low hills from the Lenches through Feckenham to Bromsgrove. He also quoted several specific declines. By the time of the *Atlas* surveys (1966 72) breeding was confined to the Severn and Avon Valley sub-regions, where birds were present in most squares. Overall, breeding was confirmed in twelve 10-km² squares (16 per cent) and suspected in a further sixteen (21 per cent).

As in Harthan's day, the Nightingale's regional stronghold is still the cluster of small woodlands centred on Himbleton, where there are probably half-a-dozen pairs in most woods and up to twice as many in some. However, the largest single "colony" appears to be in Warwickshire, at Wappenbury Wood, where there have been up to 18 singing birds in recent years. The Nightingale's numbers vary consider-

ably from year-to-year, but a BTO Census in 1976 revealed 169 singing males of which 75 were at thirty-four localities in Warwickshire, 92 were at forty localities in Worcestershire and two were in the West Midlands County, where the building of a housing estate was delayed while one pair finished nesting. The results of the 1980 census are not yet available, but the regional population is probably some 150–200 pairs, which is at the upper end of Lord and Munns' range of 20–200 pairs.

Nightingales arrive in late April and May and generally announce their presence by their superb song. The earliest recorded arrival was on April 10th (1945) and over forty-four years the average first arrival date has been April 18th, though in cold springs such as 1977 and 1978 no birds may appear until early May. Once singing ceases in mid-June, the Nightingale's unobtrusive habits make observation difficult and this period of its life remains something of a mystery. However, data for eleven years shows the average date of last departures to be August 9th, with the latest record on August 31st (1935).

Bluethroat
Luscinia svecica

Very rare vagrant. Two records.

There are only two records, both of the Red-spotted form *L. s. svecica*. One was killed near Birmingham between 1837 and 1841 (Yarrell 1871–85), but whether this was a spring or autumn occurrence is not known. The other was seen perched on a fence at Arley, Worcestershire, during the second week of April 1944 (*Brit. Birds* 38:155). Neither spring nor inland records are common in Britain (Sharrock 1974), so these occurrences of this rather shy and retiring migrant from Scandinavia were unusual.

Black Redstart
Phoenicurus ochruros

Scarce summer visitor and passage migrant. Very scarce in winter.

The Black Redstart has only become a regular breeding bird in Britain in comparatively recent times. Apart from one or two earlier isolated records, nesting really started with two pairs on coastal cliffs in Sussex in 1923 and was then regular for a few years in Cornwall, but sporadic in SE England. The first London colonisation was in 1926 on a derelict site. A few years later scattered records came from other parts of London, but it was not until the early 1940s, by when there were many bombed sites in Central London to provide plenty of suitable nesting places, that the species really became established (Sharrock).

Consequently there are few historical records for the Region. Harthan referred to one at Malvern in 1884 and Smith quoted just three, a bird at Wombourne in 1900, a possible one at Longton on November 2nd 1911 and one at Hamstall Ridware on October 18th 1915. Otherwise there were no records before 1935. Then, in November of that year and again in March 1936, what was possibly the same bird was seen at Sheriffs Lench. This was followed by an immature or female at Kings Norton Girls' Grammar School on December 9th 1937, which was the first Warwickshire record (Norris), and by passage birds at

Malvern Sewage Farm during November and December 1937 and 1938 and an adult male at Lickey on May 24th 1941.

Though the Midlands had its share of bombing during the Second World War, the Black Redstart failed to penetrate this Region in any numbers to take advantage of what appeared to be an "appropriate niche". That is apart from one pair which nested in 1943, though not on a bombed site, but on the very much intact main building of the University of Birmingham, where one, or possibly two, young were raised. The only other record in that period concerned a bird, probably of this species, that was seen in the ruined centre of Coventry in August 1944. Not until 1947 was the Black Redstart noted again in Birmingham. Then a male sang mainly from the top of the Times Furnishing Company's premises in High Street and the shell of the ruined Empire Theatre from May 20th until June 25th. Singing birds were also noted in the centre of Birmingham in 1948 and 1949, but then not again until 1952 when a pair bred. Breeding also occurred the following year, but the species was apparently not recorded again in Birmingham until 1958, since when birds have been noted most years, though breeding has not always been proved.

In London, as bombed sites were redeveloped so the Black Redstarts moved to power stations and gas works, with breeding concentrated along the Thames and Lea valleys, where there was heavy industry (Meadows 1970). This is similar to the West Midland situation, where birds favour power stations, gas works and other large, sometimes derelict, industrial installations near to water, especially canals. Breeding has occurred at power stations at Hams Hall, Meaford, Nechells and Ocker Hill; gas works at Aston, Coleshill, Nechells and Saltley; Bilston steelworks; the old Snow Hill station and canal basins. Apart from Birmingham, breeding also occurred at

Table 62: Breeding Season Records of Black Redstarts, 1958–78.

Year	Pairs Breeding	Present	Single Birds	Year	Pairs Breeding	Present	Single Birds
1958	—	—	2	1969	2	—	—
1959	—	1	1	1970	—	—	1
1960	—	—	—	1971	2	—	—
1961	—	1	—	1972	1	1	—
1962	—	—	—	1973	1	—	4
1963	—	—	1	1974	2	—	2
1964	1	1	—	1975	1	1	8
1965	—	—	2	1976	5	—	4
1966	1	—	—	1977	6	3	2
1967	2	—	—	1978	6	—	5
1968	—	1	—				

Bilston in most years from 1964–72, at Redditch in 1973, Dudley in 1974, Coventry in 1974 and 1977, when there were three pairs, the Potteries in 1976, Wolverhampton in 1977 and 1978 and Walsall in 1978. In addition, a pair was present in Worcester for much of the summer in 1961 and a nesting attempt failed at Stourbridge in 1972. The table of breeding season records shows the steady consolidation of the species in the Region since 1958.

With its many industrial sites, the West Midlands appears well suited to the Black Redstart's recently acquired habitat preference, but difficulty of access and other problems attendant with birdwatching in a gas works or power station could mean that some birds are overlooked.

Outside the breeding season the Black Redstart is a scarce, but fairly regular passage migrant. Table 63 shows the month of arrival of all birds except those breeding or holding territory during 1929–78.

Most breeding birds leave in late August or September, but there is a small influx from elsewhere in November and December. Most of these birds pass straight through or stay just a few days and only very rarely will one or two winter. However, birds stayed from December 29th 1947 until March 16th 1948 and from January 5th to April 5th 1975. There have been no new arrivals in February, but March brings a small return passage and this is followed by a second peak in May, which presumably involves breeding or potential breeding birds.

Table 63: Monthly Distribution of Arrivals of Black Redstarts, 1929–78.

	Jan	Feb	Mar	Apr	May	Jun	Jul	Aug	Sep	Oct	Nov	Dec
No. of birds	5	—	10	5	12	1	3	2	2	6	16	11

Excludes those breeding or holding territory.

Redstart
Phoenicurus phoenicurus

Widespread and fairly numerous, or numerous, summer visitor.

The *Handbook* describes the Redstart's habitat as old deciduous woods and park-lands, heaths and commons with scattered trees or copses, well timbered gardens, orchards, streams or ditch sides with pollarded willows, open hilly country with loose stone walls and to some extent, ruins, quarries and rocky localities. It seems to dislike shade and prefers open-canopy, broad-leaved woods with glades and a good shrub layer. This variety of nesting situations can be found to a greater or lesser degree throughout the Region, but there are areas which might seem suitable for this bird where it does not occur and Warwick-shire particularly appears least acceptable to its needs.

Today it shows a marked preference for high ground and is commonest in the sessile oakwoods which occur in north Stafford-shire, particularly the Churnet Valley and around the Potteries, on Cannock Chase, in Sutton Park and west of the River Severn, especially in the Wyre Forest. It is less fond of pedunculate oakwoods (Simms 1978), which goes a long way towards explaining its scarcity in much of Warwick-shire.

For Staffordshire Smith said the species was locally plentiful in well timbered low-land districts, though numbers varied greatly from year to year, and it was not uncommon among the hills, nesting in holes in walls and hollows of old trees and stumps up to a considerable elevation. He added that it was very numerous in Dove-dale and other high lying valleys. Harthan said it was common on the Malvern and Lickey Hills and plentiful in both the high country from the Wyre Forest to Tenbury and in river meadows, where it nested in pollarded willows and other hedgerow timber. He added, however, that it was un-common in the Stour Valley and eastern parts of the county. Norris said that before the First World War it was considered to be generally distributed, nesting wherever possible, but since then it had become a noteworthy event to encounter one.

Amongst a number of widely scattered localities mentioned in a special study of the Redstart in 1935, it was said to be well-established on the Bunter sands from Kinver through Enville and Highgate, where it bred in birch trees, plentiful on Cannock Chase and common in Ettington Park and along the Warwickshire Stour Valley. Several pairs were also present be-tween Droitwich and Ombersley, where they probably bred in pollarded willows, in the gravelly country around Coleshill, where they bred in old birches, and in Packington Park, whilst a few pairs were noted nesting in holes in beech and rowan in the Lickey Woods. Lord and Blake (1962) reported the Redstart as breeding in all areas of Staffordshire, but most commonly on the moors, Cannock Chase and in the Kinver–Enville–Highgate area. In 1963 a pair bred in young forestry at Hanchurch and from 1964–6 several pairs occupied nest-boxes erected in the conifer plantations on Cannock Chase (Minton 1970). The *Atlas* surveys (1966–72) largely confirmed the earlier distributions, with breeding confirmed in fifty-three 10-km squares (69 per cent) and suspected in a further seven (9 per cent). The gaps in the distribution were in and around the Conurbation, on the Birmingham plateau south-east of the city and in the Warwick-shire Avon and Stour Valleys.

There is little data on density, but nationally the population fell dramatically from a peak in 1965 to a very low level in 1973. This decline has been attributed to the Sahel drought in West Africa (Win-stanley *et al* 1974). It has certainly been evi-dent in the West Midlands, where the Red-start has been slow to recover its numbers. In Sutton Park there were 40 pairs in 1955, but only 10 in 1980 – and this latter figure was believed to be an improvement on previous years when it seemed the species was no longer present. In the Wyre Forest, counts along the Dowles Brook ranged from 15–20 singing males prior to 1966, but these had slumped to six by 1971 and in 1978 just five pairs were present. The smaller populations at Lickey and Packington have shown more marked fluctuations. Only in its favoured upland sessile oakwoods does the Redstart appear to have recovered its numbers. At Coombes Valley, numbers fell from around 30 pairs in 106 ha to just 19 pairs by 1973, but they had returned to 35 pairs again by 1977 and 1978. In the limestone dales they are quite numerous amidst the drystone walls and ashwoods and 18 singing males were located in 1977 along 14 km of the Manifold Valley between Hulme End and Waterhouses, whilst on the gritstone moors there were 10 pairs in 460 ha of the Swythamley Estate in 1978 (Yalden 1979). Sharrock suggested perhaps 30–60 pairs per 10-km square, but many squares in the West Midlands hold considerably fewer than this and the regional population is probably in the range 1,200–2,500 pairs, which is at the upper end, or slightly above, the 200–2,000 pairs estimated by Lord and Munns.

Redstarts have arrived as early as March 31st, in 1946, 1957, 1963, 1968 and 1975, but April 10th has been the average date for first arrivals over forty-two years and the main immigration usually occurs in late April and early May. Departure takes place in August, with a peak late in the month, and continues through September and over forty-one years the average date for last birds has been September 22nd. However, October records are not rare and the latest reported was on the 31st (1971). Passage birds appear at a variety of localities, especially in autumn, when small parties up to ten may be seen. Thick thorn hedges are a favoured haunt, whilst one or two have even appeared in suburban gardens. The arrival of some passage birds in autumn has coincided with "falls" on the East Coast and it seems likely that a few Scandinavian drift migrants may reach the West Midlands. However, ringing shows only the southerly movement of our resident birds, with a nestling ringed at Cheadle in 1920 and recovered in Basses Pyrenees (France) at the end of the year, and another nestling ringed in the Wyre Forest in 1972 and found in Algarve (Portugal) in September of the same year.

Whinchat
Saxicola rubetra

Fairly numerous summer visitor.

The Whinchat is a summer visitor that has declined markedly this century. Smith said that it occurred throughout Staffordshire, even in the high country, but that owing to the cultivation of former wastes it was scarcer than formerly in many districts. Tomes (1901 and 1904) said of Worcestershire that it was one of the commonest migratory birds, freely nesting in the fields and meadows, and of Warwickshire that it was common and indeed abundant. However, Harthan noted it was very scarce, except in river meadows, and added that none could be found on the Malvern Hills where forty years before it had been common. Norris, too, said a considerable decrease had taken place and described it as scarce and local.

Whinchats feed mainly on insects which they take from grasses and flowering plants, especially umbellifers, and they also nest in mowing grass. Thus, with the spraying of herbicides, cutting of roadside verges and decline in hay-making, it is hardly surprising that they should have disappeared from most lowland areas. Indeed, drainage and earlier grass cutting have even driven them from their favoured river meadows, such as the lower Avon, where there was one pair per km downstream of Pershore in 1942. There have been no recent records from railway embankments, perhaps because these are no longer subject to burning, and in the lowlands the species has even declined on commons. For example, Hartlebury Common held 10 pairs in 1951, but by 1960 there were none.

Today the Whinchat is very much a bird of the uplands, being most numerous on the northern moors and Cannock Chase. Here it frequents open moor or gorsey heath, particularly where there is a combination of rank, tussocky grass or bracken-covered slopes and drystone walls, fences or other suitably low song posts. Recently it has also held territory in young forestry plantations. Away from the uplands, the Whinchat occurs very sparingly on commons or in marshes and rough, unkempt grasslands such as those around the large industrial complexes of the Potteries, Coventry and the Tame Valley. Often it is found near water, as evidenced by recent breeding records from Alvecote, Brandon, Chasewater, Draycote and Ladywalk.

At Alvecote there was a decline from eight pairs on 225 ha in 1959 to six pairs in 1963 and none in 1974. This was the first absence in any breeding season since 1947 and unfortunately there has been only one subsequent record, in 1976. Otherwise information on density comes only from isolated samples. For example, in 1957 there was one pair to every 10 km^2 around Nuneaton and in 1960 six nests were located in 8 km^2 around Newcastle-under-Lyme. In 1968 there were six pairs on 31 ha at Ufton Fields and finally, in 1967, four pairs were found in 89 ha of favoured habitat in the Brocton–Sherbrook Valley area of Cannock Chase. Sharrock suggested 10–20 pairs per 10-km square, but admitted that whilst some squares may have 50 or even 100 pairs, there are probably few holding more than 20 or 30 and the majority have less than half-a-dozen. Thus the widespread distribution revealed by the *Atlas* surveys (1966–72), with breeding confirmed in thirty-eight 10-km squares (49 per cent) and suspected in a further nine (12 per cent), gives a misleading impression of the breeding strength. In all there are probably only nine or ten squares with 20–30 pairs, the remainder having perhaps two or three. This suggests a regional population of 250–400 pairs, which is at the lower end of Lord and Munns' range of 200–2,000 pairs.

Although birds have been recorded as early as March 31st, on Walton Heath in 1908, and April 14th (1952 and 1971), the Whinchat is one of the later summer migrants and over forty-two years the average date for first arrivals has been April 24th. In most years the main influx takes place in early May. The majority of birds leave again in August and September, when parties up to ten or a dozen may be seen, and over thirty-eight years the average date for last birds has been October 4th. Some are quite frequently seen later in October, however, and there were November records in 1968, 1972, 1975 and 1976, with the latest of these at Tillington on November 28th 1975.

The Whinchat has been rarely ringed in the West Midlands, but one that had been ringed as a nestling in Lancashire in June 1953 was found in Staffordshire in April 1955.

Stonechat
Saxicola torquata

Scarce breeding species, passage migrant and winter visitor.

As a breeding species, the Stonechat is certainly less widely distributed today than last century, when its favoured habitat of gorse, heather and bracken covered heaths and commons was considerably more extensive.

68. The Whinchat was once quite widespread, but today it is chiefly a bird of the heaths and northern moorlands, though a few pairs still frequent rough ground elsewhere.

S. C. Brown

Wood (1836) said the species was abundant in many parts of Staffordshire, frequenting extensive moors and furzey commons, but by 1938 Smith said it was mainly seen on migration in spring and autumn or during severe weather in winter. Harthan on the other hand said a few were resident and it was occasionally seen on migration, but the majority were summer visitors from March until October, whilst Norris remarked on the lack of suitable habitat in Warwickshire and described it as an occasional resident and winter visitor, with breeding reported only at intervals of several years. Tomes (1901) said the species frequented not only barren, stony places, but also cultivated fields and roadside hedges and it was said to have been abundant along the Evesham–Winchcombe road in 1916. Harthan said its disappearance from farmland and roadsides was due to changes in agriculture, the tarring of roads and the cutting of verges. As a measure of the Stonechat's decline, there were an estimated 30 pairs at Malvern around 1940 (Harrison 1941), yet none from 1943 until 1973 and only one or two since. Whilst Sharrock blamed the seven cold winters in the period 1939–63 for the currently low population, Magee (1965) concluded that habitat change and disturbance had been more important. In the West Midlands both factors have been significant, though Blake (1962) did not think habitat loss the prime cause of decline. Rather he suggested that the presence of the Stonechat as a breeding species in the West Midlands was dependent upon a surplus coastal population spreading inland and that such a situation might follow a series of mild winters or high breeding success. This certainly seems to accord with the subsequent pattern of occurrences.

None the less, there was substantial loss of habitat with the ploughing of commons during the Second World War and this was followed by an exceptionally hard winter in 1946/7. In consequence, since 1948 Stonechats have bred in only seventeen out of thirty-two years. There were no breeding records at all from 1955–9, during which time there were three cold winters, and only one in the eight years immediately following the severe winter of 1962/3. Conversely, it is worth recording that the number of breeding pairs, and incidentally passage and winter birds, increased markedly during the drought years of 1975 and 1976, when the summers were good.

In recent times breeding has been most consistent on the heaths and commons around Malvern, Kidderminster, Enville and Cannock Chase, but nesting has also occurred on the moors and on rubbish tips and industrial wasteland within the Conurbation. During the *Atlas* surveys (1966–72) breeding was confirmed in just two 10-km squares and suspected in two more, and the current regional population is probably only 3–10 pairs, which is slightly better than Lord and Munns' evaluation of up to three pairs.

From autumn to spring the Stonechat is more numerous and widespread, being seen not only on its favoured heaths and commons, but amongst rough, derelict and often marshy ground, or around the margins of reservoirs and gravel pits. During this period it might in fact be found anywhere where posts or tall, dead herbage, such as umbelliferous plants, afford perches from which it can drop onto insects. Autumn birds begin to arrive in late September, but the main arrival is in October. Some pass straight through, but others linger until driven on by hard weather. Some may even remain throughout the winter and stays of three months or more have been by no means unusual, especially during the mild winters of the 1970s. None the less numbers generally decline as the winter progresses. There is a minor peak again in late February and early March, which presumably denotes a very small return passage, but this does not

Table 64: Monthly Distribution of Arrivals of Stonechats, 1929–78.

	Jan	Feb	Mar	Apr	May	Jun	Jul	Aug	Sep	Oct	Nov	Dec
No. of birds	80	70	45	13	10	12	6	11	66	188	101	94

note: excludes breeding records.

show when averaged over a whole month. The analysis of the arrival dates of 696 birds during 1929–78 is shown in Table 64.

As Blake (1962) observed, these autumn and winter visitors appear to have no connection with the small breeding population, which does not return to its territories until late March or April and it would be interesting to know their origin.

Most passage or wintering birds appear singly or in twos or threes. Indeed parties of seven at Malvern on October 8th 1972, ten at Perton in February and March 1975 and ten in Sutton Park on October 3rd 1978 were larger than average. However, there have been two exceptional concentrations. Harthan recorded a flock of 50 collecting together at Malvern on October 28th 1938 and up to 30 gathered at Perton during hard weather in late January and early February 1976 to feed on freshly-turned earth. There has been one autumn and one winter record of birds in urban gardens.

Wheatear
Oenanthe oenanthe

Fairly numerous summer visitor and passage migrant.

As a breeding species, the Wheatear today is largely confined to the moorlands of north Staffordshire, where it is a characteristic bird of drystone walls, rocky outcrops and scree slopes. In Derbyshire, Frost (1978) found it much more numerous on Carboniferous limestone than Millstone

grit and this is the impression gained in Staffordshire as well, though detailed information is lacking. Breeding away from the moors is now mostly very localised or sporadic, but this has not always been so. Indeed hilly sheep pastures, with short turf and rabbit warrens, were once a favoured habitat, but much of this marginal land was ploughed up during the Second World War.

Smith mentioned that a few pairs nested on town slag-heaps and wastes, and cited breeding on a blast furnace tip and within 60 m of tipped molten slag in the Potteries. Thirty years later, in 1967, Wheatears were again reported as breeding on slag heaps and industrial wastes at six sites in the Potteries. Perhaps such occurrences always take place, but if so they have been overlooked both before and since, as there have been no subsequent records since 1970. Until the early 1950s breeding occurred annually in the Kinver–Enville–Highgate Common area, but there have been no nesting records from here since 1957. Its disappearance from here is probably due to a decline in the rabbit population through myxomatosis, and afforestation with conifers. Harthan said a few were still resident on the Malvern Hills, where they still maintain a tenuous hold today, and at Wolverley, though they no longer bred at former haunts on the Lickey Hills and Bredon Hill. There are few references to the Clent Hills, but breeding occurred here in 1950 at least. Norris could trace no breeding records for Warwickshire, but only one year later, in 1948, there was an unsuccessful breeding attempt at Bedworth. Following another failure in 1949, breeding was then reported from Coventry in 1965 and suspected at Kingsbury in 1968 and Draycote in 1975. Sporadic breeding also occurred at Seighford Aerodrome in 1960, Whittington Sewage Farm in 1971 and 1972, Pelsall in 1972 and Dudley Golf Course in 1977. During the *Atlas* surveys (1966–72) breeding was confirmed in seventeen 10-km squares (22 per cent) and suspected in a further eight (10 per cent). Birds were also recorded as present in many other squares, but recently-fledged juveniles sometimes appear away from the breeding grounds as early as July. It would be optimistic to apply Sharrock's assumed density of 40 pairs per 10-km square to more than the five or six squares in north Staffordshire that embrace its typical habitat. Doing so would still give a regional population of some 200–250 pairs, which is slightly above the range of 20–200 pairs suggested by Lord and Munns.

Wheatears are regular and widespread on passage, when they can be seen in such diverse habitats as short grasslands, ploughed fields, sand and gravel pits, reservoir margins, derelict or rough ground, golf courses, and parks, playing fields or allotments well within the built-up area. The species is often faithful to particular areas and has been recorded annually in the same field at Aldridge for over thirty years.

It is one of the earliest spring migrants and the first birds invariably appear in March. Indeed there have been three late February appearances, in 1968, 1977 and 1978, with the earliest on the 23rd (1968). Over forty-one years the average date for first arrivals has been March 21st. The main passage, though, is in April and early May, when some of the larger, more colourful birds have been ascribed to the Greenland race *O. o. leucorrhoa*. Birds that were trapped on April 20th 1957 and in early May 1958 were positively identified as being of this race. Especially favourable weather conditions in spring may lead to a concentrated passage. For example, on April 3rd 1966 there were 24 at Great Barr, 20 at Brandon, 12 around Great Wyrley and 11 at Hollywood Golf Course. The largest spring parties each contained 30 birds, at Bartley on May 12th 1963 and in snow at Brandon on April 1st 1973.

Wheatears begin to vacate their breeding grounds in July, but passage in autumn continues through to October or even November. It reaches its peak in late August and early September and the average date of last birds over forty-two years has been October 11th. There have been seven November records, however, with the latest in recent years on the 13th (1963 and 1976), though Smith referred to some that were still to be seen at Etruria on November 28th 1913. Autumn passage is more protracted and, despite the addition of juveniles to the population, parties are small and seldom exceed a dozen. The largest concentrations at this season were 30 at Atch Lench on September 19th 1962 and 50 at Bittell on August 16th 1952. Prior to 1954 one or two birds of the Greenland race, *O. o. leucorrhoa*, were fairly regularly noted in autumn, but there has been only one subsequently, in 1963.

Few Wheatears have been ringed or recovered in the West Midlands, but a nestling ringed on Skokholm (Dyfed) in May 1948 was recovered at Coventry in April 1949, a juvenile on Fair Isle in July 1955 was found near Stone in September 1955 and one ringed at Chasewater on April 11th 1958 was recovered at Voe (Shetland) on May 24th 1959.

There was an unusual report of a single bird on the Malvern Hills on January 29th 1967.

White's Thrush
Zoothera dauma

Very rare vagrant. Two old records.

Both records come from Warwickshire and date from last century. One concerns a bird killed at Welford in 1859 and the other a bird shot at Packington in 1895 (Norris). Despite there being many more bird-watchers, occurrences in Britain of this Asiatic vagrant have probably been fewer in number over the last fifty years than before it.

Ring Ouzel
Turdus torquatus

Scarce breeding species and passage migrant.

The Ring Ouzel's distribution in Britain is closely related to high ground and thus the only suitable nesting habitat in this Region is to be found in the northern extremities of Staffordshire, where a few pairs nest each year on the open moors. Smith stated that the high north Staffordshire moorlands formed one of the main homes of this bird in Britain and on the upland ground between Leek and the upper reaches of the River Dane, Ring Ouzels were sometimes numerous. This is certainly overstating the case, for other upland areas of Britain are equally suited to this species, if not more so.

The Ring Ouzel has certainly declined in Britain and Ireland during the last hundred years (Sharrock), though Smith's only comment to this effect was that it formerly nested in the Churnet Valley and that it no longer bred in Dovedale. In 1962 Lord and Blake's assessment was that the species was no longer as common as Smith had suggested and that it now only bred sparingly on the northern moorlands. This view was supported by the *Atlas* surveys (1966–72), which showed it to be breeding in only the three northernmost 10-km squares. With the most optimistic estimate of a maximum of 20 pairs per 10-km square (Sharrock), the regional population would be around 60 pairs, but Lord and Munn's estimate of 3–20 pairs seems more realistic.

Frost (1978) observed that in Derbyshire Ring Ouzels are currently very rare on limestone. This is equally true of Staffordshire, where the species is almost, if not entirely, confined to the Millstone grit country. Here it inhabits the steep-sided valleys, or cloughs, with their small streams, rocky outcrops and upland pastures, amongst which it forages alongside the Blackbird for earthworms and other invertebrates. Favoured localities are Blackbrook Valley, Gib Torr, Goldsitch Moss, the Roaches and Swallow Moss. Human disturbance may have caused it to foresake the limestone dales, especially Dovedale, and this could ultimately threaten its continued presence on the moors as well. After breeding, birds sometimes gather together to feed on bilberries or rowan berries and in 1970 parties of 100 and 60 respectively were noted in the Dane Valley on July 20th and the Blackbrook Valley on September 12th.

Though confined to the North Staffordshire Moors as a regular breeding bird, Ring Ouzels nested in Warwickshire on one occasion, with the *Zoologist* for 1848 recording that a pair was shot and the nest and eggs taken at Pinley about that time (Norris). It is also on record that a pair nested on the Malvern Hills in 1887, successfully rearing young, while on two other occasions, in 1849 and 1877, nesting occurred, but mishaps prevented hatching (Harthan).

As a passage bird the Ring Ouzel has been recorded annually since 1959 and it might turn up almost anywhere, though it is most regularly seen on or near uplands like Cannock Chase, the Clent Hills, the Lickeys and the Malverns. Spring passage often begins in late March, with the earliest record on the 16th (1969), and continues through to early May, with a distinct peak in the first half of April. Most spring records refer to males, which generally appear before females, and usually involve only single birds. However, parties of 15 and seven were noted at Clent in 1977 and 1978 respectively. Over twenty-two years the average date for first arrivals has been April 2nd.

In autumn birds are more widespread, sometimes even appearing in suburban gardens. However most gather in places like Cannock Chase and the Malverns, where they feed on elder or rowan berries, and there are few years when these favoured sites are not visited. Parties of a dozen or so have been seen several times and up to 35 were to be seen at Malvern in 1970. Passage may begin in late August and last until early November, with the latest records in mid-month in 1947 and on the 19th in 1976. It reaches its peak, however, between mid-September and mid-October and over sixteen years the average last departure date has been October 18th. In 1966 there was a major influx of continental Ring Ouzels to the East Coast and a party of 15 on Cannock Chase in October of that year almost certainly contained birds of continental origin. However, it is possible that late birds in other years have also come from the continent, as the autumn passage is rather protracted for that of British birds (Durman 1976). An unusual record concerned a bird in Twiland Wood on July 24th 1974.

Blackbird
Turdus merula

Abundant resident, passage migrant and winter visitor.

This most familiar of birds is an abundant breeding species which can be found in all parts of the Region from the bleak moors right into the very heart of Birmingham. The Blackbird is very adaptable and during the last two hundred years it has spread from its traditional woodland haunts to colonise hedgerows, moorland fringes and particularly suburban gardens, parks and town squares (Simms 1978).

Both Smith and Harthan commented on the Blackbird's abundance and the former said it was universally common at all seasons, adding that in the uplands it nested beneath ledges, amongst rocks and heather, or in clumps of rushes, often sharing the same habitat as the Ring Ouzel. He also observed, however, that very few remained on the northern hills in winter. Norris thought Warwickshire to be outside the normal range of birds from northern Europe and considered the species to be more truly resident than most, adding that it may stay for a life-time in a comparatively small area.

So widespread is the Blackbird that it took only the three years of the WMBC *Atlas* (1966–68) to establish breeding in all but one 10-km square. Just how numerous it is can be gauged from the fact that it was the commonest farmland species in the 1978 regional CBC, accounting for 14 per cent of all territories, and third commonest in woodland after Wren and Robin, where it accounted for 10 per cent of all territories. It outnumbered the Song Thrush by three to one. Overall, in 1978, the Blackbird's regional densities were 37 pairs per km^2 on farmland and 89 pairs in woodland, whilst its comparative national densities were 33 and 73 pairs per km^2 respectively. In suburbia, though, densities can reach 250 pairs per km^2 (Batten 1973) and

Sharrock postulated an average of 2,000 pairs per 10-km square. These figures suggest a regional population of some 120,000–160,000 pairs. Lord and Munns simply categorised it as abundant, with over 20,000 pairs. Blackbirds are less affected by hard weather than many species and most of our breeding birds are sedentary. After the 1962/3 winter its CBC index dropped by only 18 per cent, whereas that of the Song Thrush fell by 57 per cent. Consequently the Blackbird has a relatively stable population and the 40 ha of Fradley Wood, for example, held between 25–28 pairs in every year from 1970–9, except for 1976 when there were just 19 pairs.

Blackbirds feed particularly on earthworms, though in late summer and autumn they also take berries, especially haws, or feed on fallen fruit. Their habit of foraging amongst the leaf litter in hedge bottoms or on the woodland floor makes counting difficult and in common with other familiar species they often go unreported. Contrary to Norris' opinion of thirty years ago, however, ringing recoveries have since indicated that continental Blackbirds do winter in the West Midlands, coming from southern Scandinavia and the Low Countries. There has also been one breeding season recovery in central Germany. The first recovery of a bird from Finland was in 1969, but since then there have been several. This accords with the national trend of ringing recoveries and the colonisation of Finland by the Blackbird (Spencer 1975). Recoveries during the autumn migration indicate that many birds pass along the southern coast of the North Sea *en route* to the West Midlands. Once in their wintering area there is little evidence of their moving far, as most recoveries from during the same winter have been within 100 km.

Whether immigrant Blackbirds are faithful to a wintering area is difficult to assess, since resident birds use the same roosts. However, there is some evidence that birds do change their wintering areas. In particular, five Blackbirds have now been recovered in France in subsequent winters, and in three cases these were in the same month as they had previously been in the West Midlands. Some individuals, though, winter at the same site in subsequent years and one bird was caught at Hallow in both the 1975/6 and 1976/7 winters and then recaught in Germany in June 1977.

Dusky Thrush
Turdus naumanni

One record.

One visited a garden at Majors Green, Worcestershire, to feed on windfall apples and apple peelings in a compost heap during severe weather. It was seen on February 17th to 19th, 27th and 28th and March 18th, 19th and 23rd 1979. This was only the sixth British record of this rare Siberian vagrant (*Brit. Birds* 73:521).

Black-throated Thrush
Turdus ruficollis

One record.

A first-winter male was trapped at a thrush and finch roost in north Staffordshire on November 26th 1978. This was only the ninth British record of this rare vagrant from central Asia, but the date was typical (*Brit. Birds* 72:535).

Fieldfare
Turdus pilaris

Numerous or abundant passage migrant and winter visitor. Bred from 1974–7.

Chattering flocks of Fieldfare are a familiar feature of the countryside in winter. Harthan called it a regular winter

69. Large numbers of Fieldfare come from Scandinavia to winter in the Region, where they feed avidly on berries, rotting apples and soil invertebrates. *M. C. Wilkes*

visitor and Norris described it as a regular and generally distributed winter migrant in varying numbers.

There have been several September records of Fieldfare and exceptionally even August ones, the earliest of which concerned 11 seen near Leamington on August 12th 1944, but over forty-three years the average first arrival date has been October 4th. Ringing recoveries have shown that our wintering birds originate from Scandinavia, with breeding season recoveries from Sweden, Norway (two) and Finland, whilst one recovered in the Ukraine in a subsequent winter may indicate that Russian birds reach the West Midlands too. From the scanty data available it seems that both Fieldfares and Redwings may follow the North Sea coast on autumn passage.

The majority of Fieldfares do not arrive until well into October, usually slightly later than Redwings, and immigration reaches its peak in November. As with the Redwing, many of the early migrants appear to pass straight through the Region, or pause but briefly, and by December numbers have already begun to decline. This decline then continues through the winter until return passage in spring. The distribution of all reported flocks by date of arrival during the period 1929–78 is shown in Table 65.

None the less, large numbers remain throughout the winter. From the time they arrive until the crop is exhausted, Fieldfares congregate in thickets, hedges or orchards where they avidly devour berries

Table 65: Summation of all Reported Flocks of Fieldfares by Month of Arrival, 1929–78.

	Sep	Oct	Nov	Dec	Jan	Feb	Mar	Apr	May
No. of birds	15,100	33,430	48,433	35,540	23,304	20,608	60,235	28,610	750
Percentage	5	12	18	13	9	8	23	11	0

such as haws and elderberry or windfall apples. Indeed 5,000 found food in a Warwickshire apple orchard in 1962 (Simms 1978). From January onwards they turn increasingly to pastures, parklands and grasslands, where they seek out earthworms and other soil invertebrates, which at this time have often been brought nearer the surface by rising temperatures and water-tables. They are wary birds that prefer open fields and when alarmed they make straight for the highest tree-tops rather than going into low cover. Although the Fieldfare is better able than the smaller thrushes to withstand hard weather, this will often bring them into country and sub-urban gardens, especially where there are berries or apples. Numbers of both Fieldfare and Redwing vary from year-to-year as the following graph shows.

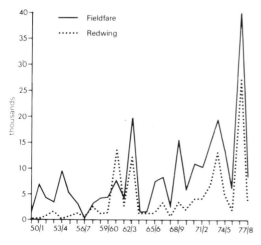

Summation of all Fieldfare and Redwing Flocks by Individual Winters, 1949/50–1977/8

Peak years often, but not always, coincide and the Fieldfare is always the dominant species. Over the period 1929–78 the total number of Fieldfares reported was almost exactly twice the total number of Redwings. There is a lack of retraps or recoveries near to ringing sites in subsequent winters, but several recoveries from Spain, Italy, Portugal and France in December and January indicate that the species does not remain faithful to one wintering area and this may partly explain its fluctuating numbers.

Flocks are most numerous in autumn, but tend to be largest in spring. The majority of records refer to small gatherings of 200–300 birds, but up to 600 are often seen together and larger parties up to 1,000 are by no means unusual. The largest concentrations to be recorded were 4,000, at Wellesbourne on December 6th 1953 and again at Coombes Valley in March and April 1973, and 15,000 at Coombes Valley in September and October of the same year.

There are also many records of heavy passage, especially in autumn, such as 1,000 an hour moving SW over Erdington on October 31st 1952 and 4–5,000 moving east in two hours at Wellesbourne on December 6th 1953.

Large roosts of Fieldfares gather in coniferous woodland, in deciduous woods that have a dense understorey of laurel or rhododendron, or in willow or hawthorn scrub. Occasionally birds roost on the ground in long grass or heather. One such was reported in long grass at Enville Common in December 1948. Most roosts are of 300–1,000 birds, but there were 9–12,000 at Sheriffs Lench on November 24th 1962 and an exceptional 25,000 at Brandon in late March 1977.

Smith commented that from mid-February to the end of April, Fieldfares were noticeable on passage and some always occurred in May. Likewise Harthan said late flocks were not infrequent up to the end of April. The spring passage actually reaches its peak in March, but birds are still very numerous and widespread throughout April and in cold springs even into early May. Over forty-two years the average date for last records has been April 30th, but 500 were still present on May 2nd in 1954 and 100 on May 4th in 1958, whilst the latest date recorded was May 17th (1974). However, there were isolated June occurrences in 1940 and 1978 and one at Belvide on July 26th 1975.

Lingering spring birds always raised speculation that the species might breed in the Region and, indeed, this was proved to be the case in 1974, when a pair in north Staffordshire reared four young and another pair may also have bred. At the time, this was the furthest south the species had bred in Britain, but it was in line with the increasing number of breeding occurrences since the first pair was found nesting in the Orkneys in June 1967. This breeding success in Staffordshire was repeated in 1975, 1976 and 1977, with at least one pair nesting each year and probably two in 1975. Unfortunately, though, none have been reported since.

Song Thrush
Turdus philomelos

Abundant resident, passage migrant and winter visitor.

The Song Thrush can be found almost anywhere where there are trees or bushes, but woodland with a good shrub layer, hedge-rows, thickets, parks and suburban gardens are its favourite haunts. Indeed, it is one of our more familiar birds, on a par with the Blackbird, Robin and Wren, though by no means so numerous as any of them. In winter it tends to foresake woodland in favour of other habitats.

Smith said Song Thrushes could be found in the lowlands throughout the year, but that they could be seen leaving the uplands very early in autumn. He added that such movements slackened in early October and not until late in the month did others appear, travelling mostly by night and mingling their cries with those of Fieldfares and Redwings. Harthan said it was common throughout the year, but also commented that most of the woodland thrushes disappear during the winter and do not return until February, whilst Norris said it was a common and generally distributed resident, but that it was part migratory.

The Song Thrush is still widespread and the *Atlas* surveys (1966–72) confirmed breeding in every 10-km square. For such a well-known species, it may surprise some people that it ranked only ninth in abundance on the regional CBC farmland plots in 1978, where the Blackbird outnumbered it by three to one, and was no higher than seventh in woodland, where it was outnumbered two to one by the Blackbird. In a Birmingham suburban garden, though, it was sixth in abundance from 1965–8. For farmland and woodland the regional CBC densities were 11 and 38 pairs per km^2 respectively. The equivalent national figures were 8 and 24 pairs per km^2. These, and Sharrock's assumed density of 1,000 pairs per occupied 10-km square, suggest a regional population of between 30,000–75,000 pairs. Lord and Munns simply placed the population as in excess of 20,000 pairs.

Nationally the Song Thrush has apparently decreased in the past twenty-five years (Simms 1978), but, apart from losses in hard winters, the recent evidence on its status in the West Midlands is conflicting. On 184 ha of mixed farm and woodland at Moreton Morrell there has been a dramatic decline, from 36 pairs in 1972, 1973 and 1975 to just eight pairs in 1978. At Fradley Wood, though, numbers have been more stable, with 11–17 pairs in 40 ha from 1966–69, when felling and re-planting took place, and 6–11 pairs subsequently. At Edgbaston Park the numbers on 16 ha have actually increased, from four or five pairs during 1966–72 to between seven and ten pairs from 1974–9.

In common with other familiar species, Song Thrushes often go unrecorded and there is surprisingly little data on their migration. However, passage usually occurs between September and November, with a peak in mid-October, and much of the movement takes place at night. Minton (1970a) said that these winter visitors come from the Low Countries or eastern England and that many pass through to destinations further south or west. At Malvern 250 were seen on September 12th 1970 and 220 stayed to feed on rowan berries from September 20th–25th the following year. Apart from these, the largest number ever reported was 400 dispersed about Whittington Sewage Farm on January 18th 1970.

Minton (*op cit*) cited a bird ringed in the West Midlands in mid-winter and recovered the following March in Holland as evidence of the continental origin of winter visitors, but no recoveries have been received since then to confirm this conclusion. Nestlings or breeding adults have been recovered to the south and west during winter, however, with birds getting to such places as Ireland, Wales, Devon, Cornwall, the Isle of Wight, north-west France, Spain and Portugal. Hence West Midlands Song Thrushes are sometimes migratory, although there are also several records of birds breeding locally being re-caught during the winter. In addition, five other birds have been recovered in the Iberian peninsula, but from the dates of

ringing and recovery these may have been birds that changed their wintering area, although it must be remembered that cold weather will cause Song Thrushes to move south.

Redwing
Turdus iliacus

Numerous or abundant passage migrant and winter visitor.

Redwings are common and widespread winter visitors to rural areas throughout the Region. Previous authors all referred to their general and widespread distribution during the winter months or commented on their movements. Smith, for example, said the first flocks almost invariably crossed the northern upland districts without alighting, even when facing a strong westerly gale; Harthan referred to the fact that in mild winters sub-song could sometimes be heard from February onwards, particularly in March and early April; and Norris commented that, as with the Fieldfare, numbers varied from year-to-year, with a good year for one species usually, but not invariably, being good for the other.

The first Redwings are sometimes noted in late September, though the major movements into the area usually occur from mid-October onwards. Over forty-four years the average date for first arrivals has been September 29th, with the earliest record on September 6th (1969). Ringing has shown that birds breeding in both Scandinavia and Russia are involved and, whilst some arrive by day, many can be heard calling as they pass overhead on nocturnal migration. As yet there is no positive proof of Icelandic bred birds visiting the West Midlands. Recoveries in the same winter as ringing show that many birds pass through the Region on their way to winter in Wales, south-west England and Iberia, but others

70.   The Redwing is a common and widespread winter visitor, which often comes to feed on windfall apples in hard weather.   *M. C. Wilkes*

stay and numbers steadily build up during November and reach their peak in December. During this time large flocks can be seen along hedgerows, where they seek out haws, blackberries, elderberries and holly berries. A count in December 1976 showed 40 per km feeding on haws from hedges alongside the M5 motorway. Once they have exhausted this food supply, many move further south and west and numbers in January are only half those of December. Those that remain turn to pastures, parklands, airfields and playing fields to feed on earthworms and soil invertebrates, which at that time of year will have been brought nearer the surface, especially in river valleys, by the rising water-table. Redwings are susceptible to cold spells and hard-weather movements sometimes occur in mid-winter. In extreme conditions most vacate the area, but others will overcome their apparent shyness and come into suburban and country gardens in search of food such as cotoneaster berries and windfall apples. They suffered considerable mortality during the 1946/7 and 1962/3 winters in particular.

During February numbers begin to rise again as return migration builds towards its March peak. Table 66 shows the distribution of all reported flocks by date of arrival during the period 1929–78.

Most have generally departed by March, but a cold spring will delay their departure and over forty-three years the average date for the last birds to be recorded has been April 16th. Exceptionally late birds were noted in early May of 1975 and 1977 though, whilst in 1978 one remained at Edgbaston from May 18th to 30th. An-

Table 66: Summation of all Reported Flocks of Redwings by Month of Arrival, 1929-78.

	Sep	Oct	Nov	Dec	Jan	Feb	Mar	Apr
No. of birds	7,201	15,256	22,035	28,285	13,585	15,550	25,360	5,482
Percentage	5	12	17	21	10	12	19	4

other late bird was seen at Henley-in-Arden on June 12th 1971, but this had a damaged wing, and it is possible that this was the same individual that was seen at Combrook on May 23rd.

Numbers vary considerably from year-to-year, presumably in relation to food supply and weather conditions (see also Fieldfare). Certainly recoveries in winters subsequent to that of ringing indicate that the Redwing is not strongly attached to a particular wintering area. Birds have been found in most continental countries, with individuals going as far south and east as southern Italy, Greece (two birds) and Iran. The nomadic existence of the Redwing is further shown by the fact that re-traps of birds at their ringing site in subsequent winters are rare, although one bird was at the same site in 1973, 1976 and 1978. Redwings were particularly numerous in the winters of 1960/1, 1962/3 before the onset of the protracted cold spell, 1973/4 mainly in autumn, and especially 1976/7. Conversely, few were reported in 1967/8, 1969/70 and 1975/6. In most years there are fewer Redwings than Fieldfares, though their exact relationship varies from year-to-year and area-to-area.

Flocks up to 500 strong are quite usual. In the main there are more and larger flocks in autumn, especially November, and least in January and February, although the most commonly reported size in every month is between 200-300. From time-to-time impressive movements or larger concentrations have occurred. For example, at least 1,000 arrived at Sheriffs Lench within an hour on October 21st

1936, 4,000 were at Coombes Valley in March and April 1973, with 7,000 in the following September and October, and 3,500 were at Stafford on March 7th 1977.

Even more impressive numbers gather at roosts. Most Redwings roost in thickets or woods with a dense understorey and they have a particular liking for laurel and rhododendron. Often they roost with Fieldfares, in which case the Redwings will occupy a lower level (Simms 1978). Most of the roosts reach their maximum numbers from December onwards, when counts up to 2,000 are by no means unusual and the largest have been 6-8,000, at Sheriffs Lench on November 24th 1962 and in Sutton Park in February 1977, and 8,000 at Brandon in late March 1977.

Mistle Thrush
Turdus viscivorus

Numerous resident.

Although the edges of woods or open country with scattered trees and bushes are its favoured habitat, the Mistle Thrush is equally at home amidst the well-timbered farmland, small woods and parklands of the West Midlands, or in its villages and large suburban parks and gardens, sometimes well inside the urban area. In autumn and winter it often visits pastures and other grasslands, or seeks out rowan, yew and holly berries.

Smith described it as occurring commonly throughout Staffordshire at all seasons, nesting somewhat sparingly in all rural districts. Both Harthan and Norris said it was a widely distributed and common resident and the former added that it was especially so in the orchard districts of Worcestershire. This widespread distribution is still apparent today and the *Atlas* surveys (1966-72) showed confirmed or suspected breeding in every 10-km square. Nowhere, though, is the Mistle Thrush very numerous. Its territories are generally large and its density consequently low. For example, 184 ha of mixed farmland and woodland at Moreton Morrell generally holds only one pair and Fradley Wood's 40 ha normally have just one or two pairs, although there were five pairs there in 1968. In recent years, there have been two or three pairs on just 16 ha in Edgbaston Park, with as many as six pairs in 1975. Overall the 1978 CBC showed a regional density of $2 \cdot 04$ pairs per km^2 on farmland and national densities of $2 \cdot 19$ pairs per km^2 for farmland and $6 \cdot 22$ pairs for woodland. Together with Sharrock's assumed range of 100-200 pairs per 10-km square, these suggest a regional population of 7,500-15,000 pairs, with the true figure probably at the lower end of this range. This agrees well with Lord and Munns' estimate of 2,000-20,000 pairs.

There have been no noticeable trends in the Mistle Thrush's status in the Region, though it does suffer from particularly severe or protracted winters, which not only reduce its numbers but also inhibit breeding as the species nests early. After breeding, birds often gather in loose flocks on pastures and grasslands. Such flocks usually contain less than 50 birds, but there have been six records of 100 or more, with the largest gathering one of 200 in Jervis Wood on September 18th 1973. Most of the larger flocks have been concentrated at good food supplies, especially rowan berries in the Lickeys. Small flocks begin to gather in July, but they are largest and most numerous in September and October, becoming smaller and less regular again in November and December. Few flocks are noted in winter, but small parties have again been reported in March. This pattern suggests some migration through the Region, particularly during August-October which is when the species disperses (Simms 1978), but Mistle Thrushes are infrequently ringed and there have been only three recoveries to substantiate this. Two involved movements between northern England and the West Midlands, whilst the third bird was ringed at Leamington Spa in April 1923 and found in West France in November of that year.

Cetti's Warbler
Cettia cetti

Very rare visitor. Five records.

Cetti's Warbler is a secretive bird which frequents thick vegetation in damp situations, such as reed-beds, marshes, ditches and willow or alder carr. Although most likely to be seen early in the morning or at dusk, even at other times it is unlikely to be overlooked because both its explosive song and distinctive call-note are far-carrying.

The species was unknown in Britain prior to 1961, but it has been expanding its range northwards in Europe for the past fifty years or so and numbers have increased since 1967, with a breeding population becoming established in southern England since 1972 (Sharrock).

In common with this trend, the first West Midland record came from Edgbaston Park, where a bird was present from May 2nd to July 12th 1975 (*Brit. Birds* 69:347) and was trapped on May 17th. This was followed by further birds in Worcestershire from May 26th to June 16th 1977 and again in the summer of 1978 and during March 1980, and in Staffordshire from January 2nd to February

11th 1978. Localities are not being disclosed lest this should prejudice the species becoming established in the area.

Grasshopper Warbler
Locustella naevia

Local, but fairly numerous, summer visitor.

Although Grasshopper Warblers are usually associated with damp, marshy habitats, they also occur in a wide variety of drier habitats including commons, rough pasture, scrub, derelict land, golf courses and forestry plantations. Indeed they might be found anywhere where there is a good cover of long grass, bramble or tangled vegetation. There has been a steady erosion of their more traditional habitats through drainage of wetlands, but to compensate there has been the creation of new habitats through extraction of minerals and more particularly afforestation. The thick undergrowth in young conifer plantations has proved very much to their liking, though this is, of course, only a transient habitat.

Smith said the species occurred locally in many parts of Staffordshire, Norris described it as scarce and local in Warwickshire and Harthan as formerly common in Worcestershire, but both Harthan and Norris considered it had greatly decreased in the first forty years of this century. Norris mentioned the war-time cultivation of land which had previously been wild as a likely cause of decline. Since then, however, numbers have apparently increased again with the creation of new habitats so that by the time of the WMBC *Atlas* (1966-8) the species was found in most areas apart from the northern moorlands and was regarded as locally numerous in favoured localities such as marshland or young conifer plantations. More thorough fieldwork for the BTO *Atlas* (1968-72)

confirmed breeding even on the northern moorland area and birds were absent only from the more industrialised areas.

Grasshopper Warbler numbers fluctuate markedly from year to year. For example, at least six pairs were on Castlemorton Common in 1941, but only one bird was "reeling" there in 1942. Subsequently there were four in 1957 and eight in 1969. Similarly characteristic fluctuations have occurred at Brandon, Sherbrook Valley and other favoured localities. To some extent these mask long-term trends, but there is some evidence of improved numbers in the last thirty years. In 1948, for example, breeding in the Tame Valley was unknown whereas today gravel workings have created new habitats and several pairs breed regularly. In south-east Worcestershire Harthan said it was more plentiful in 1954 than in any year since 1934.

More recently birds have been reported from between thirty and sixty localities in most years, but their abundance has fluctuated as the following graph shows:

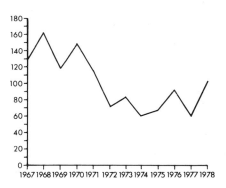

Annual mean of all reported birds = 100

Annual Index of Grasshopper Warblers, 1967-78

The high numbers from 1967-71 and low ones from 1972-7 suggest a population cycle, but data over a longer period is needed to verify this. The graph clearly shows an intriguing two-year cycle however.

There is little information on density, but six were singing on 10 ha at Brandon in 1965 and twelve on approximately 100 ha near Pershore in 1967. Such densities compare favourably with those of three pairs in 10 ha of young plantation and five in 24 ha of chalk grassland mentioned by Sharrock and suggest that the strength of Grasshopper Warblers in the West Midlands may be as good as anywhere. With breeding confirmed or suspected in seventy-one 10-km squares (92 per cent) and an assumed density of 10 pairs per square (Sharrock) a regional population of some 700 pairs is indicated. This is at the lower end of Lord and Munns' range of 200–2,000 pairs.

Over forty-two years the average first arrival date has been April 19th, with the earliest record on April 2nd (1961). The 1961 date is the third earliest British record (Hudson 1973). The average date of last departures over thirty-two years has been August 22nd, with the latest on September 26th (1976). Like several other species there has been a tendency for later departures in recent years, but whether this is indicative of a change in habit or simply a reflection of greater observer activity is not known. Grasshopper Warblers are rarely caught, but a nestling from Beckley (Oxon) was recaught at Brandon nearly a year later in May 1978.

Savi's Warbler
Locustella luscinioides

Rare summer visitor or vagrant. Six records.

The first record of Savi's Warbler in the West Midlands came from Brandon on April 21st 1968 (*Brit. Birds* 62:478). This was followed by a series of records from different localities in the Tame and Anker valleys during the spring and summer of 1972, 1976, 1977 and 1978, including the first record for Staffordshire between June 6th and July 5th 1977 (*Brit. Birds* 66:347; 70:433; 71:519).

Savi's Warbler is very much a bird of reed-beds and rank aquatic vegetation and its "reeling" song is more often than not the first indication of its presence. It began to recolonise Britain in or before 1960 after an absence of almost a hundred years (Sharrock) and these recent records reflect this recolonisation, although they have occurred at a time when the species appears to have passed its peak nationally and to be having difficulty maintaining its status. So far breeding has not been proved in the Region, but the concentration of records into a limited area may indicate prospecting and indeed a pair and a second singing male were present at one locality during 1977. The precise location of recent records has been withheld to avoid prejudicing any possible colonisation. Birds appear to arrive in late April, but there is no information about their departure.

Sedge Warbler
Acrocephalus schoenobaenus

Numerous summer visitor, widespread but local.

Although usually associated with rank, aquatic vegetation such as reed, reedmace or osiers, Sedge Warblers are often found around the drier margins of wetlands and increasingly in drier scrub, hedgerows, young forestry plantations and even arable crops.

Previous authors all described it as common, though Smith added that it was practically absent from the hill country, Harthan that it was scarce along the Teme and Norris that it was more restricted in industrial centres. Since then drainage of wetlands has led to a decline in many areas, but gravel pits have created new habitat and overall there has been little

Table 67: Population of Sedge Warblers at Alvecote, 1948–78.

	1948	1966	1967	1968	1969	1970	1971	1972	1973	1974	1975	1976	1977	1978
No. of pairs	12	12	12	12	12	20	15	10	10	12	12	12	12	12

change in distribution. Survey work for the WMBC *Atlas* from 1966–8 confirmed breeding everywhere except the moorlands and the Potteries, the West Midlands Conurbation and much of the Teme and Warwickshire Stour valleys, where the swifter flowing rivers have less suitable bankside vegetation. Even so these gaps were largely closed by more intensive field-work for the BTO *Atlas* from 1968–72, so that breeding was finally confirmed in sixty-eight 10-km squares (88 per cent) and suspected in a further six. Only on the bleakest moors, in the very heart of Birmingham and on the high ground around Clent and Lickey were birds totally absent.

Census results show great variability. In 1946 thirty-two kilometres of the River Severn yielded a mere 0·2 to 0·3 pairs per km; a navigable canal at Tardebigge held approximately eight pairs in 0·5 km in 1967 and 1970; and disused canals between 14 and 25 pairs per km. In 1965, 26 ha of disused limestone quarries at Ufton Fields held 20 pairs and in the same year at Brandon 8 ha of regenerating sallow and marsh held 30 pairs. This data and that from the CBC would suggest a total population of 7,000–8,000 pairs which is at the lower end of the range of 2,000–20,000 suggested by Lord and Munns.

Unravelling long-term population trends from short-term fluctuations is difficult. For example, at Belvide the number of singing males crashed dramatically from 26 in 1970 to 11 in 1971 and 2 in 1976 – yet only two had been noted in 1948 also. The CBC shows a similar decline, yet the index was high from 1964–8 and again in 1970 and data from the West Midlands suggests that the population at this time may have been above average. A comprehensive series of counts from Alvecote at least implies long-term stability as Table 67 shows.

Over forty-two years the average first arrival date has been April 16th, but most birds do not arrive until later in the month or in May. The earliest recorded arrival was April 1st (1972). Whilst on passage, birds sometimes turn up in built-up areas and even gardens. Return passage starts in July, but birds are fairly widespread even in September. Over thirty-six years the average last departure date has been September 27th, with the latest on October 17th (1960).

Many Sedge Warblers have been ringed as part of the BTO's *Acrocephalus* inquiry and there are several recoveries involving birds on passage. West Midland birds have been controlled during August and September along the south coast, from Cornwall to Kent, while May recoveries come from Surrey and Gloucestershire. Foreign recoveries involve two birds on Sark, one each in May and August; two birds from France in August and individuals moving north in April in Morocco and Spain.

Marsh Warbler
Acrocephalus palustris

Summer visitor. Very local, but not scarce.

The lower reaches of the Avon and Severn valleys are the stronghold of this nationally rare warbler, with more than three-quarters

71. Three-quarters of the British population of Marsh Warblers are found in the lower Severn and Avon valleys, where the Worcestershire Nature Conservation Trust has established several reserves in an effort to safeguard its preferred habitat.

A. Winspear Cundall

of the British population (Sharrock). Although seldom far from water, the Marsh Warbler is found in a wider range of habitat than the Reed Warbler, often nesting in dry situations. It prefers nettles, meadowsweet, willowherb or other rank vegetation, with a few shrubs or trees for song posts or nest anchorages. It commonly nested in osiers too, when these were still pollarded by basket-makers, but with the decline of this practice the habitat is no longer suitable and it has largely deserted such areas. The first recorded nest, in 1892, was in a stone quarry and during the 1930s and 1940s several pairs bred in bean fields (Harthan 1938). Apart from a singing bird in 1953, though, none have been reported from such a habitat since, but adults have been seen feeding young in an orchard.

Despite this diverse range of habitats, this rare warbler is confined to the lower reaches of the Avon, Severn and their tributaries. Climate is possibly an important factor in this restricted distribution, which coincides with a combination of high summer temperatures and thunderstorms. The ensuing humidity of course fosters the kind of luxuriant vegetation growth favoured by the Marsh Warbler and it is worth noting that similar conditions prevail in parts of Kent and southern England, where the species also occurs sparingly.

The Marsh Warbler is the last summer visitor to return, with the average first arrival date over twenty-nine years being June 2nd and the earliest date May 24th (1976). These dates might be distorted by observer bias, however, since few bird-watchers actually live in Marsh Warbler country and most delay their visits until they are certain of birds being in residence. Significantly some of the earliest arrivals have been at well-watched localities away from the normal areas and Green (1980) said they arrived in late May or early June. Information on their departure is scant, but Green (*op cit*) noted that the adults de-

parted at the end of July and the juveniles followed a week or so later.

Writing of the national population, Sharrock considered that numbers had undoubtedly decreased during the last twenty-five years and that loss of traditional areas through drainage or abandonment of osier-beds may have been important factors. Lord and Munns considered the species had likewise decreased in the West Midlands both in range and numbers, again probably due to habitat destruction. The full extent of any decline though is far from certain. Harthan described it as fairly common in the Avon Valley and along the banks of the Severn up to Worcester, but he knew of only one site on the River Teme. Since then the range has certainly contracted. In 1942 it was more numerous than the Sedge Warbler between Upton-on-Severn and Worcester, yet in 1970 not a single bird was found along this stretch. Elsewhere, however, it may still outnumber Sedge Warbler even today. Over the past fifty years the pattern has consistently been one of consolidation in the main breeding area, with sporadic records from elsewhere indicating either an overshooting northwards in spring or, more recently perhaps, displaced continental migrants (G. H. Green *in litt*). This has resulted in records along the Warwickshire Avon in 1918 (bred), 1938, 1945 (bred), 1952 (bred), 1961, 1965, 1975 and 1977 (bred); and the middle reaches of the Severn in 1943, 1947 (bred), and 1976–8, with breeding in the latter year. Isolated breeding also occurred in Staffordshire in 1916–8 and again in 1953; and along the Teme in 1956.

An earlier Staffordshire record concerned a pair that bred at Wood Eaton in 1914 and this and a record of a bird from the Tame Valley in 1975 are the only ones from outside the Avon and Severn catchments. Apart from recent years there are no estimates of population, but four or five sang regularly at Nafford from 1953 until

1972, since when the species has all but disappeared from this formerly favoured haunt. In 1969 there were an estimated 40–50 pairs in Worcestershire, in 1977 46 singing males were located and it seems likely the present population is in the range of 50–60 pairs. During the dry summer of 1976 it enjoyed good breeding success, but years like 1968, when there was a disastrous July flood, serve to illustrate how vulnerable such a small population can be. Certainly the Marsh Warbler can well do without the additional threats from over-zealous impatient bird-watchers. It could also do without the threats from river and drainage improvements, which not only destroy its habitat but also tend to increase the incidence of minor floods through accelerating run-off higher up the catchment area.

Reed Warbler

Acrocephalus scirpaceus

Numerous summer visitor in suitable habitats.

As its name implies, the Reed Warbler is closely associated with the common reed. Reed-beds are neither common nor extensive in the West Midlands though, so it is only locally distributed, despite nesting sometimes in osiers, reedmace, meadow-sweet, nettles, bulrushes, reed grass and even hawthorn. Of all the warblers this is the most colonial, so where suitable habitat does occur it can be quite numerous. It is also a common foster parent to the Cuckoo.

Smith said the Reed Warbler occurred very locally, but in considerable numbers in certain localities such as Aqualate and Copmere. He also referred to great numbers having frequented the Trent Valley at one time, with nests in privet and

lilac near Burton-on-Trent. Harthan said that every patch of reed by rivers, canals and pools had one or more pairs, with numbers depending on the area of reeds available. Norris too thought it fairly numerous, though rather local, and he thought there had been some decrease in the first half of this century, since Hudson and Tomes (1904) had both noted the species as common by all streams, yet by the 1940s some reed-beds remained unvisited and birds in an osier bed were exceptional. By 1951 Norris showed the highest density to be in the lower Severn Valley.

Since then progressive drainage and river improvements have destroyed much habitat, but this has been replaced by reed growth in abandoned canals and gravel-pit silt-beds. The species is very largely confined to the valleys of the main rivers and their slower-flowing tributaries, below the 100 m contour, but it does occur in suitable localities on the Birmingham Plateau, even well within the urban area, whilst a colony at Rudyard (170 m) represents the altitudinal limit. Breeding has been confirmed from the *Atlas* surveys (1966–72) in forty-five 10-km squares (58 per cent) and was suspected from a further eleven (14 per cent). More recently, ringing studies have shown numbers to be much higher than had been gauged from singing birds alone and this makes populations hard to estimate. For example, at Brandon in 1973 only nine pairs were reported as breeding, yet the 160 birds that were trapped during the summer indicated a much higher population. It is now known that there are about forty pairs at this site. Before 1970 there were few reports of any colony exceeding twenty pairs, except along the derelict Droitwich canal, but since then more than 40 pairs have been reported from Brandon (1972), 50 from Thorngrove (1977) and Ombersley (1978) and 60 from Oakley in 1978. At Thorngrove in 1978 many young perished through being tipped out of their nests by the weight of a large

Starling roost which caused the reeds to buckle. Records from Alvecote and Chillington, two of the long-established sites, show an overall increase since the late 1940s, with a peak in the early 1970s. Unfortunately there are few reliable figures on density, but 7–20 pairs per km have been recorded along the Droitwich canal, whilst at Brandon there have been 40 pairs on 68 ha. Where Reed and Sedge Warblers occur together, numbers of each often tend to be similar, though this depends very much on the habitat.

Lord and Munns placed the population within the range 200–2,000 pairs, but the density suggested by Sharrock of 50–100 pairs per 10-km square implies a considerably higher population of 3,000–6,000 pairs. Since there is probably less suitable habitat in the West Midlands than elsewhere in the Reed Warbler's range, however, a figure of 2,000–3,000 seems more probable.

Reed Warblers begin to arrive in late April, though the majority do not come until May. April 26th has been the average date for first arrivals based on thirty-eight years observations, during which the earliest record was April 10th (1977). Many leave during August, but September birds are common and over thirty-two years the average for last records has been September 19th. Since the mid-1960s there has been an increase in October records, the latest of which was October 16th (1977).

This species has been much ringed in recent years as part of the BTO's *Acrocephalus* inquiry. One result of this is the large volume of recoveries showing just how mobile the Reed Warbler is, particularly the young birds, before departing from this country. There have been recoveries right up to the northern limit of its British range, from Carnforth (Lancs.) and Otley (West Yorks.), as well as overseas recoveries from Portugal in October and November, Morocco in April and Jersey in May.

Great Reed Warbler
Acrocephalus arundinaceus

Very rare vagrant. Two records.

Just two records, both from Brandon where birds were trapped and ringed on June 12th 1977 (*Brit. Birds* 71:520) and June 10th 1979 (*Brit. Birds* 73:523). The dates are typical for this vagrant from Europe (Sharrock and Sharrock 1976).

Icterine Warbler
Hippolais icterina

One record.

The only record is that quoted by Harthan (1961), of a bird seen and heard in song at Sheriffs Lench on June 29th 1942. Both spring and inland records of this central and eastern European species are unusual (Sharrock 1974), so this occurrence was most unexpected.

Dartford Warbler
Sylvia undata

Formerly a scarce visitor, perhaps even resident, now a lost species.

The Dartford Warbler is essentially a bird of gorse-clad heaths and this habitat must once have been reasonably widespread in the West Midlands, although today it has all but disappeared. Certainly there are a few historical records from all counties, although Norris considered there to be only one authentic Warwickshire record – of a bird found at Yarningale Common on October 10th and 11th 1914 – and Harthan could find no definite record for Worcestershire, although it had reputedly occurred near Broadway and Tenbury Wells. Smith referred to breeding records from Cannock Chase and a possible sighting near Rugeley on March 22nd 1915 and considered that the species may have been overlooked in Staffordshire.

Assessing these historical records against the background of the present, much-restricted national status is difficult, but there are similar records from some surrounding counties and no doubt some at least are authentic and indicative of a wider distribution in years gone by. With the decline in Dartford Warblers nationally and the virtual absence of suitable habitat anywhere in the West Midlands today, it is hardly surprising that there are no recent records.

Barred Warbler
Sylvia nisoria

One record.

The only record is of a male, which was trapped, ringed and photographed at Brandon on June 3rd 1979. Although this species occurs regularly on the East Coast in autumn, spring records are exceedingly rare and inland ones almost unknown (Sharrock 1974). The appearance of this vagrant from eastern Europe was therefore doubly unexpected.

Lesser Whitethroat
Sylvia curruca

Numerous summer visitor.

The Lesser Whitethroat is a skulking species whose rattling song is often the first indication of its presence. Its favoured habitat is dense scrub or more especially those tall, tangled hedgerows of hawthorn and blackthorn that enclose lowland pastures.

72.   More often heard than seen, the Lesser Whitethroat nests in tall, tangled hedgerows or dense scrub.   *R. J. C. Blewitt*

Smith said the species was nowhere so numerous as the Whitethroat, being common only in lowland Staffordshire and decidedly local even there. Harthan described it as widely distributed, but much less common than the Whitethroat and Norris said it was local in distribution, not uncommon in some areas, but seldom noted in south-east Warwickshire, where incidentally many of the hedges were elm and arable farming is most prevalent.

Broadly speaking this is still the distribution, with birds most widespread in the dairy area of the Trent Valley and scarcest in the built-up parts of the Potteries, Birmingham and the Black Country and also in areas of predominantly arable farming. On the northern moors, where hedges are replaced by drystone walls, there are no breeding records, but other gaps revealed by the *Atlas* surveys (1966-72) are not conclusively due to lack of suitable habitat. Overall, breeding was confirmed in fifty 10-km squares (65 per cent) and suspected in a further twenty (26 per cent). Only in two squares were birds completely absent.

Even as early as 1948, hedgerow removal was being advanced as a reason for decline at Sheriffs Lench. The most drastic reductions in hedgerows have occurred subsequently, however, and coupled with modern cutting methods these must have led to a decline in Lesser Whitethroat numbers, though there is little data on which to base a firm assessment. In 1939, a density of 4·4 pairs per km^2 was noted around Bromsgrove, whilst there were 2·5 pairs per km^2 at Sheriffs Lench in 1947. Nationally the CBC for 1978 showed a mere 0·9 pairs per km^2 on farmland and this would indicate a regional population of 2,000–4,000 pairs, which is somewhat higher than the 200–2,000 pairs estimated by Lord and Munns. Regionally the species' appearances on CBC plots were sporadic and

numbers too small to draw any conclusions on density or long-term population change.

Most Lesser Whitethroats arrive in late April and May and leave again from July to September, with average first and last dates of April 22nd and September 19th respectively over forty-two years. The extreme dates are April 11th (1950 and 1961) and October 8th (1974). When on passage, birds are often noted in situations away from breeding areas, such as suburban gardens. One such was ringed at Dudley on August 8th 1971 and subsequently reported from Bedford on May 3rd 1972. This is the only ringing recovery.

Whitethroat
Sylvia communis

Numerous summer visitor.

In the West Midlands the Whitethroat's usual haunts are tangled hedgerows, gorse-clad heaths, scrub, wasteland and the clearings or edges of woods, particularly where there are tangled thickets of bramble or hawthorn.

According to Smith it was universally plentiful, except on the bare hills and moors, being found even on heather moor if there were a few bushes, and being quite common in the upland dales. Harthan said it was the most abundant warbler and Norris very common and widely distributed, though the former subsequently referred in 1961 to a reduction through hedge trimming.

During the *Atlas* surveys from 1966–72 breeding was confirmed in every 10-km square save one, where it was only suspected. For the first three years of the survey, however, numbers were high, with densities of six or seven pairs per km^2 on farmland, 10 pairs on 89 ha of heathland on Cannock Chase and 20 pairs in 105 ha around the gravel pits at Brandon. Such

Table 68: Changes in the Population of Whitethroats at Fradley Wood, 1966–79.

	1966	1967	1968	1969	1970	1971	1972	1973	1974	1975	1976	1977	1978	1979
No. of pairs	7	7	16	1	1	3	3	7	1	3	9	7	3	7

densities would have indicated a regional population in excess of 20,000 pairs, which conforms with the view of Lord and Munns that the species just about warranted classifying as abundant.

In 1969, though, numbers plummetted everywhere and this has been attributed to a disaster arising from the drought in their West African Sahel wintering area (Winstanley *et al* 1974). Following this crash, recovery in the West Midlands has been sporadic, with reports of a return to normal numbers in some areas by 1971, but little sign of an improvement elsewhere. This is presumed to indicate a return to optimal habitats before suboptimal ones. Numbers fell again in 1974 to an all-time low, however, and in many areas Whitethroats were outnumbered by Lesser Whitethroats. A further recovery ensued, but in 1978 there were again reports of a decline. The misfortunes of the Whitethroat are best exemplified by figures from the farmland around Sheriffs Lench, where there were 14 pairs on 142 ha in 1953, but none from 1969 until 1977, when just one pair returned. The CBC figures from Fradley Wood (40 ha) certainly show the extent of the decline and the fluctuations in the subsequent recovery, though they were somewhat distorted by habitat changes around 1968, when felling of many mature trees created an optimum habitat for the species.

From the CBC it seems the West Midlands' population may have been particularly hard hit, since the 1978 densities for farmland and woodland were only 1·2 and 3·6 pairs per km^2 respectively, compared to 2·4 and 6·7 pairs nationally.

Overall these figures indicate a current regional population of 5,000–15,000 pairs, with the true figure probably at the lower end of this range.

Most Whitethroats arrive from mid-April onwards and leave again during August and September. Observations over forty-four years have given average first and last dates of April 14th and September 24th respectively, with extremes of March 24th (1972) and October 23rd (1967). During autumn passage in particular, birds may be seen in atypical situations and records from gardens are not uncommon at this season. As examples, seven were trapped in a Nuneaton garden on August 3rd 1957 and four days later 35 were feeding in 2·5 ha of blackberry bushes at Sheriffs Lench.

The Whitethroat has provided two recoveries from the Iberian peninsula whilst on migration. The first was a bird caught at Brandon in August 1964, which was in Montes (Portugal) by the following October, and the second was a nestling from Coombe that was recovered in Vizcaya (Spain). A strange recovery is that of a juvenile ringed in Nuneaton in June 1955 and found on the Gocree lightship, in the North Sea, on June 11th 1956.

Garden Warbler
Sylvia borin

Numerous summer visitor.

Garden Warblers and Blackcaps share the same habitat of broad-leaved woodland with an open canopy and well-developed shrub layer. There is some divergence though, with the Garden Warbler more likely to be found in scrub or young plantations and the Blackcap in high forest. None the less the overlap and comparative strength of the two species has long been of interest. Smith and Norris both considered this the more numerous species, whilst Harthan thought they occurred in the same areas and numbers.

Today this is certainly not true, since there has been a definite shift in favour of Blackcap. From the available data this shift appears to have taken place quite recently as Table 69 shows.

This change is reflected nationally by the CBC, but has been more pronounced in the West Midlands. One reason for this could be the high proportion of woodland that was formerly coppiced. Whilst coppicing still took place conditions suited the Garden Warbler, but since it has ceased a higher, denser canopy more suited to the Blackcap has steadily developed. Garden Warblers used also to be common in osier beds when these were cut regularly, but again this practice has died out.

Nevertheless, the Garden Warbler is still widespread and the *Atlas* surveys (1966–72) proved breeding in fifty-six 10-km squares (73 per cent). Breeding was also suspected in a further twenty squares (26 per cent), including the moorlands, most of the Birmingham plateau and the Lias clay areas of the Avon Valley, where presumably birds were less common. In woodland, density at Randan varied between four and seven pairs in 28 ha in 1962 and 1963, but at Coombe in 1966 it was no higher than four pairs in 69 ha. On farmland around Bittell one pair bred on 40 ha in 1966. Appearances on CBC plots have been sporadic, but the species preference for open-canopied scrub is well illustrated by an increase from three to eight pairs at Fradley Wood (40 ha) when the mature trees were felled in 1968, since when it has been noted only occasionally. Such figures are broadly in line with the 30–50 pairs per 10-km square quoted by Sharrock and the 1978 national CBC densities of 0·65 pairs per km² on farmland and 4·24 pairs per km² in woodland. Between them, they suggest a regional population of 2,000–4,000 pairs, which is at the lower end of the range 2,000–20,000 postulated by Lord and Munns.

Over forty-one years the average date for first arrivals has been April 22nd, with the earliest records on April 9th, in 1967 and 1974. After the breeding season birds become unobtrusive, but data for forty years shows an average departure date for last birds of September 4th. The latest records involved birds on October 15th 1968 and the exceptionally late date of November 22nd 1975 at Acton. There have been no ringing recoveries relating to the West Midlands.

Table 69: Comparative Counts of Blackcaps and Garden Warblers, 1949–78

	1949–53	1954–58	1959–63	1964–68	1969–73	1974–78
No. of Blackcaps	15	21	12	170	180	115
No. of Garden Warblers	10	22	10	152	97	39
Blackcap as percentage of Garden Warbler	150	95	120	112	186	295

Blackcap

Sylvia atricapilla

Numerous summer visitor, scarce winter visitor.

Mature broad-leaved woodland with a rich shrub layer is the preferred habitat, though Blackcaps also occur on farmland with thick hedges or thorn scrub, in gravel-pit willow beds and the shrubberies of larger, suburban gardens. They are especially fond of rhododendrons.

Smith and Norris each described the species as fairly common, with the former noting that it was more plentiful in the lowlands. Both considered it to be outnumbered by the Garden Warbler, though, but today there are three Blackcaps to every Garden Warbler. For fuller details of their respective status see the latter species.

The *Atlas* surveys (1966–72) showed a wide distribution, with confirmed breeding in seventy 10-km squares (91 per cent) and suspected breeding in the remaining seven. Of these, however, failure to prove breeding was unlikely to have been due to lack of habitat or birds, except on the Millstone Grit country of the moors. Information on numbers is scant, but a density of $1 \cdot 2$ pairs per km^2 was recorded around Great Witley in 1948, whilst the Enville district held $0 \cdot 6$

pairs per km^2 in 1967. On smaller, more specific areas there were two pairs in 40 ha at Bittell in 1966 and 20 pairs in 69 ha at Coombe in the same year, whilst the regional CBC plots showed averages of eight pairs in 40 ha at Fradley Wood from 1966–79 (range 4–16 pairs); six pairs on 16 ha at Edgbaston Park from 1967–79 (range 3–8 pairs); and four pairs in 184 ha of mixed farm and woodland at Moreton Morrell from 1972–9 (range 1–7 pairs). Overall the regional CBC plots showed $1 \cdot 2$ pairs per km^2 on farmland and 29 pairs per km^2 in woodland in 1978, where it was the ninth commonest species. The comparable national figures were $2 \cdot 2$ and 6 pairs per km^2 respectively. Even ignoring the regional woodland density, a population of 7,000–13,000 pairs can be deduced, which is in the middle of the range 2,000–20,000 suggested by Lord and Munns.

A few Blackcaps occur during the winter, when they often visit gardens to feed from bird-tables or on cotoneaster or ivy berries. Even allowing for more observers, the incidence of these winter records has grown tremendously in recent years, as Table 70 shows.

Almost 90 per cent have occurred since 1959 and, perhaps aided by the succession

Table 70: Five-yearly Totals of Wintering Blackcaps, 1929–78

	1929–33	1934–38	1939–43	1944–48	1949–53	1954–58	1959–63	1964–68	1969–73	1974–78
No. of Wintering birds	1	3	4	4	3	5	15	21	31	102

Table 71: Monthly Distribution of Wintering Blackcaps, 1929–78

	Nov	Dec	Jan	Feb	Mar
No. of Birds	23	38	49	61	40

of mild winters, over half since 1974. The monthly distribution of these wintering birds is shown in Table 71.

The January and February peak is probably the result of birds being more obvious as they are forced into gardens to feed during hard weather. Certainly a large proportion of records at this time comes from the suburbs of the West Midland Conurbation, though there is also a bias towards the Avon and Severn valleys, which enjoy milder winters, with Kidderminster, Leamington and Malvern frequently mentioned.

These wintering birds have veiled the migration pattern in recent years, but over forty-four years the average first arrival date has been April 4th, whilst the average last departure date over forty-one years has been October 5th. A bird ringed at Packington in May 1977 was recovered in Algeria the following February; one ringed in Morocco in March 1976 was retrapped at Thorngrove on several occasions in the summer and a Warwickshire bird was reported from France in October 1976, presumably on passage. Other recoveries have come from the south of England and either show movements of juveniles around the country or migrants *en route* to or from overseas. One caught at Malvern Link in December 1968 was recaught at Truro (Cornwall) in March 1970.

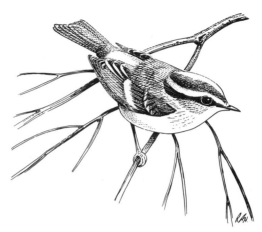

nationally, the date is typical for this wanderer from Siberia (Sharrock and Sharrock 1976).

Pallas's Warbler
Phylloscopus proregulus

One record.

The only record comes from Staffordshire, a bird being trapped and ringed during an influx of Goldcrests at Weston-under-Lizard on November 8th 1970 (*Brit. Birds* 64:361). Although this was not an exceptional year for Pallas's Warblers

Wood Warbler
Phylloscopus sibilatrix

Summer visitor, not scarce.

More than any other warbler, the Wood Warbler is a bird of high forest and is seldom seen away from those woods with a closed canopy, virtually no shrub layer and a sparse field layer. Within such woods it exploits all layers, nesting on the ground, feeding in the canopy and using the lower branches as song posts or perches when taking food to the nest. In the West Midlands only sessile oakwoods really meet its exacting requirements and, whilst birds do nest under beech or birch, there is a very high correlation between the distribution of sessile oakwoods and that of Wood Warbler.

"Hanging" woods with acidic soils on steep, well-drained slopes, such as those in the moorland cloughs, the Churnet Valley, Needwood Forest and west of the Severn, are particularly favoured; though small but regular populations are found in woods on the sandstones and gravels south of the Potteries, in Cannock Chase and Sutton Park, north Warwickshire, the Lickeys and

Table 72: Counts of Singing Wood Warblers at Lickey Woods and Wyre Forest, 1947–78

	1947	1948	1954	1961	1962	1963	1964	1965	1967	1968	1969	1972	1974	1975	1976	1977	1978
Lickey	12*	10	4	—	—	—	2/3	2/3	—	—	20	4	5	6	6	—	5
Wyre Forest	—	—	—	20	15	12	12–20	20	15	25	15	14/15	21	—	21	20	20

* Reported at the time to be more than usual.

around Redditch, Bromsgrove and Himley. Harthan noted it also occurred in coppices on the top of Bredon Hill and Norris that it used to breed regularly on Edge Hill, but today it is much less regular along the Cotswold scarp woodlands.

This basic distribution pattern has remained unchanged throughout the review period, though breeding has sometimes occurred outside it and birds regularly add confusion by singing for lengthy periods in spring without establishing permanent residency. Isolated incidents of this kind can occur well within the urban area and singing birds have been noted at both Edgbaston Park and Saltwells Wood. The *Atlas* surveys (1966–72) showed breeding confirmed in twenty-two 10-km squares (29 per cent) and suspected in a further thirty (40 per cent).

Wood Warblers are seldom censused and were virtually absent from all CBC plots, but the limited evidence available points to a relatively stable long-term population, though annual variations can be most marked. Counts of singing birds in the Lickey woods and along the Dowles Brook in the Wyre Forest, for example, are shown in Table 72.

In favoured situations this can be the commonest warbler and in 1947 it was estimated to have been twice as common as the Willow Warbler in the Wyre Forest, but there is no information on actual density either from the WM *Bird Reports* or the CBC. Sharrock, however, quoted 20–90 pairs per km² in typical oakwoods, but the West Midlands' density would almost certainly be at the lower end of this range. Forestry Commission statistics point to

little more than 7 km² of ideal habitat, so the regional population seems unlikely to be more than 100–200 pairs, which is at the upper end of the 20–200 range put forward by Lord and Munns.

Over forty-two years the average first arrival date has been April 21st, but many birds do not arrive until May, especially on the higher ground in the north. Arrivals in the second week of April are not exceptional, however, and the earliest recorded was on April 3rd (1967), excluding a very dubious claim from Solihull on March 5th 1974, which, if correct, would be a full fortnight earlier than any other British record (Hudson 1973). Most birds depart during July or early August and over thirty-three years the last date has averaged August 17th. September records are not unusual though, and the latest was September 13th (1976). To date no West Midland birds have been involved in ringing recoveries.

Chiffchaff
Phylloscopus collybita

Numerous summer visitor, scarce winter visitor.

A tangled undergrowth for nesting and tall trees for feeding and song posts are the essential ingredients for the Chiffchaff. Broad-leaved, mixed and especially old coppiced woodlands are most favoured, but well-timbered farm hedgerows, heaths, gravel pits and maturing forestry plantations are also used.

Previous authors all described the species

Table 73: Five-yearly Totals of Wintering Chiffchaffs, 1949–78

	1949–53	1954–58	1959–63	1964–68	1969–73	1974–78
No. of Birds	2	4	3	4	8	26

as common and generally distributed, though Smith regarded it as local in the lowlands and sparingly distributed in the hill country. No subsequent change in distribution has become apparent, with breeding proved in sixty-five 10-km squares (84 per cent) and suspected in a further eleven (14 per cent) between 1966 and 1972. Indeed only in the very heart of the Conurbation and on the northern moors are Chiffchaffs scarce.

Surveys in 1966 showed population densities of 18 pairs in 69 ha of woodland at Coombe, eleven pairs in 40 ha of woodland at Fradley and eight pairs in 73 ha of farmland at Frankton. As with most summer visitors, numbers tend to fluctuate from year-to-year, but the CBC points to numbers having halved between 1966 and 1978. For example, at Fradley Wood numbers had fallen from the eight mentioned above to just three pairs by 1974 and 1976, although there has been some recovery since and overall there was an average of eight pairs from 1966–79. At Edgbaston Park during the same period the average was four pairs in 16 ha, but here again numbers declined from seven pairs in 1966 to just two from 1976/8. At Moreton Morrell, however, the population changed very little from 1972–9, with an average of eleven pairs in 184 ha of mixed farm and woodland. In 1978 the CBC showed a regional density of 2·4 pairs per km² on farmland – the same as the national average – but in woodland there were thirty-one pairs per km² compared to a national average of only twelve pairs. These figures point to a regional population of 6,000–18,000 pairs. This is comfortably within the range of 2,000–20,000 suggested by Lord and Munns, though if anything the bias is probably towards the upper end.

Chiffchaffs are the earliest summer visitors to arrive and their appearance is strongly influenced by weather conditions. In a cold spring birds may be scarce until April, but over forty-four years of records the species has never failed to appear in March. Wintering birds can cause confusion, but the average first arrival date has been March 16th and during the 1960s there were records for the first week in March for five years out of ten. Until the mid-1960s the average date of last records was October 7th, but since then birds have increasingly lingered later (*cf.* Willow Warbler) and the overall last date for forty-four years is now October 14th. This spate of lingering autumn birds has been accompanied by an increase in winter records as Table 73 shows.

This recent increase parallels that of Blackcap, though as yet it has been less pronounced. The monthly distribution differs from that of Blackcap, however, with most records in late autumn (see Table 74).

Table 74: Monthly Distribution of Wintering Chiffchaffs, 1949–78

	Nov	Dec	Jan	Feb
No. of Birds	16	12	8	10

This suggests the Chiffchaff is less able to sustain itself in the hardest part of winter, perhaps because it cannot subsist on berries like the Blackcap, and that it moves further south or west, returning again in spring. Although many wintering birds are reported from gardens, the largest proportion is seen around aquatic habitats such as reservoirs and gravel pits.

There is one foreign recovery, of a bird ringed at Malvern in September 1970 and recovered in Morocco the following November. Otherwise all recoveries show returns to the same, or relatively close, sites in subsequent years. Birds showing characteristics of one of the northern or eastern races were reported as follows: one seen and heard near Evesham on September 27th 1928, one at Draycote on October 1st 1976 and one, possibly from one of these races, at Sheriffs Lench on August 9th 1955, although this is an unlikely date for such an occurrence.

Willow Warbler
Phylloscopus trochilus

Abundant summer visitor.

The Willow Warbler is the commonest of all our summer visitors and is the fourth commonest bird in West Midlands woods. With less precise habitat requirements than either the Chiffchaff or Wood Warbler, it is not confined to woodland alone, but is widespread and abundant, nesting anywhere where there are small trees or bushes. If anything it avoids a closed canopy, preferring instead clearings, rides and the margins of woods; scrub, bracken-covered heaths, large gardens, gravel pits and even hedgerows. It also frequents young forestry, up to about fifteen years old, and on the moors nests in heather as well.

Previous authors all described the species as very common and widely distributed and there has been no subsequent change in distribution, with breeding proved during the *Atlas* surveys from 1966–72 in every 10-km

square save one, where it was only suspected. Willow Warblers are even successful in exploiting suitable areas of isolated habitat in urban areas. Over the past fifteen years various comparative counts have shown it to outnumber the Chiffchaff by five to one, though since 1969 the ratio has been as high as seven to one. On the CBC plots in 1978, however, it was no more than three to one.

Figures on density have varied from 22 pairs in 69 ha of woodland at Coombe in 1966 and 25 per km^2 around Brownhills in 1969 to eight pairs in 89 ha of heathland on Cannock Chase in 1967 and a mean of 6–8 pairs per km^2 on farmland in a number of localities between 1964 and 1972. On CBC plots there was an average of 39 pairs in 40 ha of woodland at Fradley from 1966–79, whilst during the same period Edgbaston Park's 16 ha held an average of 10 pairs. At Moreton Morrell 184 ha of mixed farm and woodland held an average of seven pairs from 1972–9. There were typically very marked annual variations on each plot, but these revealed no consistent pattern. Unfortunately two of the more interesting counts do not provide accurate densities, but 140 singing birds encountered in just under one kilometre of young forestry at Bagots Park in 1968 probably represented over 200 pairs per km^2, whilst 300 in song along 15 km of the Manifold Valley probably indicates some 25 pairs per km^2 and certainly shows the strength of the species in the dales. In 1978 the regional CBC plots showed overall densities of $7·1$ pairs per km^2 on farmland and no fewer than 85 pairs per km^2 in woodland, compared to national figures of $7·7$ and 43 respectively.

From this a regional population of 35,000–50,000 pairs has been estimated, which accords with Lord and Munns' estimate of over 20,000 pairs. Sharrock assumed a density of at least 1,000 pairs per 10-km square, which would suggest a much higher population, but, notwithstanding

73.   The Willow Warbler is the most widespread and abundant of our summer visitors, nesting in woods or wherever there are small trees or bushes.   *S. C. Brown*

the high regional density of Willow Warblers in woodland, their overall density in the West Midlands is unlikely to be as high as that assumed by Sharrock because so little land is actually wooded.

The earliest Willow Warblers usually arrive in late March, but most do not come until April. Departure begins in mid-July and until 1966 most birds had left by September, with only two records beyond the first week in October. Assuming correct identification, an abrupt change then apparently occurred and subsequently birds have stayed noticeably later, with only two years lacking records beyond mid-October. Forty-four years' observations have revealed average first and last dates of March 27th and October 4th respectively,

with extremes of March 12th (1966 and 1967) and October 30th (1966 and 1972). On passage, birds can be very numerous, even in urban areas. In 1940, for example, at least 100 were in birch scrub at Edgbaston Park and in 1957 20 birds were in a small 0·25 ha allotment at Harborne. Forty passed through a Sedgley garden in one day during August 1963 and on April 16th 1977 one was watched taking insects from produce in Birmingham's Fruit Market.

There have been two foreign recoveries, one from France and one from Spain, whilst at home two birds have shown movement between the West Midlands and Dorset and another, ringed in Cheshire, was recovered at Bearley two years later.

Goldcrest

Regulus regulus

Numerous resident and winter visitor.

Throughout the Region wherever there are conifers there are likely to be Goldcrests, whether it be in large plantations of larch, pine or spruce, churchyard yews or the exotic cedars and cypress of suburban and country gardens. Hard winters have most effect on numbers and distribution. When the population is high Goldcrests will also nest in deciduous woodland, but whenever numbers are low they show a marked preference for yew and spruce.

Smith remarked that the species bred locally wherever there were large masses of conifers. He also observed that it was far commoner in winter, with birds from other regions appearing in October and November, and Wood (1836) considered that the winter flocks did not disperse until mid-February. Harthan (1961) described it as generally present and Norris as common and generally distributed. Since then the Goldcrest must have increased in numbers with the spread and growth of young forestry plantations. However, numbers were severely depleted by the hard winters of 1946/7 and 1962/3 and the latter had a strong bearing on the distribution revealed by the *Atlas* surveys. That for the WMBC, which was conducted from 1966-68, showed a very patchy distribution, with breeding confirmed or suspected in only forty-two 10-km squares (55 per cent). On the basis of this Lord and Munns placed the regional population within the range 200-2,000 pairs, which was undoubtedly correct at the time. By the time fieldwork for the BTO *Atlas* was finished in 1972, however, the CBC was showing a population more than double that of 1968 and breeding had been confirmed or was suspected in seventy-two 10-km squares (94 per cent). According to the CBC this increase was sustained until 1975, since when there has been a decline.

Information on density is scant, particularly for its favoured coniferous woodland, but there must be several hundred pairs in areas like Cannock Chase. In 1966 there were five pairs in 69 ha of woodland at Coombe, whilst Edgbaston Park held one or two pairs in 16 ha from 1966-79 and Fradley Wood from one to five pairs in 40 ha between 1966-78. The 1978 CBC data for the Region showed 8·4 pairs per km^2 in woodland, which was marginally below the national level of 8·7. At the time these figures probably implied a regional population of 3,000-13,000, but once the effect of the 1978/9 winter is fully known it is likely the true figure will be at the lower end of this range, if not below it.

During winter, when they are joined by immigrants from north-east Europe and the USSR, Goldcrests become gregarious, often consorting with wandering flocks of tits. The winter influx begins in late September or October, but flocks are most frequent and largest in November, when parties up to 50 regularly occur, and 100 or more were noted at Sherbrook Valley in 1968, Kinver in 1969 and Leamington in 1974. There is also a very small secondary peak in March which indicates a return passage, suggesting that some birds at least pass through the area to winter further south or west. Wintering parties of 20-25 are often reported, but there have been few attempts to estimate total numbers in extensive tracts of coniferous woodland. However, there were an estimated 1,000 at Enville during the winter of 1966/7 and areas like Cannock Chase probably hold several thousand. Although winter records come most consistently from places with extensive stands of conifers, such as Cannock Chase, Sutton Park, the Lickeys and the Enville area, Goldcrests have become a familiar garden bird, too, especially in the suburbs, where they regularly come to bird tables to feed on crumbs and fat. On passage they have occurred at some unexpected places, such as the top of the Post

Office tower in Birmingham in October 1967 and a redevelopment site at Aston in October 1975.

There have been several recoveries from a long distance in recent years. Birds ringed in the West Midlands have been recovered at Skegness (Lincs.), the Calf of Man and Romney (Kent), whilst one ringed in Argyll in 1976 and retrapped at Hopwas in February 1977 may give an indication of the origin of some winter visitors, although many of the winter visitors to Britain apparently come from Scandinavia and the Baltic.

Firecrest
Regulus ignicapillus

Very scarce passage migrant or winter visitor. Bred in 1975.

This diminutive little bird occurs in a wide variety of habitats. Although it prefers conifers for breeding, particularly Norway spruce, it also breeds in mixed woods and outside the breeding season has been reported from more open woodland and in scrub on commons and around reservoirs. Several have also been noted in gardens, some in suburban localities.

Harthan described the Firecrest as an occasional winter vagrant from the Continent, but Norris could quote only five Warwickshire records and there were none from Staffordshire until as recently as 1966. Since these statements there has been a definite change in status, which reflects the national spread as a breeding species.

During the thirty-five years prior to 1969 the Firecrest was very definitely a rare winter visitor, with nine of the ten birds recorded occurring in the period November to February and two-thirds of them coming in November alone. Most of the records were prior to 1950 – indeed there was only one record during the fifteen years 1950–64 – and the majority (60 per cent) came from

Worcestershire, especially the Lickey Woods.

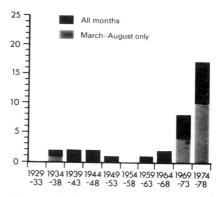

Five-yearly Totals of Firecrests, 1929–78

By comparison the ten years 1969–78 produced twenty-four records, many of which involved more than one bird, though unfortunately precise numbers are not known in every case. The general pattern of winter vagrancy continued, but with an annual average double that of earlier years. Even so it accounted for little more than a third of the records. Instead there was a marked shift towards spring occurrences, with half of the records coming in the period March to May. In Staffordshire particularly, following the first record in 1966, there were eight further records between 1972 and 1975, including six in April. Whilst birds in April could possibly be on passage, it seems more likely that they were part of a nationally expanding population prospecting new territory, especially as half of the records came from very suitable breeding territory on Cannock Chase.

The three summer records involved a solitary male in 1936, a female trapped in suitable breeding habitat in 1974 and the first positive breeding records in 1975, from Worcestershire, when three or possibly four pairs were at one locality at Lickey and one nest, found 15 m up in a larch, was subsequently destroyed, most probably by grey squirrels. Disappoint-

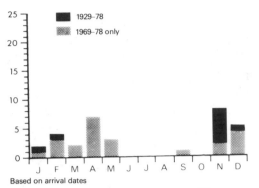

Based on arrival dates

Monthly Distribution of Firecrests, 1929–78

ingly, there have been no subsequent breeding season reports from the same locality, though two males were singing at another site near Kidderminster in 1976.

Firecrests first bred in Britain in the New Forest in 1962, but it was not until 1971 that a more extensive distribution was revealed (Sharrock). This spread continued until 1975, but was followed by a drastic reduction in numbers in 1976 (*Brit. Birds* 71:28). The West Midlands situation mirrors this exactly, with a gradual increase from 1971 to a maximum of six records in 1975, but only four in 1976 and just one November record in 1977. It remains to be seen, therefore, whether the Firecrest will be able to consolidate or even maintain its status as a breeding species or whether it will revert to being just a scarce winter visitor.

Spotted Flycatcher
Muscicapa striata

Numerous summer visitor.

The Spotted Flycatcher is an endearing bird which takes readily to nest boxes and is very faithful to traditional nest sites. Essentially it is a bird of parkland, orchards and large gardens, both rural and suburban, though it is also often found in churchyards, town parks and woodland

glades and margins. Birds build around farmsteads, too, particularly where there are cattle to attract insects, and near still water, where insects are also abundant. Indeed their ideal habitat is perhaps the combination of ornamental lakes, small woods and copses that typifies the country estate.

Previous authors all commented on its being common or widely distributed, even in the wooded valleys of the hill country, where it has to compete with the Pied Flycatcher. The *Atlas* fieldwork (1966–72) confirmed breeding in every 10-km square, including sites well within the urban area such as Edgbaston Park and Warley Park. More unusual, though, was a pair at Dudley in 1977 which bred within three metres of an operational power press in daily use.

Notwithstanding its widespread distribution, the Spotted Flycatcher has declined in numbers, particularly in recent years. In 1962 the decrease in one Birmingham suburb was attributed to large houses being demolished and their gardens destroyed. In general, though, nesting success seems to be very poor, whilst predation, particularly from grey squirrels, causes high mortality. In May 1960 at least 60 were seen in the Wyre Forest along Dowles Brook and the same number was in the Lickeys in August of that year. Such numbers have not occurred since at any locality and there have been several references to declines since 1970.

Examples of density come from Coombe, where eight pairs bred on 69 ha in 1966 and Edgbaston Park, where up to three pairs breed in 16 ha. Nationally the CBC for 1978 showed 1·44 pairs per km^2 on farmland and 3·78 pairs in woodland. These densities would indicate, even at a conservative estimate, a population regionally of 2,000–5,000 pairs, which is slightly above the 200–2,000 suggested by Lord and Munns.

Spotted Flycatchers are one of the last summer visitors to arrive and few are seen

74.   The Pied Flycatcher is a fairly scarce summer visitor to the "hanging" oakwoods of the Wyre Forest and the Churnet Valley, where it benefits from nest-box schemes.

A. Winspear Cundall

before the second week of May. Autumn passage reaches its peak in late August, when parties up to 40 strong have been noted, though 20–25 are more normal. Over forty-two years the average first and last dates have been May 1st and September 25th respectively, with extremes of April 15th (1961 and 1976) and October 17th (1970). Somewhat surprisingly there have been no ringing recoveries relating to the Region.

Pied Flycatcher
Ficedula hypoleuca

Very local, but not scarce, as a summer visitor and passage migrant.

The Pied Flycatcher is one of the speciali-

ties of the "hanging" sessile oakwoods west of the Severn and in north Staffordshire. Even in these areas though, it is somewhat local. It prefers steeply sloping valleys, where the fast-flowing streams are overhung by oaks, but the shrub layer is sparse. It feeds amongst the oaks on defoliating caterpillars. Old trees with natural holes are utilised for nesting and in the Wyre Forest it frequents old apple trees in orchards within the Forest. Breeding later than other hole nesters places the Pied Flycatcher at a disadvantage if nest sites are in short supply and there is one record from the Wyre Forest of a pair taking over a nest hole immediately it had been vacated by a brood of Nuthatches. The populations of both the Wyre Forest and Coombes Valley have increased with the provision of nest-boxes. Occasionally one or two pairs nest in mature deciduous woodland or parkland of less characteristic habitat, but these are exceptions rather than the rule.

Smith said the Pied Flycatcher probably crossed Staffordshire every year on migration, but could quote no breeding records save Whitlock's (1893) that they nested in Dovedale at one time. Harthan quoted just three breeding records, at Rhydd in 1877, Malvern Wells in 1932 and Bewdley in 1936, though he thought it had probably bred occasionally in the Wyre Forest, and Norris just one, from Sutton Park in 1882. Since about 1940, however, the species has been gradually establishing itself in the Region in line with its national expansion. By 1951 Norris referred to regular breeding in the Wyre Forest and Harthan (1961), too, subsequently mentioned its being well established here and also breeding on the Lickeys.

The Wyre Forest has become the Pied Flycatcher's main stronghold. Although the 1936 record above may have referred to this area, the first definite record came in 1943 and by 1947 six singing birds were reported. This had increased to ten in song along 1·6 km of the Dowles Brook in 1957 and to twenty in 3·2 km by 1961. Without a regular census it is hard to be certain, but since then numbers appear to have been fairly steady. In nest-boxes, for example, there were 11 pairs in 1972, and 15 in each of the years 1973–5. During 1972–5, 228 young were reared at an average of 4·1 per nest. Success varies from year-to-year, however, and in 1972 each pair reared over six young, whereas in 1974 the average barely exceeded two.

Away from the Wyre, breeding was reported from the Lickeys in 1949 and up to two pairs then bred annually until 1953, whilst in 1952 a pair bred to the north of Malvern, at Alfrick. In north Staffordshire a pair bred at Onecote, also in 1952, and occurrences in the Churnet Valley increased noticeably during the mid-1950s. In 1959 breeding occurred in west Stafford-

shire, at Maer, where it continued sporadically for ten years, and possibly also at Chillington. Back in the north a pair bred at Blackbrook Valley in 1961, whilst at Coombes Valley there was an increase from four pairs in 1967 to nine by 1973, making this the second stronghold for the species. Irregular breeding has also occurred in nest-boxes at Chillington and Shugborough, a pair bred in the Needwood Forest in 1978 (S. C. Brown *pers comm*) and birds have been seen a number of times in mid-summer at many widely scattered localities. Lord and Munns placed the regional population at between three and 20 pairs, but during the *Atlas* surveys (1966–72) breeding was proved in seven 10-km squares and suspected in a further three. Allowing for ten occupied squares and the size of the populations in the Wyre Forest and at Coombes Valley, the present regional population is certainly 30 pairs and may be as high as 50 pairs.

Pied Flycatchers begin to arrive in the last week of April, with the earliest migrants, 70 per cent of which are males, turning up almost anywhere, even in towns and suburbs. Over thirty-two years April 22nd has been the average first arrival date, with April 9th (1975) the earliest. Most breeding birds, however, do not return to their territories until May and have gone again by August. Elsewhere, though, September records are not unusual, especially from suburban gardens, but these seem likely to involve Scandinavian birds as their arrival often coincides with that of other well-known drift migrants. Over fourteen years the average last date has been September 5th, whilst a bird on October 7th 1976 is the latest ever recorded. Despite quite extensive ringing of birds from the nest boxes of the Wyre Forest there is but one recovery, of a bird ringed here in July 1972 which returned to Bromyard (Hereford–Worcester) the following May.

Bearded Tit
Panurus biarmicus

Scarce passage migrant and winter visitor.

Generally Bearded Tits are associated with pure stands of common reed, *Phragmites*, but in the West Midlands, where such habitat is scarce, they also visit other aquatic habitats with emergent vegetation such as bulrush, reedmace and willow scrub.

Smith said that Bearded Tits possibly resided in Staffordshire at one time, but that definite evidence of this was much needed. In support he quoted records from last century, including reports of breeding at Aqualate circa 1850, but rightly questioned the latter's validity in view of its dubious circumstances.

Neither Norris nor Harthan could quote any Warwickshire or Worcestershire records and there were no records at all this century from anywhere in the West Midlands until 1963. Since then there have been over 50 involving more than 200 birds and the species has occurred in every subsequent year. This remarkable change in status began shortly after Bearded Tits started an annual autumn eruption from their East Anglian and Continental breeding strongholds (Parslow 1973) and ringing recoveries have shown movements from Holland to Brandon, Minsmere to Brandon (and the reverse), Yorkshire to Brandon and Holland to Westwood Park. In addition, evidence of birds travelling in small parties is provided by three which were ringed at Goole (Humberside) and recovered at Stoke-on-Trent.

Autumn 1972 saw an exceptional influx involving some 56 birds, but over 20 were recorded in 1973, 1976 and 1977 also. Most birds arrive in autumn, half of them between late October and the end of November, and usually in small parties. Many move on again rapidly, but several parties have wintered, staying usually until the second half of March or early April and in one instance through to July. Parties

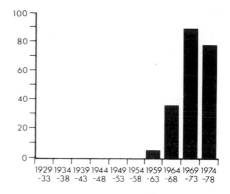

Five-yearly Totals of Bearded Tits, 1929–78

usually comprise less than ten birds (average three or four), but 35 appeared at Brandon in October 1972 and 20 of these remained into 1973. The only other parties to exceed ten were at Himley, also in October 1972, and Oakley and Berry Mound in October 1976.

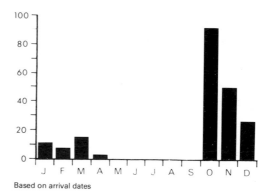

Based on arrival dates

Monthly Distribution of Bearded Tits, 1929–78

The 50 or so records are evenly distributed between the shire counties, but as yet there have been none from the West Midlands County. Birds are faithful to a few suitable habitats. For example, nearly half the Warwickshire records have come from Brandon, where birds have appeared in eight of the last twelve winters and made protracted stays on five occasions. In Worcestershire over half of the records have come from Upton Warren or West-wood Park and only in Staffordshire does there appear to be no particularly favoured locality, although Belvide has been visited more than anywhere else.

Long-tailed Tit
Aegithalos caudatus

Numerous resident.

Certain differences set the Long-tailed Tit apart from the true tits. To begin with it does not nest in holes, but builds a beautiful structure of moss, lichen and feathers, though one eccentric individual in the West Midlands used pieces of crisp packets and silver foil for adornment. It is thus less dependent than the other tits on trees during the summer and tends to avoid the closed canopy of mature woodland in favour of more open situations. Thorn hedges, gorse bushes and blackthorn thickets on farmland and heaths are its typical breeding habitat, though young plantations are occasionally chosen and a few nest around woodland margins, rides and glades. In winter though, when the other tits disperse more widely, Long-tailed Tits often congregate in woodland, where they feed in the upper shrub and lower tree layers of both broad-leaved and coniferous woods. However, similar parties can also be found in hedgerows, scrub and along alder-lined streams.

Previous authors said little about distribution, but Norris (1951) showed it to be general and Lord and Blake (1962) said it bred fairly commonly except in the Black Country, but was sparingly distributed in south-east Staffordshire. The species still has a wide distribution and the *Atlas* surveys from 1966–72 confirmed breeding in seventy-one 10-km squares (92 per cent)

and showed it as suspected in a further three. Only on the bleakest parts of the moors and the most densely built-up parts of Birmingham were birds completely absent.

As they are insectivorous, Long-tailed Tits are susceptible almost to the point of extermination to very severe winters such as those of 1917, 1940, 1947 and 1963. Fortunately their numbers recover quickly and in recent years, with the succession of mild winters, their population has been high. The 40 ha of Fradley Wood usually hold one or two pairs, whilst 184 ha of mixed wood and farmland at Moreton Morrell hold up to five pairs. The CBC for 1978 showed $2 \cdot 04$ pairs per km^2 on farm-land and $9 \cdot 6$ pairs per km^2 in woodland for the limited number of regional plots and national averages of $1 \cdot 35$ and $5 \cdot 12$ pairs respectively. Overall this suggests the Long-tailed Tit might be above average strength in the West Midlands and points to a regional population of 5,000–9,000 pairs. This is again higher than the 200–2,000 suggested by Lord and Munns, but coming so soon after the 1962/3 winter their estimate was probably correct at the time.

In winter Long-tailed Tits consort with other tits in mixed feeding flocks, but they tend to move faster and further than other species and frequently become detached. Such flocks commonly number 20–30 birds and up to 50 are not unusual. Both Smith and Harthan mentioned flocks of 80, whilst at least 100 have been recorded twice – at Hanbury on March 12th 1957 and Dowles Brook on February 6th 1972. On the whole Long-tailed Tits are sedentary and recoveries show only limited move-ment, although in some cases several birds of the same flock have made the same journey. The longest movement was from Temple Grafton to the Wirral (Cheshire), but the origin of the birds that caused a major influx in the autumn of 1960 remains unknown.

75. The Marsh Tit has a preference for mature, deciduous woodland, where it spends much of its time feeding in the shrub layer. *R. J. C. Blewitt*

Marsh Tit

Parus palustris

Numerous resident.

The name Marsh Tit is something of a misnomer since this species is by no means confined to damp areas. It does, however, have a close affinity with broad-leaved woodland, especially ash, beech and oak, and seems to prefer more mature situations than the Willow Tit. Seldom in the canopy or on the ground, it spends much of its time searching for seeds where there is a rich shrub layer of hazel, elder, guelder rose, honeysuckle, bramble or yew. In autumn it also attacks oak buds and thistle-heads and, like the Coal Tit, it also stores food. Occasionally it will visit orchards or rural gardens, but records from urban situations are rare though one or two are seen quite regularly in Edgbaston Park.

Previous authors held some contradictory views on status. In Staffordshire, Smith said that Marsh Tits were found in most wooded parts and that they were commoner than any of the other tits in Dovedale, though in the Trent Valley they were largely replaced by Willow Tits. Lord and Blake (1962) said it bred fairly commonly, with most in the Trent Valley, the north-west, west and south-west. Of Worcestershire, Harthan said it was fairly common in all woods and country districts, but in Warwickshire Norris thought it scarce and rather local. Fincher (1955) said that it appeared to breed throughout the West Midlands, but nowhere commonly, nesting mainly in deciduous woods. From

1966–72 the *Atlas* surveys showed breeding confirmed in fifty-nine 10-km squares (77 per cent) and suspected in eight (10 per cent). The remaining ten squares, where birds were present but not suspected of breeding, are most intriguing. They included the northern hills, for example, which embrace Dovedale; parts of the Conurbation and Tame Valley, where suitable habitat is scarce, but also the Coal Measures north-west of Coventry which are heavily timbered with oak; and most of Bredon Hill, where the species was reputedly the commonest tit in 1965 and was again said to be common in 1977, especially above the 200 m contour.

In view of the conflicting statements on status it is a pity there is so little data on density. However, the 40 ha of Fradley Wood have regularly held up to two pairs and 184 ha of mixed farmland and woodland at Moreton Morrell up to three pairs. Nationally the CBC showed 4·5 pairs per km² in 1978 and this suggests a regional population of 2,000–4,000 pairs, which is just above the 200–2,000 range postulated by Lord and Munns.

Outside the breeding season Marsh Tits are less gregarious than the other species of tit and Barnes (1975) regarded a flock of 20 as exceptional. A flock of this size occurred at Hanbury on December 27th 1957, however, whilst the same number was noted in dispersed flocks in 1954 and twice in 1964. The most reported was 25 along Dowles Brook on February 18th 1968.

Willow Tit

Parus montanus

Numerous resident.

It was not until the turn of the century that the Willow Tit was distinguished as a separate species from the Marsh Tit, so it is hardly surprising that its distribution has only recently begun to be clarified. Even now the distribution of both species is confused by misidentifications. There is considerable geographical overlap between the species, but there are some marked differences in preferred habitat. Whilst the Marsh Tit is closely associated with mature deciduous woodland, the Willow Tit prefers wet or scrubby birch woods, willow and alder fringed streams, gravel pits and hedgerows with small saplings. Its main requirement is plenty of soft, rotting timber into which it can excavate its nest-hole. In winter the Willow Tit wanders more widely in search of haws and honeysuckle berries in particular, and at this time it may even visit gardens to feed. As a rule it associates less with the other species of tit and seldom joins their feeding flocks.

Smith regarded the Trent Valley and its neighbourhood as the Willow Tit's Staffordshire home, Harthan said it was widely distributed in Worcestershire, but nowhere common, and Norris described it as scarce and very local in Warwickshire. A distinct spread was noted in the Birmingham area from 1949, however, and in the Avon Valley in 1950. In 1951 Norris reported gaps in the lower Severn Valley, the Feldon area of Warwickshire and the Black Country, whilst Lord and Blake (1962) added to this the moorlands and the Potteries. A similar, but slightly wider, distribution was shown by the *Atlas* surveys from 1966–72, when breeding was proved in sixty 10-km squares (78 per cent) and suspected in a further ten (13 per cent).

Though not entirely absent, birds were still found to be scarce on the northern hills and breeding did not occur at Coombes Valley, for example, until 1973. They were also scarce in Birmingham and the Black Country and throughout an extensive area of the lower Severn Valley, where the rejuvenated rivers are contained by deeper banks so the overhanging trees have their roots above water and do not rot so readily.

Throughout the period under review evidence has conflicted as to whether the Marsh Tit or the Willow Tit is the commoner. Observations have tended to do no more than reflect variations in habitat, though. For example, in 1966, nine Willow Tits were trapped and ringed compared to eight Marsh Tits at Coombe, where there is both extensive mature woodland and a tree-fringed lake, whereas at Brandon, which is an open gravel pit with no mature woodland, there were seven Willow Tits but no Marsh Tit. A census at Sheriffs Lench in 1950 revealed six pairs on 162 ha.

More recently Edgbaston Park has consistently held one pair in an area of 16 ha and Fradley Wood held up to three pairs in 40 ha until 1970, when felling took place. These samples are too small to be a reliable guide to numbers on their own, but Sharrock assumed 40–80 pairs per occupied 10-km square and this would point to a regional population of 3,000–6,000 pairs. A range of 2,000–4,000 seems more likely, however, compared with the 200–2,000 suggested by Lord and Munns. Outside the breeding season there are few reports on numbers, but 11 at Belvide on August 24th 1970 were the most noted.

Coal Tit
Parus ater

Widespread and numerous resident and winter visitor.

Though small numbers occur in broad-leaved woodland, especially oakwoods, the Coal Tit's long, thin bill is perfect for feeding amongst pine needles and it differs from other members of the tit family in the Region by showing a marked preference for conifers. In plantations it may well be amongst the commonest species, but even a few conifers in mixed woods, landscaped parklands, urban parks, gardens, cemeter-ies and churchyards will hold a few pairs. In most respects it shares the same habitat as the Goldcrest. In winter small numbers join with other tits and finches in feeding flocks and Smith mentioned their accompanying Bramblings to feed on beech mast. They are also regular visitors to garden bird-tables, often making repeated visits and carrying away their food to a safe cache.

The Coal Tit's distribution has expanded this century. Tomes (1904) had no proof of nesting in either Warwickshire or Worcestershire, Harthan said there were very few Worcestershire breeding records and Norris regarded it as fairly common, but local in Warwickshire. In Staffordshire Smith considered it fairly common in most wooded districts, though possibly preferring high-lying plantations and scrub-clad valleys to open lowland country, whilst Norris (1951) and Fincher (1955) noted its absence from the south of both Warwickshire and Worcestershire. *Atlas* fieldwork between 1966 and 1972, however, proved breeding in sixty-three 10-km squares (82 per cent) and birds were present or suspected of breeding in the remaining fourteen squares. This expansion has coincided with a noticeable increase in conifer planting, with the area in 1965 being three times that of 1924.

The present distribution is largely dictated by suitable habitat, with gaps only in the bleaker, tree-less parts of the moors; the more densely built-up parts of the Conurbation; and the Lias clay area of the Avon Valley, including the Cotswold scarp, where the soils are too heavy or calcareous to suit conifers.

Numbers are affected by severe winters such as those in 1946/7 and 1962/3 but from the CBC would appear to recover quickly afterwards, with the national population having nearly trebled since 1966. Even where the population is densest, like the conifer plantations of Cannock Chase, the Coal Tit is more reluctant than

76.   The Coal Tit's long, thin bill is well adapted for feeding in conifers and this species
often stores its food in crevices and holes.   *M. C. Wilkes*

either the Blue or Great Tits to forsake
natural nest sites in favour of nest-boxes
and the Cannock Chase nest-box scheme
from 1962–6 produced only 42 nests, or 13
per cent of all occupied boxes. The CBC
has shown up to five pairs in 40 ha at
Fradley (1966–79) and usually one pair in
16 ha at Edgbaston Park during the same
period. In 1978 the very limited regional
CBC showed 12 pairs per km² of woodland, compared to 15·5 pairs nationally. In
both cases most census plots are in deciduous woodland, however, and Sharrock
quotes up to 100 pairs per km² for more
favoured areas such as coniferous woodland. Overall, a regional population of
5,000–15,000 pairs is indicated, which conforms well with the range of 2,000–20,000
given by Lord and Munns.

In winter the resident population is augmented by varying numbers of continental
immigrants. More than usual arrived in the
Region in both 1957 and 1971, when birds
were reported from localities where they
had not hitherto been recorded. Most
winter reports refer to loose flocks of less
than 50, though an estimated 100 were at
Enville on March 17th 1973. As with Goldcrest, there have been few attempts to
census extensive areas of coniferous woodland, but a survey of Enville in November
1967 revealed an estimated 1,000 birds and
Cannock Chase must similarly hold large
numbers. As yet ringing recoveries tell us
nothing about the origins of winter immigrants to the Region, the only recovery
being of a nestling ringed at Kineton and
recovered in Sutton Coldfield.

Blue Tit

Parus caeruleus

Widespread and abundant resident.

The Blue Tit breeds wherever a few trees or shrubs provide suitable nest holes and is absent only from the barest hills and moors. It prefers broad-leaved trees, especially oak and beech, but has proved very versatile in coming to terms with a human environment, exploiting nest-boxes, drain-pipes, letter-boxes and lamp-posts as well as natural holes. After breeding, birds usually resort to woods, orchards and gardens, where they feed on insects, often high in the canopy. In winter, small flocks mix with other tits in hedges and woods to feed on birch seeds, beech mast, elder-berries, blackberries or rose hips. The Blue Tit is the most likely tit to feed in osier, reed and sallow beds and it also comes regularly to bird-tables, even in the heart of built-up areas. Norris even referred to a party invading a butcher's to feed on mutton, sirloin and beef!

The familiar Blue Tit has always been one of our commonest birds. Smith described it as numerous everywhere except the barest hills and moors, Harthan considered it more abundant at all seasons than any other tit and Norris said it was widespread and common. In overall abundance, it ranked fourth amongst species caught in the Bartley Heligoland trap in 1950, fifth in a Birmingham suburban garden from 1965–8, sixth in the woodland CBC plots in 1978 and fifth on the farmland plots the same year. During the *Atlas* surveys (1966–72) breeding was confirmed in every 10-km square. Numbers are depleted by severe winters, however, and after that of 1946/7 Blue Tits were found to be outnumbered by Great Tits at Sheriffs Lench. A similar set-back followed the 1962/3 winter, but since then numbers have steadily increased.

Most reports show Blue Tits to be two to four times more numerous than Great Tits and this is confirmed by an analysis of nest-box records over the past ten years (see Great Tit). The same analysis of 233 nests also shows an average of 5·49 fledglings, although in the conifer plantations of Cannock Chase (1962–6) there were only 3·55 fledglings per nest. Breeding success varies from year-to-year, with 1957 and 1975 being good years and 1958 and 1961 poor ones. The density of Blue Tits varies widely. At Fradley Wood between 1966–9, before the oaks were felled, there were on average 34 pairs in 40 ha, whereas from 1970–8 in what was then largely birch scrub the average fell to 16 pairs. At Edgbaston Park – no more than 3 km from the centre of Birmingham – there was an average of 16 pairs in 16 ha during 1966–79, whilst 184 ha of mixed farmland and woodland at Moreton Morrell held an average of 20 pairs from 1972–9.

The regional CBC plots in 1978 gave an average of 15·9 pairs per km^2 on farmland and 62·5 pairs in woodland, and the national plots 16·5 and 57·4 pairs respectively. These densities suggest a regional population of 80,000–100,000, which is less than might be extrapolated from Sharrock. Lord and Munns simply estimated more than 20,000 pairs.

Winter flocks usually comprise less than 100 birds, but 300 were noted at Brandon Hall on December 23rd 1974 (see also Great Tit) and the same number were present each winter from 1974–6 at Edgbaston Park. During winter, birds come more into the urban area and in the centre of Birmingham four were seen in 1950 and several in 1972. Some of these are immigrants and autumn movements were revealed by a ringing scheme at West Bromwich in 1949, whilst in 1957 there was evidence of the irruption that was so marked in eastern and south-eastern Britain, with more records than usual from the inner suburbs and industrial areas. Despite this being one of the most ringed species, however, recoveries still tend to come from near the ringing site. Where movement is shown, most travel less than 30 km, but a few go as far as 100 km. Frequently it is young birds that make these movements. The one exception is a Blue Tit caught at Titchwell (Norfolk) in October 1974 and recaught in Birmingham in December 1975.

Great Tit
Parus major

Widespread and abundant resident.

During the breeding season the Great Tit frequents a variety of habitats that afford suitable nest-holes. Broad-leaved woodland is its favoured haunt, but parks, gardens and well-timbered hedgerows are used too and it takes readily to artificial sites such as nest-boxes, drain-pipes and letter-boxes. Outside the breeding season it forms small flocks along with other tits, often feeding on beech mast or yew berries, and it is more inclined to feed on the ground than the other tits. At this season bird-tables and tit-feeders are also regularly visited, especially in hard weather when milk bottles are also raided – a habit first noted in the West Midlands in the severe winter of 1946/7.

Its ability to adapt to, and exploit, man's environment has aided its success and, after the Blue Tit, the Great Tit is the commonest tit. Smith said it occurred universally, though being scarcer in the hill country than the lowlands during summer; Harthan described it as common and noted that it was more often seen around inhabited places than in country woods and hedges during winter; and Norris said it was widespread and common. During the *Atlas* surveys (1966–72) breeding was recorded in every 10-km square and its general status is indicated by its being the eighth commonest species to be trapped at Bartley in 1950, the seventh commonest in

a Birmingham suburban garden from 1965–8, the ninth commonest in CBC woodland plots in 1978 and the seventh commonest on farmland plots in the same year.

Most reports indicate two to four times more Blue Tits than Great Tits and nest-box returns over the past ten years have revealed 126 Great Tit nests against 233 Blue Tits' – a ratio of 1:2. The Great Tits' fledging success from these was 3·95 young per box. These figures exclude the experimental nest-box scheme on Cannock Chase, where in the conifer plantations between 1962 and 1966 Great Tits occupied 211 nest-boxes against the Blue Tits' 69 – a ratio of 3:1. Here fledging success was lower, with an average of 3·07 young per box. Amongst the more unusual breeding records were ones of double broods in 1954 and 1957. Reduced numbers were reported after the hard winter of 1962/3, but since then the population has returned to full strength. At Fradley Wood from 1966–9, before the oaks were felled, an average of 16 pairs was recorded in 40 ha, but from 1970–8, in largely birch scrub, this had fallen to four pairs. At Moreton Morrell, 184 ha of farmland and woodland held an average of 12 pairs during 1972–9, whilst at Edgbaston Park from 1966–79 there was an average of eight pairs in 16 ha. The 1978 CBC showed regional densities of 8·6 pairs per km² on farmland and 31·3 pairs in woodland and national ones of 8·2 and 36·2 respectively. These densities point to a regional population of 40,000–50,000 pairs, which, as with Blue Tit, is somewhat lower than would be extrapolated from Sharrock. A population of over 20,000 was suggested by Lord and Munns.

Winter flocks of Great Tits are usually less than 50 and one of this size was considered noteworthy in 1949, but since 1974 over 100 have occurred more than once in Edgbaston Park and an exceptional 500 were at Brandon Hall on December 23rd 1974. Again a colour ringing scheme at West Bromwich in 1949 showed a definite autumn movement, but the species is largely sedentary and despite much ringing has shown little movement overall. As with Blue Tits, the majority of recoveries are less than 30 km from the ringing site, but two birds have moved more than 100 km and the longest recorded movement was of 115 km.

Nuthatch
Sitta europaea

Fairly numerous resident.

At all seasons the Nuthatch shows a marked preference for mature, deciduous timber, especially oak, beech and sweet chestnut, and is consequently less widely distributed than the Treecreeper. It is mostly confined to small, mature woods, old orchards, shelterbelts, avenues and parkland, where it nests in the holes of trees and feeds along their trunks and main branches. It will nest readily in nest-boxes, but seemingly with mixed success judging from the fact that only one pair out of six was successful in raising young in 1978 at both Compton Verney and the Wyre Forest.

Smith described the Nuthatch as a very local resident, rare in north Staffordshire, but Lord and Blake (1962) noted an increase and assessed it as breeding in all areas except the moors and south-eastern districts. Harthan too described it as local, observing that it was found in groups or avenues of old trees, especially elms, being far less common in large woods. Norris also observed it was largely local, being encountered with some regularity in south Warwickshire, but less frequently in the north, where in many areas it was something of a rarity.

The *Atlas* surveys (1966–72) confirmed breeding in fifty-five 10-km squares (71 per cent) and revealed suspected breeding in a further sixteen (21 per cent), thereby confirming a generally wide distribution. Proved breeding came on the moors only during the latter years of the surveys, but whether this represented recent colonisation or just more intensive searching is not certain. Providing there is suitable habitat, Nuthatches occur well inside built-up areas and they breed regularly, for example, along the Kenilworth Road, Coventry, where seven pairs were noted in 4 km in 1970, and at Edgbaston and Warley Parks. They are most scarce in the Avon and Severn valleys, where there is noticeably less mature woodland, and in the more heavily developed areas.

Numerically its fortunes appear to have fluctuated over the years, with declines regularly reported from Sutton Park, but overall it seems that the gradual national increase noted by Parslow (1967) has been reflected in the West Midlands. At Edgbaston Park, for example, just one pair was recorded sporadically on 16 ha prior to 1975, but there have since been two or three pairs every year. The species was not recorded on any other regional CBC plot in 1978, but nationally in woodland there were $4 \cdot 9$ pairs per km^2. It seems likely from this that the regional population is between 700–1,400 pairs, which is in the middle of Lord and Munns' range of 200–2,000 pairs.

Nuthatches are largely sedentary and faithful to traditional haunts, even to the extent of having nested in the same tree between Habberley and Kidderminster for fifteen consecutive years. It is surprising, therefore, to read Smith quoting Mosley as saying that "at least one hundred came to feed on insect-infested currant bushes at Rolleston on August 16th 1846". This record seems very dubious, but if correct would be quite exceptional, since Nuthatches seldom congregate in other than family parties and more than ten is unusual. The most seen in recent times have been 15 at Atherstone on November

29th 1952 and 12 at Arbury on January 29th 1964. There are several notes of birds wedging nuts, often hazel, into crevices in the bark of trees in order to hammer them open, but one observer recorded this behaviour with soft fruit too. More unusual records refer to one feeding on a dead Woodpigeon at Chillington in 1958 and two at Hewell Park in 1976 feeding on an angler's maggots.

Treecreeper
Certhia familiaris

Widespread and numerous resident.

The Treecreeper is one of those species that is rather taken for granted and information is scarce, perhaps because its unobtrusive habits make it hard to census. Smith thought it probably far more plentiful than is generally supposed and Norris also echoed this opinion, whilst Harthan said it was common in woodlands throughout Worcestershire.

This latter statement is still true and Treecreepers are likely to be found wherever there are trees, be they conifers or broad-leaved. They nest behind the loose bark of trees like birch and willow, or in ivy, and Smith said they had been persuaded to breed at Cheadle by securing a piece of bark against a tree, but with sufficient clearance around to take a nest. During the *Atlas* surveys (1966–72) breeding was proved in sixty-seven 10-km squares (87 per cent), including much of the West Midland Conurbation, and suspected in a further nine squares (12 per cent). Where breeding was not proved, poor coverage is as likely a cause as the absence of birds. It is clear from the gaps in distribution revealed by the WMBC *Atlas* alone, however, that Treecreepers are sparsely distributed on the northern moors, in the Conurbation and in the lower Avon and Severn valleys, where woodland is also sparse. Density is hard to gauge, but 30 were found in two-and-a-half hours in the Wyre Forest on June 25th 1950, two or three pairs were regularly present in 40 ha of Fradley Wood until the main trees were felled in 1969, and the 16 ha of Edgbaston Park held an average of almost two pairs a year from 1966–79. Nationally the CBC in 1978 showed $1 \cdot 09$ pairs per km^2 on farmland and $5 \cdot 02$ pairs in woodland and from this a regional population can be postulated of 4,000–6,000 pairs, which is somewhat higher than the 200–2,000 suggested by Lord and Munns.

In winter Treecreepers often roost in crevices in the bark of Wellingtonias. At this season too they sometimes join foraging parties of tits and small flocks up to ten are occasionally reported, mostly between January and March. Autumn influxes have been noted more than once at Edgbaston Park, but there is no ringing evidence to substantiate a movement into the area and it seems probable that these are just local birds flocking to a good feeding area.

There is conflicting information on the effects of the hard winters in 1946/7 and 1962/3, with marked decreases reported from some areas, but no obvious changes noted in others. At Worcester in 1963 the population was said to be down by 25 per cent, however, and it is quite clear that Treecreepers are susceptible to heavy losses in severe weather. Unfortunately the CBC provides no base line prior to 1966, but the subsequent increase with the succession of mild winters indicates that numbers soon recover. In the West Midlands this recovery has been aided by a wealth of nest sites afforded by diseased elms with peeling bark.

Golden Oriole
Oriolus oriolus

Rare passage migrant.

Up to 1979 there had been 26 acceptable records of Golden Orioles involving 29 birds – seven from Warwickshire, including pairs twice; eleven from Worcestershire, including one pair; and eight from Staffordshire. Seven of these date from the last century, including a pair at Kyre in 1868, which probably nested (Prescott 1931). Of those where sex was specified, 18 were males and only four females.

Since 1960 Golden Orioles have appeared in ten out of nineteen years, with two birds in 1960 and three in 1975. Records have come from widely scattered localities, but the Wyre Forest area has been visited more than once during this period. Several have been reported near reservoirs, but this may merely reflect the whereabouts of bird-watchers.

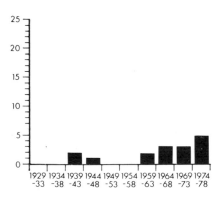

Five-yearly Totals of Golden Orioles, 1929–78

Most are spring records, with three in April; eight in May, mostly in the last ten days; seven in June and one in early July. There are just three autumn records, one of a bird which stayed from early September 1960 to early October at Enville, and two in October, one as late as October 28th (1970).

Little information is available on habitats, but birds are invariably well-concealed in the canopy, frequently in oak, though one was seen feeding on ripe pears at South Littleton in 1892.

Red-backed Shrike
Lanius collurio

Rare passage migrant or vagrant. Formerly bred.

The Red-backed Shrike was once a characteristic bird of dry, gorse-clad heaths with patches of hawthorn scrub, particularly those along the Severn Valley from Enville to Malvern. Sadly it has declined everywhere during this century and today it appears only rarely, but whether on passage or as a vagrant is unknown. Some idea of its former strength is given by Hudson, who in 1903 counted 12 birds in 16 km of roads around Stratford, although even he admitted that frequently he could cover four times that distance and see but one or two.

With the decline, the range of the Red-backed Shrike receded steadily southwards. Mosley (1863) said it nested uncommonly about Burton-on-Trent, but that some years before it was so regular that "every juvenile robber of birds' nests knew its eggs". In 1887 it was reputedly increasing around the Shropshire border and birds frequented the Blymhill – Weston-under-Lizard area until at least 1903. At one time it nested in Dovedale and bred occasionally along the Churnet, with a pair at Oakamoor in 1918 (Smith). The same author also quoted breeding records from the Trent Valley until 1912. In Warwickshire, Norris said the Red-backed Shrike was locally distributed throughout the rural parts of the county until 1915, but that thirty years later it had become scarce and greatly reduced in numbers. In Worcestershire the decline came later and Harthan, writing in 1946, said it was local, but still breeding fairly regularly on the slopes of the Lickeys and from the Teme Valley south to the Malverns. At the latter locality there were still an estimated thirty pairs in 1941, (Harrison 1941), but by 1961 only four remained.

The catalogue of recent occurrences

Table 75: Decline of the Red-backed Shrike, 1950–57

	1950	1951	1952	1953	1954	1955	1956	1957
No. of pairs	7	5	2	2	0	1	0	2

makes dismal reading. A nest at Stourton in 1948 was the last confirmed breeding record for Staffordshire, though a pair was seen at Enville in the summer of 1951, then in the 1950s the population declined as shown in Table 75.

Subsequently a pair bred at Lillington in 1960 – the last Warwickshire record – and finally the last positive proof of breeding to be recorded in the WM *Bird Reports* came from Malvern in 1962, although birds were seen in their characteristic habitat of lowland heath near Kidderminster during the summers of 1963 and 1964. However, Teagle (1978) mentioned Saltwells Claypit as a breeding site until 1966 and a passage bird appeared here in 1969 (D. Smallshire *in litt*).

Since 1964 there have been only eight other sightings, at Sutton Park in April 1965, Upton Warren in September 1966, Sutton Park in June 1968, Brandon in May 1971, Wyre Forest in August 1972, Chapel Chorlton in June 1976, Droitwich in September 1976 and Upton Warren in June 1977. Of these only the Sutton Park birds were in typical breeding habitat and it seems certain that the records refer only to passage birds or wandering individuals.

Great Grey Shrike
Lanius excubitor

Scarce passage migrant and winter visitor.

Though scarce, the Great Grey Shrike is an annual visitor to the West Midlands. It might appear almost anywhere where there are scattered thorn bushes or trees for

perches, though it often shows a liking for barren, bleak areas of heathland or rough grassland and is particularly faithful to a few favoured localities. Tomes (1904) said that Warwickshire records were too many to enumerate, yet Norris knew of none between 1904 and 1947, whilst Harthan listed six Worcestershire records prior to this century and Smith at least half-a-dozen for Staffordshire, including one at Ellastone on the improbable date of July 10th 1921.

During this century there have been 227 records, all but nine of them since 1949, with 1966 marking a very pronounced upturn. Prior to then no more than four birds had been reported in any one year, yet since there has never been less than seven, whilst the period 1972–6 was especially rich, with peaks of 24 in 1972 and 26 in 1976. The quinquennial totals for the last fifty years are shown in Table 76.

Worcestershire, with 16 per cent, has the fewest Great Grey Shrike records, except of course for the new West Midlands County, whilst Staffordshire, with 48 per cent, has the most. To a large extent this is because

Table 76: Five-yearly Totals of Great Grey Shrikes, 1929–78

	1929–33	1934–38	1939–43	1944–48	1949–53	1954–58	1959–63	1964–68	1969–73	1974–78	Total
No. of birds	0	0	0	0	7	6	14	38	76	77	218

Cannock Chase is so favoured by Great Grey Shrikes, with records in twelve of the fifteen years 1964–78. In all there have been 31 records (14 per cent) from here, with two-thirds of the birds staying longer than a day. Records have come from all parts of the Chase, but mostly from the Brocton Field – Oldacre Valley area, where up to two birds have been seen on several occasions, sometimes together. Nearby, Chasewater – Brownhills Common is another popular area, with 20 records and three-quarters of its birds having stayed longer than a day, whilst Brandon and Sutton Park have each had 12 records. Other areas visited with some regularity include the Tame Valley, Bittell, Castlemorton Common and the heaths around Kidderminster. Away from Cannock Chase, bird are usually solitary.

The monthly distribution of arrival dates is shown in the histogram.

An unusually early arrival was at Blithfield on September 6th 1970, but otherwise arrival takes place mainly during October and November, with a peak in late October. First dates continue through from December to April, or exceptionally May, but some of these possibly relate to earlier arrivals that have remained undetected. However, the secondary peak in late March and April is clearly indicative of a return passage. This is reinforced by departure dates, with many of the earlier arrivals passing through during October, November and December. There is then a marked exodus in late March and early April. The latest records were both on May 14th, from Enville in 1951 and Blackbrook Valley in 1972.

Woodchat Shrike
Lanius senator

One record.

The only record was that quoted by Harthan of a pair at Weatheroak Hill on May 14th 1893 – a typical date for this vagrant from Southern Europe (Sharrock 1974).

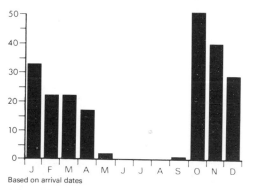

Monthly Distribution of Great Grey Shrikes, 1929–78

Jay
Garrulus glandarius

Numerous resident.

The Jay is an arboreal and wary bird with a marked preference for coppiced woodland with standard oaks, or for conifer plantations, though it can also be found in overgrown hedgerows and dense shrubberies in suburban parks and gardens. There are no

401

records of its having nested in Birmingham's streets, as it has in London, but a pair did attempt to nest in the middle of Leamington in 1951.

The Jay is widespread, occurring wherever there is suitable habitat. Smith said it occurred more or less commonly in all well-wooded districts, Harthan that it was common wherever there were woods and Norris that it could be seen in practically every suitable area. During the *Atlas* surveys (1966–72) breeding was proved in sixty-six 10-km squares (86 per cent) and suspected in a further nine (12 per cent), with the lowest population on the northern moors, in the more intensively built-up areas of the Conurbation and in the less-wooded districts of the Avon and Severn valleys.

The numerical status of the Jay depends on habitat, food and persecution. It feeds principally on invertebrates and fruit from trees, having a highly developed, symbiotic association with oak. During the summer it feeds on young oaks, whilst the nestlings are fed on oak-defoliating caterpillars, and in autumn and winter it relies heavily on a diet of acorns, many of which it buries for later retrieval. Inevitably some are never found and this helps propagate the oak. Whenever there is a poor acorn crop, autumn wanderings of Jays are noticeable. Unfortunately the Jay is also partial to soft fruit, eggs, nestlings and game-chicks, so not surprisingly it is scarce in horticultural and well-keepered areas, with Smith quoting as many as 150 taken in a single winter near Oakamoor. With a more enlightened attitude in recent years, however, its numbers have increased and it has been quick to exploit the relative safety of the suburbs. In 1947 its density in south Warwickshire woods was said to be ten pairs per km², whilst at Fradley Wood, before the felling of its favourite oaks in 1969, there were five pairs in 40 ha. The 16 ha of Edgbaston Park have also held up to five pairs, though the average from 1966–79

was 2·5 pairs. Such densities are high and help to explain why, in 1978, the few regional woodland CBC plots showed 9·6 pairs per km² compared to only 4·98 pairs nationally. Probably the regional population is 2,000–4,000 pairs, which is somewhat above the range of 200–2,000 pairs put forward by Lord and Munns.

Small parties are sometimes reported in autumn, when up to 20 are quite frequent and Harthan once saw 40 disturbed from an oakwood by foxhounds. In 1957 there was an influx in eastern and south-eastern England and some birds penetrated as far as the West Midlands, with up to 30 at Bartley on September 28th and 19 flying north at Cannock on November 11th.

Magpie
Pica pica

Widespread and numerous resident.

The Magpie occurs wherever there are isolated trees, thickets or scattered shrubs in which it can build its domed nest. It is found throughout the Region on farmland and along woodland margins and has increasingly spread into suburban gardens and parks. None the less it remains commonest in areas of grassland enclosed by tall, thick hedges and dotted with copses or shelterbelts.

Smith said it was common locally in lowland Staffordshire, but perhaps its true home was amongst the northern hills, where it abounded in the high-lying woods and scrubby tracts of the limestone hillsides. Harthan said its population varied in proportion to the number of gamekeepers, being very abundant in 1943, whilst Norris quoted Tomes' (1904) reference to a general decrease in Warwickshire, but reported a subsequent increase with certainly no shortage in the southern part of the county, especially in the hawthorn roughs at Snitterfield, Wilmcote and Lighthorne. Although it has since lost much suitable habitat through clearance of scrub, hedges and trees, this has to some extent been offset by the further fragmentation of woodland and overall the Magpie's distribution has not changed significantly, with breeding confirmed in every 10-km square between 1966–72.

The Magpie enjoys a varied diet, which, in addition to innocuous items like snails, insects, fruit and berries, unfortunately includes eggs, nestlings and young birds. Consequently it has been heavily persecuted in the past and its recent increase in numbers reflects reduced persecution as much as anything else. In particular it has benefited from the safe sanctuary of the suburbs and two or three pairs have regularly nested in 16 ha at Edgbaston Park. Nationally the CBC showed a population increase of nearly 30 per cent from 1966–78, by when there were $3 \cdot 76$ pairs per km^2 on farmland compared to a regional figure of $3 \cdot 05$ pairs. Little is known about density in the Region's woodlands, but nationally there were $7 \cdot 01$ pairs per km^2 in 1978. Such densities would imply a regional population of 8,000–13,000 pairs, which is within the range of 2,000–20,000 pairs suggested by Lord and Munns.

Magpies are gregarious and parties up to 20 are commonplace, even in the suburbs, whilst 30 is not unusual and 60 were seen at Sheriffs Lench in 1944. They also roost communally. In 1935, 70 were seen going to roost in a larch coppice near Evesham; up to 100 roosted at Shirley in 1949 and at Edgbaston Park in 1974 and 1976, whilst the maximum reported was 150, also at Edgbaston Park, in 1973.

Birds have been seen to follow foxes more than once in recent years and on one occasion a fox was observed eating a dead Magpie, surrounded by a noisy gathering of 12 further birds. Others have been seen on the backs of bullocks, pecking at the ears and roots of the tail, and one was seen attacking a grass snake. Despite being common, the Magpie is rarely ringed and the only known recovery is of a bird ringed at Bournville in 1975 and shot at Wythall in 1977.

Nutcracker
Nucifraga caryocatactes

One record.

The only record is of one in a garden at Stapenhill, Burton-on-Trent, on October 27th and 28th 1968 (*Brit. Birds* 63:372). In that year an invasion of the slender-billed Nutcracker, *N. c. macrorhynchos*, from Siberia reached unprecedented proportions and perhaps the most remarkable feature was the absence of any records from Warwickshire, Worcestershire, Gloucestershire or Oxfordshire, whilst birds appeared in all surrounding counties (Sharrock and Sharrock 1976).

Chough
Pyrrhocorax pyrrhocorax

One dubious record.

Hastings (1834) mentioned "a bird killed at Lindridge, near Tenbury, in November 1826 when perched on a building where it was resting after a long flight". In the absence of a full knowledge of the status and distribution of the Chough at this time, it is impossible to put this quite exceptional record into perspective.

Jackdaw
Corvus monedula

Widespread and numerous resident.

The Jackdaw occurs throughout the Region, nesting colonially in crevices in trees, rocks and buildings, and feeding extensively on insects, which it takes from around livestock and pastures with long grass. Occasionally it feeds on grain stubbles as well and it commonly scavenges at rubbish tips. In the countryside, woods, old orchards and scrub are all frequented, but its preference for lush pastures and gnarled old trees are best met by parkland and this is perhaps its optimal habitat. Large colonies also occur amongst the cliffs and caves of the limestone dales and Smith observed that the great bluff of Beeston Tor harboured so many that it was locally known as "the Jackdaw". It is not just a bird of open country, however, but also inhabits church towers and house chimneys in village and town alike.

Previous authors all described the Jackdaw as plentiful and this is still true today, with breeding proved in all but two 10-km squares (97 per cent). Only in the heart of the Black Country was breeding not even suspected, though Lord and Blake (1962) observed that its density on Cannock Chase was low and this is still true. Conversely the highest densities are found around ruined buildings, with Kenilworth Castle, for example, believed to have held the largest colony in Warwickshire in 1949; limestone caves, with a colony of 50 pairs at Wolfscote Dale in 1973; and parklands such as those along the Avon Valley. In the last mentioned area, however, the population might be affected by the felling of dead elms, as a colony of 200 was discovered in dead trees at Compton Verney in 1975. In recent times the Jackdaw has colonised the new or rapidly expanding stone quarries of north Staffordshire, Rowley Regis, the Nuneaton district and south-east Warwickshire. There is little information on density, but 16 ha at Edgbaston Park hold about 15 pairs and up to two pairs have bred on the 184 ha CBC plot at Moreton Morrell. In 1978 the national farmland density averaged $1 \cdot 97$ pairs per km^2. Considering these figures, and allowing also for the high densities in western Britain, the figure of $1 \cdot 6$

pairs per km^2 overall (Sharrock) is probably too high for the West Midlands and a regional population at the lower end of a range 5,000–15,000 pairs seems most likely. This is considerably less than the estimate of 20,000 plus pairs made by Lord and Munns.

In winter, flocks of feeding Jackdaws 300 to 500 strong are common and the species also roosts communally, often in the company of Rooks. There have been four reports of flocks exceeding 1,000, with the most being 2,000, at Hanbury on November 16th 1978, and 2,500 flying SW towards their roost at Patshull within ten minutes at Bilbrook on January 2nd 1967. According to Smith Jackdaws could be seen on migration, passing westwards along the Churnet Valley in autumn and appearing in increased numbers in the hill country about mid-October. Ringing evidence points to such movements being very local, however, since of the eight recoveries the longest journey was only from Alcester to Burton-on-Trent and in the main the Jackdaw moves very little. Judging from the numbers trapped and ringed, this was the commonest corvid found scavenging on rubbish tips in Warwickshire and Worcestershire (Green 1978).

Rook
Corvus frugilegus

Widespread and abundant resident, but has declined significantly since the mid-1950s.

The Rook is primarily a bird of farmland, where it nests in hedgerow trees, copses and small woods and probes for invertebrates amongst pastures, grass leys and stubbles. It is a widespread and fairly abundant species, absent only from the most densely-wooded and most heavily-industrialised areas of the Region.

Previous authors all described the Rook as common, or widespread and numerous.

77.   The Rook is a bird of farmland with tall trees, and surveys in 1973 and 1975 revealed around a thousand rookeries in the Region.   *M. C. Wilkes*

A comprehensive survey during 1973 (Dean 1974) revealed 1,022 rookeries containing 19,079 nests in an area of 7,100 km². The average density of nests was 2·7 per km², or 2·9 if inadequately censused areas were climinatcd. The average rookery size was 18·7 nests, while three-quarters of all rookeries contained less than twenty-five nests. At the other end of the scale, the largest rookeries were 161 nests, in the Manifold Valley, and approximately 150 nests, near Warwick. Four species of tree – elm, oak, beech and ash – held over 90 per cent of the nests for which the host tree was identified, with elm alone accounting for more than 50 per cent. Rookeries in elms, however, were on average only half the size of those in oaks.

The density of nests in different areas varied widely, from zero at the centre of Birmingham to 9·7 nests per km² in the Manifold Valley. In general, densities were highest in the valleys of the rivers Avon, Severn, Penk, Sow and Manifold, which are largely mixed agricultural areas supporting good numbers of hedgerow trees and small copses. Interestingly, high densities were often found in areas of heavy soils, perhaps because these have a high water table that brings invertebrates nearer the surface. Densities were lowest amidst the industrialised parts of the Birmingham plateau and the Potteries. The heavily wooded Cannock Chase and Wyre Forest also supported few rookeries. Birds do not shun built-up areas entirely, however, and there is a small rookery in the main shopping street of Leamington Spa, one in Worcester and another in the centre of Nuneaton.

A repeat census was conducted as part of the 1975 national BTO enquiry (Sage and Vernon 1978) and this revealed a regional total of 22,478 nests. Sixty-eight fully-censused 10-km squares contained 21,475 nests, giving a mean density of $3 \cdot 16$ nests per km^2 for the Region, compared with a national average of $3 \cdot 9$. Fifty-eight 10-km squares were fully censused in both years and the comparative figures for these squares were as shown in Table 77.

Although these figures imply an increase of $8 \cdot 6$ per cent between the two years, the indications are that the number of Rooks has declined significantly since the mid-1950s. In 1946 Norris and Hawkes found a density of $11 \cdot 5$ nests per km^2 in an area of 555 km^2 of southern Warwickshire, which compares with a density of only $4 \cdot 5$ in a similar area during 1973. In 1945 A. J. Harthan and H. J. Tooby found a density of $7 \cdot 5$ nests per km^2 in an area of 948 km^2 centred on southern Worcestershire, while G. H. Green discovered a similar density of nests in an area of 334 km^2 in 1952 (Dean *op cit*). By 1973 the densities in comparable areas had fallen to $3 \cdot 0$ and $4 \cdot 0$ respectively. Although based on limited sections of the Region, these figures suggest that the Rook population of the West Midlands has declined by about 60 per cent since the early 1950s. This decline has, in fact, been nationwide, with the 1975 BTO Census revealing a decrease of about 45 per cent in the English population since 1945/6. Increasing urbanisation, agricultural changes and climatic factors have all been suggested as possible factors, but no conclusive reasons for the decline have yet been demonstrated. However, it may be significant that the national population increased by 20 per cent between the 1930s and the 1945/6 census, so perhaps the Rook is subject to population cycles.

In the West Midlands, where the elm is a very important host species, the outbreak of Dutch elm disease has added a new and potentially very serious dimension to the pressures exerted on Rooks. The indications at present, however, are that it continues nesting in dead trees until these are felled, when it moves to the next tallest tree available. Recently nests have been noticed in much smaller trees and at much higher densities than normal, with twenty-five nests in a single small oak at Draycote in 1980. Further adaptability in the choice of nest sites is demonstrated by occasional instances of birds nesting on man-made structures such as girders and pylons at power stations. Such incidents have been recorded at Stourport power-station from 1953 to at least 1960, Hams Hall in 1958, Birch Coppice colliery in 1965, Gratton in 1975 and Whateley in 1978.

Outside the breeding season, large gatherings of Rooks are regularly reported, frequently involving pre-roost assemblies. Over 1,000 birds have been noted on several occasions, with the largest flock comprising 1,500 birds at Hanbury Park on November 16th 1978. Like all corvids, the few recoveries available indicate little movement. There is, however, a record of a

Table 77: Comparison between the Rook Censuses of 1973 and 1975.

Year	Area in km^2	No. of Rookeries	No. of Nests	Average No. of Nests per Rookery	Density of Nests per km^2	Density of Rookeries per km^2
1973	5,800	918	16,954	$18 \cdot 5$	$2 \cdot 92$	$0 \cdot 16$
1975	5,800	1,033	18,902	$18 \cdot 3$	$3 \cdot 26$	$0 \cdot 18$

Rook ringed at Spurn (Humberside) in November 1953 and recovered in February 1954 near Tamworth.

Carrion Crow
Corvus corone

Widespread and numerous or abundant resident.

More than any other corvid, the Carrion Crow has managed to adapt to a variety of new environments and there are few places where it cannot be seen. It is an omnivorous species, feeding on insects, earthworms, small mammals, eggs and birds as well as carrion. Less typically it has been noted taking fish from water or dropping walnuts and freshwater mussels from a height to crack them open. More than anything, though, it is the scavenger of farmland and is commonest in pastoral landscapes with fragments of woodland, shelterbelts and copses. None the less it spans a wide range of habitats from moors and heaths to town gardens, parks and streets and is regularly seen in some numbers around waste ground, reservoirs and rubbish tips. Carrion Crows nest singly, usually in trees or bushes, though man-made structures such as electricity pylons have been utilised as well. Sometimes they feed and roost communally.

Until the beginning of this century, the Carrion Crow was mercilessly persecuted by keepers. Smith, however, said it was widely distributed in Staffordshire and plentiful in many districts where game was not preserved, adding that in the northern hill country it was exceedingly common, with many nesting in thorn bushes. In Worcestershire, Harthan said it was widely distributed, but most abundant in the river valleys. Norris quoted decreases in 1895 and again in 1904, but thought that a relaxation in keepering during two world wars had led to a subsequent increase. Even

so, he still described it as a resident only in small numbers, but acknowledged that it was more common in south Warwickshire, towards the Cotswolds, where it had increased in recent years. There has been no subsequent change to this distribution, with breeding confirmed in every 10-km square bar one during the *Atlas* surveys (1966–72).

Numbers have undoubtedly increased, however, with the size of flocks rapidly building-up in the early 1960s and becoming a feature of summer as well as winter. An indication of the degree of persecution can be gleaned from the fact that 103 were killed on the Arbury estate even as late as 1960. Harthan found winter flocks of up to 30 in riverside meadows worthy of mention, and prior to the mid-1950s the most reported were two gatherings of 50, one feeding on freshwater mussels along the shoreline at Bittell in October 1952 and the other at Pattingham on December 31st 1955. In 1963, 70 were counted around Blithfield Reservoir, whilst by 1966 the numbers at Bittell had risen to 100 and the following year topped 150. Since then counts at this latter locality have been more stable, with 100 again in 1977 and 170 in 1978. Some of the CBC plots also show an increase. For example, 184 ha of farmland and woodland at Moreton Morrell held none from 1973–5, five pairs in 1976 and 1978 and nine pairs in 1979. The 1978 CBC also showed farmland density to vary from one pair on 95 ha at Wellesbourne to five pairs on 64 ha at Willey. Overall there were $3 \cdot 87$ pairs per km^2 regionally and $4 \cdot 35$ pairs nationally. From these figures, and the national average of $6 \cdot 45$ pairs per km^2 for woodland, it seems the regional population probably lies between 15,000–25,000 pairs, which is at the upper end or slightly above Lord and Munns' estimate of 2,000–20,000 pairs.

In addition to those mentioned already, even larger flocks have occurred elsewhere, especially in the Tame Valley, where from

1967 to 1970 gatherings of 500 were noted annually and as many as 800 were at Little Packington on June 11th 1973. Some impressive numbers have occurred at Whittington Sewage Farm too, with 250 in 1968, 190 in 1971 and 600 in December 1975. Indeed flocks up to two hundred are quite common in some places today.

Similar increases have been noted at roosts, where numbers are even larger. In 1956 a roost of 70 at Sheriffs Lench was noteworthy, but by 1962 150 were roosting in alders in Sutton Park and then in 1966 as many as 500 roosted on the banks of the fly-ash lagoons at Hams Hall Power Station. The largest roost of all was 3,000 at Bunkers Hill during the 1972/3 winter, but this seems to have been exceptional and numbers up to 500 are much more typical of the present-day pattern. There have been only five recoveries of ringed birds, all nestlings which showed no significant movement.

The Hooded Crow *C. c. cornix* is a rare winter visitor and passage migrant of erratic appearance, though a few individuals have over-wintered from time-to-time and breeding has occurred at least twice. Norris acknowledged that the earliest Warwickshire records were most confused, but there was definite proof of breeding in Sutton Park in 1883 and it is possible that nesting also occurred here in other years about this time. In those days, and for the first two decades of this century, the Hooded Crow was slightly commoner than today, being a regular, if scarce, visitor in winter months. From Staffordshire, Smith said that a very few

appeared along river valleys, such as the Blithe, Dove and Churnet, in October or early November and again in spring; and that birds were present every winter at Whittington, in the Stour Valley. In the hill country, though, he said they were rare, yet a pair was proved to have bred in Dovedale in 1915 and others may have done so previously. After 1920 occurrences declined markedly. Between 1923 and 1933 there was not a single record and at the end of the Second World War the Hooded Crow was described by Harthan as an occasional winter visitor and by Norris as a very rare winter visitor.

Hooded Crows have occurred in twenty-five of the last fifty winters. The erratic nature of their appearances is well demonstrated by the fact that four winters is the longest period of either continuous absence (1941/2 – 1944/5) or continuous presence (1970/1 – 1973/4). Furthermore the frequency with which five or more birds have appeared in a single winter is the same as that for just single-bird winters. Thus the pattern is one of small annual influxes interspersed with blank years. Overall, between 1929–78 there were 79 records involving 95 individuals, distributed as shown in Table 78.

The European population of Hooded Crows is migratory (Busse 1969) and passage along the east coast of Britain is regular. The above accords well with a westerly displacement of this passage and it seems very probable that the appearance of Hooded Crows in the West Midlands largely coincides with easterly winds. The geographical distribution of records also

Table 78: Five-yearly Totals of Hooded Crows, 1929–78.

	1929–33	1934–38	1939–43	1944–48	1949–53	1954–58	1959–63	1964–68	1969–73	1974–78
No. of records	1	4	2	5	13	9	4	7	13	21
No. of birds	1	4	2	5	22	9	4	8	16	24

fits this theory of westerly displacement, with 44 per cent of birds seen in Warwickshire, 38 per cent in Staffordshire and only 18 per cent in Worcestershire. Apart from this, there is little evident pattern as to where Hooded Crows occur, with reservoirs, derelict sites, sewage farms, rubbish tips and hilly country all being frequented.

There is more consistency over the monthly distribution of records. Coombs (1978) says the first arrivals on the east coast come in early October and passage continues throughout the month. Evidence of return passage is more scant, but it appears to occur in late March and early April.

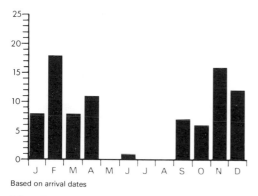

Based on arrival dates

Monthly Distribution of Hooded Crows, 1929–78

In the West Midlands, as the above histogram shows, autumn passage begins in October, or exceptionally even September, and peaks in late November, after which there is a marked tailing-off to the end of the year. Records then steadily increase again to a peak in mid-February, after which they slowly decline until all birds have left by the end of April. There is also a marked second peak in early April. An exceptional record refers to a party of six seen on stubble near Coventry on September 6th 1950. Not only was this the largest single party to occur, but it was also an unusually early date. Most birds (76 per cent) are seen only on one day, but the remainder stayed for periods ranging from seven to 113 days, with a mean stay of 46 days. Half of these protracted stays spanned dates from January to March, suggesting that early migrants pass straight through, but that later arrivals may over-winter.

Raven
Corvus corax

Scarce wanderer, mostly in spring and autumn, but occasionally in winter. Very scarce breeding species.

The West Midlands lies just to the east of the Raven's Welsh stronghold and wandering birds appear most years in the hilly country west of the Severn, especially around Malvern. One or two pairs occasionally stay to breed, but as yet the species is not firmly re-established.

In former times of course Ravens were commoner and more widespread, as many place names signify. Smith said they were once common in Staffordshire, not only in the northern hills where they bred on the Roaches and around Swythamley, but in many lowland districts too. In 1844 they were not considered rare, with Dovedale, Cheadle and Ramshorn favoured localities, but by 1863 they had deserted Needwood Forest. Hastings (1834) said that Ravens had become rare in Worcestershire, but that they still nested at Stanford-on-Teme in the late 1840s, near Upton-on-Severn in 1856 and Bewdley in 1859. These were the last breeding records and Harthan thought it unlikely ever to nest again owing to lack of carrion on which to feed. However, in 1949 and 1950 a pair did breed – at Stanford-on-Teme again – whilst in the latter year another pair bred at Malvern. In 1951–2 pairs also attempted to breed in Staffordshire, at Enville. The Raven has never been so common in Warwickshire and Norris could quote only two known breeding records, both from the eighteenth century, plus three sightings, of which the most recent came from Warwick Park in the 1850s. Since then there have been just two records, of single birds at Brandon and Tysoe, both in 1972.

Since the war records have been distributed as shown in Table 79.

Almost three-quarters of these records came from Worcestershire, with the Malvern area contributing half, but birds were also seen more than once in the Teme Valley, Wyre Forest, the Lenches and over Bredon and Clent Hills. The remaining quarter came from Staffordshire and particularly from the Enville area. However individuals have appeared on or near the moors as well, at Goldsitch in 1958 and 1960, Wolfscote Dale in 1969, Coombes Valley in 1971 and Ramshorn in 1977. Also in 1977 there were parties of seven at Flash in August and October (these being the largest ever recorded in the West Midlands) and four at Gun Hill in August. These would seem to suggest breeding somewhere in the southern Pennines and might hopefully portend future colonisation.

As Table 80 shows, Ravens have appeared in every month, though most have occurred either between late December and late April or during August and September. The records also show two concentrated peaks – one in mid-March and the other in late October. August records are more evenly spread.

Starling
Sturnus vulgaris

Abundant resident, passage migrant and winter visitor. Widespread at all seasons.

The status of this well-known species appears to have changed little over the years. It nests wherever it can find a suitable hole, from trees to buildings or crev-

Table 79: Five-yearly Totals of Ravens, 1944–78

	1944–48	1949–53	1954–58	1959–63	1964–68	1969–73	1974–78	Total
No. of Birds	2	26	21	10	10	42	29	140

Table 80: Monthly Distribution of Ravens, 1944–78.

	Jan	Feb	Mar	Apr	May	Jun	Jul	Aug	Sep	Oct	Nov	Dec
No. of Birds	14	17	29	17	6	4	7	22	15	19	4	12

Note: Breeding records have not been included.

ices in quarries and rocks; and from the centre of Birmingham to the wilds of the moors, there can be few places where none can be seen. It is even better known in winter, however, when huge flocks feed on grasslands and enormous numbers gather to roost in woods or on city centre buildings. At this season many also feed on suburban lawns or squabble over scraps on the bird-table and in urban areas the Starling is a characteristic bird of chimney pots and television aerials. It is also common on rubbish tips at all seasons.

Smith thought the Starling's numbers had increased greatly in the fifty years prior to the Second World War and described it as very common everywhere. Harthan said it was widely distributed, but with very few nesting in towns, whilst Norris considered that it more than doubled its strength in winter months. There has been no change in this universal distribution since and the *Atlas* surveys from 1966–72 proved breeding in every 10-km square. Today, however, there are more than a very few in urban areas, so there has obviously been an increase since Harthan's time, perhaps consequent upon the passing of legislation in 1956 to control atmospheric pollution. The 1978 CBC showed a regional density of 2·65 pairs per km^2 on farmland and 24·01 pairs per km^2 in woodland, compared to national figures of 9·61 and 19·36 respectively. So, making allowance for substantial urban and suburban populations, the regional population probably lies between 50,000–150,000 pairs. Lord and Munns merely assessed the Starling's population as being in excess of 20,000 pairs. Of passing interest is an exceptionally early breeding record concerning unfledged young which were discovered at Selly Oak in January 1935.

Surprisingly there are few flock counts of Starlings, but of those available 2,000 at Trescott on March 9th 1952 and 3,000 at Knighton on October 29th 1973 were the largest. By comparison there is much information on roosts. After breeding, young Starlings often resort to osier or reed-beds to roost and their combined weight on the fragile reeds has been known to topple Reed Warblers from their nest. The largest roosts, however, occur in late autumn and winter after the arrival of the winter visitors. In the centre of Birmingham, flocks of birds swirling around in the night sky before huddling together for the night, are a familiar sight. Old buildings, with cornices, pediments and ledges are preferred, but thermal insulation and the texture of the building material are also influential. Numbers are so great that their droppings cause serious erosion to some of the soft sandstone buildings characteristic of the Midlands and in Birmingham various attempts have been made to dissuade Starlings from roosting on buildings. Mostly these have met with little success and the city centre roost has varied in size from 20,000 in 1949/50 and 1951/2 to 11,600 in 1974/5 and 10,300 in 1975/6. The latter two winters were mild, however, and during the exceptionally cold weather of 1962/3 as many as 31,600 Starlings took advantage of the warmth retained by the city centre at night. In country districts numbers are even more impressive, with six-figure roosts by no means uncommon. Frequently observers have simply referred to "millions" or "impossible to estimate", but counts have been made of 200,000 and 400,000 at Sheriffs Lench, in January 1950 and January 1960 respectively; 400,000 at Church Lench in the autumn of 1962; between a quarter and one million at Sandon in 1964/5; 2–3 million in conifers at Drakes Broughton on November 6th 1970 and one million at Codsall on December 9th 1972. Perhaps the largest concentration of all, however, was at Shustoke on January 3rd 1967, when "the sky was thick with birds from horizon-to-horizon, up to fifty birds deep and passing continuously for four minutes." Conifers and thorn thickets are favoured roost sites, but deciduous woods are also used if there is a dense under-storey. Birds remain faithful to certain sites and at Hopwas Wood they repeatedly returned each year despite annual shoots to disperse them. At other sites, though, a roost will suddenly disband and reform elsewhere, perhaps because the birds destroy their own habitat. For example, a roost in rhododendron at Newcastle-under-Lyme in 1961/2 peaked at two million birds in March and the deposit of droppings, which covered the bushes to a depth of 15 cm in places, caused much damage. In June 1960 a roost in an orchard at Sheriffs Lench destroyed an entire cherry crop. Roosting birds are themselves vulnerable and many were killed by torrential rain in Birmingham on June 18th 1968 and 300 were found dead around the cathedral on August 19th and 20th 1970 after similar weather. In foggy weather birds flighting to roost occasionally become disorientated and in 1975 1,000 roosted on the street lights of the M5 service station at Strensham.

Immigrant Starlings begin to arrive during September, but passage usually peaks in the last week of October and the first week of November. Counts of 200 per hour in the Tame Valley on November 11th 1952 and 100 per hour at Caldecote on November 5th 1950 are indicative of the strength of movement. In his study of diurnal migration at Walsall, Jenkins (1953) found this to be the commonest species, accounting for 25 per cent of all birds. Birds return in March, peaking late in the month. Ringing has shown locally bred Starlings to be largely sedentary, with these winter immigrants originating from the continent. Analysis of recoveries from the West Midlands shows their summering zone to be mostly between 53° and 61° N and 4° and 40°E. This is the area from Belgium north-eastwards to the western USSR and includes the Netherlands, northern Germany, Denmark, southern Norway and Sweden, south-west Finland

and northern Poland. Recoveries during the migration periods show the most common route to be along the southern shores of the North Sea, in particular the coast of the Netherlands. Some birds probably come to the West Midlands directly across the North Sea. Ringing also tells us something of the Starling's numerical status. Of the birds trapped in a Birmingham suburban garden from 1965–8, for example, over half were Starlings, making this easily the commonest species (Edwards 1969), whilst at Warwickshire and Worcestershire rubbish tips it was second only to the Black-headed Gull in abundance, accounting for 32 per cent of all birds ringed (Green 1978).

Rose-coloured Starling
Sturnus roseus

Rare vagrant. Five old records.

None has been recorded this century, but the following are admissable records from last century of this vagrant from south-east Europe and southern Asia: an adult male shot in an orchard at Barton in the summer of 1854; a female killed at Powick in August 1855; one seen at Rushton Spencer in 1875; an adult male shot at Haselor on January 20th 1890; and an immature female shot in Sutton Park on November 10th 1890. Tomes (1904) also referred to an adult male shot near Bidford, but gave no date, and Norris considered this might have been the Haselor bird. All of these records accord with the pattern of vagrancy outlined by Sharrock and Sharrock (1976).

House Sparrow
Passer domesticus

Abundant and widespread resident.

Like most ubiquitous species, the House

Sparrow tends to be ignored and most reports refer only to some aberrant behaviour or plumage. It occurs wherever there is human habitation, right from the very heart of Birmingham and the other large towns and cities down to the tiniest hamlet. Only where settlements are sparse, such as the northern moors, is it less than abundant. Harthan remarked that the rural population varied with the number of poultry, in which case, with modern battery houses, numbers may have declined. With the rapid growth of towns and cities over the past fifty years, however, the urban population has probably grown tremendously and it is likely that numbers today are higher than at any time. Breeding was proved in every 10-km square during the *Atlas* surveys from 1966–72 and Sharrock's assumed density of 10–20 pairs per km^2 would give a regional population of 75,000–150,000 pairs. Lord and Munns simply placed the population at over 20,000 pairs.

In late summer House Sparrows often flock together to feed on ripening corn and they can cause considerable damage to crops at this time. In August 1952, 1,500 were noted in growing wheat near Handsworth and the same number was seen in corn at Bittell in 1976. Later in autumn and winter they move to stubbles, plough or waste land, but flocks are then smaller and seldom exceed 700, although 1,200 were noted with Woodpigeons at Lower Penn in December 1964. Such flocks sometimes mob intruders and once a workman was actually struck on the shoulder when he disturbed birds feeding on grain. On another occasion a flock drove off a pair of marauding rats for fifty metres. This aggressive tendency is also manifest in the way the species will evict House Martins and usurp their nests, apparently on one occasion just to shelter from the rain. The House Sparrow is an opportunist, too, and it has learnt to emulate the tits by taking cream from milk bottles and feeding from suspended tit-feeders.

As a sedentary species, House Sparrows are seldom ringed, but an adult that was ringed in 1957 was still alive in 1961, when it must have been at least five-and-a-half years old.

Tree Sparrow
Passer montanus

Widespread and numerous resident.

Colonies of Tree Sparrows occur in a variety of situations throughout the Region. Most are likely to be encountered where there are old, decaying trees, whether in woodland, parkland, orchard or hedgerow; but pollarded streamside willows are another favoured haunt, Sand Martin burrows are occasionally purloined and the species takes readily to nest-boxes. Nests in the cavities of stone or brickwork, station lamps and the insulators on electricity pylons are a further indication of the Tree Sparrow's versatility. In 1969 they were recorded as sub-tenants of the herons at Gailey and three pairs bred in the straw of the hide at Belvide. Outside the breeding season, large flocks of Tree Sparrows often consort with finches and buntings to feed on weed seeds along hedgerows, or amongst stubble, crops such as kale, or the rough margins of reservoirs.

Tree Sparrows are well-known for fluctuations in numbers. Neither Dickenson (1798) nor Pitt (1817) considered it even bred and Garner (1844) thought it rare. Smith, however, described it as a local breeder, but with many roaming the countryside in autumn and winter, Harthan (1961) said it was local, but generally distributed and Norris (1947) termed it scarce and local in distribution. Reports in the late 1950s and early 1960s indicated an increase, however, and the *Atlas* surveys (1966–72) showed a much wider distribution, with breeding confirmed in seventy-

78. The Tree Sparrow is well-known for its fluctuating population and in good years winter flocks may exceed a thousand birds. *M. C. Wilkes*

six 10-km squares (99 per cent) and suspected in the remaining one. Since then numbers have been fairly constant.

During 1966-8, before felling and replanting, the 40 ha of Fradley Wood held an average of 17 pairs, but subsequently numbers declined with the change in habitat and today the wood is all but deserted. Conversely, at Edgbaston Park, where the habitat has changed little, Tree Sparrows have increased from around one pair on 16 ha until 1972 to four or five pairs since. Other examples of density in 1978 were nine pairs on 73 ha of farmland at Flyford Flavell and three pairs in 4 ha at Tocil Wood. Overall, the CBC regional plots in 1978 showed 10·8 pairs per km^2 in woodland, whilst on farmland there were 6·92 pairs per km^2 compared to 4·76 nationally. Such densities point to a regional population well in excess of the 200-2,000 pairs estimated by Lord and Munns and probably between 9,000-20,000 pairs.

More information is available on winter flocks, the commonest size of those reported being between 50-150 (52 per cent) and the mean size 164.

Table 81 clearly shows a peak from 1959-63 coincident with an increase in breeding numbers. Whilst the mean flock size has not increased since, the number of flocks reported has nearly trebled. In this respect, 1976 was a particularly good year with large numbers at many localities. The largest concentration of Tree Sparrows ever reported was up to 1,500 at Blithfield in the winter of 1961/2, but other noteworthy totals have been 1,300 at the same locality in January 1960 and 1,000 at Draycote in December 1976. Flocks have been reported in all months from August to April inclusive, but are largest and most frequent during December and January. There is no evidence of autumn or spring passage through the area and ringing recoveries show mainly only local movements. However, one ringed at Middleton in July 1974 was recaught at Farnham (Surrey) in November 1974 and another ringed at Packington in September 1976 was at Warminster (Wilts) in February 1978.

Chaffinch
Fringilla coelebs

Widespread and abundant resident, passage migrant and winter visitor.

The Chaffinch ranges from the suburbs to the moors, breeding wherever there are woods, including conifer plantations, scattered trees, orchards and hedgerows. A favourite site for its characteristic nest is in a tree, but bushes and hedges are also freely used. Post-breeding flocks resort to oakwoods to feed on defoliating caterpillars, then in winter birds congregate in flocks to feed on beech mast or around farmyards, manure heaps, stubble or ploughed fields. Parks and gardens are often visited as well in winter.

There has been no apparent change in the distribution of the Chaffinch from that recorded by previous authors and breeding was proved in every 10-km square between 1966-72. There has however been a definite fluctuation in numbers. Harthan described the Chaffinch as the commonest bird in

Table 81: Five-yearly Means of all Reported Flocks of Tree Sparrows, 1954-78.

	1954–58	1959–63	1964–68	1969–73	1974–78
Mean flock size	147	280	158	167	159

Worcestershire throughout the year, but this cannot now be true. In 1950, though, it was certainly the commonest species in the Heligoland trap at Bartley, where one-in-five of the birds caught was a Chaffinch. By 1952 a decline was being reported and this appears to have been associated with poor hatching success widely attributed to the use of toxic chemical dressings. The population appeared to reach its lowest ebb from 1960–4, since when there has been some recovery.

There are several indications of the Chaffinch's status within different habitats in the Region. A study of visible migration at Walsall in 1952, for example, showed it to be the third commonest migrant (13 per cent), behind Starling and Skylark, whilst ringing in a Birmingham suburban garden from 1965–8 showed it to be the ninth commonest species. Again in 1952, it was found to be the eleventh commonest breeding species in the mixed habitats of the Sandwell Valley, though only at a density of four pairs per km². In more rural situations, the 40 ha of Fradley Wood held 28 pairs prior to felling of the standard oaks in 1969, since when there has been an average of only 14 pairs, and 184 ha of mixed farm and woodland at Moreton Morrell held an average of 15 pairs from 1972–9. There were 16 pairs in 89 ha of heathland on Cannock Chase in 1967; 15 pairs per km² on farmland around Brandon in 1966; 14 pairs in 74 ha of conifer plantations on the moors in 1978 and an average of two pairs in 16 ha at Edgbaston Park from 1966 to 1979. These densities tend to be at the lower end of the ranges quoted by Sharrock, suggesting that the Chaffinch might be less numerous in the West Midlands than in other parts of the country, though almost everywhere it seems to rank amongst the top ten species. On the regional CBC plots in 1978 it was the third commonest farmland species and the eighth commonest woodland species, with densities of 19·5 and 31·2 pairs per km²

respectively as against 23·4 and 48·8 pairs per km² nationally. Such densities point to a regional population of 60,000–120,000 pairs. Lord and Munns merely estimated 20,000 pairs plus.

The largest numbers of Chaffinches occur in winter roosts or flocks, when resident birds are joined by continental immigrants. Again fortunes have been mixed, with a large roost at Sheriffs Lench numbering 2,000 in 1950 and 3,000 subsequently, but declining to a mere 100 by 1962. Likewise a flight of 2,300 was seen going to roost at Enville in 1957, but ten years later the entire roost numbered no more than 400. Other sizeable roosts have been 1,000 at Sutton Park in 1977 and 800 at Maer in 1978, both perhaps indicative of some recovery in numbers. Most roosts are in woods with a dense shrub layer and rhododendron is especially favoured. Feeding birds disperse more, so flocks are smaller, with 100–200 commonplace and up to 500 by no means unusual. The largest flocks recorded were 700 at Whittington Sewage Farm in December 1972 and 1,000 in Edgbaston Park during the autumn of 1976. Overall there has been an increase of one-third in the mean size of reported flocks over the past ten years. Other finches and buntings frequently associate with feeding Chaffinches and another phenomenon is the oft-quoted tendency for flocks to comprise almost entirely one sex. The only data to support this, however, concerns 180 birds on November 18th 1944 that were all females.

In a roost at Maer, however, Emley (1980) noted that males predominated at the start of the winter, but by March, when males had presumably left to take up their territories, females were predominant. Smith also considered that females were more ready than males to desert the hostile upland environment in winter and this might help to explain why flocks of predominantly one sex are sometimes encountered. It also points to birds migrat-

Table 82: Summary of Chaffinch Movements to and from the Region by Country and Month of Ringing or Recovery.

	Jan	Feb	Mar	Apr	May	Jun	Jul	Aug	Sep	Oct	Nov	Dec	Total
Norway	—	—	—	—	2	1	—	1	—	—	—	—	4
Sweden	—	—	—	2	1	—	1	—	2	—	—	—	6
Poland	—	—	—	—	—	—	—	—	1	—	—	—	1
Denmark	—	—	1	2	—	—	—	—	—	1	—	—	4
Germany	—	—	—	1	—	1	—	—	—	1	1	1	5
Holland	1	—	—	—	—	—	—	—	—	5	1	—	7
Belgium	1	1	1	—	—	—	—	—	—	8	9	—	20
France	—	—	—	—	—	—	—	—	1	—	—	—	1

ing and a southerly or south-westerly passage is noted most autumns.

This passage usually peaks in late October, when some impressive numbers have been noted. On October 23rd 1952, for example, 199 passed through Walsall within an hour, whilst on October 25th 1955 there were 500 per hour moving north-westwards over Wylde Green and on October 22nd 1958 400 per hour passed south-westwards across north Birmingham. Return passage in spring is less strong and even more variable in direction, though a north-westerly movement peaking in mid-March is usually discernible, with 64 per hour passing through Erdington in March 1951 and 200 flying NNW at Alvecote on March 1st 1953. Table 82 shows the country and month of all foreign recoveries. All birds were caught in the West Midlands during the winter and show how Scandinavian Chaffinches join our resident birds for the winter, migrating to us via the Benelux countries, but returning direct. The relatively few recoveries within Britain show movement of these winter visitors within this country and there is no evidence that birds bred in this area are anything other than sedentary.

Brambling
Fringilla montifringilla

Regular winter visitor and passage migrant, variable in number but usually fairly numerous or numerous.

Bramblings are ground feeders and are usually seen in company with Chaffinches feeding on beech mast, or in mixed finch flocks feeding in stubble, ploughed fields, around manure heaps or hop waste, or amongst the weeds of waste ground. They also roost with other finches in woods with a dense shrub layer such as rhododendron. Less usually they have been noted in orchards, blackcurrant and kale fields or feeding on rowan berries. During hard weather like that experienced early in 1979 they will freely visit gardens to feed from bird tables, but mostly they are faithful to certain favoured woodland haunts on the

Birmingham and central Staffordshire plateaux, especially at Lickey, around Atherstone and Stoke-on-Trent, and in the Churnet Valley. Generally fewer are seen in the three main river valleys, though large flocks have occurred at Wilden in recent years.

Like Siskin and Crossbill, Brambling numbers fluctuate markedly from one year to the next, indicating that its appearances are related to availability of food and severity of weather. For example, over 8,000 were reported in 1973/4 and again in 1975/6, but only a mere 200–300 in the intervening winter. A similar fluctuation occurred in the three winters 1966/7 – 1968/9. Nevertheless, birds invariably occur somewhere in the Region every year and the sparseness of records prior to the 1950s seems more likely to be the result of fewer observers and wartime difficulties than to any lack of birds.

The earliest Brambling arrival was on September 3rd 1956, at Sheriffs Lench, but the average first arrival date over forty-three years has been October 9th. Most birds do not begin to arrive until mid-October, however, and even then flocks are few and small and it seems certain that many of these earlier arrivals move on through the Region. This passage continues until mid-November. Numbers then begin to increase from mid-December onwards and it is likely this is when most of the wintering birds arrive, reaching their maximum numbers in January and February. Most of the reported flocks comprise less than 50 birds, but flocks up to 300 are not uncommon and larger numbers have occurred on occasions. Smith referred to numbers at Hilderstone that "could only be calculated in thousands" and Norris to a flock of nearly 1,000 in December 1904. The largest reported flocks in recent times were at Wilden, where there were 700 in January 1974 and 600 feeding on *Chenopodium* seeds in February 1976; around Brewood, where the regular flocks

that feed amongst blackcurrant bushes reached 1,500 in February 1973 and 2,000 in February 1974; at Worcester rubbish-tip, where 1,000 were present in mid-February 1979; and near Summerfield, where 1,000 were present in a huge finch flock, again in mid-February 1979.

Numbers begin to decline in early March, when many of the wintering flocks disperse as birds start to return to their northern breeding grounds, but there is another marked passage which peaks in late March and early April. Over forty-two years the average last departure date has been April 13th, but birds have been recorded into May on seven occasions, with the latest on May 12th (1973). The one bird to be recovered in April, on the 24th near Vara (Sweden), was still south of the breeding range (Newton 1972) and may still have been on migration, so there have been no breeding season recoveries of West Midland Bramblings. Other movements confirm the similarity of this species to Chaffinch, however, with autumn recoveries from the Benelux countries. There have been four October reports from that region, two each from Holland and Belgium. There have also been three reports of birds that may have chosen to over-winter in different areas – one in each of Belgium (December), France (December) and West Germany (February). Summer records are rare, but a male was in song at Beaudesert on July 14th 1979 and a male remained at Belvide from August 3rd to 9th 1980.

Serin
Serinus serinus

One record.

At least two, possibly three, were seen near Evesham on June 17th and 18th 1978, one of which was singing (*Brit. Birds* 72:541). Serins bred in England for the first time in

1967 and it was confidently expected that they would colonise Britain as part of their spread through Europe, but records have since declined (Sharrock) and this record from Worcestershire would appear to have been no more than an isolated occurrence.

Greenfinch
Carduelis chloris

Widespread and abundant resident.

The Greenfinch is equally at home in town or country and small colonies nest just as readily in city parks, churchyards and large gardens with thick, evergreen shrubs as they do in more conventional situations such as thickets, straggling hedges and the margins and glades of woods. In recent times it has taken to young conifer plantations as well. Greenfinches feed on seeds which are taken either from the plant itself or from the ground and its heavy bill enables it to tackle those too large for other finches apart from the Hawfinch. In the West Midlands, birds used to feed on elms in summer and it is yet uncertain whether the loss of so many elms will adversely affect the Greenfinch population. During autumn it often resorts to hornbeam or yew trees, but in winter it wanders more widely to farmland or closed woodland in its search for food and often takes haws, cracking the seeds like a Hawfinch. It was once a common bird of stackyards, but since the advent of combine-harvesting it has substituted this source of food with that of rural and suburban bird-tables, having quickly mastered the art of hanging from suspended bags to obtain nuts.

Smith described the Greenfinch as universally common, except in the northern hill country, where hedges are mostly displaced by drystone walls. He commented that the hawthorn hedges of the lowlands supplied favoured nesting places, but in the limestone country many bred in stunted bushes on the steep hillsides. Harthan said it was common everywhere, even in towns, and that a considerable movement of wandering flocks took place in February and March; whilst Norris recorded it as a common resident, which was generally distributed and frequently noted outside the intensely industrial area, with considerable numbers roaming the countryside in large, mixed finch-flocks in winter.

These statements are still true today, though the Greenfinch now has become much commoner in suburban areas. During the *Atlas* surveys from 1966–72 breeding was proved in seventy-five 10-km squares (97 per cent) and suspected in the other two. Information on density is scant, but in 1978 there were 11 pairs on 95 ha of farmland at Wellesbourne and at Moreton Morrell 184 ha of farm and woodland usually holds from two to five pairs. The 1978 regional CBC plots showed this to be the eleventh commonest farmland species, with a density of $5 \cdot 49$ pairs per km^2. This compares favourably with the national farmland density of $5 \cdot 37$ pairs per km^2, but there is no reliable regional figure for woodland to compare with the national density of $9 \cdot 96$ pairs per km^2. Overall though, it seems likely from these figures that the regional population is some 25,000–50,000 pairs, which is above the range of 2,000–20,000 estimated by Lord and Munns.

Outside the breeding season, the Greenfinch forms feeding flocks, often consorting with other finches to seek out seeds from ploughed fields and stubbles. In particular, from July to September it frequently joins Linnets to feed on oilseed rape. In 1950 it was the commonest bird in the Heligoland trap at Bartley, accounting for 36 per cent of all birds ringed, but as rickyards such as that at Bartley have become scarcer, so it has exploited new feeding areas in the suburbs. At the same time the Chaffinch has declined as a suburban bird. The Greenfinch was the first species to emulate the tits in feeding from suspended nut-bags (WM *Bird Report* 1959) and some idea of its strength in suburban areas is given by ringing totals, with 44 trapped in a single day in a Foleshill garden on January 3rd 1960, 143 ringed by April 18th of the same year in a garden at nearby Nuneaton, and a staggering 260 ringed in a garden at Wyken during the first three-months of 1966. From 1965–8 the Greenfinch was the fourth commonest species to be trapped in a Birmingham garden, accounting for 5 per cent of all birds caught. During 1977–8, however, there have been reports of a decline in suburban areas, but it is too early yet to tell whether this is a permanent or temporary setback.

Prior to 1950 there was little information on flock size, except for 200 at Bittell in September 1934, but since then there has been an even distribution of reported flocks between 50 and 250, but few larger ones. The most reported were 750 at Huntington on February 2nd 1974 and 700 at Arbury in late December 1978, the latter feeding on mustard sown for Pheasants. Analysis of winter flock-size by five-yearly intervals has revealed no significant change over time. Post-breeding flocks up to 250 have been reported as early as July, but numbers do not usually build-up until September and October. They then remain fairly high throughout autumn and winter, but begin to decline after the end of February as birds start returning to their breeding territories. However, in cold springs they will remain until April and there is one record of 300 at Chasewater as late as May 7th (1950). Roosts are larger, seldom being less than 200 and having reached 1,000 on three occasions – at Sheriffs Lench on January 15th 1950, Sutton Park from December 1955 to January 1956 and Trickley Coppice on December 20th 1964. In general the larger roosts form later in the year than flocks, with the largest ones appearing in December and finally dispersing in March.

The Greenfinch is one of the most ringed species and there have been many recoveries, yet they show no consistent pattern. Birds have been recovered in all compass directions from the Region and, whilst most move less than 50 km, a substantial minority move quite long distances with individuals having reached Northumberland in the north and Devon, Hampshire and Surrey in the south. There have been just two foreign recoveries. The first concerned a nestling ringed at Cheadle on June 20th 1910 and picked up dead in France on April 2nd 1911 and for more than twenty years this remained the only foreign recovery of a British ringed Greenfinch. The second West Midland recovery came from Spain in 1956.

Goldfinch
Carduelis carduelis

Widespread and numerous resident.

Fortunately the Goldfinch is now much commoner than in the past and today it can be found nesting in trees and bushes on farmland, in orchards and parklands, along woodland edges and rides or around open glades, and in garden shrubberies, even in suburban districts. Outside the breeding season small, twittering "charms" of this delightful finch search open and waste land for weed seeds, especially thistles.

From its position of "not rare" (Garner 1844), Smith remarked that the Goldfinch had greatly decreased during the previous half-century, although he could still quote a flock of 200 as recently as October 1919. Harthan said that small numbers could be seen throughout the year and that the species was fairly common in orchard districts, but nowhere abundant. He noted that larger flocks fed upon thistle seed in weedy fields in winter and quoted one of 90 in December 1935. Norris considered the

79. In many districts Goldfinches are now much commoner than in the past and in autumn flocks gather to feed on seed heads.   *M. C. Wilkes*

species resident in small numbers, but rather local, and thought it doubtful whether there had been any change in the preceding twenty years, though there had been a decline in the fifty years before that. Fincher (1955) reported an absence from the northern third of Staffordshire, but summarised the distribution elsewhere as widespread, but seldom common. Since then there has been a resurgence in numbers and the Goldfinch has spread into suburban districts. During the *Atlas* surveys (1966–72) breeding was confirmed in seventy-two 10-km squares (94 per cent) and suspected in the remaining five, which embraced the bleakest parts of the moors, the intensive arable areas south of Lichfield, the Stour Valley and an area south of Cannock. In all these districts birds were presumably scarcer than elsewhere. In 1978

the CBC showed a regional farmland density of 1·02 pairs per km² compared to a national figure of 2·76, but whether this difference is significant or not cannot be determined as the species was present on very few plots. However, allowing for the suburban population, it is likely that there are some 6,000–10,000 pairs in the Region compared to the 200–2,000 pairs estimated by Lord and Munns.

The spread into suburban areas began in the early 1950s and by 1975 the Goldfinch was reportedly second only among finches to the Greenfinch in abundance, having ousted the Chaffinch into third place. As an indication of its strength, six singing birds were counted along 0·75 km of road in Nuneaton in 1967 and a flock of 50 was seen feeding on weeds on an inner-city redevelopment site in Birmingham the same year. Small parties have been noted in the very centre of Birmingham on several occasions, 15 fed from a single teasel-head in a suburban garden at Hall Green in 1976 and in 1978 up to 200 were seen around Foleshill Gas Works.

Outside the breeding season, Gold-finches form small flocks and Norris referred to parties of 5–20 being occasional in Warwickshire. Such flocks have been reported in every month of the year, but are largest and most frequent in September and October, when the seed-heads of thistles, teasels and devils-bit scabious are most abundant. In 1976, when Siskins were conspicuously absent, Goldfinches also fed on alders in large numbers. There is a secondary flocking peak in April, but numbers in winter are usually lower, in-dicating some evacuation of the Region during the intervening months. During the 1950s and early 1960s flocks up to 100 were commonplace, but 200 at Enville on April 11th 1953 was unusual. Since 1968, how-ever, there has been a noticeable increase in flock size and today a flock of 200 is by no means infrequent and up to 600 were at Wilden on October 5th 1974, and 300 at Holt on October 16th 1977. Little is known about the Goldfinch's roosting habits in the Region, but it would appear to roost in small numbers of little more than 50. The only ringing recovery concerns a bird which had moved from Holt in September 1977 to Spain by the following November.

Siskin
Carduelis spinus

Fairly numerous or numerous winter visitor. Scarce or very scarce breeding species.

The Siskin is best known as a winter visitor, when small acrobatic parties feed on the seeds of alders or, less commonly, birch and larch. In spring though, many resort to plantations of spruce, pine or larch to feed on cones and in recent years birds have tended to linger later and return earlier in the autumn. Singing males have been heard more than once and the first breeding re-cords came in 1973, from the Stour Valley and Cannock Chase, followed by one from the Wyre Forest in 1974. Somewhat surprisingly, there have been no subsequent breeding records, but males have held territory and one or two pairs almost certainly breed each year in the Wyre Forest or on Cannock Chase. This tentative colonisation is in line with the recent national spread (Sharrock) and has been assisted in the West Midlands, as else-where, by the post-war spread of conifer plantations. During spring and early summer Siskins feed largely on conifer seeds and as these plantations mature so their seed crop becomes more reliable.

Smith said birds appeared in greater quantities in a severe winter, but that they had probably been more frequent in former times. The largest flock he mentioned was nearly 60, at Trentham in January 1893, but he was able to quote a breeding record – of a pair which nested in a spruce at Wightwick in 1916. This isolated occur-rence constitutes the first breeding record for the Region. Harthan, referring to a statement by Curtler (1853) that the Siskin was resident, pointed out that there was no evidence of its having nested and classed it as an irregular winter visitor. Norris re-garded it as a sporadic winter visitor, which appeared most years, but he considered a flock of 50 to be larger than average.

Subsequent data clarifies the Siskin's winter status. Birds may begin to arrive in September, but do not usually appear until October. Over forty-one years the average first arrival date has been October 6th, with the earliest record on August 15th (1966). Numbers build-up in November and reach a peak in December, but then decline again during January and February. There then follows a small secondary peak in March,

Table 83: Monthly Distribution of Arrivals of Siskins, 1929–78.

	Sep	Oct	Nov	Dec	Jan	Feb	Mar	Apr
No. of birds	25	601	2,342	3,929	3,126	1,392	1,523	1,193

Note: Records of less than three birds are not included.

suggesting that some birds at any rate pass through the Region in late autumn and return in spring. Support for this theory comes from a bird ringed in Warwickshire in November 1975 and recovered in West Germany later in the same winter and one caught at Birmingham on April 4th 1978 and retrapped at Epping (Essex) eleven days later. Birds are still numerous even in April and over thirty-nine years the average last date has been April 14th, with the latest record, excluding those of possible breeding birds, on May 19th (1969). One at Selly Oak on July 10th 1936 appears to be an isolated summer record. The number of birds reported by month of arrival is as shown in Table 83.

When Siskins first arrive in autumn they invariably feed on alder, or less often birch, seed and they are faithful to favoured localities such as Bittell and the Lickey Woods. Birches are deserted fairly early in autumn, however, but birds remain in alders for most of the winter before moving into conifers in March. Flocks up to 100 are quite common, though most only involve 25–50 birds, and they frequently associate with Redpolls. Larger flocks are scarce and occur more often in spring than in autumn, presumably because conifer plantations are less widely distributed than alders. Staffordshire, with its plantations at Enville and on Cannock Chase, has the predominance of spring flocks, though good numbers also occur in the conifers of Wyre Forest. The most recorded were 400 in Sutton Park on December 14th 1955, 300 in the Sherbrook Valley on April 7th 1971 and well over 400 in

larches at Beaudesert during March and April 1980 (D. Smallshire *in litt*). Birch, alder, larch and spruce are all present at favoured localities such as Bittell, Cannock Chase, Enville and the Lickeys and in these situations birds are often present throughout the winter.

Siskins appear most years, but their numbers fluctuate. Variations are not erratic or entirely related to weather, however, but follow a definite cycle as the graph shows.

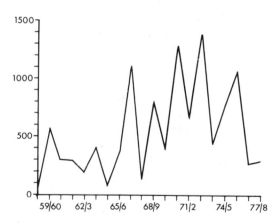

Summation of all Siskin Flocks by Individual Winters, 1958/9–1977/8

The biennial cycle of peaks and troughs is very evident and conforms to the pattern of abundance for the spruce cone-crop in the breeding quarters (Haapanen 1966), which controls the overall Siskin population. Within this, troughs can still be above, or peaks below, average numbers and 1966/7; 1970/1; 1972/3 and 1975/6 were good Siskin winters, but 1967/8;

1976/7 and 1977/8 bad ones. This pattern is typical of an irruptive species seeking out the most abundant food source each winter and probably correlates with the abundance of birch and alder seed within the Region itself. That birds do winter in widely different areas is evidenced by one recovery of a bird ringed at Shenstone in 1973 and recovered in Spain in March 1974. Newton (1972) showed a scarcity in the Siskin's natural food supply during March and in recent years birds have increasingly turned to garden bird-tables at this time to feed from suspended nut-bags. Why they should show a marked preference for red bags, however, remains a mystery.

Few Siskin had been ringed before birds began to visit gardens, but there is one old recovery, of a bird ringed in Holland in October 1931 and recovered in Worcester in February 1932. Another interesting recovery concerns a bird ringed in Latvian SSR on September 15th 1975 and found dead on Cannock Chase just fourteen days later.

Linnet
Carduelis cannabina

Widespread and numerous or abundant as a resident, summer visitor and passage migrant.

The Linnet is widespread on farmland and

80.   The Linnet is a widespread breeding species, which often gathers in large flocks to feed on weed seeds in autumn.   *S. C. Brown*

most abundant on gorse-clad hillsides and heaths. A few pairs breed in heather on the moors and sometimes even in suburban gardens, whilst small flocks feed on open ground in the suburbs or on the weed-strewn redevelopment sites of the inner city. It nests in small colonies in hedges, bushes, saplings and young plantations alike and an indication of the variety in nest sites is provided by three nests within a small area at Blithfield Reservoir in 1972 – one each in gorse, hawthorn and a small pine.

Smith said Linnets occurred universally, but that the numbers found nesting were considerably less than in former times, especially in certain parts of north Staffordshire. In winter he found them absent, or nearly so, from the high ground, but plentiful in the lowlands at all seasons, especially autumn. Harthan observed that it was far more abundant in some years than others, with flocks up to a hundred or more frequenting weedy stubble fields in winter, whilst Norris noted it in comparatively small numbers in most areas, but probably a good deal less commonly than formerly.

Its distribution has not changed since and during the *Atlas* surveys (1966–72) breeding was proved in seventy-six 10-km squares (99 per cent) and suspected in the remaining one, namely the Millstone Grit country. Breeding densities have included six pairs on 81 ha of farmland at Wilnecote in 1971; 15 pairs in 6 ha of gorse at Wilden in 1974 and an average of 2·3 pairs in 40 ha of woodland at Fradley from 1966–79, though this varied from one to seven pairs in different years. In 1978 the regional CBC data showed the Linnet to be the eighth commonest farmland species at a density of 7·94 pairs per km² compared to only 5·54 pairs nationally, but this difference was largely due to the high density of 25 pairs on 95 ha at Wellesbourne. There is no reliable regional density for woodland, but nationally it was 5·72 pairs

per km² and overall these figures suggest a regional population of 15,000–30,000 pairs, which is at the upper end or above the range of 2,000–20,000 pairs postulated by Lord and Munns.

In summer Linnets feed directly from herbaceous plants, but in winter they feed from the ground (Newton 1972). Outside the breeding season they often congregate in large flocks to feed on thistles, fat hen or *Persicaria* amongst kale or cabbage fields and wastelands. Increasingly since 1974 it has also fed on stands of oilseed rape up to 2 m high. Reservoirs with low water-levels are also favoured sites. The mean size of all reported flocks is about 200, but numbers up to 500 are not unusual and there are a dozen records of even larger flocks, with the largest concentration being flocks totalling 1,200 at Draycote in September 1976. Flocks increased steadily both in frequency and mean size to a peak between 1969 and 1973, since when they have declined slightly. They have occurred in every month except June, but are largest and most widespread during September and October, after which there is a decline in November, another peak in December, then a further decline through the winter before a final build-up to a spring peak in April. The peaks are clearly indicative of both spring and autumn passage and indeed during the late 1940s and early 1950s the species' apparent desertion of the Region in winter aroused much interest. Ringing evidence supports this migration as Table 84 shows.

Several of the recoveries in the table relate to nestling and juvenile birds, showing that Linnets bred in the Region frequently winter in France and Spain. However there are also records of birds bred and wintering locally.

A few winter roosts have been reported, mainly in December and January, with the largest being 1,000 at Sheriffs Lench on January 15th 1950 and 800 at Old Hills on September 18th and 19th 1970.

Table 84: Summary of Linnet Movements to and from the Region by Country and Month of Ringing or Recovery.

	Jan	Feb	Mar	Apr	May	Jun	Jul	Aug	Sep	Oct	Nov	Dec	Total
France	3	—	1	—	—	—	—	—	—	13	5	5	27
Spain	—	2	—	—	—	—	—	—	—	1	—	1	4
British Isles													
0–100 km	2	—	—	1	—	1	—	—	—	—	—	—	4
101–150 km	—	—	—	—	—	1	—	—	—	1	—	—	2

Twite

Carduelis flavirostris

Not scarce but very local, both as a summer and winter visitor.

There are two distinct habitats where Twite might be found in the West Midlands. During summer a few pairs breed amongst the heather of the northern moors, usually in the vicinity of upland pastures to which they resort for feeding (Orford 1973). In winter small numbers are occasionally encountered on waste ground around reservoirs or on derelict slag heaps, though only at Chasewater, where both habitats are in close proximity, are they at all regular.

Garner (1844) said Twites nested in Staffordshire and were common in hilly places, whilst Smith said a few pairs nested sparingly in the Roaches and Axe Edge districts and on the Warslow Moors. He also observed that after the breeding season flocks appeared fitfully on the whole moorland area and that birds frequented lowland Staffordshire well nigh universally in autumn and winter, with bird-catchers taking their toll of singles or small parties that occurred at Longton from September and October onwards. The last breeding records mentioned by Smith were from near the source of the Manifold in 1925 and 1927, after which there was no confirmed breeding until a nest with eggs and subsequently young was found on Morridge in 1967. Strong evidence of a few other scattered pairs was received at the same time. During the intervening period the Twite population of the southern Pennines had declined to a low ebb, but since the mid-1960s it has undergone a strong resurgence (Frost 1978).

Atlas survey work during 1966–8 showed confirmed breeding in one 10-km square and suspected breeding in another, but by 1972 birds had spread and breeding was confirmed in three squares, with birds also present in a fourth. Numbers were increasing too. In 1968 two or three pairs bred, but by 1971 there were an estimated twenty pairs; 1973 brought a large increase to an estimated forty pairs and 1974 a peak of over fifty pairs. Since then the increase has ceased and in 1978 there were again some forty pairs. Birds begin their return to the moors in April each year, but most come back in May and some as late as June. As in Smith's day, small flocks may still be encountered until mid-October, with the largest in recent times being 80 at Thorncliff in 1972 and 50 around Flash in 1973. One on November 12th (1961) is the latest autumn record from the moors.

Outside the breeding season the Twite remains something of an enigma. At Chasewater birds have appeared every winter since 1948/9, but elsewhere its appearances are no more than erratic. The first of the Chasewater birds arrive during October, but many do not appear until November. They leave again during March or early April, with extreme dates of September 29th (1957) and April 24th

(1965). During the course of a winter numbers appear to fluctuate markedly and it seems that birds either disperse over a wide area or else remain well-concealed in tall herbage such as fat hen. For example, only five were found on December 27th 1963, but 42 were present the following day; whilst 75 were noted on December 11th 1976, but only 30 the next day. However, identification is a problem and there is always the possibility of some Redpolls and Linnets having been included in error. In recent years much of the favoured feeding ground has been reclaimed and this has been blamed for apparently reduced numbers since 1974. As yet, though, the evidence of a real decline rather than the typical fluctuations, is slender and, if all were correctly identified, the annual maximum of 200 on November 25th 1977 was in fact the highest ever as Table 85 shows.

Away from Chasewater, small numbers were noted on Cannock Chase during winter and spring from 1950–3, though a flock of 200 reported from a birch wood there on January 3rd 1951 should be treated with some scepticism, since Redpolls may have been involved. One or two were then seen at Blithfield Reservoir in 1959, 1962–6 and 1969, but none subsequently until 1978. Similarly the species appeared annually at Belvide from 1968–73, with a party of 25 after heavy snow on March 4th 1970, but has not been present since.

Outside Staffordshire, Tomes (1901 and 1904) regarded the Twite as a rare winter straggler to both Warwickshire and Worcestershire, but Harthan knew of only two definite records in Worcestershire, to which can be added three more – of singles in 1949 at Sidemoor and in 1973 at Bittell and Halesowen. Likewise Norris could quote only one Warwickshire record, although there have been several since, with small flocks up to 40 apparently noted fairly regularly in winter along the Tame Valley during the 1970s. Of the sporadic, scattered records from elsewhere, many were in autumn or spring and clearly involved birds on passage. The extreme dates for such records are September 11th (Bittell 1973) and May 4th (Blithfield 1968). Whilst ringing data provides no evidence about the origin of passage and wintering birds, some of the autumn records have coincided with major influxes on the east coast.

Redpoll
Carduelis flammea

Widespread and numerous as a resident, passage migrant and winter visitor.

Fifty years ago the Redpoll was little more than a winter visitor to birch woods in the Region, though historical evidence points

Table 85: Maximum Counts of Twite at Chasewater, 1948/9 – 1977/8.

Winter	Max	Winter	Max	Winter	Max	Winter	Max	Winter	Max
1948/9	9	1954/5	31	1960/1	14	1966/7	40	1972/3	70
1949/50	30	1955/6	15	1961/2	45	1967/8	90	1973/4	90
1950/1 no count		1956/7	35	1962/3	20	1968/9	70	1974/5	55
1951/2	10	1957/8	50	1963/4	42	1969/70	80	1975/6	40
1952/3	11	1958/9	35	1964/5	15	1970/1	50	1976/7	75
1953/4	10	1959/60	8	1965/6	19	1971/2	95	1977/8	200

to its once having been more numerous. During the past twenty years, however, it has enjoyed an unprecedented resurgence, accompanied by a considerable expansion in its breeding range, and today the Redpoll is not only a regular winter visitor to birches and alders, but is also familiar as a breeding species in young conifer plantations, birch scrub and, to a lesser extent, hawthorn and rank hedgerows.

Smith said it nested in the Dove and Trent valleys as well as locally in much of lowland Staffordshire and that large flocks, sometimes several hundred strong, appeared from other regions in autumn and winter, with a maximum of 500 at Longton in October 1918. Both Harthan and Norris, however, considered the species primarily a winter visitor, with just a few resident, though Harthan did refer to half-a-dozen or so that used to breed in damson trees at Evesham until 1910. The largest flock mentioned by Norris was 50, in Sutton Park in April 1915, and another flock of similar size fed under rowans on the Lickeys in 1934. By the end of the second war, however, even flocks of 15 and 17 had become noteworthy and it was 1954 before a flock of 50 was again reported.

Even as recently as the 1950s breeding was largely confined to northern Staffordshire (Norris 1951 and Fincher 1955), but by 1959 the Redpoll had been taken off the list of winter visitors as its remarkable population explosion began to gather momentum (see graph).

The WMBC *Atlas* survey from 1966-8 showed confirmed breeding in eight 10-km squares (10 per cent) and suspected breeding in a further twelve (16 per cent) – all in north or west Staffordshire and the West Midlands Conurbation – but by the conclusion of the BTO *Atlas* survey in 1972 these figures had increased to twenty-seven (35 per cent) and twenty-one (27 per cent) respectively, with consolidation throughout the whole of Staffordshire and the West Midlands Conurbation, a spread into

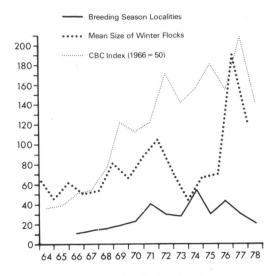

Fluctuations in Redpoll, 1964-78

north Warwickshire and initial, sporadic colonisation of both the Avon and Severn valleys. Since 1974 Redpolls have spread increasingly into urban parks and suburban gardens and in 1977 nests were located in both lilac and privet in a Quinton (West Midlands) garden. In the same year a pair also bred in heather at Highgate Common, despite an abundance of more conventional habitat, and this may have been an indication of numbers having reached a saturation threshold. Certainly many observers consider numbers to be past their peak and the CBC at Fradley Wood (40 ha) has shown this trend (see Table 86).

This is the only data on regional density, but nationally there were 1·92 pairs per km² in all habitats in 1978, whilst Sharrock postulated a density of 100 pairs per 10-km square. These figures would imply a regional population of 2,000–5,000 pairs, which is well above the range 20–200 suggested by Lord and Munns in 1970.

The Redpoll's population increase is typical of an irruptive species and shows a three-yearly cycle of peaks and troughs (*cf.* Siskin), which probably correlates with the abundance of birch seed, which is its main food source in winter. Newton (1972) has

Table 86: Population of Redpoll at Fradley Wood, 1973–79.

	1973	1974	1975	1976	1977	1978	1979
No. of pairs	4	2	5	8	15	4	6

shown that birds from northern Britain migrate as far south as the Continent in search of a good food supply, but once they have discovered a plentiful crop of birch seed they will often remain in large numbers throughout the winter and if the habitat is suitable many will stay to breed. In this respect post-war afforestation, with conifers planted alongside natural birch scrub, has been a great boon, since Redpolls nest readily in young conifers of 2 m or so in height. Ringing evidence shows that West Midland birds conform to the national pattern, with three Warwickshire-ringed birds being recovered in Belgium during the November of a subsequent winter and a bird ringed in Staffordshire in September 1970 being recovered in France in October 1971. Within the British Isles, however, recoveries show considerable movements in varying directions.

Prior to the recent population explosion, flocks only occurred from October to April, but since 1969 there have been more reports in September and May also. Most flocks occur in December, but the mean size is greatest in January. There is then a decline during February and March followed by an increase in April and May, which indicates a return spring passage. Redpoll were especially abundant in the 1976/7 winter, when one or two flocks up to 500 strong were noted on Cannock Chase between December and March, but in the main flocks in excess of 150 are scarce and most reports refer to parties of less than 100.

The Mealy Redpoll, *C.f. flammea*, is a rare winter visitor that has occurred seventeen times. The only historical reference is that of Harthan, who said that Mealy Redpoll were caught occasionally by bird-catchers during the last century. Smith said that it perhaps had been overlooked, but the only definite record was of three taken by bird-catchers at Longton in December 1913. Two more were reported in 1943/4, from Randan and Powick, but then none until 1971/2. Since then there have been reports in four years out of seven, with up to five at Hampton-in-Arden twice during March 1972 and ten at Shirley in late January and early February 1976. In fact the winter of 1975/6 was quite exceptional, with reports from seven localities involving 17 birds. Apart from one at Hatton on September 22nd 1976, all records have fallen within the period December to March inclusive.

Two-barred Crossbill
Loxia leucoptera

One record.

A male consorted with Crossbills at Beaudesert, Cannock Chase, from December 16th 1979 until April 1st 1980, much to the delight of many of the country's bird-watchers (*Brit. Birds* 73:527). The species is a very rare vagrant to the British Isles from northern Europe. This bird appeared to be paired with a female Crossbill and mating was observed.

Tomes (1901) also mentioned a female killed near Worcester in 1838 and this was originally included by Harthan in his *Birds of Worcestershire*. However, in his 1961 review, Harthan said "Professor Newton examined this skin at Cambridge and a

label attached to it states that it seemed in all respects to resemble the American race *L.l. leucoptera* (White-winged Crossbill). It is doubtful if this was a genuine wild bird. *The Handbook* does not admit this sub-species to the British list.''

Crossbill
Loxia curvirostra

Scarce and erratic visitor, mainly in late summer. Very scarce breeding species.

Crossbills occur in small flocks in many suitable localities during irruption years, but are very scarce or absent in intervening ones. Their nomadic life is geared to the fruiting cycle of spruce and larch and each summer they search out the best cone crops and remain there until food runs out (Newton 1972). So long as food is sufficient, birds will remain in the same area and one or two pairs may breed after major irruptions.

Smith found Crossbills irregular visitors and evidence of breeding only circum-stantial. Birds appear to have been plentiful in 1838, 1888/9 and 1909/10, when a nest without eggs was found near Penkridge. However, Harthan referred to a record of nesting at Enville, which had been erroneously attributed to Worcestershire. Both he and Norris also described the Crossbill as an irregular visitor, with birds usually arriving in July to feed on larch and pine cones. Flocks were noted in 1821, 1869 and 1909/10 in Worcestershire and 1845, 1855, 1909/10 (large numbers) and 1927 in Warwickshire.

During the period 1929–78 the largest irruption occurred in 1935 (at least 250 birds), but substantial numbers also arrived in 1958/9 (128 birds), 1962/3 (185 birds) and 1972 (172 birds), with smaller influxes in 1953, 1956 and 1966. Conversely 1961, 1968, 1969, 1975 and 1976 were all blank years – the only ones since 1956, though fifteen of the preceding twenty-seven years were also blank. This reflects both the increase in observations and the gradual maturing of post-war larch and spruce plantations. The latter has certainly aided breeding, which occurred at Lickey and Enville in 1959 and Enville again in 1963 and 1964. Birds were also present during the breeding season at Enville in 1967, on Cannock Chase and along Coombes Valley in 1973 and in the Wyre Forest in 1977 and 1978. Breeding may also have occurred on Cannock Chase in 1980 following a small irruption in 1979.

During irruption years Crossbills begin to arrive about the second week in July and peak in the last week of July. These early immigrants may be augmented by further arrivals during late summer and autumn, but many also pass on, and overall numbers decline from late July until October, after which they remain fairly constant until February. There then follows a slight March peak and a further decline, with June having fewest records of all. Even allowing for some birds passing through to other areas, it is likely that the tailing-off of numbers during August and September is because some of the larger flocks disperse and remain partially undetected. Whether the slight March peak indicates a small return passage or merely an enlarged, post-breeding population, however, remains unanswered.

Crossbills feed almost exclusively on spruce and larch cones and are seldom seen away from conifer plantations, where the first indication of their presence is often the twittering flight calls of excited flocks. For this reason most records come from Cannock Chase, Sutton Park, Lickey, Enville and the Wyre Forest. On October 6th 1963, however, two were apparently feeding on rowan berries in Sutton Park, whilst a few frequented a plum orchard at Sheriffs Lench in July 1956. Two-thirds of the records refer to small parties of less than ten birds, but parties up to 50 are not unusual in good years and flocks of 100 – perhaps the same birds – were reported from Enville on August 31st 1935 and March 15th 1936.

Bullfinch

Pyrrhula pyrrhula

Widespread and numerous resident.

The Bullfinch feeds on a variety of flowers, buds, berries, soft fruit and seeds taken directly from plants such as oak, ash, birch, rowan, fruit trees, hawthorn, clematis and honeysuckle. Consequently it is seldom seen far from the cover of woods, copses, garden shrubberies and thick hedgerows, though it does resort to orchards at blossom time, sometimes in considerable numbers.

According to Smith Bullfinches were met with universally at all seasons, being common in wooded localities and even penetrating the upland valleys wherever there is scrub to provide a nest site. He also referred to small flocks in autumn

and winter, but observed that the species suffered severely in continued hard weather and that in heavy snow it would resort to open fields to feed on "hard heads". Harthan noted it as common in woods and referred to its habit of destroying the buds of early-flowering fruit trees – an important consideration in the Vale of Evesham, where fruit growing was of great commercial importance. Nevertheless Harthan considered the Bullfinch did negligible damage to commercial orchards as the woodlands, especially where clematis grew wild, supplied them with plenty of food. In general, birds only resort to orchards in any number when ash fruit is scarce or absent (Summers 1979). Norris described it as a widely distributed resident in small numbers, often appearing in twos and threes in winter and being more common in less intensely agricultural areas where hawthorn and bramble prevailed.

The same widespread distribution was revealed by the *Atlas* surveys from 1966–72, when breeding was proved in seventy-three 10-km squares (95 per cent) and suspected in a further three (4 per cent) that embraced the Millstone Grit moors, the Potteries and the Stour Valley. There has been a steady increase in numbers, which began in the south of the Region during the late 1950s and spread gradually northwards. By 1962 there were reports of serious damage to orchards in the Vale of Evesham and in that year 400 were shot and many more trapped in the Church Lench, Harvington and Norton parishes, whilst 163 were trapped on a single fruit farm in 1963. This general increase in numbers was accompanied by a spread into suburban areas, with birds at Edgbaston Reservoir in 1955 and Hall Green in 1962 being the first at these localities for eleven and ten years respectively. In 1955 also, a party of 14 appeared in town gardens at Stourport, whilst in 1960 eight nests were located within 40 m of each other in a Malvern garden. The CBC showed just one or two pairs at Edgbaston Park from 1966–73, but four or five subsequently, whilst at Moreton Morrell numbers in 184 ha of farmland and woodland have varied from one to eight pairs and Fradley Wood (40 ha) has held from one to five pairs. In 1978 the overall regional densities were $3 \cdot 46$ pairs per km^2 on farmland and $10 \cdot 8$ pairs km^2 in woodland, compared to national figures of $2 \cdot 06$ and $7 \cdot 79$ respectively. Such densities suggest a regional population of 10,000–15,000 pairs, which is well in excess of the 200–2,000 estimated by Lord and Munns.

Some indication of winter densities is given by 50 males in 13 km^2 at Exhall, near Coventry, in February 1961 and over 100 in 10 km^2 at Brandon and Baginton in November 1964. During autumn and winter Bullfinches gather in small parties, keeping in touch with one another by their characteristic piping calls. In Norris' time twos and threes were normal, but since the mid-1950s 15 or 16 has been the average size of those notified and 50 have been reported on six occasions, mostly between 1970 and 1973. Since then there have been no reports of parties larger than 25 which suggests a decline, though most reporters are still referring to good numbers and increases. Parties are most frequently seen in December (30 per cent) and January (27 per cent), but occur throughout from September to February. March records, however, are very few. Parties are largest in November, when the mean size of those reported was 23, but gradually diminish during the course of the winter.

In Britain the Bullfinch is largely sedentary and the longest movement involving a West Midland bird was 32 km, from Banbury (Oxon) to Rugby. The northern race, *P. p. pyrrhula*, migrates regularly, however, but seldom penetrates as far south as this Region, though three males and a female were claimed at Lickey on January 19th 1949.

Hawfinch
Coccothraustes coccothraustes

Scarce resident.

Its shy, elusive behaviour makes the Hawfinch difficult to detect, but nowhere in the West Midlands is it common. Since it feeds mainly on the fruits of trees like hornbeam, beech, cherry and holly, it is largely confined to well-wooded districts, parklands or orchard areas, but over the years its distribution and numbers have fluctuated. In particular it has disappeared from many of the old orchard areas with the demise of cherry growing. When its favourite fruits are scarce, it moves into sub-optimal habitats in search of haws or ash seeds.

Plot (1686) considered the Hawfinch scarce and Garner (1844) irregular in its appearance, yet subsequently they were more widely seen. Smith said Hawfinches bred locally practically throughout Staffordshire, except on the bare hills, and that considerable numbers appeared in autumn and winter, when small parties roamed the countryside and often appeared in districts where few bred, where they frequently did considerable damage in gardens.

Harthan also said that the Hawfinch was considered a rare winter visitor in the early nineteenth century, with large flocks at Worcester in 1833 (Hastings 1834), but none at Malvern until 1854 or Evesham until 1869 (Tomes 1901). It then increased rapidly, with breeding at Malvern Wells in 1867, Evesham in 1879, Stourport in 1887 and Dudley in 1893. At Barnt Green the species was still uncommon in the 1880s, but by 1920 flocks of up to 50 were not infrequent and 45 were seen in the air together in 1935. Subsequently the Hawfinch again became scarce in Worcestershire, though it still nested regularly at Malvern and doubtless at some other localities too.

In Warwickshire, Tomes (1904) noted that the species was more abundant than

81. The shy, elusive Hawfinch is nowhere common, but can be seen regularly at a few favoured localities.   *R. J. C. Blewitt*

formerly, though according to Norris this was a peak year in the south of the county. Norris also recorded a slight increase in the Birmingham area until the First World War, after which it decreased and became a rarity, though perhaps more plentiful than records suggested due to its secretive habits.

By 1950 though, Norris (1951) could show breeding only in the Enville, Bromsgrove, Martley, Evesham and north Warwickshire areas. Twenty years later the *Atlas* surveys from 1966–72 confirmed breeding only in west Staffordshire and the Teme Valley, though in addition breeding was suspected around Evesham, Enville and Sutton Park. In all breeding was confirmed or suspected in just seven 10-km squares (13 per cent), but there were birds present during the breeding season in a

further fifteen squares (19 per cent), including Malvern, Bredon, the Avon Valley, Wyre Forest, Needwood Forest, Churnet Valley and west Staffordshire. In the absence of survey data the total breeding population is hard to estimate, but it seems likely to be at the upper end of, or above, the range 3–20 pairs suggested by Lord and Munns.

Hawfinches have been seen in every year since the war, with the five-yearly distribution shown in Table 87.

This indicates marked fluctuations, but no long-term trends. Most records have come in winter or early spring, with almost half of them in March and April and fewest in September. The monthly distribution of dated records is shown in Table 88.

However, Hawfinches are most conspicuous in March and April, when they

Table 87: Five-yearly Totals of Hawfinches, 1949–78.

	1949–53	1954–58	1959–63	1964–68	1969–73	1974–78
No. of birds	124	97	139	69	113	157

Table 88: Monthly Distribution of Hawfinches, 1929–78.

	Jan	Feb	Mar	Apr	May	Jun	Jul	Aug	Sep	Oct	Nov	Dec
No. of birds	69	69	171	162	23	37	26	21	8	26	33	49

indulge in noisy display flights, but tend to disappear in the canopy once the trees have burst into leaf and nesting starts. Even so, this distribution has not been consistent over time, with the majority of October records, for example, being concentrated into the 1974–8 period.

Birds appear in small parties of up to ten in winter and spring. Larger groups have been observed on half-a-dozen occasions, with up to twenty-five intermittently at the Lower Avenue, Chillington, from the autumn of 1976 through to early April 1977 and up to eighteen at Lickey in March 1960 the largest parties. Smith said he had seen small numbers seemingly migrating over the moorlands in October and November and cited a mixed concourse of thrush-like birds and Hawfinches proceeding westwards up the Churnet Valley against a gale on November 24th 1917, but there is no ringing evidence to throw any light on the movements of this elusive species.

then elapsed before the next record, of what was believed to be a first winter bird at Chasewater from December 5th to 8th 1948. One flew across Belvide on October 28th 1956 and another was seen and heard at the same locality on November 22nd and 29th 1959. The fourth Staffordshire record came from Chasewater on November 3rd 1973, whilst 1978 brought the second Warwickshire record when three were discovered in a large finch and bunting flock at Kingsbury Water Park on January 8th. The following week a fourth bird was discovered and at least two of these stayed until February 26th. During this time, what was assumed to be one of these birds visited nearby Ladywalk. This unexpected occurrence followed a major influx to the east coast in the preceding autumn. Finally, during the severe weather in early 1979, a male and a female were seen in meadows alongside the Trent near Burton from January 21st to 28th, where they fed with other passerines in hay put out for cattle; and a female or immature was seen at Bartley, also on January 21st 1979.

Lapland Bunting
Calcarius lapponicus

Rare winter visitor. Eight records.

There were no records until October 21st 1904, when a male was caught in a clap net at Acocks Green (Norris). Over forty years

Snow Bunting
Plectrophenax nivalis

Regular, but scarce autumn passage migrant or winter visitor in varying numbers.

Smith said Snow Buntings were occasionally reported in autumn and winter and that in severe times they had been known to frequent rick-yards along with other small birds. Harthan described it as a rare winter visitor and Norris as very rare. Between them these authors listed fourteen admissable records involving fifteen birds prior to 1929.

The period 1929–78, however, has shown the Snow Bunting to be a more regular, though still somewhat erratic, autumn passage migrant and winter visitor. During this period there have been ninety-one records involving 162 birds, distributed as follows:—

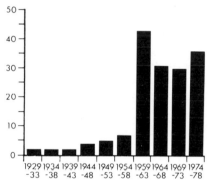

Five-yearly Totals of Snow Buntings, 1929–78

Since 1947/8 birds have failed to appear in only three winters and there has been no blank year since 1958/9. Occurrences peaked in the early 1960s and again in the late 1970s, with 1959/60, 1961/2 and 1976/7 being particularly good winters. The majority of records referred to single birds or small parties up to half-a-dozen strong, but seven have been recorded on three occasions – at the Malvern Hills in November 1961, Draycote in the 1970/1 winter and Blithfield in November 1977 – whilst the most ever reported was thirteen at Chasewater on October 17th 1976.

Interestingly, half-a-dozen localities have produced three-quarters of all records, involving over 80 per cent of all birds, namely:—

Table 89: Records of Snow Buntings at Principal Localities, 1929–78

	No. of Records	No. of Birds
Chasewater	20	42
Blithfield	16	28
Bittell	9	17
Draycote	9	17
Bartley	9	12
Malvern Hills	4	16

The first five of these are all reservoirs with both hard and natural shorelines which, when water levels are low, offer an abundance of seeds along their open, sparsely vegetated margins. As such they are the Region's closest resemblance to the Snow Bunting's typical coastal wintering haunts. Equally the Malverns satisfy the species summer preference for high ground. A few birds have been recorded on or near the moors, too, but not with the frequency or numbers reported from Derbyshire (Frost 1978).

The earliest record was on October 9th (1976), but most birds have arrived in November (46 per cent) or December (18 per cent). Many were noted only once, so presumably passed on to other areas, but there is no evidence of any return spring passage. Some birds, however, made protracted stays. Significantly, though, apart from two birds at Oldbury in 1962 which stayed for one and two months respectively, none have stayed longer than a week except at the five principal reservoirs.

Yellowhammer
Emberiza citrinella

Abundant and widespread resident.

The Yellowhammer is a widespread species likely to be encountered anywhere in open country that affords elevated song posts. It

435

is one of the commonest birds on farmland, especially where there are hedgerows
and trees from which it can deliver its
familiar song, but it is equally at home
along roadsides and railways, on bracken-
covered hillsides and heaths, or in the edges
of woods. Recently it has even colonised
forest clearings and young plantations,
especially on Cannock Chase. Only in the
most industrialised or densely developed
districts is it entirely absent. Yellowhammers feed throughout the year on
grain, seeds and invertebrates, which they
obtain from the ground, and outside the
breeding season they are frequently found
in small flocks, often accompanied by larks
or finches.

Smith remarked that the Yellowhammer
was common at all seasons in well-
cultivated, lowland districts, but added
that it was scarce in the bare moorlands
and hilly areas of the north, though it must
have been more numerous there when corn
was more freely grown. Harthan and
Norris respectively described the species as
abundant and common, but both remarked
on the decline in birds breeding along roadside verges due to disturbance from increasing traffic and verge-cutting. Today,
of course, traffic has reached a volume that
these authors could scarcely have imagined
thirty years ago, but at least verge-cutting
has been reduced as an economy in recent
years and the Yellowhammer remains one
of the commoner roadside species.

There appears to have been little change
in distribution over the years, with the
Atlas fieldwork from 1966–72 showing
breeding as confirmed in seventy-five 10-
km squares (97 per cent) and suspected in
the remaining two. Failure to prove breeding in these two squares may have been due
to difficulties in observer coverage, since
they both include plenty of suitable habitat. Consequently the species can be said to
breed everywhere, although it is sparse and
local in the Conurbation and on the moors.
In 1950 it ranked fourth amongst species

caught in a rick-yard at Bartley, accounting
for 12 per cent of all birds ringed, whilst on
the regional CBC farmland plots in 1978 it
ranked fifth, accounting for 6 per cent of
all territories. The Yellowhammer appears,
therefore, to have maintained its relative
status, but its numbers have fluctuated,
with reported declines of between a third
and a half following the severe winters of
1946/7 and 1962/3 and references to local
declines about 1960. These declines seemed
to arouse interest in the species and,
encouraged by the ease with which it could
be censused, there is much valuable
information on farmland densities from
the 1960s. At Wilnecote, for example,
density dropped from seven to nine pairs
on 81 ha prior to the 1962/3 winter to three
or four pairs thereafter; whilst at Cofton
Richards there was a decline from six pairs
on 81 ha in 1966 to only half this level the
following year. Conversely, density at
Brandon rose from five pairs on 105 ha in
1965 to 12 pairs in 1966, whilst at Arley
(Warwickshire) there was an increase from
six pairs on 61 ha in 1966 to 10 pairs in
1967. The CBC data shows great variability. For example, numbers at Fradley
Wood ranged from one to nine pairs on
40 ha, with a mean of five pairs, and at
Moreton Morrell there were between six
and twenty-five pairs on 184 ha of farmland and woodland, with a mean of sixteen
pairs. In 1978 the Yellowhammer was the
commonest species on a farmland plot at
Flyford Flavell, with twenty-one pairs in
73 ha, and overall the index showed $15 \cdot 9$
pairs per km^2 regionally and $9 \cdot 4$ pairs
nationally. The respective 1978 densities
from the woodland plots were $9 \cdot 6$ and $12 \cdot 8$
pairs per km^2. Thus it would seem that the
Yellowhammer population of the Region is
particularly strong at the present time, with
perhaps some 30,000–50,000 pairs. Lord
and Munns estimated over 20,000 pairs.

Outside the breeding season, Yellowhammers often flock together on stubble or
around livestock feeding stations. They

begin to congregate in October and flocks then persist until March or early April, being most frequent and largest during December and January, when the average size of those reported was eighty. Only a few flocks exceed one hundred, but 300 have been noted on three occasions – at Frankley in December 1957, Kings Bromley in January 1968 and Coombe in October 1975 – whilst a total of 350 was present in several small flocks at Draycote in December 1970. In winter some birds wander into urban areas. Ten were seen at Moor Street station, Birmingham, in 1970 and birds have also been noted in Edgbaston Park from time-to-time. There is little information on roosting habits, except for 300 which roosted in dead bracken at Islandpool on February 4th 1971 and 100 which roosted in low conifers at Brandon Wood on January 14th 1973. Since 1975 there has also been an increasing tendency for small numbers up to twenty to roost with Reed Buntings at Brandon (Potter 1977).

Little is known about the movement of Yellowhammers either, since, despite fairly frequent ringing, the only recovery is of a bird ringed at Wellesbourne and found five days later some 14 km hence.

Cirl Bunting
Emberiza cirlus

Formerly a very scarce resident, now no longer present.

The distribution of the Cirl Bunting in the West Midlands has always been restricted and its population small. Smith said that failing further evidence it would hardly be advisable to place this species definitely on the Staffordshire list. Subsequently, however, there has been one record of a female at Kinver in May 1951. In Warwickshire, Norris said it was formerly very local and uncommon, but that there had been no records since 1931, when a nest was found near Rugby. With one or two exceptions all historical records came from south of the Avon and particularly from the foot of the Cotswold fringe. In recent times single males were seen at Bartley in February 1950 and Shipston-on-Stour in July 1959.

In Worcestershire a small, relict population lingered in the lower Severn Valley for many years, though this seems now to have finally died out as only one site has been known since 1972 and this appears to have been deserted since 1977. Harthan described the Cirl Bunting as scarce and local. A few pairs nested regularly south of Upton-on-Severn and it was fairly common until 1910 around Malvern. In 1938 the species reappeared at Malvern and this remained its stronghold until 1972, when birds were last seen. Even this population was small, however, with only one or two pairs and, apart from an exceptional flock of about twenty near Bushley on January 2nd 1936, there is no record referring to more than three birds. Bannister (1941), though, considered it might be commoner than it appeared, since it is easily overlooked. He also observed that it was a local wanderer, often deserting old haunts for many months, with pairs keeping close company throughout the winter.

Harthan considered the Cirl Bunting to be fond of pasture land with plenty of hedgerow timber, but birds have occurred in other typical habitats as well. On the Malvern Hills, for example, one sang for many years from a disused quarry hewn into the hillside, whilst the last known pair regularly visited a country garden. In the Mediterranean, Cirl Buntings favour warm, sheltered valleys (Sharrock), so perhaps the warm summers enjoyed by the Severn Valley were instrumental in sustaining this population and the recent succession of cooler summers a factor in its decline.

Little Bunting
Emberiza pusilla

One record.

Norris provided the sole record, of a male in winter plumage caught with bird lime at Pailton early in October 1902 and kept in a cage for nearly fifteen months. The date conforms well with the pattern demonstrated by Sharrock and Sharrock (1976) for vagrancy in this species from Fenno-Scandia and northern Russia.

Reed Bunting
Emberiza schoeniclus

Numerous resident and passage migrant.

Although traditionally a bird of wet or damp situations, with progressive drainage since the last war, the Reed Bunting has increasingly turned to drier areas. Today it is not only a common breeding species in waterside vegetation, but can also be found on farmland and even in young conifer plantations, where it sometimes breeds alongside Yellowhammers. Outside the breeding season Reed Buntings congregate in small flocks to feed from the ground on grain, seeds and invertebrates and some birds wander into urban situations, where they are often seen on wasteland and may occasionally resort to bird-tables.

Smith said that Reed Buntings abounded in suitable localities, but that they bred only sparingly in the hill country. Harthan

said that they were resident in marshy places, especially osier beds, and Norris that they were few in number and local in distribution, with the banks of the Avon and the sides of streams and lakes commonly frequented. All three authors commented on a decrease in numbers in winter.

With its spread into drier habitats, the Reed Bunting now has a much wider distribution and the *Atlas* surveys from 1966–72 showed confirmed breeding in seventy-four 10-km squares (96 per cent) including the whole of the Conurbation. Only along the Stour Valley in Warwickshire and around the Wyre Forest was breeding not proved. Where reliable census data is available it seems the species has increased in numbers, but this may be because birds tend to congregate in the diminishing remnants of their most favoured habitat. Certainly Edgbaston Park held two pairs in 1944 and one or two pairs from 1966–72, but this increased to five or six pairs between 1973–6, though it dropped again to three pairs in 1979. At Alvecote, the population increased by half, from 8 to 12 pairs, between 1950 and the 1970s and then peaked at 20 pairs in 1970; whilst that at Brandon more than trebled, from 24 to 80 pairs, between 1968 and 1973. The latter figure represents a density of 76 pairs per km^2, which accords closely with Sharrock's figure of 50–70 pairs per km^2 for typical wetland situations, but even higher densities have been recorded from small areas of eminently suitable habitat. In 1966, for example, Ufton Fields held 30 pairs on 32 ha and the Leasowes five pairs on 2 ha. In drier situations there were four pairs on 81 ha of farmland at Cofton Richards in 1967, whilst 40 ha at Fradley Wood held just one pair from 1967–75, but this had increased to five pairs by 1979. In 1975, eleven pairs were recorded nesting in 56 ha of oilseed rape and birds have also bred and are quite common and widespread in winter wheat. Breeding has also been recorded in spring beans, gorse, a conifer hedge and soft fruit in the Vale of Evesham. A CBC census along part of the River Avon in Worcestershire showed the Reed Bunting to be the third commonest species. Overall, in 1978, the CBC showed a regional density of 4·1 pairs per km² on farmland and a national density of 4·7 pairs per km². Between them, these densities point to a regional population of at least 15,000 pairs, which is considerably in excess of the 200–2,000 pairs estimated by Lord and Munns.

Outside the breeding season Reed Buntings can be gregarious and loose flocks are often reported. Three-quarters of those reported have involved less than 100 birds, with an average flock size of 60. However there are four records of more than 200, with 250 in a mixed finch flock at Thorn Hill on January 29th 1972 and 300 at Belvide on April 4th 1958 the most ever reported. In general, flocks are largest in January and February, when hard weather tends to bring birds together more, but most numerous in December, March and April, when there is a noticeable passage through the Region. Late March and early April is also the typical time for birds to return to their breeding grounds.

Reed Buntings often roost in reed beds or similar vegetation. Numbers are usually less than 100, but there have been several reports of gatherings up to 400, whilst the most recorded was 1,000 at Brandon from January to March 1974. The Brandon roost has been well studied and numbers have been found to be very variable. The roost peaks during November, after which numbers fall and then rise again towards a second peak in February. During these peaks males outnumber females. In mild weather there may be several satellite roosts rather than one composite one (Fennell and Stone 1976).

The Reed Bunting has been extensively ringed and at Brandon no fewer than 449 individuals were trapped during 1973, two of which proved to be over eight years old. There have been a large number of movements between suitable sites in the Midlands and the surrounding counties and in recent years there has been evidence of longer migrations, with three birds that were in Hampshire during the winter being in the West Midlands Region during the summer.

Corn Bunting
Miliaria calandra

Fairly numerous, or numerous, resident in arable districts.

As its name implies, the Corn Bunting is closely associated with arable farmland, though it also occurs in grasslands and commons. Above all it seems to avoid trees, perhaps because its colonial and polygamous behaviour requires an abundance of food and in this way it can reduce competition from other species for the food available. Instead it has exploited a largely vacant niche in open country, where it feeds on grain, seeds and insects taken from the ground, and uses hedges, bushes, fences and overhead wires as perches from which to deliver its characteristic song. In winter it roosts communally, often in reeds or reedmace, but occasionally in conifer plantations.

With a patchy distribution in which birds are often absent from seemingly suitable areas, the Corn Bunting has long intrigued the Region's ornithologists. During the past hundred years it has also undergone a significant change in status. Centuries ago strip-farming amid an open, pre-enclosure landscape must have been to the Corn Bunting's liking. As the enclosure landscape matured, so it would have suited Corn Bunting less, but even so it is perhaps not surprising that Tomes (1901 and 1904) considered it common and well distributed and Smith quoted several earlier writers

82. The Corn Bunting is a bird of treeless, arable country and has a strong affinity with barley growing.   *M. C. Wilkes*

who had described it as abundant. The agricultural depression, which began late last century and lasted until the Second World War, however, saw a decline in arable farming and a consequent reduction in Corn Buntings. Even as early as 1903 Hudson noted that 50 Yellowhammers and 10 Reed Buntings could be seen for one of this species and so great was the decline that Harthan found the species had almost disappeared from Worcestershire and Norris that recent Warwickshire records were almost non-existent. With the war-time resurgence of arable farming, the Corn Bunting has returned again, however. Two or three additional factors have aided its recovery. Firstly the 1950s and 1960s brought an emphasis on barley growing and this is a crop with which the Corn

Bunting has a close affinity. Indeed, as the accompanying graph shows, there is a strong correlation between the increases in barley growing and Corn Bunting numbers. The post-war removal of hedges and more particularly trees also expanded its preferred habitat and in recent years the outbreak of Dutch elm disease has extended this still further. Today Corn Buntings are commonly found in grass leys as well as cereals (J. A. Hardman *pers comm*).

The recovery began in the Avon and Stour valleys south-west of Stratford in the late 1940s and was remarked on by both Norris (1951) and Fincher (1955), although their respective descriptions of regular breeding only in the Shipston area and the Cotswold fringe were geographically

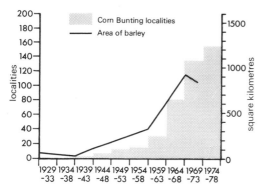

Five-yearly Totals of Corn Bunting Localities compared with the increase in Barley Growing

slightly inaccurate. During the next decade birds spread up the Avon Valley as far as Coventry and re-established themselves in the lower Severn Valley and around both Nuneaton and Tamworth. This position was consolidated in the early 1960s and was followed by a further expansion that embraced the Severn and Worcestershire Stour valleys, the lower Avon Valley, the Feldon area of Warwickshire, the Cotswold fringe, the Lichfield district and one or two isolated areas of west Staffordshire. This was the distribution found during the WMBC breeding survey of 1966–8. During the next five years Corn Buntings also spread into the arable districts adjoining the Needwood Forest, the Trent Valley and the Penk basin, so that by the time *Atlas* survey work was completed in 1972, breeding had been confirmed in twenty-seven 10-km squares (35 per cent) and suspected in a further twenty-one squares (27 per cent). Although breeding was confirmed in both the moorland area and central Staffordshire, the distribution in these districts is patchy and, along with the Birmingham plateau and the hilly district of west Worcestershire, they remain largely uncolonised. In each of these areas the farming pattern is predominantly dairying or stock rearing and, as the accompanying map shows, the distribution of the Corn

Bunting correlates very closely with areas of arable or horticultural production, particularly on light, sandy soil.

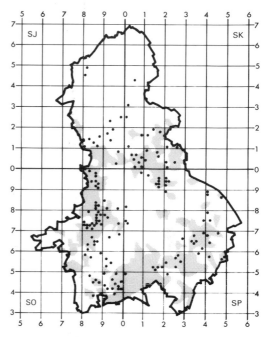

Predominantly arable or horticultural areas

· Corn Bunting records 1974–78

Distribution of Corn Bunting, 1974–78, and Types of Farming

In favoured areas density can be high. For example ten singing birds were found along 4 km of the Leam Valley in 1978, whilst the density at Hammerwich in 1971 was 4 pairs per km². In its Avon stronghold, the Corn Bunting ranked fifth in abundance in 1976 on a CBC plot at Wellesbourne, with a density of 17 pairs on 95 ha. Compared to the 1978 national CBC average of 2·3 pairs per km², this is high indeed and, notwithstanding the patchy distribution, it seems the regional population must be well in excess of the 20–200 pairs estimated by Lord and Munns in 1970 and possibly in the range of 1,000–5,000 pairs.

Table 90: Roosts of Corn Buntings at Brandon, 1972–76.

	1972	1973	1974	1975	1976
Max. size of roost	62	100+	80	50	100
Date	Nov 26th	Jan 27th	Jan–Apr	Feb	Dec

Outside the breeding season Corn Buntings feed in small flocks on cultivated land or waste ground. For example, 17 were noted amongst sprouts at Wellesbourne in October 1953 and 30 fed on *Chenopodium sp* at Wilden in October 1978. The largest feeding flock was up to 100 at Kingsbury in the winter of 1978. As a rule such flocks disperse during March or April, but as many as 60 have been seen together as late as May 13th. During the winter months small parties have even occurred in inner Birmingham, at Nechells and Saltley. The larger roosts are usually in reed-beds or other emergent vegetation and seldom occur before December, but they then persist through to March or early April. In 1963 even a modest roost of ten, at Chesterton, was considered noteworthy, but their size has grown along with the spread in distribution and increase in population, and nowadays up to 100 is by no means unusual. The largest recorded roosts were 150, at Eathorpe in 1975 and Brandon in 1978, and 500 at Exhall, near Alcester, on December 31st 1973 (J. A. Hardman *pers comm*).

Some idea of the variability of roosts is given by the data from Brandon shown in Table 90 (Potter 1977).

Corn Buntings are not noted for long movements and one ringed at Far Cotton (Northants) on September 22nd 1974 and recaught at Alcester on February 8th 1976 is, at 63 km, the longest known movement within the British Isles. Other recoveries have involved birds moving between Eathorpe and either Brandon or Ufton, both distances of 7 km.

10 Breeding Distribution Maps

These maps have been compiled by amalgamating those in the WMBC *Atlas of Breeding Birds of the West Midlands* (Lord and Munns 1970) with the BTO *Atlas of Breeding Birds in Britain and Ireland* (Sharrock 1976). As such, they show the breeding distribution by 10-km squares during the seven consecutive years 1966–72, denoted as follows:

●   Definite breeding
●   Probable breeding
·   Possible breeding

The following species appear in the above works, but are known to be very localised or sporadic and have not been mapped:

Black-necked Grebe	Goshawk
Bittern	Ringed Plover
Greylag Goose	Dunlin
Egyptain Goose	Marsh Warbler
Wood Duck	Bearded Tit
Mandarin	Red-backed Shrike
Hen Harrier	

Likewise no maps are shown for the following species for which breeding was confirmed in every 10-km square:

Mallard	Willow Warbler
Moorhen	Spotted Flycatcher
Woodpigeon	Blue Tit
Swallow	Great Tit
House Matin	Magpie
Dunnock	Starling
Robin	House Sparrow
Blackbird	Chaffinch
Song Thrush	

Data from the BTO *Atlas* is reproduced by kind permission of T. & A. D. Poyser and the maps are based on the National Grid by permission of the Controller of Her Majesty's Stationery Office, Crown Copyright reserved.

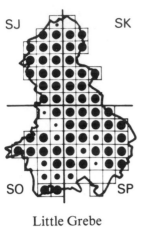

Little Grebe

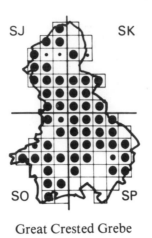

Great Crested Grebe

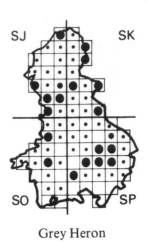

Grey Heron

Mute Swan

Canada Goose

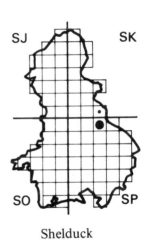

Shelduck

Gadwall

Teal

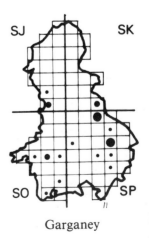

Garganey

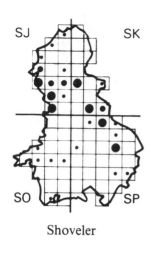

Shoveler

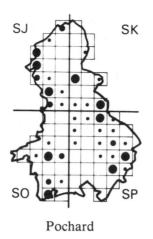

Pochard

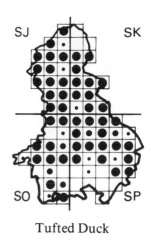

Tufted Duck

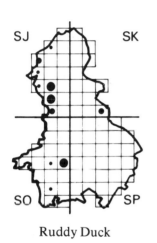

Ruddy Duck

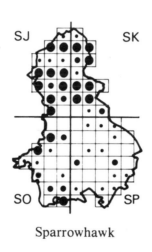

Sparrowhawk

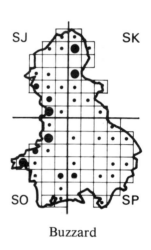

Buzzard

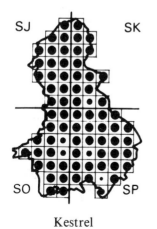

Kestrel

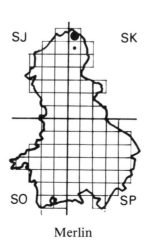

Merlin

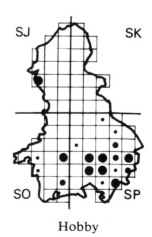

Hobby

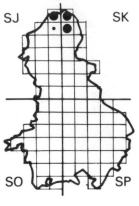
Red Grouse

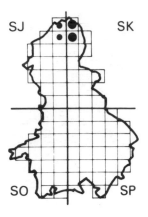
Black Grouse

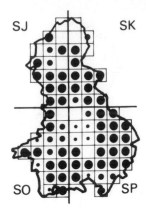
Red-legged Partridge

Grey Partridge

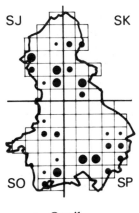
Quail

Pheasant

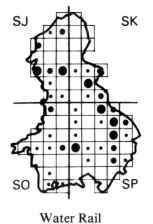
Water Rail

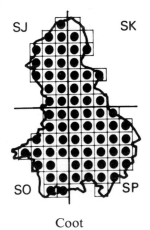
Corncrake

Coot

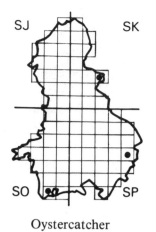

Oystercatcher

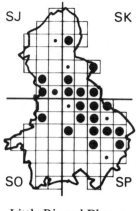

Little Ringed Plover

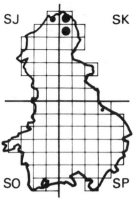

Golden Plover

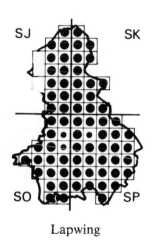

Lapwing

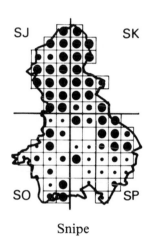

Snipe

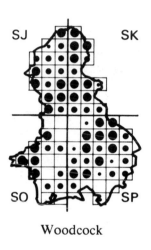

Woodcock

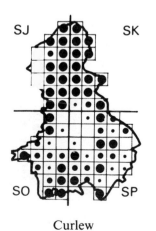

Curlew

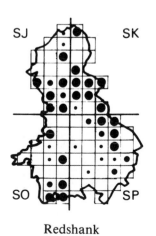

Redshank

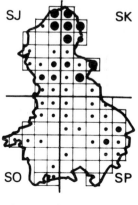

Common Sandpiper

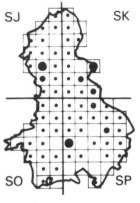

Black-headed Gull

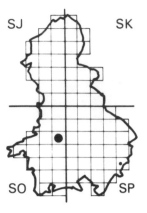

Herring Gull

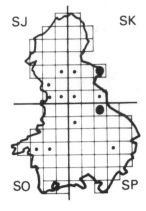

Common Tern

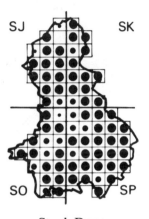

Stock Dove

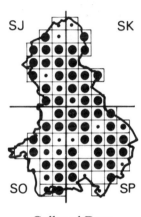

Collared Dove

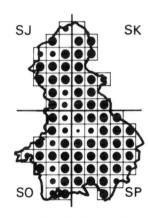

Turtle Dove

Cuckoo

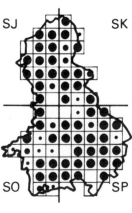

Barn Owl

Little Owl

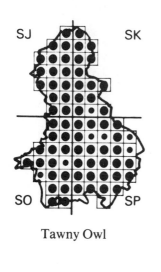

Tawny Owl

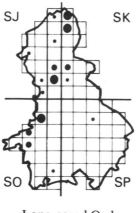

Long-eared Owl

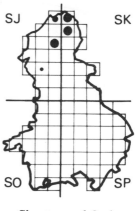

Short-eared Owl

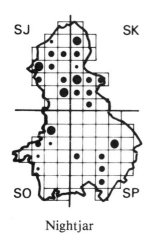

Nightjar

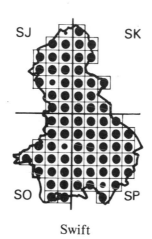

Swift

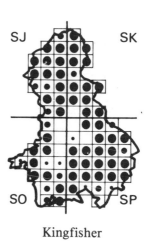

Kingfisher

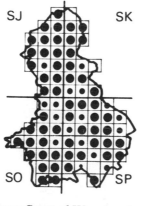

Green Woodpecker

Great Spotted Woodpecker

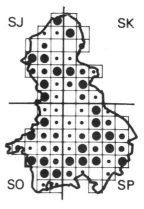

Lesser Spotted Woodpecker

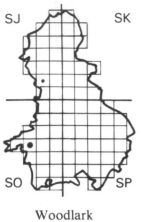

Woodlark

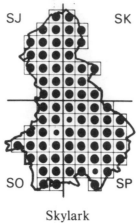

Skylark

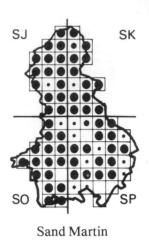

Sand Martin

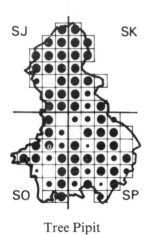

Tree Pipit

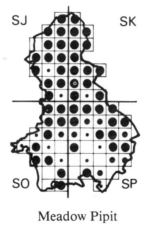

Meadow Pipit

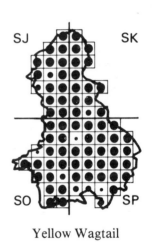

Yellow Wagtail

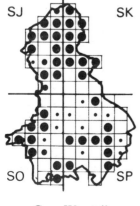

Grey Wagtail

Pied Wagtail

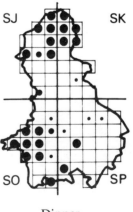

Dipper

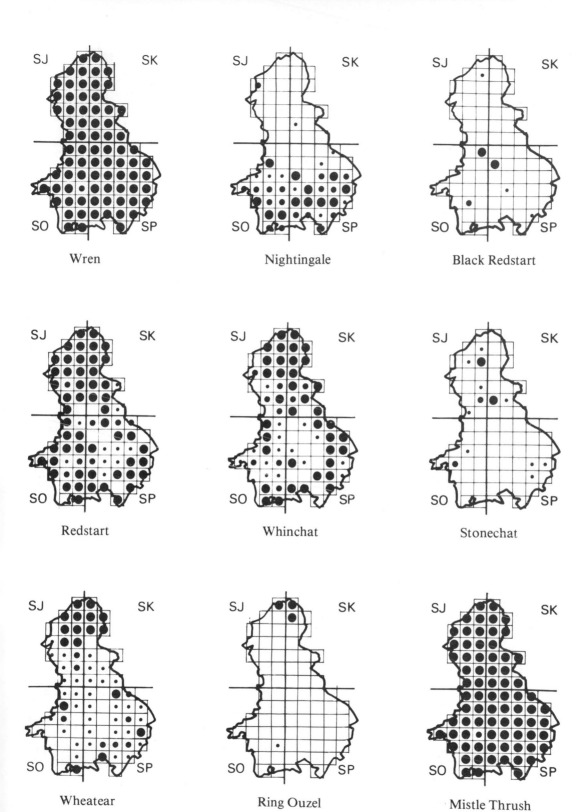

Wren

Nightingale

Black Redstart

Redstart

Whinchat

Stonechat

Wheatear

Ring Ouzel

Mistle Thrush

451

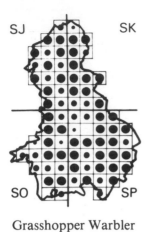

Grasshopper Warbler

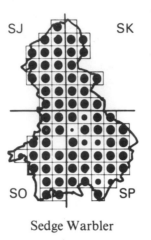

Sedge Warbler

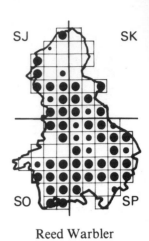

Reed Warbler

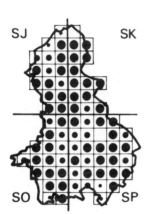

Lesser Whitethroat

Whitethroat

Garden Warbler

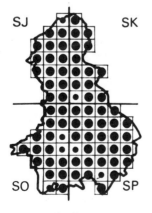

Blackcap

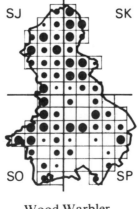

Wood Warbler

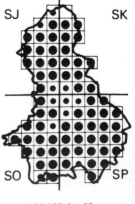

Chiffchaff

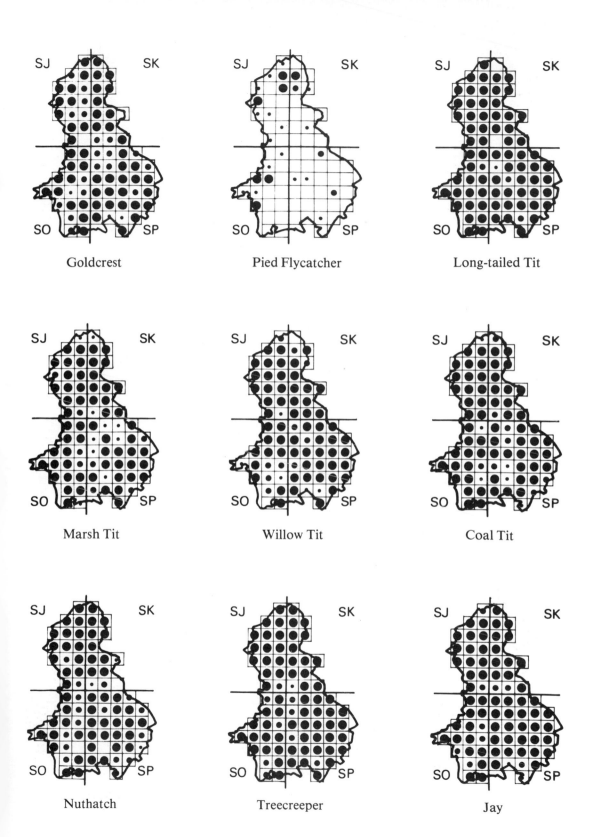

Goldcrest

Pied Flycatcher

Long-tailed Tit

Marsh Tit

Willow Tit

Coal Tit

Nuthatch

Treecreeper

Jay

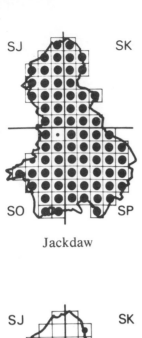

Jackdaw

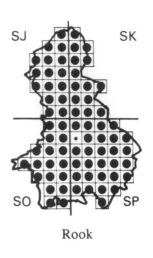

Rook

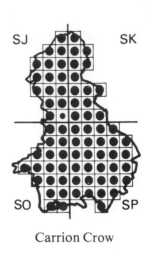

Carrion Crow

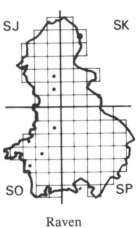

Raven

Tree Sparrow

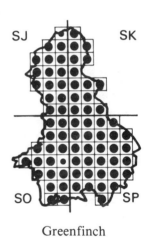

Greenfinch

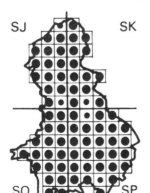

Goldfinch

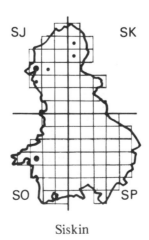

Siskin

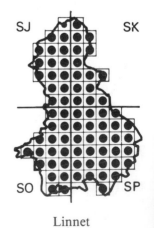

Linnet

454

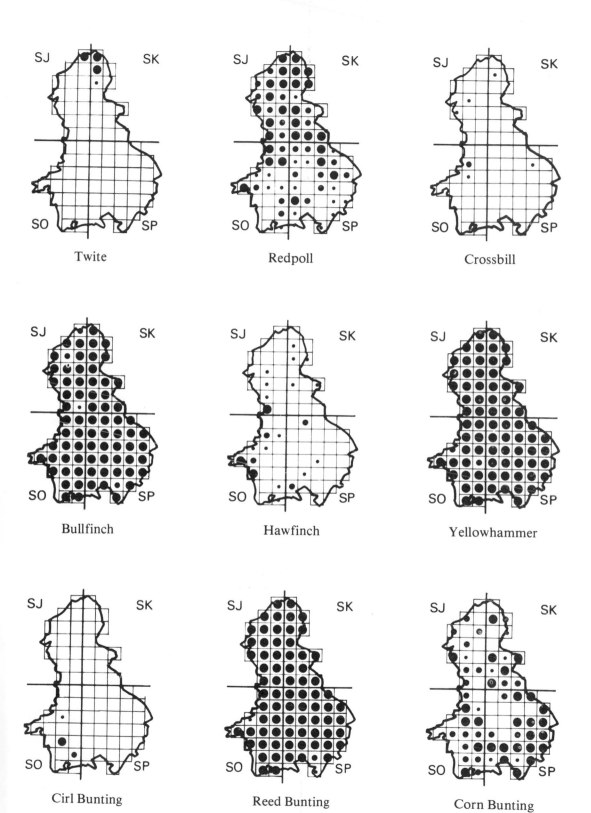

Twite

Redpoll

Crossbill

Bullfinch

Hawfinch

Yellowhammer

Cirl Bunting

Reed Bunting

Corn Bunting

Appendix I

Species not fully admitted to the British List

The official British and Irish list, maintained jointly by the British Ornithologists' Union and the Irish Wildbird Conservancy, includes under Category D species which, though reliably identified within the past fifty years:—

   i   may not have occurred in a wild state
   ii have certainly arrived with ship assistance
  iii have only ever been found dead on the tideline or
  iv belong to feral populations which may or may not be self-supporting.

Six such species have occurred in the West Midlands. Of these, White Pelican, Greater Flamingo and Red-headed Bunting belong to category (i) and Wood Duck, Bobwhite and Ring-necked Parakeet to category (iv).

White Pelican

Pelecanus onocrotalus

One, undoubtedly an escape, was observed at Leamington Spa Reservoir on October 26th 1975.

Greater Flamingo

Phoenicopterus ruber

There are two early and six recent records, all presumed to be of feral birds: at the Manifold Valley in September 1881, Warley on December 22nd 1909, Wormleighton Reservoir on October 14th 1962, Ladywalk during April and May 1968, Kingsbury in early June 1968, Brandon in August 1968, Bittell in late November 1968 and Blithfield in December 1968. Several, if not all, of the 1968 records may have involved the same individual. With the increasing number of escapes, it is possible that some of these individuals were of the American race, *P.r. ruber*, or the Chilean Flamingo, *P. chilensis*.

Wood Duck

Aix sponsa

The following records of this introduced North American species have been published in the WM *Bird Reports* since 1974, but there are known to be several undocumented occurrences: two at Knypersley in February 1974, a drake in Sutton Park from October to December 1975, one at Kingsbury in February, May and September 1977, one at Brocton on December 26th 1977, and a female at Chillington on June 24th 1979. It is noted in the BTO *Atlas* that adults with ducklings were seen at Bishops Offley during 1972–4 and it was suspected that breeding took place on a private estate upstream.

Bobwhite

Colinus virginianus

Sainter (1878) included this species in his list for north Staffordshire, but gave no further details.

Ring-necked Parakeet

Psittacula krameri

This is a popular species with aviculturists, which frequently escapes and nationally may yet consolidate its feral breeding population. Records were first published in the WM *Bird Reports* in 1975, when an escaped pair bred successfully in a hollow tree near Pelsall, one or two more were noted in the Stafford area and another raided peanuts in a Streetly garden. The only subsequent records were singles at Bartley in January 1977 and Belvide on November 26th 1978, and one flying around gardens at Wombourne on October 13th 1979, although some of the six records of Parakeets that were not specifically identified during 1976–8 were probably of this species.

Red-headed Bunting

Emberiza bruniceps

One at Draycote on May 6th 1976, with possibly the same bird at Barston on May 11th and 12th, and a male at Wheaton Aston between July 19th and August 3rd 1977, during which time it was heard singing, are the only known records.

Although the following is strictly a Category C species on the British and Irish list, this is on the strength of its self-supporting Scottish population and its history in the West Midlands is more appropriately recorded here.

Capercaillie

Tetrao urogallus

A male and two females were introduced onto Cannock Chase in the early 1970s, but the introduction has almost certainly been unsuccessful.

Appendix II

Exotic Species

This list includes those species which are not included on the British Ornithologists' Union's official British and Irish list, but which are known to have occurred in the West Midlands as escapes from captivity or introductions. The list is not necessarily exhaustive.

The order and English names follow the *List of Recent Holarctic Bird Species* (Voous 1977), *A Coloured Key to the Wildfowl of the World* (Scott 1965) and *A Check List of the Birds of the World* (Gruson 1976), in that order of priority.

Pelican sp	*Pelecanus* sp
Marabou	*Leptoptilos crumeniferus*
Sacred Ibis	*Threskiornis aethiopica*
Chilean Flamingo	*Phoenicopterus chilensis*
Flamingo sp	Phoenicopteridae
Red-billed Whistling Duck	*Dendrocygna autumnalis*
White-faced Whistling Duck	*D. viduata*
Black Swan	*Cygnus atratus*
Black-necked Swan	*C. melancoryphus*
Bar-headed Goose	*Anser indicus*
South African Shelduck	*Tadorna cana*
New Zealand Shelduck	*T. variegata*
Chiloe Wigeon	*Anas sibilatrix*
Cape Teal	*A. capensis*
Australian Black Duck	*A. superciliosa*
Chinese Spotbill	*A. poecilorhyncha zonorhyncha*
Red-billed Teal	*A. erythrorhynchos*
Chilean Pintail	*A. georgica*
Bahama Pintail	*A. bahamensis*
Argentine Red Shoveler	*A. platalea*
Marbled Duck	*Marmaronetta angustirostris*
Baer's Pochard	*Aythya baeri*
New Zealand Scaup	*A. novaeseelandiae*
Indian White-backed Vulture	*Gyps bengalensis*
Lanner	*Falco biamarcus*
Red-headed Falcon	*F. chiquera*
Lagger	*F. jugger*
California Quail	*Lophortyx californicus*
Silver Pheasant	*Lophura nicthemera*
Peafowl	*Pavo cristatus*

Sarus Crane	*Grus antigone*
Demoiselle Crane	*Anthropoides virgo*
Crowned Crane	*Balearica pavonina*
Barbary Dove	*Streptopelia "risoria"*
Diamond Dove	*Geopelia cuneata*
Sulphur-crested Cockatoo	*Cacatua galerita*
Cockatiel	*Nymphicus hollandicus*
Budgerigar	*Melopsittacus undulatus*
Parrot sp	Psittacidae
Parakeet sp	Psittacidae
Macaw sp	probably *Ara* sp
Eagle Owl	*Bubo bubo*
Peking Robin	*Leiothrix lutea*
Red-billed Blue Magpie	*Urocissa erythrorhyncha*
Common Mynah	*Acridotheres tristis*
Pin-tailed Whydah	*Vidua macrocoura*
Paradise Whydah	*V. paradisea*
Black-headed Weaver	*Ploceus cucullatus*
Weaver sp	probably *Ploceus* sp
Canary	*Serinus canaria*
Zebra Finch	*Poephila guttata*

Unacceptable Records

The following records gained some measure of acceptance through being published in earlier works, although none is now considered admissible. Records which have been submitted, but not officially accepted and published are not included.

Greater Yellowlegs

Tringa melanoleuca

Three noisy birds, one of which was shot, were reported from Powell's Pool, Sutton Park, on November 22nd 1907. Coburn purchased the corpse and identified it as a Greater Yellowlegs. He had previously shot the species in Canada, but, as this skin was not examined independently and exhibited formally, the record was viewed with some incredulity (*Brit. Birds* 4:109). Both Witherby *et al* (1938) and Norris included the record in square brackets.

Scops Owl

Otus scops

One near Fladbury prior to 1834 and another brought to a Worcester taxidermist about 1860 were regarded as dubious by Harthan, who placed both in square brackets. A third bird in the Chillington collection, which was said to have been obtained locally probably in the nineteenth century, was included without question by Smith, but there are no supporting details. None of these records appeared in *The Handbook*, (Witherby *et al* 1938).

Snowy Owl

Nyctea scandiaca

Smith mentioned one killed near Burton-on-Trent sometime before 1881 (which may actually have been a Derbyshire record) and another seen near Pipe Ridware on December 28th 1917, but doubted both and placed them in square brackets. *The Handbook* (Witherby *et al* 1938) contained no reference to any Staffordshire records.

Crested Tit

Parus cristatus

A record of one on Cannock Chase on April 11th 1954 was placed in square brackets in the WM *Bird Report* as there was only one observer and this was felt to be inadequate for such an unusual occurrence.

Scarlet Rosefinch

Carpodacus erythrinus

One, said to have been shot at Powick in December 1855 and preserved at Worcester, was placed in square brackets by Harthan.

Pine Grosbeak

Pinicola enucleator

Hastings (1834) included this species in his list for Worcestershire, but gave no description or details of capture.

Rustic Bunting

Emberiza rustica

One was reported at Chasewater on August 17th 1958, but this record appeared in square brackets in the WM *Bird Report* and was subsequently rejected by the *British Birds* Rarities Committee (*Brit. Birds* 53: 157).

Appendix IV

Names of Plants mentioned in the Text

These are listed in alphabetical order of vernacular names. In general both the vernacular and scientific names follow Clapham, A. R., Tutin, T. G. and Warburg, E. F. (1962) *Flora of the British Isles*, but Mitchell, A. (1974) *A Field Guide to the Trees of Britain and Northern Europe* has been used for exotic trees not included in the *Flora*, Hubbard, C. E. (1968) *Grasses* for cereals and Hillier and Son's Annual *Catalogue of Trees and Shrubs* for garden plants.

Alder	*Alnus glutinosa*
Apple	*Malus sylvestris*
Ash	*Fraxinus excelsior*
Mountain	*Sorbus aucuparia*
Barley	*Hordeum vulgare/distichon*
Bean	*Vicia faba/Phaseolus* spp
Beech	*Fagus sylvatica*
Beet, Sugar	*Beta vulgaris* ssp
Bilberry	*Vaccinium myrtillus*
Hybrid	*V. myrtillus* x *vitis-idaea*
Birch	*Betula pendula/pubescens*
Downy	*B. pubescens*
Silver	*B. pendula*
Blackberry	*Rubus fruticosus* agg
Blackthorn	*Prunus spinosa*
Bluebell	*Endymion non-scriptus*
Bracken	*Pteridium aquilinum*
Bramble	*Rubus* spp
Buckthorn	*Rhamnus catharticus*
Bulrush	*Schoenoplectus lacustris*
Bur-reed	*Sparganium* spp
Cabbage	*Brassica oleracea*
Cedar	*Cedrus atlantica/deodara/libani*
Cherry	*Prunus avium/cerasus*
Chestnut	*Castanea sativa/Aesculus hippocastanum*
Sweet	*C. sativa*
Clematis	*Clematis vitalba*
Clover	*Trifolium* spp
Cotoneaster	*Cotoneaster* spp

Cotton-grass	*Eriophorum* spp
Cowberry	*Vaccinium vitis-idaea*
Cranberry	*Vaccinium oxycoccos*
Crowberry	*Empetrum nigrum*
Currant, Black	*Ribes nigrum*
Cypress	*Chamaecyparis/Cupressus* spp
Damson	*Prunus domestica insititia*
Dogwood	*Thelycrania sanguinea*
Elder	*Sambucus nigra*
Elm	*Ulmus* spp, usually *U. procera*
Wych	*U. glabra*
Fat Hen	*Chenopodium album*
Fir, Douglas	*Pseudotsuga menziesii*
Firethorn	*Pyracantha* spp
Flag, Yellow	*Iris pseudacorus*
Fumitory	*Fumaria* spp
Gooseberry	*Ribes uva-crispa*
Gorse	*Ulex europaeus*
Guelder Rose	*Viburnum opulus*
Hair-grass, Tufted	*Deschampsia cespitosa*
Hawthorn (Haw)	*Crataegus monogyna/oxyacanthoides*
Hazel	*Corylus avellana*
Heather	*Calluna vulgaris*
Hemlock	*Tsuga* spp
Hip	*Rosa* spp
Holly	*Ilex aquifolium*
Honeysuckle	*Lonicera periclymenum*
Hop	*Humulus lupulus*
Hornbeam	*Carpinus betulus*
Horsetail, Water	*Equisetum fluviatila*
Ivy	*Hedera helix*
Kale	*Brassica oleracea acephala*
Larch	*Larix decidua/leptolepsis*
Laurel	*Prunus laurocerasus/lusitanica*
Lettuce	*Lactuca sativa* vars
Lilac	*Syringa vulgaris*
Lime	*Tilia cordata/x europaea/platyphyllos*
Small-leaved	*T. cordata*
Ling	*Calluna vulgaris*

Meadowsweet	*Filipendula ulmaria*
Mercury, Dogs	*Mercurialis perennis*
Mistletoe	*Viscum album*
Moor-grass, Purple	*Molinia caerulea*
Mountain Ash	*Sorbus aucuparia*
Myrtle, Bog	*Myrica gale*
Nettle	*Urtica dioica/urens*
Oak	*Quercus* spp
Pedunculate	*Q. robur*
Sessile	*Q. petraea*
Oat	*Avena sativa/strigosa/byzantina*
Orchid, Southern Marsh	*Dactylorhiza praetermissa*
Osier (Common)	*Salix viminalis*
Pea	*Pisum sativum* var
Pear	*Pyrus communis*
Pine	*Pinus* spp
Corsican	*P. nigra larico*
Lodgepole	*P. contorta latifolia*
Scots	*P. sylvestris*
Plum	*Prunus domestica domestica*
Poplar	*Populus* spp
Black	*P. nigra*
Lombardy	*P. nigra italica*
White	*P. alba*
Potato	*Solanum tuberosum*
Primrose	*Primula vulgaris*
Privet	*Ligustrum vulgare/ovalifolium*
Rape, Oilseed	*Brassica napus arvensis*
Redwood	*Sequoia sempervirens*
Reed	*Phragmites communis*
Reed-grass	*Glyceria maxima/Phalaris arundinacea*
Reedmace	*Typha latifolia/angustifolia*
Rhododendron	*Rhododendron ponticum*
Rose	*Rosa* spp
Guelder	*Viburnum opulus*
Rowan	*Sorbus aucuparia*
Rush	*Juncus* spp
Sallow	*Salix caprea/cinerea*
Scabious, Devils-bit	*Succisa pratensis*
Sedge	*Carex* spp
Greater Tussock	*C. paniculata*
Service Tree, Wild	*Sorbus torminalis*
Spindle Tree	*Euonymus europaeus*

Sprout	*Brassica oleracea bullata gemmifera*
Spruce	*Picea* spp
Sitka	*P. sitchensis*
Sugar Beet	*Beta vulgaris* ssp
Sundew, Round-leaved	*Drosera rotundifolia*
Sycamore	*Acer pseudoplatanus*
Teasel	*Dipsacus fullonum*
Thistle	*Carduus/Cirsium* spp
Thorn	*Crataegus monogyna/Prunus spinosa*
Tomato	*Lycopersicum esculentum*
Turnip	*Brassica rapa rapa*
Walnut	*Juglans regia*
Wayfaring Tree	*Viburnum lantana*
Wellingtonia	*Sequoia gigantea*
Wheat	*Triticum aestivum/turgidum*
Willow	*Salix* spp
Willowherb	*Epilobium* spp
Yew	*Taxus baccata*

Appendix V

Names of Animals (other than Birds) mentioned in the Text

These are listed in alphabetical order of vernacular names. Both the vernacular and scientific names follow current usage.

Ant	Hymenoptera: Formicidae
Cattle	*Bos taurus* (domestic)
Chub	*Leuciscus cephalus*
Dace	*Leuciscus leuciscus*
Deer	Cervidae
Earthworm	Annelida: Oligochaeta
Fox	*Vulpes vulpes*
Gudgeon	*Gobio gobio*
Horse	*Equus* spp (domestic)
Moth, Pine-looper	*Bupalus piniaria*
Mouse,	
Wood	*Apodemus sylvaticus*
Yellow-necked Field	*Apodemus flavicollis*
Mussel, Freshwater	Lammellibranchia
Newt, Warty	*Triturus cristatus*
Pig	*Sus* (domestic)
Poultry	*Gallus gallus* (domestic)
Rabbit	*Oryctolagus cuniculus*
Rat,	*Rattus norvegicus/rattus*
Brown	*R. norvegicus*
Roach	*Rutilus rutilus*
Sheep	*Ovis aries* (domestic)
Snail	Mollusca
Snake, Grass	*Natrix natrix*
Squirrel, Grey	*Neosciurus carolinensis*
Trout	*Salmo* spp
Vole,	
Bank	*Clethrionomys glareolus*
Short-tailed	*Microtus agrestis*

Gazetteer

This gazetteer lists alphabetically all places within the West Midlands Region that are referred to in this book, together with their respective county as follows: S=Staffordshire, W=Warwickshire, WM=West Midlands and Wo=Worcestershire. For precise localities a four-figure National Grid reference is given, but for some rivers, canals and large tracts of land a two-figure reference is used. In the case of the longest rivers and canals or very general areas, no reference is included, but these are all shown on the various maps. There is also an index to the pages on which the more important habitats are described.

Place names generally conform to those used in the *Ordnance Survey 1:50,000 Second Series*, but where part appears in parenthesis this denotes it has been abbreviated in the text *e.g.* Kingsbury (Water Park) is referred to simply as Kingsbury. Where localities straddle a county boundary within the Region, both counties are shown, though for recording purposes the WMBC has generally ascribed them to a single county *e.g.* Alvecote (Warwicks) or Chasewater (Staffs).

Localities outside the Region are not listed, but throughout the text those in Great Britain are followed by their respective county (shown in parenthesis), whilst for those abroad the respective country is similarly denoted.

Place	County	Grid Ref.	Page	Place	County	Grid Ref.	Page
Abberley	Wo	SO7567	—	Ashley	S	SJ7636	—
Abbots Bromley	S	SK0824	—	Ashow	W	SP3170	—
Acocks Green	WM	SP1283	—	Astley	Wo	SO7867	—
Acton	S	SJ8241	—	Aston(-by-Stone)	S	SJ9131	—
Adbaston	S	SJ7627	—	Aston	WM	SP0888	—
Alcester	W	SP0957	—	Aston Mill	Wo	SO9434	96
Aldridge	WM	SK0500	—	Atch Lench	Wo	SP0350	—
Alfrick	Wo	SO7453	—	Atherstone	W	SP3097	—
Allesley	WM	SP2980	—	Avon, River	W/Wo	—	75–96
Alne, River	W	SP15/16	89–90				
Alrewas	S	SK1715	—	Baddeley Edge	S	SJ9150	—
Alton	S	SK0742	126	Badsey	Wo	SP0743	—
Alvechurch	Wo	SP0272	—	Baggeridge	S	SO8992	46
Alvecote	S/W	SK2504	114–115	Baginton	W	SP3474	79
Alveston	W	SP2356	—	Bagots Park	S	SK0927	—
Amington	S	SK2304	—	Bagots Wood	S	SK0727	128
Anker, River	S/W	SP39/SK20	114–115	Barford	W	SP2760	—
Aqualate (Mere)	S	SJ7720	139–140	Barlaston	S	SJ8938	—
Arbury (Park)	W	SP3389	58	Barnt Green	Wo	SP0073	—
Arden, Forest of	W	SP16	—	Barr	WM	SP09	—
Arley	W	SP2890	—	Bartley (Res.)	WM	SP0081	41–42
Arley	Wo	SO7680	—	Barton	W	SP1051	—
Arrow, River	W/Wo	SP05/06	89–90	Barton-under-			
Arrow Valley Pk.	Wo	SP0665	—	Needwood	S	SK1818	—

Place	County	Grid Ref.	Page	Place	County	Grid Ref.	Page
Bearley	W	SP1760	—	Bow Brook	Wo	SO94/95	92
Beaudesert				Bow Wood	Wo	SO9355	—
(Old Park)	S	SK0313	—	Brackenhurst			
Beckford	Wo	SO9735	96	Covert	S	SK1422	—
Bedworth	W	SP3587	—	Bradley Green	Wo	SO9861	—
Bedworth Slough	W	SP3487	81	Brailes Hill	W	SP2938	—
Beeston Tor	S	SK1054	—	Brandon Hall	W	SP4076	—
Belvide (Res.)	S	SJ8610	103–104	Brandon (Marsh)	W	SP3875	77–78
Bentley Woods	W	SP2895	57–58	Brandon Wood	W	SP3976	79
Berkeley				Branston (Gravel Pit)	S	SK2120	116
presume Byrkley	S	SK1623	—	Bredon	Wo	SO9136	—
Berkswell	WM	SP2479	58	Bredon Hill	Wo	SO9539	146
Berrow Hill	Wo	SO9962	—	Brereton	S	SK0516	—
Berry Mound	Wo	SP0977	—	Bretford	W	SP4277	—
Bescot	WM	SP0096	—	Brewood	S	SJ8808	—
Betley	S	SJ7548	137–138	Bridgtown	S	SJ9708	—
Bewdley	Wo	SO7875	64	Brierley Hill	WM	SO9187	—
Bickenhill	WM	SP1882	—	Brierley Hill			
Biddulph	S	SJ8857	—	Pools	WM	SO9188	45
Biddulph Moor	S	SJ9058	—	Brinklow	W	SP4379	—
Bidford-on-Avon	W	SP1051	—	British Camp	Wo	SO7640	—
Bilbrook	S	SJ8703	—	Broadway	Wo	SP0937	—
Bilston	WM	SO9496	—	Brocton Coppice	S	SJ9819	51
Binton	W	SP1454	—	Brocton Field	S	SJ9817	50
Birch Coppice				Bromsgrove	Wo	SO9570	—
Colliery	W	SK2500	—	Bromwich Wood	WM	SO9981	—
Birchfield	WM	SP0690	—	Broome	Wo	SO9078	—
Birchley Wood	W	SP4078	79	Brown Edge	S	SJ9153	—
Birdingbury	W	SP4368	—	Brownhills	WM	SK0405	—
Birmingham	WM	SP0686	—	Bubbenhall	W	SP3672	—
Airport	WM	SP1783	—	Budbrooke	W	SP2565	—
Plateau	S/W/Wo/WM	—	—	Bunkers Hill	S	SO8782	—
University	WM	SP0483	—	Bunsons Wood	WM	SP3184	57
Bishops Offley	S	SJ7829	101	Bunster Hill	S	SK1451	—
Bishops Wood	S	SJ7531	138	Burmington	W	SP2637	—
Bittell (Res.)	Wo	SP0174	55	Burnt Wood	S	SJ7434	—
Blackbrook				Burslem	S	SJ8749	—
Sewage Farm	S	SO8398	46–47	Burton Green	WM	SP2675	—
Blackbrook Valley	S	SK0064	133	Burton Hastings	W	SP4189	—
Black Country	WM	SO98/99	—	Burton(-on-Trent)	S	SK2423	—
Blackdown	W	SP3168	—				
Blakedown	Wo	SO8878	—	Caldecote	W	SP3594	—
Blakeshall	Wo	SO8381	—	Caldon Canal	S	SJ95/SK04	—
Blithe, River	S	SK02/03	—	Cannock	S	SJ9710	—
Blithfield (Res.)	S	SK0623	106–108	Cannock Chase	S	SJ91/SK01	48–51
Blymhill	S	SJ8112	—	Cannon Hill Park	WM	SP0683	—
Blythe, River	W/WM	SP17/28	110–111	Carrant Brook	Wo	SO93	—
Bockleton	Wo	SO5862	—	Castle Bromwich	WM	SP1489	—
Bodymoor Heath	W	SP1996	112–114	Castle Hill	Wo	SO8181	—
Bonehill	S	SK1902	—	Castle Hill,			
Bosley Cloud				Dudley	WM	SO9490	—
presume The Cloud	S	SJ9063	—	Castlemorton			
Bourne Brook	W	SP28/29	—	Common	Wo	SO7839	146
Bournville	WM	SP0481	—	Catshill	Wo	SO9573	—

Place	County	Grid Ref.	Page	Place	County	Grid Ref.	Page
Feckenham	Wo	SP0061	—	Hampton-in-Arden	WM	SP2080	—
Feldon	W	—	—	Hams Hall	W	SP2092	109–110
Fenton	S	SJ8944	—	Hamstall Ridware	S	SK1019	—
Fillongley	W	SP2887	—	Hanbury	Wo	SO9663	68
Finham Brook	W	SP27/37	—	Hanch Reservoir	S	SK1013	106
Finham Sewage				Hanchurch (Hills)	S	SJ8340	125
Works	W	SP3374	—	Handsworth	WM	SP0490	—
Fladbury	Wo	SO9946	—	Handsworth Wood	WM	SP0590	—
Flash	S	SK0267	—	Hanley	S	SJ8747	—
Flyford Flavell	Wo	SO9854	—	Harborne	WM	SP0284	—
Foleshill	WM	SP3582	—	Harbury Plateau	W	SP35	—
Ford Green	S	SJ8950	121	Hartlebury Common	Wo	SO8270	65
Forest Banks	S	SK1228	—	Hartshill	W	SP3293	—
Four Oaks	WM	SP1099	—	Hartshill Hayes			
Fowlea Brook	S	SJ8646	—	(Country Park)	W	SP3194	57–58
Fradley Wood	S	SK1513	106	Harvington	Wo	SP0549	—
Frankley (Res.)	Wo	SP0080	—	Haselor	W	SP1257	—
Frankton	W	SP4270	—	Hatherton	S	SJ9510	—
Freeford	S	SK1307	—	Hatton	W	SP2367	—
Froghall	S	SK0247	—	Hawksmoor	S	SK0344	126
Fullmoor Wood	S	SJ9411	—	Hay Mills	WM	SP1185	—
				Hearsall Common	WM	SP3178	80
Gailey (Res.)	S	SJ9310	103	Hednesford	S	SK0012	—
Gallows Green	Wo	SO9362	—	Henley-in-Arden	W	SP1565	—
Gaydon	W	SP3654	—	Hewell (Grange)	Wo	SP0069	—
Gerrards Bromley	S	SJ7734	—	Highgate Common	S	SO8489	141
Gib Torr	S	SK0264	133	High Green	Wo	SO8745	—
Gnosall	S	SJ8220	—	Hilderstone	S	SJ9434	—
Goldsitch Moss	S	SK0164	133	Hillmorton	W	SP5273	—
Goosehill Wood	Wo	SO9360	93–94	Himbleton	Wo	SO9458	—
Gorcott Hill	Wo	SP0968	—	Himley (Hall)	S	SO8891	46
Grafton Wood	Wo	SO9756	—	Holden Lane	S	SJ8950	121
Gratton (Hill)	S	SJ9356	—	Hollywood	Wo	SP0877	—
Great Alne	W	SP1159	—	Holt	Wo	SO8262	66–67
Great Barr	WM	SP0494	—	Holt Fleet	Wo	SO8263	—
Great Chatwell	S	SJ7914	—	Hopton Heath	S	SJ9526	—
Great Haywood	S	SJ9922	—	Hopwas (Hayes)	S	SK1705	116
Great Witley	Wo	SO7566	—	Hornhill Wood	Wo	SO9558	93
Great Wyrley	S	SJ9907	—	Hulme End	S	SK1059	—
Greaves Wood	S	SK1527	128–129	Hunthouse Wood	Wo	SO7070	144
Greenway Bank	S	SJ8855	124	Huntington	S	SJ9713	—
Greys Mallory	W	SP2961	—	Huntley Wood	S	SJ9941	—
Griff	W	SP3588	—	Hurcott (Pool)	Wo	SO8577	63
Grimley	Wo	SO8360	66–67				
Grove End	WM	SP1695	—	Idlicote	W	SP2844	—
Gun Hill	S	SJ9761	—	Ilmington	W	SP2143	—
				Inkberrow	Wo	SP0157	—
Habberley	Wo	SO8077	—	Ipstones	S	SK0249	—
Hagley	Wo	SO9180	—	Isbourne, River	Wo	SP03/04	—
Halesowen	WM	SO9683	—	Islandpool	Wo	SO8580	—
Hall Green	WM	SP1081	—	Itchen, River	W	SP36/46	—
Ham Bridge	Wo	SO7361	—				
Hammerwich	S	SK0607	—	Jephson Gardens	W	SP3265	85
Hamps, River	S	SK05	—	Jervis Wood	S	SJ8639	—

Place	County	Grid Ref.	Page	Place	County	Grid Ref.	Page
National Agricultural				Pirton	Wo	SO8747	72
Centre	W	SP3271	—	Polesworth	W	SK2602	—
National Exhibition				Potteries, The	S	SJ84/85	—
Centre	WM	SP1983	—	Potters Cross	S	SO8484	—
Nechells	WM	SP0989	—	Powick	Wo	SO8351	—
Needwood Forest	S	SK12	128–129	Preston Brook	W	SP16	89
Netherton	WM	SO9388	—	Preston-on-Stour	W	SP2049	—
Newbold Comyn	W	SP3365	—	Preston Pastures			
Newcastle-under-				presume Preston			
Lyme	S	SJ8445	—	Fields	W	SP1766	—
New Close Wood	W	SP4077	79	Priors Hardwick	W	SP4756	—
New Invention	WM	SJ9601	—	Priory Wood	WM	SP0291	—
Norbury	S	SK1242	—	Pulley			
Northfield	WM	SP0279	—	see Oakley	Wo	SO8960	—
North Staffordshire-							
Moors	S	—	131–135	Quinton	W	SP1847	—
Northwick Marsh	Wo	SO8357	—	Quinton	WM	SO9984	—
Norton	Wo	SO8751	—				
Norton	Wo	SP0447	—	Radford	S	SJ9320	—
Nuneaton	W	SP3691	—	Radway	W	SP3648	—
				Ragley (Hall)	W	SP0755	90
Oakamoor	S	SK0544	126	Ramshaw Rocks	S	SK0162	—
Oakley	Wo	SO8960	70–71	Ramshorn	S	SK0845	—
Ocker Hill	WM	SO9793	—	Randan Wood	Wo	SO9272	54
Ockeridge Wood	Wo	SO7962	67	Ravenshill Wood	Wo	SO7353	144
Offley Moss pre-				Redditch	Wo	SP0467	—
sume Loynton Moss	S	SJ7824	—	Red Hill	W	SP1356	—
Okeover	S	SK1548	—	Rednal	WM	SP0076	—
Oldacre Valley	S	SJ9718	51	Rhydd	Wo	SO8345	—
Oldbury	WM	SO9889	—	Rhydd Covert	Wo	SO8075	—
Old Hills	Wo	SO8248	73	Ribbesford	Wo	SO7873	65
Oliver Hill	S	SK0267	—	Rickerscote	S	SJ9320	—
Ombersley	Wo	SO8463	67–68	Ripple	Wo	SO8738	—
Onecote	S	SK0455	—	Roaches, The	S	SK0063	134
Orslow	S	SJ8015	—	Rocester	S	SK1139	—
Overbury	Wo	SO9537	—	Rock Coppice	Wo	SO7673	143
Oxford Canal	W	—	—	Rolleston	S	SK2327	—
				Romsley	Wo	SO9679	—
Packington (Park)	W	SP2284	111	Rough Close	S	SJ9239	—
Pailton	W	SP4781	—	Rough Hill	WM	SO9296	—
Park Hall	S	SJ9244	123	Rough Knipe	S	SK0049	—
Patshull	S	SJ8000	140	Rough Wood	WM	SJ9800	—
Pattingham	S	SO8299	—	Roundhill Wood	Wo	SO9858	93
Peak District	S	—	131–135	Round Oak	WM	SO9287	—
Pelsall	WM	SK0103	—	Rowley Regis	WM	SO9687	—
Penk, River	S	SJ91/92	101–105	Rubery	Wo	SO9877	—
Penkridge	S	SJ9214	—	Rudyard (Res.)	S	SJ9459	124
Pennines	S	SK05/06	—	Rugby	W	SP5075	—
Perry Barr	WM	SP0791	—	Rugeley	S	SK0418	—
Pershore	Wo	SO9445	—	Rushton Spencer	S	SJ9362	—
Perton	S	SO8598	46	Ryton Gravel Pit	W	SP3772	79
Piddle Brook	Wo	SO94/95	92	Ryton Wood	W	SP3772	79
Pinley (Green)	W	SP2066	—				
Pipers Hill	Wo	SO9565	68	Saltley	WM	SP0987	—

Place	County	Grid Ref.	Page	Place	County	Grid Ref.	Page
Trench Wood	Wo	SO9258	94	Westlands	S	SJ8344	—
Trent, River	S	—	99–117	West Midlands			
Trent and Mersey				Conurbation	WM	—	—
Canal	S	—	—	West Midland			
Trentham Gardens	S	SJ8640	100	Safari Park	Wo	SO8075	—
Trentham Park	S	SJ8540	125	Westminster Pool	Wo	SO9980	—
Trent Vale	S	SJ8643	—	Weston-on-Avon	W	SP1551	—
Trescott	S	SO8497	—	Weston Park	S	SJ8010	140
Trickley Coppice	W	SP1599	—	Weston-under-			
Trimpley (Res.)	Wo	SO7778	64	Lizard	S	SJ8011	—
Trittiford Park	WM	SP0980	—	Weston Wood	W	SP2735	79
Tunstall	S	SJ8651	—	West Park	WM	SO9099	—
Turners Hill	WM	SO9688	—	Westport (Lake)	S	SJ8550	122–123
Tutbury	S	SK2129	—	Westwood (Park)	Wo	SO8763	70
Tutnall	Wo	SO9970	—	Wetmore	S	SK2524	—
Twiland Wood	Wo	SO9780	—	Wetton Mill	S	SK0956	—
Two Gates	S	SK2101	—	Whateley	W	SP2299	—
Tysoe	W	SP3444	—	Wheaton Aston	S	SJ8512	—
				Whichford Wood	W	SP3034	—
Ufton Fields	W	SP3861	148	Whiston	S	SJ8914	—
Ufton Hill	W	SP3961	—	Whitacre	W	SP2191	—
Ufton Wood	W	SP3862	148	Whitacre Heath	W	SP2192	—
Upton House	W	SP3645	—	White Sitch	S	SJ7912	—
Upton-on-Severn	Wo	SO8540	—	Whitmore	S	SJ8141	—
Upton Warren	Wo	SO9367	68–70	Whitmore Heath	S	SJ7940	—
Uttoxeter	S	SK0933	—	Whitnash	W	SP3263	—
				Whittington			
Victoria Park,				Sewage Farm	S	SO8783	62–63
Leamington Spa	W	SP3165	—	Wick	Wo	SO9645	—
				Wickhamford	Wo	SP0641	—
Walsall	WM	SP0198	—	Wightwick	WM	SO8798	—
Walton (Hall)	W	SP2852	87	Wilden	Wo	SO8272	63
Walton Heath	S	SJ8932	—	Willey	W	SP4984	148
Walton Hill	Wo	SO9479	—	Willoughbridge	S	SJ7439	—
Walton-on-Trent	Derbys	SK2118	—	Willoughby	W	SP5167	—
Wannerton	Wo	SO8678	—	Wilmcote	W	SP1658	—
Wappenbury Wood	W	SP3770	79	Wilnecote	S	SK2201	—
Warley Park	WM	SP0186	46	Winnington	S	SJ7238	—
Warslow	S	SK0858	—	Witley (Court)	Wo	SO7664	66
Warwick	W	SP2864	—	Wolf Edge	S	SK0267	—
Warwick Park	W	SP2863	87	Wolford Wood	W	SP2333	147
Waterhouses	S	SK0850	—	Wolfscote Dale	S	SK1357	—
Water Orton	W	SP1791	—	Wolston	W	SP4175	—
Waverley Wood	W	SP3571	79	Wolverhampton	WM	SO9198	—
Weatheroak (Hill)	Wo	SP0574	—	Wolverley	Wo	SO8279	—
Weaver Hills	S	SK0946	—	Wombourne	S	SO8692	—
Wednesbury	WM	SO9895	—	Woodbury Hill	Wo	SO7464	—
Wednesfield	WM	SJ9400	—	Wood Eaton	S	SJ8417	—
Weeford	S	SK1403	—	Woodgate Valley	WM	SP0083	42
Welches Meadow	W	SP3265	84	Woodseaves	S	SJ7925	—
Welford-on-Avon	W	SP1452	—	Wootton Wawen	W	SP1563	89
Welland	Wo	SO7940	—	Worcester	Wo	SO8454	—
Wellesbourne	W	SP2755	85–86	Worcester and	Wo/		
West Bromwich	WM	SP0091	—	Birmingham Canal	WM	—	—

Place	County	Grid Ref.	Page	Place	County	Grid Ref.	Page
Wordsley	WM	SO8887	—	Wylde Green	WM	SP1294	—
Wormleighton				Wyre Forest	Wo	SO77	142–143
(Reservoir)	W	SP4451	147	Wythall	Wo	SP0775	—
Wychall	WM	SP0379	—				
Wyche Quarry	Wo	SO7744	—	Yarningale			
Wyken	WM	SP3680	—	Common	W	SP1865	—
Wyken Slough	WM	SP3683	—	Yoxall	S	SK1419	—

Bibliography

ALEXANDER, H. G. 1929. The Birds of the Lickey Hills and Bittell Reservoirs. *Proc. Birm. Nat. Hist. and Phil. Soc.* Vol XV:pt VIII:197–212.

AMPHLETT, J. and C. REA. 1909. *The Botany of Worcestershire.* Birmingham.

ATKINSON-WILLES, G. L. 1963. *Wildfowl in Great Britain.* Nature Conservancy Monograph No. 3. London.

AUDEN, G. A. (Ed.). 1913. *A Handbook for Birmingham and the Neighbourhood.* Birmingham.

BANNISTER, C. W. B. 1941. Cirl Buntings. WM *Bird Report* 7:6–7.

BARNES, J. A. G. 1975. *The Titmice of the British Isles.* Newton Abbot.

BATTEN, L. A. 1973. Population dynamics of suburban Blackbirds. *Bird Study* 20:251–258.

BATTEN, L. A., R. H. DENNIS, I. PRESTT and the RARE BREEDING BIRDS PANEL. 1979. Rare Breeding Birds in the United Kingdom in 1977. *Brit. Birds* 72:363–381.

BIRD, V. 1974. *Staffordshire.* London.

BIRD, V. 1979. *Portrait of Birmingham.* London.

BLAKE, A. R. M. 1955. Buzzard Survey, 1954. WM *Bird Report* 21:18–20.

BLAKE, A. R. M. 1962. The Stonechat in the West Midlands. WM *Bird Report* 28:13–17.

BLAKER, G. B. 1934. *The Barn Owl in England and Wales.* RSPB London.

BLURTON-JONES, N. G. 1956. Census of breeding Canada Geese, 1953. *Bird Study* 3:153–170.

BOOTH, A. and K. 1979. The Birds of Betley Mere and the surrounding area (unpublished).

BOYD, A. W. 1929–39. Reports on Belvide Reservoir. *Brit. Birds* Vols 19–29.

BRANDON MARSH CONSERVATION GROUP. 1976. *Report for 1972–76* (printed for private circulation).

BRANSON, N. J. B. A., E. D. PONTING and C. D. T. MINTON. 1978. Turnstone Migrations in Britain and Europe. *Bird Study* 25:181–187.

BRITISH ASSOCIATION FOR THE ADVANCEMENT OF SCIENCE. 1913. G. A. Auden (Ed.). *A Handbook for Birmingham and the Neighbourhood.* Birmingham.

BRITISH ASSOCIATION FOR THE ADVANCEMENT OF SCIENCE. 1950. M. J. Wise. (Ed.). *Birmingham and its Regional Setting – A Scientific Survey.* Birmingham.

BRITISH ORNITHOLOGISTS' UNION. 1971. *The Status of Birds in Britain and Ireland.* Oxford.

BUSSE, P. 1969. Results of Ringing European *Corvidae. Acta. Orn. Warsz.* XI 8:263–328.

CADBURY, D. A., J. G. HAWKES and R. C. READETT. 1971. *A Computer-mapped Flora – A Study of the County of Warwickshire.* London.

CAMPBELL, B. 1960. The Mute Swan Census in England and Wales, 1955–56. *Bird Study* 7:208–223.

CAMPBELL, B. 1965. The British Breeding distribution of the Pied Flycatcher, 1953–62. *Bird Study* 12:305–318.

CHANCE, E. P. 1922. *The Cuckoo's Secret.* London.

CHANDLER, R. J. and K. C. OSBORNE. 1977. Scarce Migrants in the London Area, 1955–74. *London Bird Report* 41:73–99.

CHASE, R. W. 1886. Birds of the Birmingham District. *British Association Handbook of Birmingham.* Birmingham.

CHERRY, G. E. 1975. The West Midlands. In *Britain's Planning Heritage.* R. Taylor and M. Cox (Eds.). London.

CHRISTIE, D. 1975. October Reports. *Brit. Birds* 668:85–88.

CLAPHAM, A. R., T. G. TUTIN and E. F. WARBURG. 1962. *Flora of the British Isles.* Cambridge.

COLEMAN, A. E. and C. D. T. MINTON. 1979. Ringing and breeding of Mute Swans in relation to natal area. *Wildfowl* 30:27–30.

COLEMAN, A. E. and C. D. T. MINTON. 1980. Mortality of Mute Swan progeny in an area of south Staffordshire. *Wildfowl* 31:22–28.

COOMBS, F. 1978. *The Crows.* London.

COOPER, D. and R. JUCKES. 1977. Common Bird of Prey Census. *BTO News* 85: 3.

COULSON, C. J. 1974. Kittiwake *Rissa tridactyla* pp 134–141 in S. Cramp, W. R. P. Bourne and D. R. Saunders. *The Seabirds of Britain and Ireland.* London.

CRAMP, S. and K. E. L. SIMMONS (Eds.). 1977–79. *The Birds of the Western Palearctic.* Oxford.

CURTLER, M. 1853. In *Stanley's Guide to Worcester.* Worcester.

DAY, J. J. and S. W. WALKER (Eds.). 1979. *Progress Report on the Habitats, Flora and Fauna of Grimley Brickpits from December 1978 to August 1979.* (unpublished).

DAY, J. C. U. and J. WILSON. 1978. Breeding Bitterns in Britain. *Brit. Birds* 71:285–300.

DEAN, A. R. 1974. The Rookeries of Warwickshire, Worcestershire and Staffordshire. WM *Bird Report* 40:11–21.

DEAN, B. R. 1971. The Status of Sea-ducks in the West Midlands. WM *Bird Report* 37:14–18.

DENNIS, R. H. 1964. Capture of moulting Canada Geese on the Beauly Firth. *Wildfowl* 15:71–74.

DICKENSON, J. H. 1798. in *History and Antiquities of Staffordshire.* S. Shaw (Ed.). London.

DUNN, E. 1981. Roseates on a lifeline. *Birds* Vol 8 No 6:42–45.

DURMAN, R. F. 1976. Ring Ouzel Migration. *Bird Study* 23:197–205.

EDEES, E. S. 1972. *Flora of Staffordshire.* Newton Abbot.

EDWARDS, A. T. 1969. The Birds of a Suburban Garden. WM *Bird Report* 35:10–12.

EMLEY, D. W. 1980. Notes on Maer Hills. *WMBC Bulletin* 283:4.

ENGLISH TOURIST BOARD. 1978. *Staffordshire Moorlands* – Mini-guide.

FENNELL, J. F. and D. A. STONE. 1976. A winter roosting population of Reed Buntings in Central England. *Ringing and Migration* 1:108–114.

FERGUSON-LEES, I. J. 1971. Studies of less familiar birds 164, Wood Sandpiper. *Brit. Birds* 64:114–117.

FINCHER, F. 1955. Notes on the Distribution of certain of the Birds Breeding in the Club's Area. WM *Bird Report* 22:10–13.

FISHER, J. 1952. *The Fulmar.* London.

FLEGG, J. 1981. Swifts on the Wing. *The Living Countryside* 2:349–351.

FORESTRY COMMISSION. 1928. *Report on the Census of Woodlands and Census of Production of Home Grown Timber, 1924*. HMSO.

FORESTRY COMMISSION. 1952. *Census of Woodlands, 1947-9*. HMSO.

FROST, R. A. 1978. *Birds of Derbyshire*. Buxton.

GARNER, R. 1844. *Natural History of the County of Stafford*, with supplement, 1860. London.

GILL, C. and C. G. ROBERTSON. 1938. *A Short History of Birmingham*. Birmingham.

GLUE, D. E. and R. A. MORGAN. 1972. Cuckoo hosts in British habitats. *Bird Study* 19:187-192.

GREEN, G. H. 1977, 1978. *Wintering Gulls in Worcestershire and Warwickshire, Progress Reports 1 and 2* (printed for limited circulation).

GREEN, G. H. 1980. Worcestershire's Marsh Warblers. *Worcs. Nature Cons. Trust Newsletter*, May 1980.

GRIBBLE, F. C. 1976. A Census of Black-headed Gull colonies in England and Wales in 1973. *Bird Study* 23: 135-145.

GRUSON, E. S. 1976. *A Check List of the Birds of the World*. London.

HAAPANEN, A. 1965, 1966. Bird fauna of the Finnish forests in relation to forest succession. *Ann. Zool. fenn.* 2:153-196; 3:176-200.

HALE, W. G. 1980. *Waders*. London.

HARDMAN, J. A. 1974. Biology of the Skylark. *Ann. Appl. Biology* 76:337-341.

HARDMAN, J. A. and D. R. COOPER. 1980. Mute Swans on the Warwickshire Avon – a study of decline. *Wildfowl* 31:29-36.

HARDMAN, J. A., D. R. COOPER and R. J. JUCKES. 1977. Bird Ringing in a Changing Woodland Habitat. *Journ. of Forestry* LXXI:2.

HARDY, A. R. and C. D. T. MINTON. 1980. Dunlin migration in Britain and Ireland. *Bird Study* 27:81-92.

HARRISON, C. 1975. *Field Guide to the Nests, Eggs and Nestlings of British and European Birds*. London.

HARRISON, J. G. 1941. *Handbook of Birds of the Malvern District*. London.

HARTHAN, A. J. 1946. *The Birds of Worcestershire*. Worcester.

HARTHAN, A. J. 1953. Recent Changes in Bird Life of the Lenches. WM *Bird Report* 18:10-11.

HARTHAN, A. J. 1961. A Revised List of Worcestershire Birds. *Transactions of the Worcestershire Naturalists' Club* XI:167-186.

HASTINGS, Sir Charles. 1834. *Illustrations of the Natural History of Worcestershire*. London.

HAVINS, P. J. N. 1974. *Portrait of Worcestershire*. London.

HAVINS, P. J. N. 1976. *The Forests of England*. Newton Abbot.

HAWKER, D. M. Status of the Shelduck in the West Midlands, 1934-65. WM *Bird Report* 33:8-12.

HAWKER, D. M. 1970. Common Scoters inland. *Brit. Birds* 63:382-384.

HEWSON, R. 1967. Territory, behaviour and breeding of the Dipper in Banffshire. *Brit. Birds* 60:244-252.

HICKMAN, D. 1979. *Warwickshire*. London.

HILLIER and SON. 1971. *Hillier's Manual of Trees and Shrubs*. Winchester.

HOSKINS, W. G. 1970. *The Making of the English Landscape*. Harmondsworth.

HOSKINS, W. G. and L. D. STAMP. 1963. *The Common Lands of England and Wales*. London.

HUBBARD, C. E. 1968. *Grasses*. Bungay.

HUDSON, R. 1965. The spread of the Collared Dove in Britain and Ireland. *Brit. Birds* 58:105–139.

HUDSON, R. 1973. *Early and Late Dates for Summer Migrants*. BTO Guide 15.

HUDSON, R. 1976. Ruddy Ducks in Britain. *Brit. Birds* 69:132–143.

HUME, R. A. 1976. Inland flocks of Kittiwakes. *Brit. Birds* 69:62–63.

HUME, R. A. 1978. The Status of the Rarer Grebes. WM *Bird Report* 44:19–21.

HUME, R. A. 1978 a. Variations in Herring Gulls at a Midland roost. *Brit. Birds* 71:338–345.

HUME, R. A. and P. J. GRANT. 1974. The upperwing patterns of Common and Arctic Terns. *Brit. Birds* 67:133–136.

HUNT, A. E. 1977. Lead poisoning in Swans. *BTO News* 90:1–2.

HUTCHINSON, C. D. and B. NEATH. 1978. Little Gulls in Britain and Ireland. *Brit. Birds* 71:563–582.

JONES, B. E. (in prep.). Mortality rates and causes of death of British Canada Geese (*Branta canadensis*) with particular reference to the West Midlands.

JONES, B. E. and C. D. T. MINTON. (in prep.). Movements of British Canada Geese with particular reference to the West Midlands.

LAPWORTH, C. 1913. The Birmingham Country – Its Geology and Physiography. In G. A. Auden (Ed.) *A Handbook for Birmingham and the Neighbourhood*. Birmingham.

LEACH, I. H. 1981. Wintering Blackcaps in Britain and Ireland. *Bird Study* 28:5–14.

LEATHERBARROW, J. S. 1974. *Worcestershire*. London.

LEES, E. 1828. *The Strangers' Guide to the City of Worcester*. Worcester.

LORD, J. 1962. Analysis of Common Scoter records. West Midlands, 1934–60. WM *Bird Report* 28:17–19.

LORD, J. 1976. Duck Counts at Blithfield Reservoir, 1955–74. WM *Bird Report* 42:11–12.

LORD, J. and A. R. M. BLAKE. 1962. *The Birds of Staffordshire*. West Midland Bird Club.

LORD, J. and D. J. MUNNS (Eds.). 1970. *Atlas of Breeding Birds of the West Midlands*. London.

LORD, J. and A. J. RICHARDS. 1965. Kestrel Enquiry, West Midlands, 1964. WM *Bird Report* 31:15–17.

LORD, P. 1972. *Portrait of the River Trent*. London.

LOVENBURY, G. A., M. WATERHOUSE and D. W. YALDEN. 1978. The Status of Black Grouse in the Peak District. *Naturalist* 103:3–14.

MACALDOWIE, A. M. 1893. The Birds of Staffordshire. *Transactions of the North Staffordshire Field Club* 27.

MACKNEY, D. and C. P. BURNHAM. 1964. *The Soils of the West Midlands*. Harpenden.

MAGEE, J. D. 1965. The breeding distribution of the Stonechat in Britain and the causes of its decline. *Bird Study* 12:83–89.

MASEFIELD. J. R. B. 1908. Birds. In the *Victoria County History of Staffordshire*. London.

MASON, C. F. 1969. Waders and terns in Leicestershire and an index of relative abundance. *Brit. Birds* 62:523–533.

MAYR, E. 1951. Speciation in birds. *Proc. X Int. Orn. Congr., Uppsala, 1950*. 91–131.

MEADOWS, B. S. 1970. Breeding distribution and feeding ecology of the Black Redstart in London. *London Bird Report* 34:72–79.

MEEK, E. R. and B. LITTLE. 1977. Ringing studies of Goosanders in Northumberland. *Brit. Birds* 70:273–283.

METEOROLOGICAL OFFICE. 1952. *Climatological Atlas of the British Isles*. HMSO.

METEOROLOGICAL OFFICE. 1963. *Average of Temperatures for Great Britain, 1931–60*. HMSO.

METEOROLOGICAL OFFICE. 1968. *Monthly and Annual Average of Rainfall for the United Kingdom, 1931–60*. HMSO.

MILLWARD, R. and A. ROBINSON. 1971. *Landscapes of Britain – the West Midlands*. Basingstoke.

MINISTRY OF AGRICULTURE, FISHERIES AND FOOD. 1979. *Agriculture in the West Midlands*.

MINTON, C. D. T. 1968. Pairing and Breeding of Mute Swans. *Wildfowl* 19:41–60.

MINTON, C. D. T. 1968a. Ringing Secretary's Report. WM *Bird Report* 34:8.

MINTON, C. D. T. 1970. Cannock Chase Tit Nest Box Study. WM *Bird Report* 36:10–12.

MINTON, C. D. T. 1970a. Movement of Thrushes to and from the West Midlands. WM *Bird Report* 36:15–24.

MINTON, C. D. T. 1971. The Gailey Reservoir Heronry 1960–70. WM *Bird Report* 37:24–29.

MINTON, C. D. T. 1975. The Waders of the Wash – ringing and biometric studies. *Wash Feasibility Study*. Wash Wader Ringing Group.

MITCHELL, A. 1974. *A Field Guide to the Trees of Britain and Northern Europe*. London.

MOORE, N. W. 1957. The past and present status of the Buzzard in the British Isles. *Brit. Birds* 50:173–197.

MORLAND, B. 1978. *Portrait of the Potteries*. London.

MOSLEY, Sir Oswald. 1863. *Natural History of Tutbury*. London.

MURTON, R. K. 1971. *Man and Birds*. London.

MURTON, R. K. and N. J. WESTWOOD. 1974. Some effects of agricultural change on the English avifauna. *Brit. Birds* 67:41–69.

NATURE CONSERVANCY COUNCIL. 1977. *Chaddesley Woods – National Nature Reserve*. Shrewsbury.

NEWTON, I. 1972. *Finches*. London.

NORMAN, R. K. and D. R. SAUNDERS. 1969. Status of Little Terns in Great Britain and Ireland in 1967. *Brit. Birds* 62:4–13.

NORRIS, C. A. 1947. *Notes on the Birds of Warwickshire*. Birmingham.

NORRIS, C. A. 1951. *West Midland Bird Distribution Survey, 1946–50* (printed for private circulation).

OFFICE OF POPULATION CENSUS AND SURVEYS. *Census of England and Wales, 1931; 1951; 1961; 1971*. HMSO.

OGILVIE, M. A. 1969. The status of the Canada Goose in Britain 1967–69. *Wildfowl* 20:79–85.

OGILVIE, M. A. 1981. The Mute Swan in Britain, 1978. *Bird Study* 28:87–106.

OGILVIE, M. A. and A. K. M. ST. JOSEPH. 1976. Dark-bellied Brent Geese in Britain and Europe, 1955–76. *Brit. Birds* 69:422–439.

ORFORD, N. 1973. Breeding Distribution of the Twite in central Britain. *Bird Study* 20:51–62, 121–126.

O'SULLIVAN, J. and the RARITIES COMMITTEE. 1977. Report on rare birds in Great Britain in 1976. *Brit. Birds* 70:405–453.

PALMER-SMITH, M. 1978. *Birds of the Malvern District*. Malvern.

PARSLOW, J. L. F. 1967. Changes in status among breeding birds in Britain and Ireland. Pt. 1. *Brit. Birds* 60:2–47.

PARSLOW, J. L. F. 1973. *Breeding Birds of Britain and Ireland*. Berkhampsted. (This is mainly a reprint of papers that appeared in *Brit. Birds* in 1967 and 1968.)

PIKE, G. V. 1979. Kestrels in the Birmingham Area. WM *Bird Report* 45:23–25.

PITT, W. 1817. *Topographical History of Staffordshire*. Newcastle-under-Lyme.

PLOT, R. 1686. *Natural History of Staffordshire*. Oxford.

POTTER, C. H. 1977. *Brandon Marsh Conservation Group Report 1972–76* (printed for private circulation).

POTTS, G. R. 1969. The influence of eruptive movements, age, population size and other factors on the survival of the Shag. *J. Anim. Ecol.* 38:53–102.

PRATER, A. J. 1973. The wintering population of Ruffs in Britain and Ireland. *Bird Study* 20:245–250.

PRATER, A. J. 1974. The population and migration of Knot in Europe. *Proc. IWRB Wader Symposium, Warsaw 1973:* 99–113.

PRATER, A. J. 1975. The wintering population of the Black-tailed Godwit. *Bird Study* 22:169–176.

PRESCOTT, F. E. 1931. List of Birds in Bockleton and the Neighbourhood. In *Tenbury*. F. Joyce. Oxford.

PRESTT, I. 1965. An enquiry into the recent breeding status of some of the smaller birds of prey and crows in Britain. *Bird Study* 12:196–221.

PYMAN, G. A. 1959. The status of the Red-crested Pochard in the British Isles. *Brit. Birds* 52:42–56.

RATCLIFFE, D. A. 1963. The status of the Peregrine in Great Britain. *Bird Study* 10:56–90.

RATCLIFFE, D. A. 1977. *A Nature Conservation Review*. London.

READERS DIGEST. 1965. *Complete Atlas of the British Isles*. London.

ROBSON, R. W. and K. WILLIAMSON. 1972. The breeding birds of a Westmorland farm. *Bird Study* 19:202–214.

ROGERS, H. B., R. C. J. BURTON, D. W. SHIMWELL and J. L. BOSTOCK. 1977. *Cannock Chase Country Park* (Report to the Countryside Commission).

ROGERS, M. J. and the RARITIES COMMITTEE. 1979. Report on rare birds in Great Britain in 1978. *Brit. Birds* 72:503–549.

ROOTH, J. 1971. The status of the Greylag Goose *Anser anser* in the western part of its distribution. *Ardea* 59:17–27.

SAGE, B. L. and J. D. R. VERNON. 1978. The 1975 National Survey of Rookeries. *Bird Study* 25:64–86.

SAINTER, J. D. 1878. *Scientific Rambles round Macclesfield*. Macclesfield.

SCOTT, P. 1965. *A Coloured Key to the Wildfowl of the World*. Wildfowl Trust.

SEAGO, M. J. 1977. *Birds of Norfolk*. Norwich.

SEARS, J. N. and A. W. CUNDALL. 1955. *A Survey of the Birds of Earlswood* (unpublished).

SEVERN-TRENT WATER AUTHORITY. 1975. *River Avon Basin Study – Interim Survey of Water Services*. Birmingham.

SEVERN-TRENT WATER AUTHORITY. 1976. *River Tame Basin, Report of Survey*. Birmingham.

SHARROCK, J. T. R. 1969. Grey Wagtail passage and population fluctuations in 1956–67. *Bird Study* 16:17–34.

SHARROCK, J. T. R. 1974. *Scarce Migrant Birds in Britain and Ireland*. Berkhampsted.

SHARROCK, J. T. R. (Ed.). 1976. *The Atlas of Breeding Birds in Britain and Ireland*. BTO/IWC. Berkhampsted.

SHARROCK, J. T. R. (Ed.). 1980. *The Frontiers of Bird Identification*. London.

SHARROCK, J. T. R. and E. M. SHARROCK. 1976. *Rare Birds in Britain and Ireland*. Berkhampsted.

SHOOTER, P. 1970. The Dipper population of Derbyshire, 1958–68. *Brit. Birds* 63:158–163.

SIMMS, E. A. 1949. An Overland Migration Route. WM *Bird Report* 15:10–11.

SIMMS, E. 1971. *Woodland Birds*. London.

SIMMS, E. 1978. *British Thrushes*. London.

SMALLSHIRE, D. and A. J. RICHARDS. 1976. *The Birds of Belvide Reservoir*. West Midland Bird Club.

SMITH, F. R. and the RARITIES COMMITTEE. 1972. Report on rare birds in Great Britain in 1971. *Brit. Birds* 65:322–354.

SMITH, F. R. and the RARITIES COMMITTEE. 1974. Report on rare birds in Great Britain in 1973. *Brit. Birds* 67:310–348.

SMITH, T. 1938. *The Birds of Staffordshire*. North Staffs. Field Club.

SMOUT, C. 1969. Long-tailed Duck. In *Birds of the World*. J. Gooders (Ed.). London.

SPENCER, K. G. 1969. Overland migrations of Common Scoter. *Brit. Birds* 62:332–333.

SPENCER, R. 1975. Changes in the distribution of recoveries of ringed Blackbirds. *Bird Study* 22:176–190.

STAFFORDSHIRE COUNTY COUNCIL. 1977. *Some of Staffordshire's Woods*.

STAMP, L. D. (Ed.). 1937–46. *The Land of Britain – Report of the Land Utilisation Survey of Britain*. (Vols. for Staffordshire, Warwickshire and Worcestershire.) London.

STAMP, L. D. and S. H. BEAVER. 1958. *The British Isles – A Geographic and Economic Survey*. London.

STANLEY, P. I. and C. D. T. MINTON. 1972. The unprecedented westward migration of Curlew Sandpipers in autumn 1969. *Brit. Birds* 65:365–380.

STOKE-ON-TRENT, CITY OF. 1974. *Land Reclamation*.

STOKE-ON-TRENT, CITY OF. 1979. *Study Tour of the Potteries*. (Prepared for the Annual Conference of the Royal Town Planning Institute.)

STUTTARD, P. and K. WILLIAMSON. 1971. Habitat requirements of the Nightingale. *Bird Study* 18:9–14.

SUMMERS, D. D. B. 1979. Bullfinch dispersal and fruit bud damage. *Brit. Birds* 72:249–263.

TAVERNER, J. H. 1959. The spread of the Eider in Great Britain. *Brit. Birds* 52:245–258.

TAVERNER, J. H. 1967. Wintering Eiders in England during 1960–65. *Brit. Birds* 60:509–515.

TEAGLE, W. G. 1978. *The Endless Village*. Nature Conservancy Council, Shrewsbury.

THOROLD, H. 1978. *Staffordshire*. London.

TOMES, R. F. 1901. Birds. In the *Victoria County History of Worcestershire*. London.

TOMES, R. F. 1904. Aves. In the *Victoria County History of Warwickshire*, 1. London.

TOOBY, H. J. 1945. Birds of the Old Hills, near Malvern. WM *Bird Report* 11:1–3.

TOOBY, H. J. and A. J. HARTHAN. 1943. A survey of Sand Martin colonies in Worcestershire. WM *Bird Report* 9:3–5.

VOOUS, K. H. 1977. *List of Recent Holarctic Bird Species.* BOU. London.

WALKER, A. F. G. 1970. The moult migration of Yorkshire Canada Geese. *Wildfowl* 21:99–104.

WALKER, S. W. 1979. *The Birds of Oakley Pool – report 1978/9* (unpublished).

WARWICKSHIRE COUNTY COUNCIL. 1978. *County Landscape Plan.*

WAUGH, D. R. The diet of Sand Martins during the breeding season. *Bird Study* 26:123–128.

WESTMACOTT, R. and T. WORTHINGTON. 1974. *New Agricultural Landscapes.* Countryside Commission, Cheltenham.

WEST MIDLAND BIRD CLUB. 1934–79. WM *Bird Reports* 1–46.

WEST MIDLAND BIRD CLUB. 1952–80. *Bulletins* 1–288.

WEST MIDLAND GROUP. 1948. *Conurbation – A Planning Survey of Birmingham and the Black Country.* London.

WHITLOCK, F. B. 1893. *Birds of Derbyshire.* London and Derby.

WILLIAMSON, K. 1967. The bird community of farmland. *Bird Study* 14:210–226.

WILLIAMSON, K. 1969. Habitat preferences of the Wren on English farmland. *Bird Study* 16:53–59.

WILLIS-BUND, J. W. 1891. *List of Worcestershire Birds.* Worcester.

WILSON, J. 1973. Wader populations of Morecambe Bay, Lancashire. *Bird Study* 20:9–23.

WINSTANLEY, D., R. SPENCER and K. WILLIAMSON. 1974. Where have all the Whitethroats gone? *Bird Study* 21:1–14.

WISE, M. J. (Ed.). 1950. *Birmingham and its Regional Setting – A Scientific Survey.* Birmingham.

WITHERBY, H. F., F. C. R. JOURDAIN, N. F. TICEHURST and B. W. TUCKER. 1938–44. *The Handbook of British Birds.* London.

WOLTON, A. W. 1951. The Heligoland Trap. WM *Bird Report* 17:6–8.

WOOD, N. 1836. *British Song Birds.* London.

WORCESTERSHIRE NATURE CONSERVATION TRUST. 1978–81. *Reserves Handbook.*

WOUTERSEN, K. 1980. Migrating Little Gulls in the Netherlands. *Brit. Birds* 73:192–193.

WRIGHT, L. and J. PRIDDEY. 1973. *Heart of England.* London.

YALDEN, D. W. 1972. The red grouse (*Lagopus lagopus scoticus* (Lath.)) in the Peak District. *Naturalist* 922:89–102.

YALDEN, D. W. 1979. Birds of the Swythamley Estate. WM *Bird Report* 45:14–22.

YALDEN, D. W. 1979a. An estimate of the number of Red Grouse in the Peak District. *Naturalist* 104:5–8.

YARRELL, W. 1871–85. *A History of British Birds.* (Fourth edition). London.

Index to Birds

Vernacular names are indexed under the last word. Scientific names are indexed under the generic name and are shown in italics. Numbers in Roman type refer to the main entry for each species. Those in italics refer to the breeding distribution maps. (For lists of breeding species not mapped see page 443.) The index does not include references to birds in the first eight chapters or to the exotic species listed in Appendix II. For an index to the principal localities see the Gazetteer in Appendix VI.